UHPLC in Life Sciences

RSC Chromatography Monographs

Series Editor:
R.M. Smith, *Loughborough University of Technology, UK*

Advisory Panel:
J.C. Berridge, *Sandwich, UK*, G.B. Cox, *Indianapolis, USA*, I.S. Lurie, *Virginia, USA*, P.J. Schoenmakers, *Eindhoven, The Netherlands*, C.F. Simpson, *London, UK*, G.G. Wallace, *Wollongong, Australia*

Titles in this Series:
0: Supercritical Fluid Chromatography
1: Chromatographic Integration Methods
2: Packed Column SFC
3: Chromatographic Integration Methods, Second Edition
4: Separation of Fullerenes by Liquid Chromatography
5: Applications of Solid Phase Microextraction
6: HPLC: A Practical Guide
7: Capillary Electrochromatography
8: Hyphenated Techniques in Speciation Analysis
9: Cyclodextrins in Chromatography
10: Electrochemical Detection in the HPLC of Drugs and Poisons
11: Validation of Chromatography Data Systems: Meeting Business and Regulatory Requirements
12: Thin-layer Chromatography: A Modern Practical Approach
13: High Temperature Liquid Chromatography: A User's Guide for Method Development
14: High Performance Chelation Ion Chromatography
15: Protein and Peptide Analysis by LC-MS: Experimental Strategies
16: UHPLC in Life Sciences

How to obtain future titles on publication:
A standing order plan is available for this series. A standing order will bring delivery of each new volume immediately on publication.

For further information please contact:
Book Sales Department, Royal Society of Chemistry, Thomas Graham House, Science Park, Milton Road, Cambridge, CB4 0WF, UK
Telephone: +44 (0)1223 420066, Fax: +44 (0)1223 420247, Email: books@rsc.org
Visit our website at http://www.rsc.org/Shop/Books/

UHPLC in Life Sciences

Edited by

Davy Guillarme and Jean-Luc Veuthey
School of Pharmaceutical Sciences, University of Geneva, University of Lausanne, Switzerland
Email: Davy.guillarme@unige.ch; jean-luc.veuthey@unige.ch

RSC Publishing

RSC Chromatography Monographs No. 16

ISBN: 978-1-84973-388-5
ISSN: 1757-7055

A catalogue record for this book is available from the British Library

Published by The Royal Society of Chemistry,
Thomas Graham House, Science Park, Milton Road,
Cambridge CB4 0WF, UK

Registered Charity Number 207890

For further information see our web site at www.rsc.org

Preface

Since the initial development of modern liquid chromatography in the 1970s, this technique has received strong interest and has seen tremendous improvements in equipment (such as columns, stationary phases, pumping systems and injection valves). In addition, there has been a consistent trend towards analyses with either faster separations or enhanced resolution. During the last century, these goals were generally attained to the detriment of sensitivity, robustness and both system complexity and stability. In 2004, there was a breakthrough with the commercial introduction of columns packed with sub-2 μm fully porous particles and chromatographic systems able to withstand pressures of up to 1000 bar. Chromatography under such extreme conditions is generally described as UHPLC [ultra-high-pressure (or -performance) liquid chromatography]. Since 2004, more than 2500 papers have been published on this topic, including theoretical investigations of both the technique and its applications, from industrial, governmental and academic laboratories.

The aim of this book is to provide a single source of information with an extensive overview of UHPLC and its applications in the life sciences. The book is divided in two major sections: Chapters 1 to 7 are dedicated to the theoretical basis of UHPLC, and Chapters 8 to 14 describe its application to life science analyses.

Chapter 1 gives a general background of UHPLC and a theoretical comparison between it and other existing strategies for increasing the speed or resolution of the separation. In Chapter 2, the importance of suitable instrumentation and columns for UHPLC is critically discussed. Important aspects of transferring a method from HPLC to UHPLC conditions are discussed in Chapter 3 and include the rules used and certain inherent

---

RSC Chromatography Monographs No. 16
UHPLC in Life Sciences
Edited by Davy Guillarme and Jean-Luc Veuthey

Published by the Royal Society of Chemistry, www.rsc.org

problems encountered. Chapter 4 explains the interest in UHPLC with mobile phase temperatures higher than the ambient temperature. Indeed, high temperatures provide some obvious advantages, such as reducing the mobile phase viscosity and polarity. In 2006, a new generation of columns that were packed with sub-3 μm superficially porous particles (also known as core-shell particles) was introduced to the market. This promising technology is described in detail in Chapter 5. A performance comparison between columns packed with sub-3 μm core-shell particles and sub-2 μm porous particles is presented in the same Chapter. Another recent trend in HPLC is the rapid development of hydrophilic liquid chromatography (HILIC) for the analysis of polar compounds and to provide an alternative selectivity for ionizable compounds. Because UHPLC-HILIC columns packed with porous, sub-2 μm particles are now commercially available, the advantages and limitations of the UHPLC-HILIC approach are described in Chapter 6. Finally, the last Chapter in the first section of the book is devoted to both the coupling of UHPLC to MS devices and the technological constraints that make this coupling slightly more difficult than traditional HPLC-MS. It is noteworthy that approximately 60% of the papers published on UHPLC deal with MS detection.

The second section of the book is dedicated to the application of UHPLC to life science analyses. During the first stages of drug discovery, the number of potential drug candidates is quite large. For this reason, it is important to develop high-throughput screening (HTS) methods for the rapid determination of physico-chemical properties (such as log *P*, solubility and permeability). Chapter 8 presents the attempts that were made over the last few years to achieve the HTS screening of drugs using UHPLC and UHPLC-MS technologies. The analysis of biological fluids is of prime importance for drug development, therapeutic drug monitoring and clinical/forensic/toxicological analyses. The number of biological samples that require rapid, quantitative analysis is generally large, and thus UHPLC has been widely employed in this field, which is discussed in Chapter 9. Chapter 10 provides more specific details for the doping control analysis of urine samples by UHPLC-MS and describes the well-accepted two-step strategy, which involves a screening step and a confirmation step. Seized drugs of abuse are generally in powder form, and the strategy for the rapid screening and profiling of these specific drugs is considered in Chapter 11. The need for the determination of drugs in environmental samples is shown in Chapter 12. In this case, a UHPLC-MS strategy was selected because of the combination of improved resolution from UHPLC and high selectivity and sensitivity from the MS devices. Other complex living organism of interest in drug analysis is plant material. As reported in Chapter 13, various approaches can be required, namely, fingerprinting, profiling and metabolomics of plant constituents. UHPLC is useful for all of these approaches because of its ability to provide either rapid or high-resolution analysis. Finally, there has been an important development of the metabolomic/metabonomic approaches for biomarker discovery, and UHPLC-MS is again used as a valuable tool for the determination of biomarkers in biological fluids and tissues. This topic is extensively discussed in Chapter 14.

We would like to conclude by warmly acknowledging the different authors, who are all recognized experts in their fields, for their contributions towards making this book a reference source for UHPLC for people in both academia and industry.

Dr. Davy Guillarme

Prof. Jean-Luc Veuthey

Contents

RSC Chromatography Monographs No. 16
UHPLC in Life Sciences
Edited by Davy Guillarme and Jean-Luc Veuthey

Published by the Royal Society of Chemistry, www.rsc.org

CHAPTER 1

General Overview of Fast and High-resolution Approaches in Liquid Chromatography

DEIRDRE CABOOTER* AND GERT DESMET

Vrije Universiteit Brussel, Department of Chemical Engineering, Pleinlaan 2, 1050 Brussels, Belgium
*E-mail: dcaboote@vub.ac.be

1.1 Trends in HPLC

Over the past 50 years, column packings have evolved from irregularly shaped silica particles with sizes of 30–100 µm to spherical particles with diameters of 3–5 µm and even less than 2 µm.[1] The tendency to develop smaller particle sizes has in essence been driven by the urge to obtain more efficient columns and faster separations. As column efficiency—especially at high velocities—mainly depends on the rate of mass transfer in the stationary phase, decreasing the diffusion distances of solutes in the stationary phase by reducing the particle size is an effective way to reach higher efficiencies. Moreover, the band spreading originating from non-uniformities in the packed bed is proportional to the particle size and will hence decrease with decreasing particle size. A drawback of using smaller particles is that the pressure drop per unit plate increases with the square of the decreasing particle size, making high operating pressures necessary to take full advantage of their performance.[2,3]

Sub-2 µm particles have been developed in the last decade as an answer to the increasing demand from industry to shorten analysis times and increase sample throughput.[4–6] In order to operate these sub-2 µm particles at or above

RSC Chromatography Monographs No. 16
UHPLC in Life Sciences
Edited by Davy Guillarme and Jean-Luc Veuthey

Published by the Royal Society of Chemistry, www.rsc.org

their optimal flow rate, instrumentation capable of delivering pressures higher than the conventional 400 bar (up to 1300 bar) has become commercially available under the name of ultra-high performance liquid chromatography (UHPLC).[7–9] Moreover, new techniques to manufacture monodisperse sub-2 μm packing material with a suitable mechanical strength, such as bridged ethyl hybrid (BEH) particles and high-density silica particles, have been developed.[7] In different fields of analysis, such as drug and metabolite analysis and bioanalytical and environmental separations, it has been demonstrated that sub-2 μm particles can offer an increase in analysis time of 5–20 compared to columns packed with 5 μm particles, while maintaining a comparable efficiency.[9] A drawback of working at ultra-high pressures is the occurrence of viscous heating. Viscous heating is caused by the friction between different fluid layers inside the column at high flow rates and pressures. The generated heat leads to a number of unwanted effects when it is poorly dissipated: the temperature of the mobile phase increases, leading to axial and radial temperature gradients inside the column. These temperature gradients affect the viscosity of the mobile phase and the retention factor of the analytes, and cause changes in band broadening.[10,11] An effective way to reduce the effect of viscous heating involves reducing the diameter of the column to 2.1 or 1.0 mm (the viscous-heating-induced additional plate height contribution increases according to the sixth power of the column radius!).

Another way to speed up the analysis time of a separation is by working at elevated temperatures. High-temperature HPLC (HT-HPLC) operations benefit from an enhanced mass transfer due to the decreased mobile phase viscosity and increased analyte diffusivity at high temperatures.[12,13] This essentially leads to an increase of the optimal linear velocity, implying that separations can be performed at higher flow rates without a significant loss in efficiency. The decreased viscosity of the mobile phase, moreover, leads to a decreased column back pressure, further allowing the flow rate to be increased. It also permits the use of longer columns and smaller particles, making ultra-high efficiency separations possible in very short analysis times.[14,15] Working at elevated temperatures, the surface tension and dielectric constant of water, moreover, decrease, allowing a large amount of the organic solvent in the mobile phase to be replaced by water—while maintaining the retention capacity—and hence performing more environmentally friendly analyses.[16] In addition to this, the peak shape of basic compounds has been reported to improve at elevated temperatures due to reduced secondary interactions with free silanol groups.[17,18] Important considerations to make when performing HT operations are the stability of the analyte and the stationary phase at elevated temperature. Alternative silica-based stationary phases able to withstand temperatures up to 100°C are available,[13,19] as well as non-silica based materials, such as zirconia, carbon and polymer packings, which are stable up to temperatures of 150–200 °C.[13,20,21]

Another means to obtain higher separation efficiencies by increasing the mass transfer rate is the development of alternative particle designs such as superficially porous particles and monolithic columns.

Superficially porous particles are composed of a solid silica core with a thin porous shell. They were, in fact, the first supports made available for HPLC separations in the late 1960's. At that time, the average particle size was 40 μm. In 2006, superficially porous particles with average particle sizes of 2.6–2.7 μm were re-introduced into the market. The diffusion of analytes is restricted to the thin porous shell that has a thickness of a maximum of 0.5 μm, making a very fast mass transfer possible. The average size of the entire particle, on the other hand, leads to back pressures that are much lower than the ones obtained in sub-2 μm columns of the same length. The combination of these two effects results in efficiencies that rival those of totally porous sub-2 μm particles, but at only one-half to one-third of the column back pressure, hence at conventional HPLC pressures.[22] In addition to this, superficially porous particles have a very narrow particle size distribution compared to fully porous particles. Whether or not this can explain the exceptionally low A-term constants and minimum plate heights (h_{min}) is still a question of debate.[23,24] Very recently, Phenomenex (Torrance, CA, USA) also developed sub-2 μm superficially porous particles with a 1.25 μm solid core surrounded by a 0.23 μm porous shell.[25]

In contrast to packed beds of spherical particles, monolithic columns can be described as a continuous piece of macroporous material with relatively large flow throughpores and without interparticular voids. Although they yield higher plate heights than the best performing porous particles, the high permeability of the monoliths generates a much smaller back pressure compared to packed columns, allowing them to be used in very long columns and therefore making them extremely suited for ultra-high efficiency separations.[26–28] Monolithic columns can be subdivided into two main categories: polymer-based and silica-based monoliths.

Polymer monoliths are polymerized *in situ* in a plastic or stainless steel tube. They are prepared from a mixture of monomers, a cross-linker, a free radical initiator and a porogenic solvent. The polymerization is initiated either thermally or by irradiation with UV light.[29,30] A large variety of monomers (*e.g.* styrene, divinylbenzene, methacrylate and acrylamide) can be used for the preparation of the monoliths. This leads to a wide variety in chemistries.[31–33] Monoliths can also be prepared with reactive functionalities that can be modified to produce different stationary phases.[28] Organic polymer monoliths consist of clusters of little organized microglobules with large throughpores (generally 0.5–1.0 μm). Polymer monoliths are mostly suited for the separation of larger molecules such as proteins, nucleic acids and synthetic polymers.[34–36]

Silica monoliths are prepared using the classical sol-gel process. Silane compounds, such as tetramethoxysilane (TMOS) or tetraethoxysilane (TEOS), are successively hydrolysed and polycondensated in the presence of a porogen.[37] The concentration of the porogen [e.g. polyacrylic acid (HPAA) or polyethylene oxide (PEO)] has a large influence on the morphology of the macroporous network.[38] As the condensation reaction evolves, the solubility of the formed polymers decreases until phase-separation takes place. After the phase-separation, the monolith is aged and dried very carefully.[38] Because a porous silica rod shrinks during the drying step, it may not be able to stick to the wall of its mould. The monolith is therefore removed from the mould and encased in a PEEK tube after its preparation. The moderate mechanical strength of PEEK limits the operation of the columns to 200 bar. Finally, the monoliths are derivatised, typically with C_{18}, to obtain a column suitable for reversed-phase applications. Silica monoliths have a bimodal structure: the large through-pores (1–3 μm) lead to high permeabilities, while the mesopores in the skeleton (10–25 nm) provide the monolith with a large surface area. Because of the small size of the mesopores, silica monoliths are particularly suited for the separation of small molecules, such as peptides and drug candidates.[39,40]

1.2 Comparison of Chromatographic Techniques and Supports

For manufacturers, as well as users, of HPLC instrumentation and columns it is very important to know which support and/or technique is best suited for a particular application. The recent developments in liquid chromatography (LC) therefore require good and independent comparison techniques. In the past, a number of simple quality parameters have been proposed to assess the performance potential of a system. The most simple and well known is the minimal plate height H_{min} (m) of a support. This parameter, however, does not give any information on the permeability of a packing and can therefore not be used to compare, for example, a packed bed and a monolithic column. In this case, the information on the flow resistance and the accompanying possibility to apply higher flow rates or use longer column lengths cannot be accounted for. When using reduced plate heights ($h_{min} = H_{min}/d_{ref}$), another very popular approach to compare different support types, this problem remains.[41,42] In this case, a second difficulty arises, as it is very hard to find a general reference length d_{ref} (m) valid for any type of support.[43] The minimum separation impedance $E_{0,min}$, defined as $E_{0,min} = H_{min}^2/K_{v0}$, is another frequently used quality parameter that combines information on the permeability K_{v0} (m^2) and minimal plate height of a support.[44] Being a dimensionless parameter, however, the main shortcoming of this approach is that it does not give any information on the speed of analysis.[45]

The answer to the question which technique is the "best" is also never unique and almost always depends on the desired efficiency or resolution and on the nature of the sample being analysed. Moreover, the answer relies on the evaluation criterion being used. A system yielding a fast critical pair analysis can, for example, perform badly when one aims for the maximum peak capacity in a certain time.

To meet with the shortcomings existing in most current comparison techniques, the kinetic plot method (KPM) has been developed as a data representation method that allows comparing the separation performance of different packing structures and operation methods in a uniform and standardized way.[45,46] Using the KPM, the performance of a chromatographic system is no longer expressed in terms of absolute or reduced plate heights (as is done in a van Deemter curve), but in a more universal and practically relevant unit: the time required to obtain a certain efficiency, resolution or peak capacity.

1.3 Constructing Kinetic Plots

Kinetic plots are as easy to establish as normal van Deemter curves and are based on a simple transformation of the different equations that determine the performance of a chromatographic system.[45,46] These equations are the well-established pressure drop equation:[47]

$$u_0 = \frac{\Delta P}{\eta} \frac{K_{v0}}{L} \tag{1.1}$$

wherein u_0 is the mobile phase linear velocity (m s^{-1}), η the mobile phase viscosity (Pa s^{-1}), L the column length (m) and ΔP the column pressure drop (Pa), and the basic equation that relates the plate height H (m) and plate number N with the column length:

$$L = N \,.\, H \tag{1.2}$$

Eqns (1.1) and (1.2) can be combined to obtain the following expression:

$$N = \left(\frac{\Delta P}{\eta}\right) \left[\frac{K_{v0}}{u_0 H}\right]_{exp} \tag{1.3}$$

Wherein the subscript "exp" refers to experimentally obtained data points.

Inserting eqn (1.3) into the equation that determines the dead time t_0 (s) of a chromatographic separation:

$$t_0 = \frac{L}{u_0} \tag{1.4}$$

it is found that:

$$t_0 = \left(\frac{\Delta P}{\eta}\right) \left[\frac{K_{v0}}{u_0^2}\right]_{exp} \tag{1.5}$$

For a certain value of ΔP and η, eqns (1.3) and (1.5) allow the transformation of any experimental (u_0, H)-couple, obtained in a column with permeability K_{v0}, into a (N, t_0)-couple. This (N, t_0)-couple represents the efficiency, N, that can be obtained in a certain time, t_0, would the same support be used in a column that is exactly long enough to generate the pressure, ΔP, at the given velocity, u_0. The calculations needed to obtain these kinetic plot data can very easily be implemented into a spreadsheet calculator such as Microsoft Excel.[46] To obtain the ultimate performance limits of the columns or supports under consideration, the plots should be constructed for the largest possible ΔP_{max} value (either the maximum pressure that the instrument can deliver or the maximum allowable pressure of the column). The kinetic plots can also be used to investigate the effect of the different pressures by recalculating the same set of plate height data for a series of different ΔP_{max} values. Care should, however, be taken when extrapolating low pressure data into the range of pressures above 400 bar. In this case, effects such as the occurrence of radial temperature gradients and changes in viscosity that take place under ultra-high pressure conditions could become influential.[48,49] Downscaling the pressure is never a problem, as it can only lead to an underestimation of the performance.[50] The viscosity, η, in eqns (1.3) and (1.5) can be treated as a normalization variable, using the same reference value for all the data series represented in the same kinetic plot. The effect of viscosity differences in experiments conducted with a different mobile phase will in this way be eliminated. The actual viscosity of the mobile phase used in each experiment can, however, also be used. This approach allows to account for the fact that more open-porous reversed phase systems need mobile phases with a higher percentage of water and hence a higher viscosity to obtain a sufficiently large retention factor. Once N and t_0 are known, it is straightforward to also calculate some important derivative quantities such as the retention time t_R (s), the peak capacity n_p and the separation resolution R_s:

$$t_R = t_0\,(1 + k) \tag{1.6}$$

$$R_s = \frac{1}{2}\cdot\frac{(k_2 - k_1)}{\left(\frac{1+k_1}{\sqrt{N_1}}\right)+\left(\frac{1+k_2}{\sqrt{N_2}}\right)} \tag{1.7}$$

$$n_p = 1+\sum_{i=1}^{n}\frac{t_{R,i}-t_{R,i-1}}{4\cdot\sigma_{t,i}} \tag{1.8}$$

where k is the retention factor, the subscripts "1" and "2" refer to the first and second compound of a critical pair respectively, σ_t is the peak standard deviation (s), $t_{R,i}$ is the retention time of the ith compound (with $t_{R,0} = t_0$) and n is the number of compounds in the sample.

Figure 1.1 provides an example of how the comparison between a commercial monolithic silica support and a sub-2 μm particle column that is usually made *via* a van Deemter plot (Figure 1.1a) can be transformed into a practically more informative plot of t_0 *versus* N (Figure 1.1b) using eqns (1.3) and (1.5). As can be noted by following the numbered data points for the sub-2 μm particle column, each (u_0, H)-data point in Figure 1.1(a) is transformed

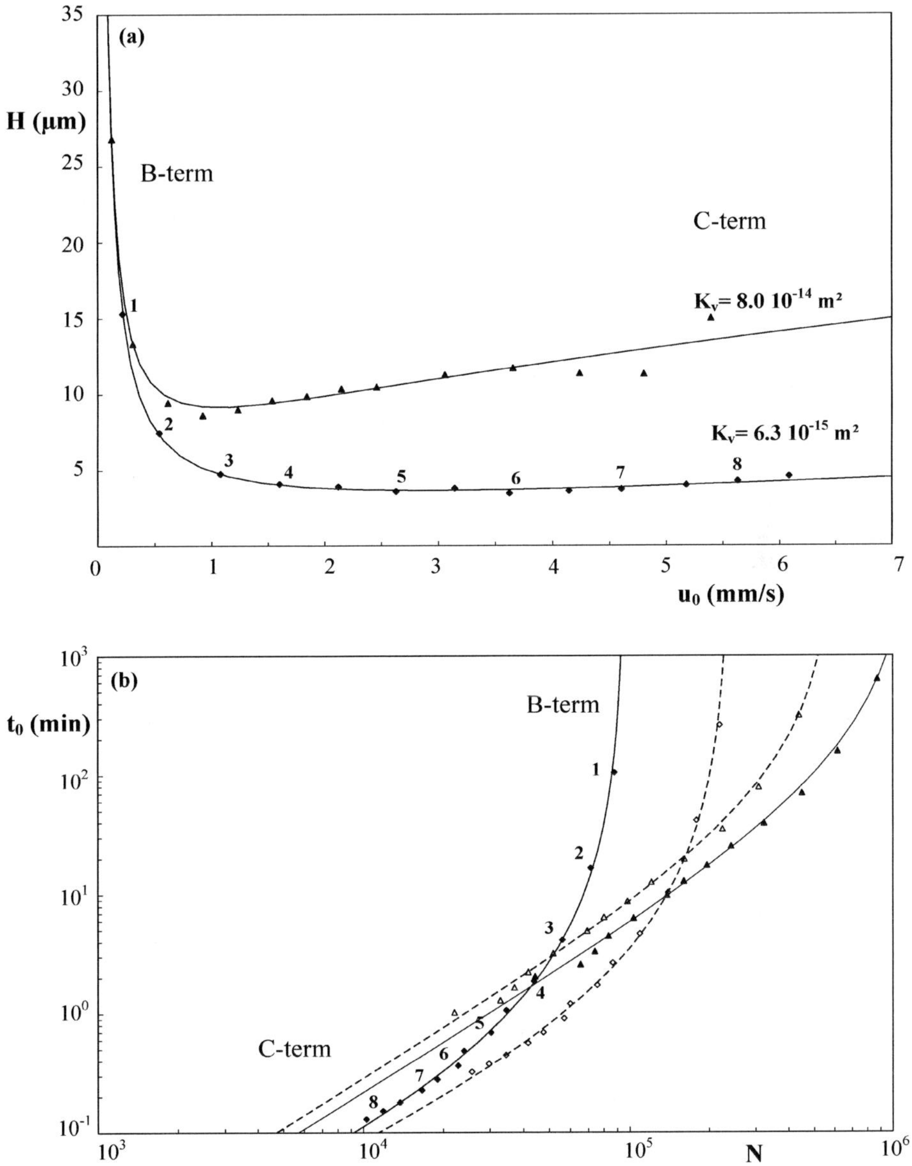

Figure 1.1 Transformation obtained using eqns (1.3) and (1.5) of (a) experimental van Deemter data into (b) a kinetic plot for the case of a fully porous sub-2 μm particle column (◆) and a commercial silica monolith (▲). The full line data in (b) compare both systems at identical pressure (P = 400 bar), and the data indicated by the dashed line compare each support type at its proper operation limit (P = 200 bar for the monolith and P = 1000 bar for the sub-2 μm column). Experimental conditions: T = 30 °C, test compound: 10 ppm methylparaben (k = 2), mobile phase: 35:65 ACN/H_2O for the sub-2 μm column (Agilent Zorbax SB 2.1 mm × 50 mm, d_p = 1.8 μm) and 27:73 ACN/H_2O for the monolith (Phenomenex Onyx 2.1 mm × 100 mm, d_{dom} = 2 μm), V_{inj} = 2 μL, λ = 254 nm.

into a unique (t_0, N)-data point in Figure 1.1(b). As low velocities allow the use of long columns at the maximum pressure, the data points corresponding to the B-term dominated region of the van Deemter curve transform into the long analysis time and high efficiency end of the kinetic plot curve (upper right of Figure 1.1b). Inversely, the data points originating from the C-term dominated region transform into the short analysis time and low efficiency end of the plot (lower left of Figure 1.1b). The latter plot readily reveals that monolithic supports are superior to sub-2 μm packed bed columns when large efficiencies (N >50 000) need to be pursued. On the other hand, when smaller efficiencies are needed, the packed bed format clearly outperforms the considered monolithic column format. This difference in behaviour is essentially due to the much more open structure and the accompanying low flow resistance of the monolithic support (allowing the use of very long columns at a sufficiently high linear velocity) on the one hand, and due to the smaller diffusion distances and the better packing homogeneity of the packed bed column (leading to high separation efficiencies in relatively short columns) on the other hand.[51] Whereas the two solid line curves in Figure 1.1(b) compare the two support types for the same pressure (thus providing a view on the intrinsic differences originating from their different geometry), the two dashed lines compare the two materials on the basis of their proper pressure limit (200 bar for the monolithic column and 1000 bar for the particulate column, respectively). The former comparison is more relevant for the monolithic column manufacturers (it gives clues on the suitability of the structures they are synthesizing) and the latter is more relevant for the column users (it tells them what support performs best in a given range of desired efficiencies).

From the above it is clear that the information obtained from a kinetic plot is economically and practically more relevant than the information obtained from a van Deemter curve. The information is at once also maximally generalized. The situation is not limited to a column with a certain length or to a certain amount of plates, but yields the conditions for all possible plate counts and column lengths. An additional advantage is that the kinetic plots can be established without having to define a reference diameter and hence allow comparison of the performance of differently shaped and sized LC supports without the need for a common reference length.

1.4 History of the Kinetic Plot Method

The approach of displaying time against efficiency to compare different chromatographic supports is certainly not new and was actually already used in the very beginning of modern chromatographic history. Calvin Giddings presented this type of plot in 1965 when he compared the performance limits of LC with those of gas chromatography (GC) on a system-independent basis.[52] Later, Knox and Saleem, and Guiochon, used the same approach to compare the performance of packed bed columns with open-tubular columns.[53,54] In 1997, Hans Poppe proposed the plotting of t_0/N *versus* N instead of t_0 *versus* N

to obtain a more expanded view on the kinetic performance in the C-term dominated range.[55] Later, in 2006, the concept was extended to gradient-based separations by Carr and co-workers.[56] However, the representation of chromatographic performance data in terms of analysis time *versus* plate number found little followers among the more practically oriented scientists in the field, mainly because the establishment of the plots required a rather unwieldy numerical optimization procedure.[45]

The simple data transformation given in eqns (1.3) and (1.5), however, yields exactly the same curve as that obtained *via* any computer-assisted kinetic optimization procedure and opened the way to a simpler and more user-friendly means of plotting experimental plate height data directly into a kinetic plot, which in turn lead to a more widespread use of kinetic plots.[45,24,57–68]

Variants to eqns (1.3) and (1.5) have recently been proposed by Carr *et al.* and Neue,[59,61] writing the relation between H and u_0 in the denominator of eqn (1.3) in an explicit form using, for example, the Knox or the van Deemter plate height expression.

According to its definition, the kinetic optimum of a given support or operating condition is achieved when a given desired efficiency or peak capacity is reached in the shortest possible time, or, equivalently, when a maximal efficiency or peak capacity is reached during a given allotted analysis time. This kinetic optimum is achieved if and only if the employed chromatographic support is used in a column that is precisely long enough to reach a given user-specified pressure limit (note that each column length reaches this limit at a different flow rate).[69] This maximal pressure condition is automatically satisfied in eqns (1.3) and (1.5), whereas it had to be calculated iteratively in the traditional approaches.[52,55,56] Figure 1.2(a) illustrates that the best kinetic performance is, indeed, always obtained at the maximal pressure end of the fixed length kinetic plot curves.[70,71] In Figure 1.2(a), these are the dashed curves, established by measuring N and t_R at various flow rates on a column with a fixed length: it is impossible to find a better combination of efficiency and time than those situated on the kinetic performance limit (KPL) curve (continuous curve) connecting the high pressure end points of all fixed length curves. The KPL curve envelopes the complete area of kinetic performances that can be achieved in practice with the tested material and thus represents a unique "signature" of its kinetic quality.

Comparing different materials or operating conditions on the basis of this "signature" automatically guarantees that the comparison occurs on a fair basis (each system is compared at its optimum, thus preventing that one system is tested under less optimized conditions than the other one) and is fully comprehensive (all possible values of N or n_p are addressed at the same time).[45,46] The curve represents the composite effect of the band broadening and the column permeability on the kinetic performance and thus allows the assessment of the quality of the compromise that sometimes needs to be made between a packing with a high flow resistance but small band broadening (for example, a packed bed column containing sub-2 μm particles) and its opposite (for example, a monolithic column with wide through-pores).

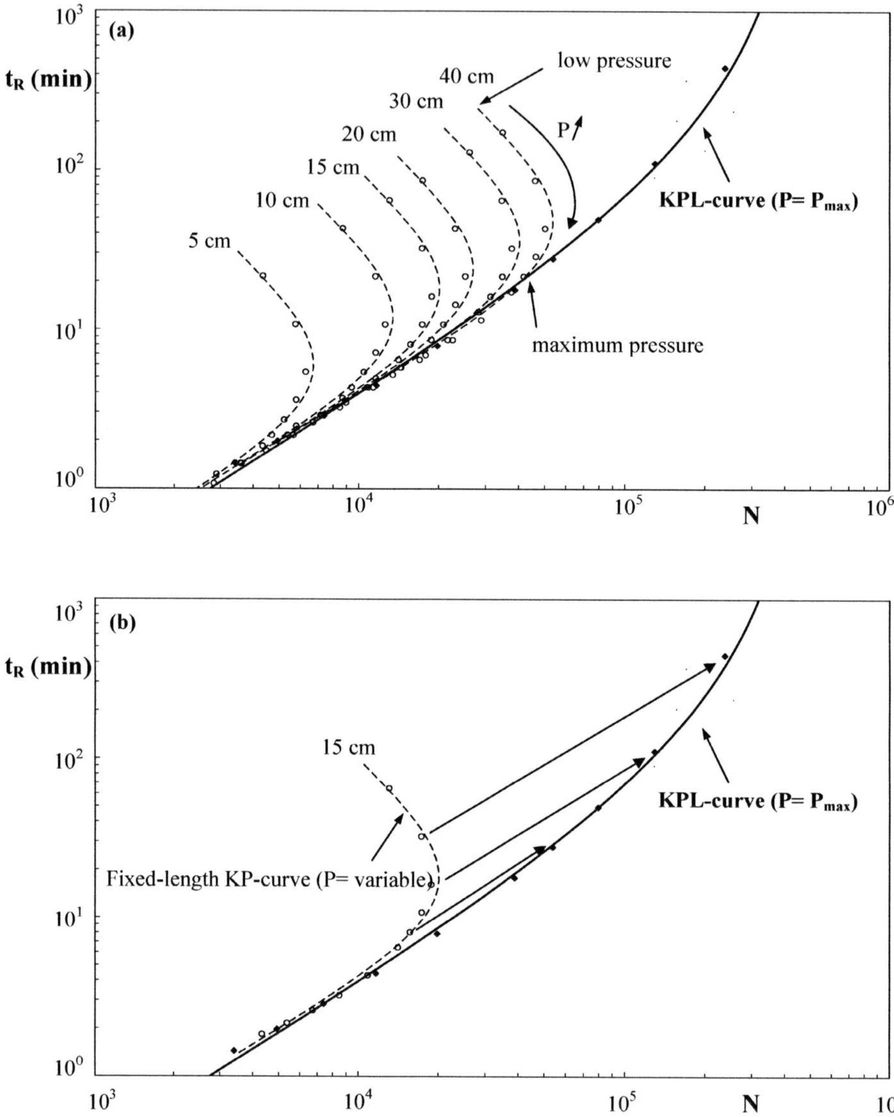

Figure 1.2 (a) Experimental fixed length kinetic plot curves (dashed curves) on columns with a different length containing the same particle type and the enveloping KPL curve (continuous curve). The arrow on the fixed length curve for L = 40 cm indicates the direction of the pressure change. This direction is the same for the other fixed length curves as well. (b) Direct data transformation from a fixed length kinetic plot curve into the KPL curve as obtained when using either eqns (1.3)–(1.5) or eqns (1.9)–(1.12). Experimental conditions: T = 30 °C, test compound: 100 ppm of a pharmaceutical compound (MW = 674 with two amide functions) (k = 10), mobile phase: 60:40, ACN/0.1% formic acid; column: superficially porous (HALO 2.1 mm × 150 mm, d_p = 2.7 μm), V_{inj} = 0.5 μL, λ = 254 nm.

1.5 Unification of the Isocratic and Gradient Kinetic Plot Method

Whereas eqns (1.3) and (1.5) are straightforward to apply under isocratic separation conditions, this no longer holds under gradient-elution conditions because in this case the (average) plate height, H, is more difficult to determine exactly (it requires the exact knowledge of the retention factor experienced by the analytes when eluting from the column).[72,73] It has, however, been shown that the use of plate heights to establish the KPL curve is only an unnecessary detour.[69] In fact, the KPL curve can also be directly established using a set of t_R- (or t_0-) and n_p- (or N-) data read out from a series of chromatograms produced at different flow rates on a column with a given length, and subsequently transforming these data using a length elongation factor, λ, to implement the aforementioned condition that the kinetic optimum is only obtained when $\Delta P = \Delta P_{max}$:

$$t_{R,KPL} = \lambda \cdot t_{R,exp} \tag{1.9}$$

$$n_{p,KPL} = 1 + \sqrt{\lambda} \cdot \left(n_{p,exp} - 1\right) \tag{1.10}$$

$$N_{KPL} = \lambda \cdot N_{exp} \tag{1.11}$$

with λ given by:

$$\lambda = \frac{P_{max}}{P_{exp}} \tag{1.12}$$

The subscripts "exp" and "KPL" in eqns (1.9)–(1.12) refer to the experimental data points (as collected on a single fixed length column for a series of different flow rates) and the corresponding data points on the KPL curve, respectively. In eqn (1.12), P_{exp} is the maximal pressure reached during the gradient run and P_{max} is the reference pressure for which the KPL curve is established (typically, P_{max} would be the maximally affordable instrument or column pressure).

In physical terms, eqns (1.9)–(1.12) directly represent the data transformation presented in Figure 1.2(b), based on the fact that an experimental kinetic plot can be calculated by extrapolating data obtained on a single column length to an imaginary set of columns, each with a different length but all operated at the same pressure.[74] The major advantage of eqns (1.9)–(1.12) is that they hold under both isocratic- and gradient-elution conditions.[69] A necessary condition for the application of eqns (1.9)–(1.12) under gradient conditions is that all data are collected using the same mobile-phase gradient history. This is achieved by maintaining a fixed ratio of t_G/t_0 and t_{dwell}/t_0 at all considered flow rates (and on all column lengths). Figure 1.3 shows that when this condition is fulfilled, the same KPL curve is always obtained, independent

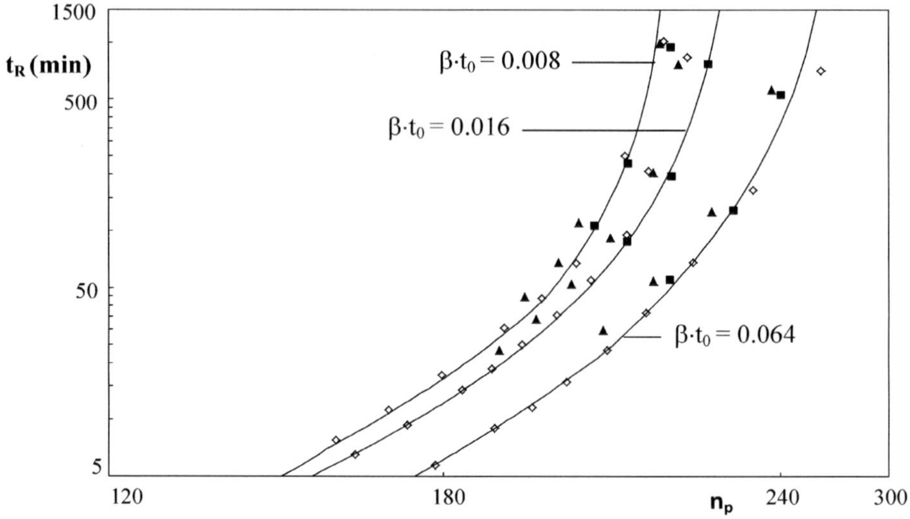

Figure 1.3 Verification of the overlap of the KPL curves originating from experiments conducted on columns with a different length (15 cm, ◇; 30 cm, ▲; 60 cm, ■) for various degrees of gradient steepness ($\beta \cdot t_0$ = 0.008, 0.016 and 0.064, with $\beta = \Delta\phi/t_G$). The gradient program was adapted to ensure a constant t_G/t_0 and t_{dwell}/t_0. Experimental conditions: T = 30 °C; test compounds: 0.02 mg mL^{-1} uracil, 0.1 mg mL^{-1} benzene, 0.05 mg mL^{-1} naphthalene and 0.05 mg mL^{-1} phenanthrene; mobile phase: ϕ_0 = 50:50 ACN/H_2O and ϕ_{end} = 100:0 ACN/H_2O; column: superficially porous (HALO 2.1 mm × 150 mm, d_p = 2.7 μm), V_{inj} = 1 μL, λ = 210 nm (adapted from reference 69).

of the length of the column used to collect the experimental kinetic performance data. The latter condition represents the fact that, as shown in Figure 1.2(a), for example, there is only one KPL curve enveloping the entire array of possible fixed length columns. On the other hand, the obtained KPL curve can depend significantly on the actual value of the gradient steepness (especially when considering a t_R *versus* n_p kinetic plot as is the case in Figure 1.3), as can be noted from the fact that the different considered t_G/t_0 cases lead to different KPL curves. This observation hints at the fact that a kinetic plot representation is also ideally suited to kinetically optimize the gradient conditions, for example, to maximize the gradient peak capacity in a given time.[56,65]

1.6 Relation between the Kinetic Performance under Isocratic and Gradient-Elution Conditions

It is sometimes claimed that although the isocratic performance of certain chromatographic columns is poorer than others, this is of no consequence for the gradient-elution performance. This, however, suggests that the performance under gradient and isocratic conditions are unrelated, which in turn would be in contradiction with the generally accepted notion that the basic band-broadening process in the gradient and the isocratic mode can be described by the same dependency on the particle diameter and mobile-phase velocity:

$$H = H_{eddy} + B\,(k)\cdot\frac{D_{mol}}{u} + C\,(k)\cdot u\cdot\frac{d_p^{\,2}}{D_{mol}} \tag{1.13}$$

where H_{eddy} (m) is the contribution of eddy diffusion to the plate height, B and C are dimensionless van Deemter constants, d_p (m) is the particle size of the chromatographic support and D_{mol} ($m^2\ s^{-1}$) is the molecular diffusion coefficient.

Whereas the analytes experience only one k value and one D_{mol} value in the isocratic mode, they experience a whole series of different k and D_{mol} values when travelling through the column in the gradient mode. Conceptually, this, however, does not change much compared to eqn (1.13).[75] In the gradient mode, the B and C constants, as well as the D_{mol} value appearing in eqn (1.13), need to be calculated by averaging them over the mobile phase history measured, but they can anyhow be expected to remain constant for different flow rates or column lengths provided the t_G/t_0 and t_{dwell}/t_0 ratio is kept constant, which is, nevertheless, the "*condition sine qua non*" for a fair column comparison and for a valid kinetic plot construction.[69]

This also implies that if one would compare an isocratic separation using a mobile phase that has a composition close to the average mobile-phase composition experienced during a gradient separation, the gradient and the isocratic mode should display very similar band broadening properties. Obviously, the gradient-elution mode might additionally benefit from the peak compression effect.[72,76,77] This effect is, however, to a first approximation independent of the degree of band broadening,[76] so that the advantage given to the gradient separation should be the same for all support types. In other words, the peak compression effect cannot be expected to alter the order in the performance comparison of different particle types when switching from the isocratic to the gradient-elution mode.

The only case wherein this order might be significantly affected occurs when comparing columns with significantly different retention behaviour. From the theory of gradient elution it follows clearly that the peak widths of the compound are determined not only by the column efficiency, but also by its retention factor at the point of elution k_e.[69,75,78] In addition, the achievable peak capacity (which is the best measure for column performance in gradient elution)[75] is also determined by the retention times of the compounds [see also eqn (1.8)], which are proportional to the average retention factor. Therefore, comparing columns which have largely different k and k_e values (as could, for example, be expected when comparing C_{18} packed bed and polymer monolithic columns), a good performance in gradient elution could be found for some specific samples, even when the isocratic performance is poor. These cases are, however, outside the scope of this Chapter.

To support the argumentation based on eqn (1.13), Figure 1.4 compares two fully porous and two porous-shell materials with a similar (C_{18}) surface chemistry in gradient mode and in isocratic mode. The isocratic experiments were performed using uracil as t_0 marker and dibenzothiopene sulfoxide to determine the column performance. For the gradient experiments a more extensive set of seven waste-water pollutants was used. The isocratic mobile-phase composition was adjusted for each column type in such a way that dibenzothiopene sulfoxide eluted with a retention factor $k \cong 7$, *i.e.* the average of the retention factors of the first and last eluting compound in the gradient runs (see below) and therefore yielding the most representative "average" retention behaviour of the considered sample in gradient elution.

To determine the separation performance in gradient elution, a constant ratio of gradient time t_G over column dead time t_0 was maintained ($t_G/t_0 = 12$) for all columns at all flow rates. The initial (ϕ_0) and final (ϕ_{end}) mobile-phase compositions of the gradient were adjusted for each column type independently to yield $k \cong 2$ for the first (2-napthoic acid) and $k \cong 12$ for the last compound (fluorine) in the sample, following a recently presented approach.[79] The separations on the different columns were thus compared for the same k-based elution window. For the column pressure, the highest value encountered during the gradient run was taken (corresponding to the mobile-phase composition having the highest average viscosity).

For both the isocratic and gradient runs, the peak capacity n_p was calculated using eqn (1.8). For the isocratic experiments, eqn (1.8) reduced to $n_p = 1 + (t_R - t_0)/(4 \times \sigma_t)$. On each column, the performance was measured at 11 different flow rates, ensuring that the column pressure drops were within the limits set by the manufacturers (*i.e.* 400 bar for the columns with fully porous particles and 600 bar for the superficially porous particles).

Figure 1.4(a) compares the isocratic kinetic performance of the four considered particle types, whereas Figure 1.4(b) shows the same for gradient-elution conditions. In both cases, the kinetic performance is reported as a total analysis time *versus* peak capacity, as this is, nevertheless, the most convenient one for the gradient-elution case. To compare all material types on the same basis, the same maximum column pressure (400 bar in the presently considered case) was used for all of them, regardless of the fact that some of the particle types have a higher operating limit [600 bar for most of the superficially porous columns (with $d_p > 2$ μm), and even 1000 bar for the 2.1 mm ID Kinetex columns].

Comparing the isocratic (Figure 1.4a) and the gradient (Figure 1.4b) kinetic performance of the different particle types, a striking similarity between the two operating modes can be noted. In both cases, the superficially porous particles clearly outperform the fully porous ones over most of the accessible n_p range, and also the subtle differences between the two different particle types belonging to the same group of materials (superficially porous or fully

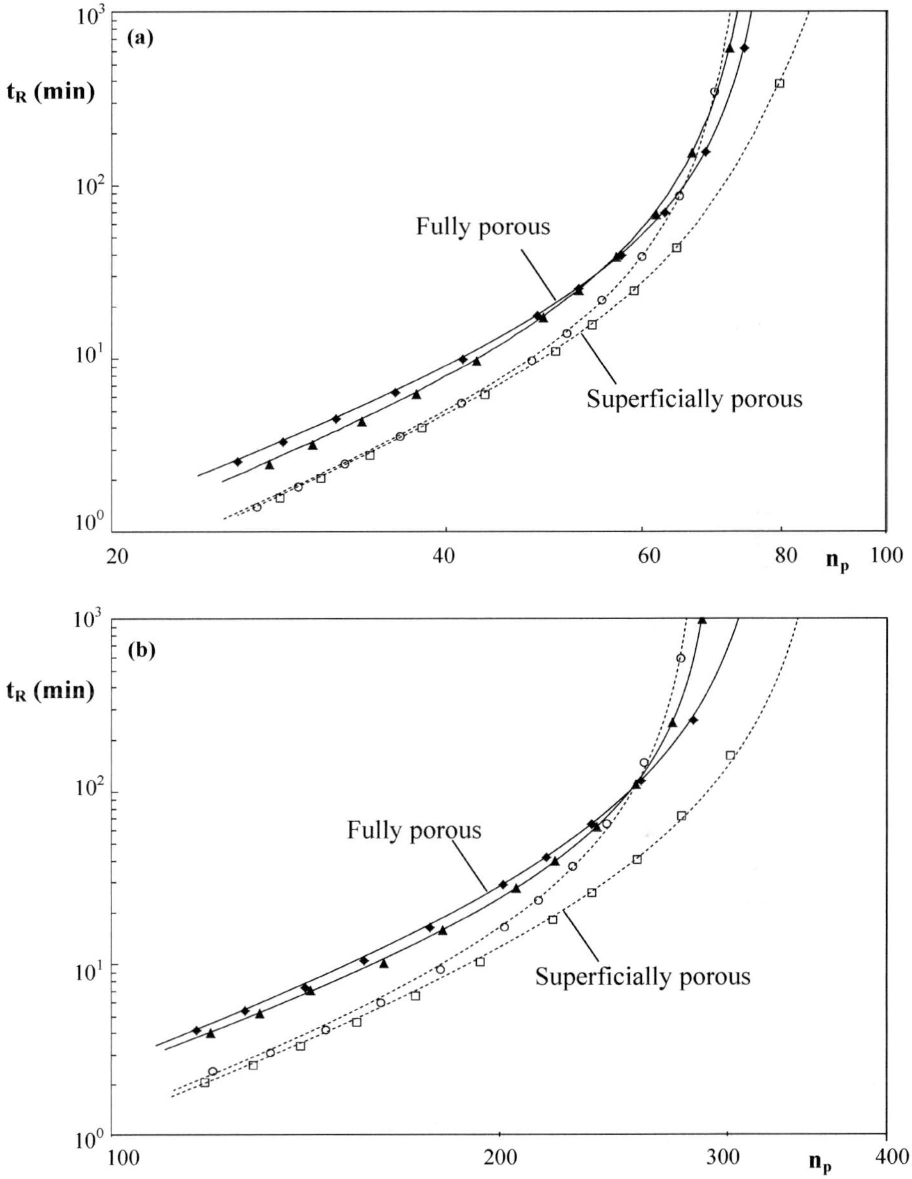

Figure 1.4 Comparison of the kinetic performance limit of columns packed with fully porous [continuous curves, ●: Xbridge C_{18} (100 mm × 4.6 mm, d_p = 3.5 μm) and ▲: ACE C_{18} (100 mm × 4.6 mm, d_p = 3.0 μm)] and superficially porous particles [dashed curves, ○: Kinetex C_{18} (100 mm × 4.6 mm, d_p = 2.6 μm), □: HALO C_{18} (100 mm × 4.6 mm, d_p = 2.7 μm)] in (a) isocratic elution, ϕ_{iso} = 32.5 ± 1.0% can, and (b) gradient elution with ϕ_0 = 14.0 ± 0.5% ACN, ϕ_{end} = 71.8 ± 3.0% ACN and t_G/t_0 = 12. Experimental conditions: aqueous phase; 10 mM ammonium acetate; T = 30 °C; test compounds: 0.04 mg mL^{-1} uracil, 0.1 mg mL^{-1} 2-napthoic acid, quinoline, dibenzothiopene sulfoxide, benzofuran, indene, indane and fluorine; V_{inj} = 1 μL, λ = 210 nm.

porous) are largely maintained when switching from the gradient to the isocratic elution mode and *vice versa*.

For the fully porous columns, the ACE column yielded a better kinetic performance than the XBridge column for $t_R > 0.5$ h in isocratic elution and $t_R > 1.5$ h in gradient mode. This behaviour is expected from the difference in particles size (3.0 μm *vs.* 3.5 μm). When comparing the superficially porous particles, the tested HALO column provided a better performance in the B-term dominated part of the KPL curve (high n_p end) in both elution modes (as previously observed),[80] whereas the difference in performance between the two tested columns was negligible around the optimum velocity and in the C-term regime (low n_p end).

1.7 Influence of the Test Conditions on the Obtained Kinetic Performance Limit Curve

Just as a fair comparison based on a van Deemter curve should preferentially occur using the same compounds and the same mobile phase (and/or the same retention factors), this also holds for a kinetic plot-based comparison. This is especially true when considering kinetic plots involving one or more parameters that depend heavily on the retention factors of the compounds, such as the total residence time t_R or the peak capacity n_p.[81]

When columns or conditions with widely differing retention properties are to be compared, the use of a constant mobile-phase composition or gradient profile could lead to widely differing retention factors. In this case, given the significant impact the value of k can have on the observed peak capacity or efficiency,[82] it might be preferable to adapt the mobile phase (gradient) for each different system individually, so that they are all compared under the conditions of (approximately) equal retention factors (as illustrated in Section 1.6 and reference 79). In some extreme cases [for example, when comparing a non-porous and a fully porous particle column, or when comparing a reversed-phased and a hydrophilic interaction liquid chromatography (HILIC) system], some radically different mobile-phase compositions will have to be employed to pursue conditions producing similar retention windows on the different systems. In such cases, it cannot be excluded that the selection of the sample compounds can lead to a bias, favouring or disfavouring some of the tested columns or operating conditions more than the others. A possible way out of this problem is to select a sample containing a good variety of the typical compounds one expects in a given application, and optimize the mobile phase (gradient) of each considered system individually. In this way, each different system is represented by its best possible performance, providing a suitable and practically relevant comparison basis. It is, however, advised to always use test compounds that are as relevant as possible for the type of applications one is mostly running as the observed kinetic performance of a chromatographic support depends strongly on the properties of the analyte used to perform the kinetic evaluations.[62,67]

In addition to the considerations one needs to make when establishing gradient kinetic plots (requiring that all measurements are conducted for the same t_G/t_0 and t_{dwell}/t_0 ratios so that the compounds always have the same retention factor), special attention should also be paid to the elimination of extra-column band broadening (ECBB) and pressure drop sources. As described by Heinisch *et al.*,[83] this can relatively easily be done for isocratic separations. For gradient separations, the correction for ECBB is more tedious because the bands may undergo a different focusing and defocusing effect at the column inlet and outlet, so that the ECBB one would measure *via* the classical method (replacement of column via a union piece) is no longer representative for the true ECBB in gradient elution. Indeed, the isocratic correction expression overestimates the contribution of the ECBB since it lumps both the pre- and post-column contributions.[69,83] Whereas the latter is independent of the elution mode (isocratic or gradient), the contribution to the observed peak width of the former is much smaller in gradient elution due to the focusing effect on the front of the column (where the retention is very high at the start of the gradient). Both contributions should therefore be considered separately, but this requires elaborate procedures. The only elegant solution for the ECBB problem hence consists of reducing it as much as possible and measuring the data such that they need as little correction as possible (preferably the column peak variance should be $>>$10 times the extra-column variance).

When kinetic plots are used to select the best possible chromatographic system (*i.e.* particle size, flow rate, column length *etc.*), and when this would turn out to have a significantly different length than that of the column used to generate the kinetic plot data, it is important to know the conditions wherein this column length extrapolation can be made without introducing a significant error (when kinetic plots are used to assess the column packing quality this is less relevant because the length extrapolation occurs for all different tested packing materials in the same way, *i.e.* by assuming a length-independent plate height). Unfortunately, the increasing operating pressures that accompany the use of longer columns, introduce several, often non-linear, effects on column performance and analyte retention.

An increase in operating pressure typically results in an increase in retention (except in some rare cases).[84,85] In addition, the mobile-phase viscosity increases (η) with increasing pressure and as a result the diffusion coefficients (D_{mol}) decrease. The equations underlying the KPM [see, for example, eqns (1.3) and (1.5)], however, assume that column performance (*H*), retention (*k*) and mobile-phase viscosity are independent of column length and hence also of column pressure. In addition, heat is generated inside the column due to viscous heating, causing both radial and axial temperature gradients that in turn affect D_{mol}, k, η and H in an inverse way than pressure. For systems operating at moderated pressures ($<$ 400 bar) or when using capillary columns, these effects are in general small and the experimental performance agrees almost perfectly with the kinetic plot predictions,[86] as can be seen from Figure 1.5.

At higher operation pressures the viscous-heating effects become more pronounced, although the amplitude of the errors on the kinetic plot predictions depends on the thermal environment in which the columns are operated. It has been shown that for a near-adiabatic system (still-air oven) the error between predicted and measured values is relatively small (on average only 3%). For a near-isothermal system (forced-air oven) the average error is larger (with a maximum of 13%).[87]

1.8 Some Reflections on Recent Trends in Liquid Chromatography using the Kinetic Plot Method

In the last few years, the KPM has been used on numerous occasions to assess and compare the true kinetic performance of newly developed chromatographic supports and operation modes. Kinetic plots have been used for two purposes: (i) to evaluate the intrinsic quality of columns (see Section 1.8.1) and (ii) to select the support type and/or operating condition that is most appropriate for a particular application (see Section 1.8.2).

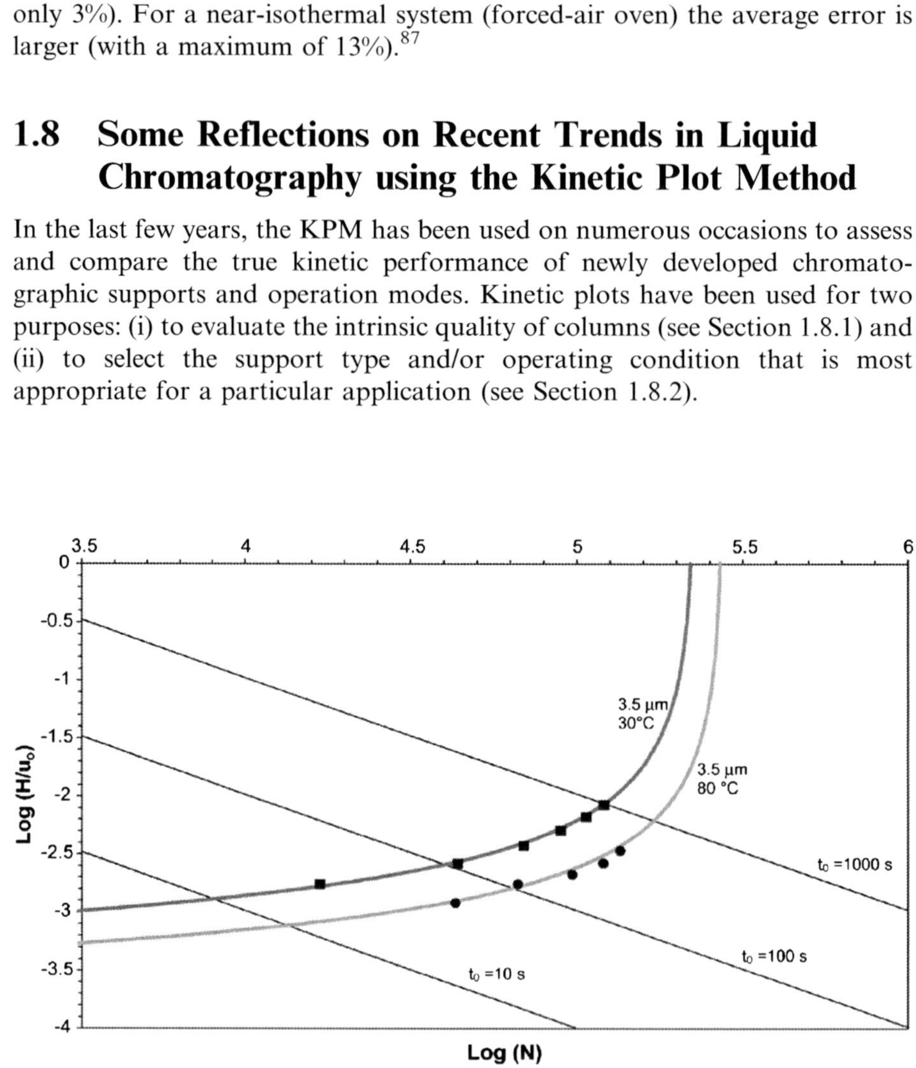

Figure 1.5 Kinetic performance of 3.5 μm Stable Bond Zorbax C_{18} at 30 °C and 80 °C for the isocratic elution of phenol with a 40:60 (%, v/v) ACN/water mobile phase (continuous lines, theoretical prediction; solid data points, experimental verification) (reprinted with permission from reference 86).

1.8.1 Intrinsic Column Quality Evaluation

The most suited kinetic plots for the evaluation of the intrinsic quality of a column are plots that are minimally affected by the actual retention factors of the analytes, for example, using t_0 instead of t_R and using N- or a t_G-based peak capacity (where the end of the elution window is determined by the breakthrough of the end of the gradient profile) to quantify the separation efficiency instead of a t_R-based peak capacity (where the end of the elution window is determined by the retention factor of the last eluting compound). If desired, a magnified view of the difference between the kinetic plot curves corresponding to the different tested columns can be obtained by plotting the data as t_0/N^2 *versus* N instead of as t_0 *versus* N. The former type of plot furthermore offers the additional advantage that it produces curves that pass through a minimum at exactly the same velocity as the corresponding van Deemter curve. Plotting the N axis in the reverse direction furthermore also puts the B-term and the C-term dominated parts of the curve in the same position with respect to the curve minimum as in a van Deemter plot.

Figure 1.6, for example, shows how t_0/N^2 *versus* N plots have been used to evaluate the kinetic performance of a series of newly synthesized silica-monolithic columns.[57,88] The plot in Figure 1.6 shows that these second-generation monolithic silica capillary columns yield a better kinetic performance than a particulate column packed with 2 μm particles as soon as efficiencies of more than 25 000–30 000 theoretical plates are needed.[89]

Recently, insights obtained using the KPM described by Miyamoto *et al.*[28] to produce ultra-long monolithic silica–C_{18} capillary columns (1140–1240 cm) yielding record efficiencies of over 1 000 000 theoretical plates. An example of such a high-resolution separation for alkylbenzenes is shown in Figure 1.7. A kinetic plot analysis was also used to examine the chromatographic performance of a new type of silica rod column, indicating that the monolithic columns operated at 300 bar can provide faster separations than a column

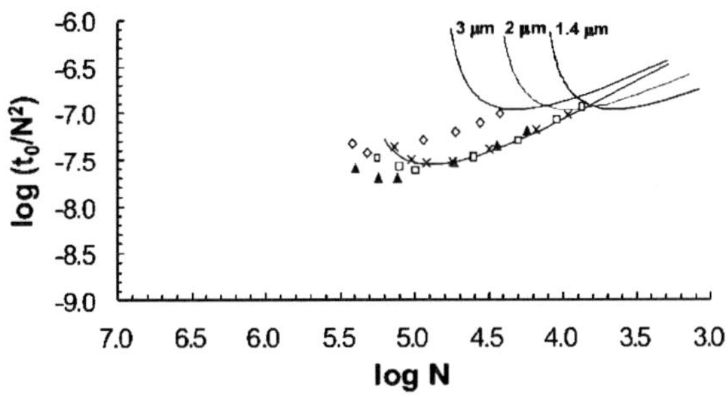

Figure 1.6 Comparison of the performance of packed bed columns with different particles sizes (assuming $\eta = 4.6 \times 10^{-4}$ Pa s, $\phi_0 = d_p^2/K_{v0} = 700$, $D_{mol} = 2.22 \times 10^{-9}$ m^2 s^{-1}, and h calculated using the following Knox equation: $h = 0.65 \times \nu^{1/3} + 2/\nu + 0.08 \times \nu$) for different types of silica-monolithic columns (symbols) and a maximum pressure P_{max} of 1000 bar (reprinted with permission from reference 57).

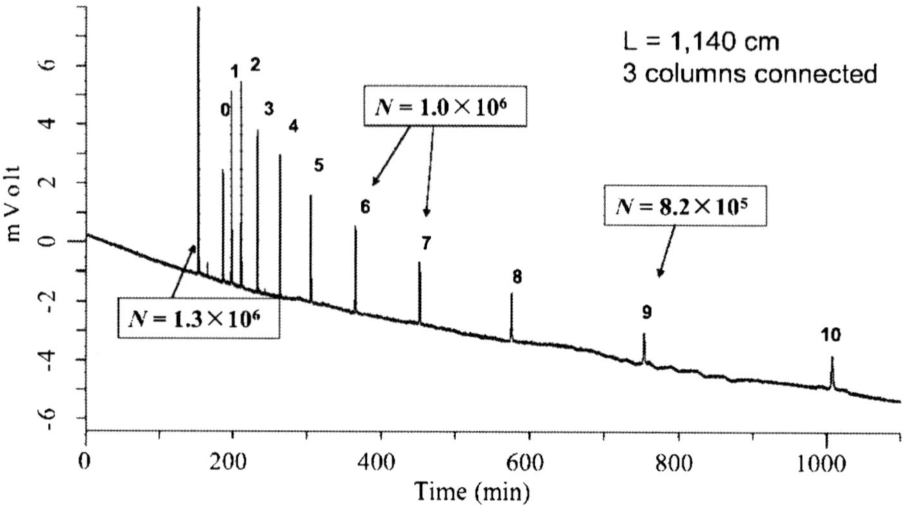

Figure 1.7 Separation of alkylbenzenes. Mobile phase: acetonitrile/water (80:20); column: three monolithic silica C_{18} columns connected in series, MS-100H-C18; effective length: 1140 cm; ΔP= 35.4 MPa; u= 1.24 mm s^{-1}; detection: 210 nm, temperature: 30 °C; sample: thiourea and alkylbenzenes ($C_6H_5C_nH_{2n+1}$, n = 0–10) (reprinted with permission from reference 28).

packed with totally porous 3 μm particles operated at 400 bar in a range where the number of theoretical plates (N) is greater than 50 000.[90] These observations are expected to lead to the commercialization of a second generation of more robust and efficient monolithic columns that will rival the efficiencies of the current small particle columns.

Using the kinetic plot method, it has also been demonstrated that the presence of particle fines in sub-2 μm high-performance columns leads to an inferior kinetic performance.[91] The effect of these particle fines is especially apparent when using a so-called reduced kinetic plot, wherein both the x- and y-axis are made dimensionless so that the effect of any difference in average particle size (which is inevitable when particle fines are present) is excluded. To obtain a dimensionless y-axis, it is sufficient to multiply the t_0/N^2 value with the ratio of $\Delta P/\eta$, so that one readily obtains Knox's separation impedance parameter:[61,92]

$$E_0 = \frac{\Delta P_{max}}{\eta}\frac{t_0}{N^2} \qquad (1.14)$$

A dimensionless x-axis can be obtained by plotting the ratio of N_{opt}/N instead of purely plotting N.[91]

The same correlation between a narrow particle size distribution (PSD) and a superior kinetic performance (in terms of h_{min}, A and E_{min}) has been observed for superficially porous and 3.0–3.5 μm fully porous columns, where especially the superficially porous particles demonstrated an extremely narrow PSD [standard deviation (σ) = 0.05–0.10 compared to σ= 0.15–0.20 for the fully porous particles] and consequently outperformed all investigated fully porous particles.[23] These findings suggest that reducing the PSD of the current generation of fully porous particles could lead to a significant improvement of their performance.

1.8.2 Selection of the Most Suitable Chromatographic System

For the selection of the kinetically most advantageous chromatographic system (support type, stationary phase type, mobile-phase propagation method, *etc.*) for a specific application, kinetic plots containing the practically most relevant parameters are mostly suited, preferring t_R over t_0 to quantify the separation speed and preferring a t_R-based peak capacity over the t_G-based peak capacity or the plate number N to quantify the separation efficiency. Using a t_R-based peak capacity, the resulting kinetic plot curve can depend significantly on the employed gradient slope and starting composition, thus offering the possibility to kinetically optimize these parameters. When it specifically comes to selecting a system offering the best resolution (R_s) between a given critical pair, it is even possible to directly make a plot of t_R versus R_s.[68]

Finding the optimal particle size for a given range of desired efficiencies has probably been one of the most frequently adopted applications of the KPM in the past few years.[56,69,93,94] In each case, observations were made that are fully similar to those shown in the different panels of Figure 1.8, compiling results on different C_{18} columns for the isocratic elution of propylparaben (k = 10). Figure 1.8 shows that, when employed at the same pressure limit, small particles are to be preferred in the range of relatively easy separations (range of small N), whereas large particles (3 and, even, 5 μm) yield better performances when very high efficiencies are needed. A fully similar observation is made under gradient-elution conditions.[69]

Recently, several studies have also assessed the difference in kinetic properties between fully porous and porous-shell particles.[25,63,65,79] It has been demonstrated that novel sub-2 μm superficially porous particles, which can be operated up to 1000 bar, exhibit a significantly higher kinetic performance than fully porous sub-2 μm particles (at the same pressure limit) and 2.7 μm superficially porous particles (at 600 bar pressure limit) for both small and large molecules in isocratic elution and over the entire range of analysis times.[25] Comparing fully porous sub-2 μm columns with 2.7 μm superficially porous particles, it has been shown on several occasions that superficially porous particles outperform fully porous particles when compared at the same operating pressure.[25,63,79] This shows that expanding the allowable operating pressure of superficially porous particles to those of sub-2 μm particles (*i.e.* 1000 bar and above) would result in previously unsurpassed separation performance.

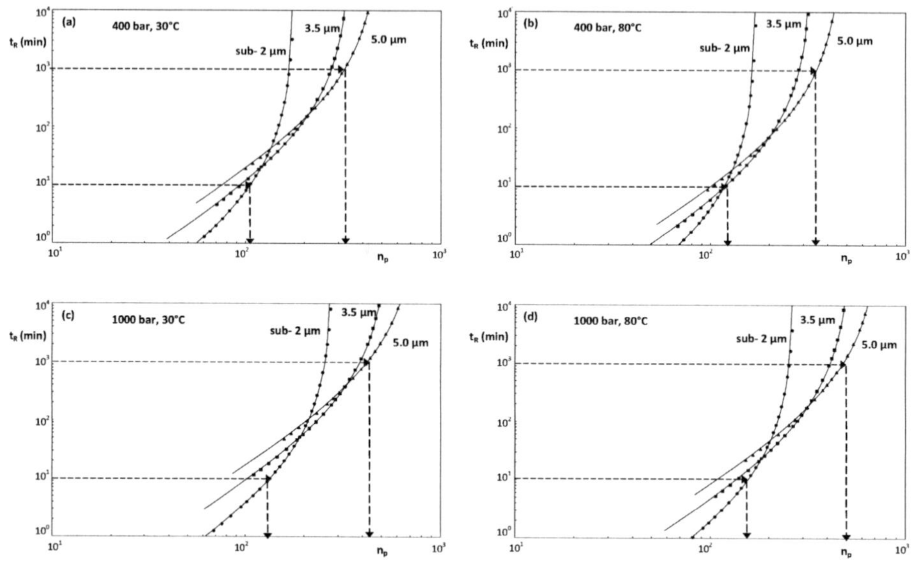

Figure 1.8 Plot of t_R *versus* n_p for columns with different particle sizes [sub-2 μm (●), 3.5 μm (■) and 5 μm (▲)] and at different combinations of pressure and temperature: (a) $T = 30$ °C and $\Delta P = 400$ bar; (b) $T = 80$ °C and $\Delta P = 400$ bar; (c) $T = 30$ °C and $\Delta P = 1000$ bar and (d) $T = 80$ °C and $\Delta P = 1000$ bar. The t_R values relate to a compound eluting with $k = 10$ (adapted from reference 98).

Kinetic plots are also ideally suited to quantify the potential advantage of HTLC. Because the main parameters affected by temperature (η, k, H, ΔP) are incorporated in the kinetic plot equations [eqns (1.3), (1.5) and (1.6)], the KPM allows the effect of an elevated temperature operation on different chromatographic performance parameters to be combined into one single plot. In several independent studies, it has been demonstrated that the gain in analysis time by moving from a 30 °C to an 80 °C operation is of the order of 2 to 2.5 if both conditions are compared for the same achieved efficiency.[86,95] This gain factor is clearly large and significant, but not as large as the factors of 10 to 20 that are sometimes claimed. The latter is due to the fact that in a kinetic plot the two systems are always compared on the basis of an optimized length, whereas in a pure one-to-one comparison the column length is usually only optimized for one condition but not for the other. Figure 1.9 shows an example of how the separation of a pharmaceutical mixture, performed on sub-2 μm columns, can be expedited with a factor of 2 by switching from 30 °C to 80 °C while maintaining the separation resolution. For larger temperature increases, for example, a temperature increase from 30 °C to 120 °C, this kinetic gain is even larger with a gain factor of approximately 3 to 4.[86,95] In fact, the kinetic gain factor that can be expected over most of the C-term dominated range when working at elevated temperatures largely parallels the decrease in viscosity of the mobile phase induced by the temperature

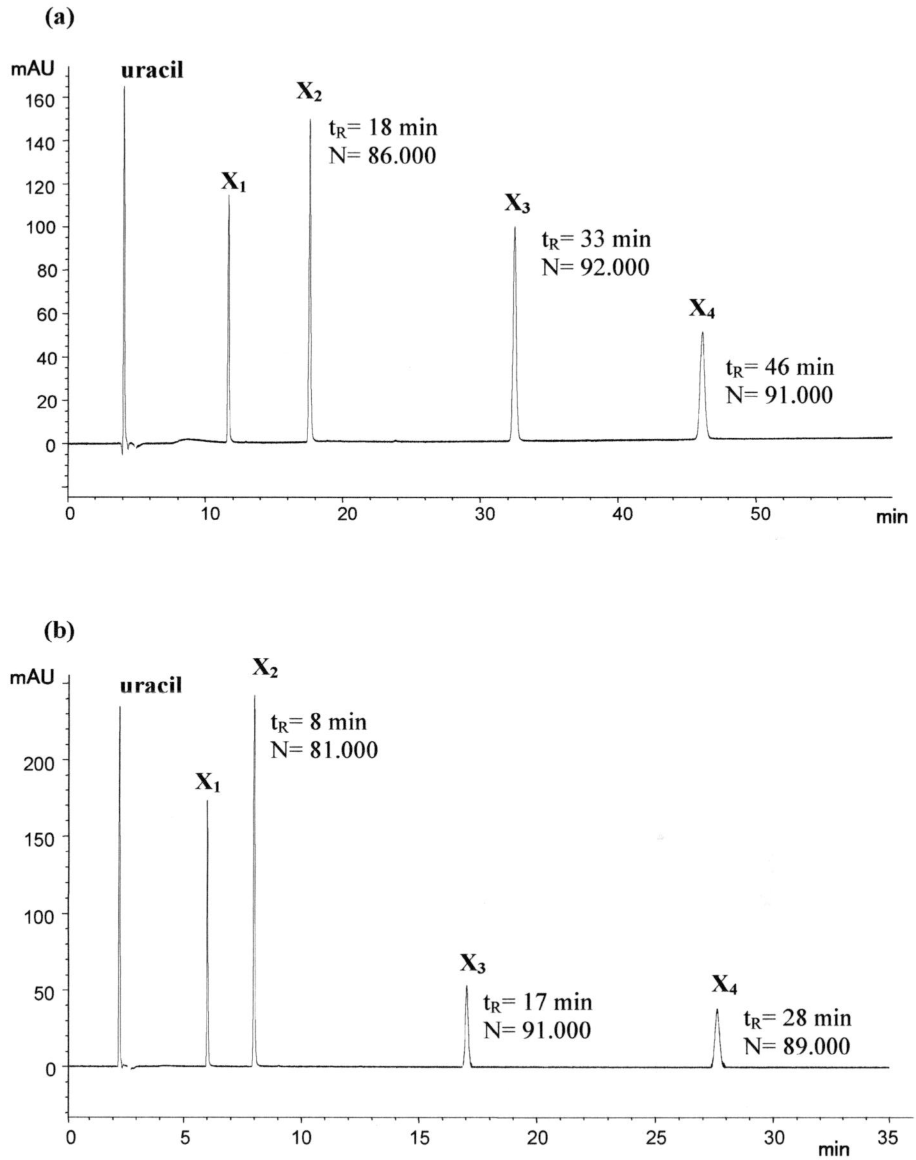

Figure 1.9 Chromatograms obtained for a pharmaceutical mixture (X_1–X_4) on a 40 cm Acquity C_{18} column, operated at maximum pressure and at: (a) T = 30 °C (F = 0.2 mL min^{-1}; P = 960 bar; mobile phase: 59:41 ACN/0.1% HCOOH) and (b) T = 80 °C (F = 0.35 mL min^{-1}; P = 925 bar; mobile phase: 53:47 ACN/0.1% HCOOH). The retention times and plate counts of each compound are indicated on the chromatograms (adapted from reference 99).

increase.[95] The same observations have been made in gradient elution.[96] As is evident from Figure 1.8 [compare (a) with (b) and (c) with (d)], the effect of an elevated temperature on the chromatographic performance of a support is mostly pronounced in the C-term region, hence for fast separations requiring relatively low efficiencies. Elevated operating temperatures are less suited to speed up ultra-high resolution separations ($N > 150\ 000$), because of the negative effect of high temperature on the kinetic performance in the B-term region, caused by the increased diffusivity at elevated temperatures.

The effect of an elevated inlet pressure, on the other hand, is clearly present in the B-term region and therefore mainly beneficial for separations requiring large peak capacities.[45,59,69,72,86,97] This is also what is observed in Figure 1.8 [compare (a) with (c) and (b) with (d)]. The KPL curve shifts to the right with increasing ΔP_{max}, providing a better separation performance and/or shorter analysis times. The gain for fast (low efficiency) analyses is limited: for example, to reach a separation performance of $n_p = 120$ at 30 °C on the 3.5 μm support [Figure 1.8 (a) and (c)], a 2.5-fold increase in pressure (400 *vs.* 1000 bar) only results in a decrease of analysis time with a factor of 1.4 (15 *vs.* 21 min). In the high efficiency region, however (for example, for $n_p = 250$), increasing the operating pressure by the same factor reduces the required analysis time by a factor of more than 3 (125 *vs.* 430 min).

These results hence show that the beneficial effect of elevated pressures (essentially improving the performance at high efficiencies) is complementary to that of high temperatures (essentially improving the performance in the low efficiency range).[95,97]

1.9 Conclusions

Kinetic plots are ideally suited to assess and compare the separation performance of new support structures and operation modes under isocratic and gradient-elution conditions in an unbiased way. Moreover, due to the availability of a very simple set of equations for their construction (*cf.* eqns 1.9–1.12), kinetic plot-based comparisons or quality tests can be directly carried out on real experimental data, collected with the sample and mobile-phase conditions of a particular interest.

Using kinetic plots for the evaluation of the recent trends in LC, it has been demonstrated that supports with an open porous structure and a high permeability, such as monolithic columns and fully porous particles with a particle size of 3.5–5.0 μm, are the best choice for high-resolution separations. This is because the low flow resistance of these supports allows them to be used in very long columns at reasonable flow rates.

Supports with small diffusion distances, such as sub-2 μm fully porous particles and superficially porous particles, on the other hand, are to be preferred for fast and easy separations. Due to the small diffusion distances leading to high efficiencies, only very short column lengths are needed for most routine separations, significantly reducing the analysis time. The high backpressure emanating from the small particle sizes, however, puts an upper limit on the column length, making these supports less suited for high-resolution separations requiring long columns lengths. However, the higher the available instrument pressure, the higher the range of efficiencies over which small particles perform better than larger particles.

Comparing the efficiency of the recently (re-)introduced superficially porous particles to those of the sub-2 μm fully porous particles, the former seem to perform equally well as the latter at much lower operating pressures. Therefore, increasing the operating pressure range of the superficially porous particles to 1000–1300 bar will definitely extend the range of high-throughput and high-resolution separations for these particles in the future.

Finally, a complementarity has been observed for high-temperature and high-pressure operations. Due to the negative influence of elevated temperature on the kinetic performance in the B-term region, high temperatures are mainly suited for fast and easy high-throughput separations. Ultra-high pressure, on the other hand, allows the column length to be increased while maintaining the separation speed and is therefore mainly suited for high-resolution separations.

Acknowledgements

D. Cabooter is a fellow of the Research Foundation Flanders. K. Broeckhoven is thanked for his contribution to this chapter.

References

1. K. K. Unger, R. Skudas and M. M. Schulte, *J. Chromatogr. A*, 2008, **1184**, 393.
2. I. Halasz, R. Endele and J. Asshauer, *J. Chromatogr.* 1975, **112**, 37.
3. A. D. Jerkovich, J. S. Mellors and J. W. Jorgenson, *LC GC Eur.*, 2003, **6a**, 20.
4. J. E. MacNair, K. D. Patel and J. W. Jorgenson, *Anal. Chem.*, 1999, **71**, 700.
5. J. S. Mellors and J. W. Jorgenson, *Anal. Chem.*, 2004, **76**, 5441.
6. N. Wu, J. Lippert and M. Lee, *J. Chromatogr., A*, 2001, **911**, 1.
7. J. R. Mazzeo, U. D. Neue, M. Kele and R. S. Plumb, *Anal. Chem.*, 2005, **77**, 460A.
8. M. E. Swartz, *LC-GC N. Am.*, 2005, **22**, 8.
9. M. Verstraeten, K. Broeckhoven, M. Dittmann, K. Choikhet, K. Witt and G. Desmet, *J. Chromatogr. A*, 2011, **1218**, 1170.
10. U. D. Neue and M. Kele, *J. Chromatogr. A*, 2007, **1149**, 236.
11. G. Desmet, *J. Chromatogr. A*, 2006, **1116**, 89.
12. T. Greibrokk and T. Andersen, *J. Chromatogr. A*, 2003, **1000**, 743.
13. C. Zhu, D. M. Goodall and S. Wren, *LC-GC Eur.*, 2004, **10**, 530.

14. D. T-T. Nguyen, D. Guillarme, S. Heinisch, M-P. Barrioulet, J-L. Rocca, S. Rudaz and J-L. Veuthey, *J. Chromatogr. A*, 2007, **1167**, 76.
15. Y. Xiang, B. Yan, B. Yue, C. V. McNeff, P. W. Carr and M. L. Lee, *J. Chromatogr. A*, 2003, **983**, 83.
16. D. Guillarme and S. Heinisch, *Sep. Purif. Rev.*, 2005, **34**, 181.
17. D. V. McCally, *J. Chromatogr. A*, 2000, **902**, 311.
18. S. Heinisch, G. Puy, M.P. Barrioulet and J.L. Rocca, *J. Chromatogr. A*, 2006, **1118**, 234.
19. S. Heinisch, J-L. Rocca, *J. Chromatogr. A*, 2009, **1216**, 642.
20. J. W. Li and P. W. Carr, *Anal. Chem.*, 1997, **69**, 2193.
21. T. P. Weber, P. T. Jackson and P. W. Carr, *Anal. Chem.*, 1995, **67**, 3042.
22. J. J. DeStefano, T. J. Langlois and J. J. Kirkland, *J. Chromatogr. Sci.*, 2008, **46**, 254
23. D. Cabooter, A. Fanigliulo, G. Bellazzi, B. Allieri, A. Rottigni and G. Desmet, *J. Chromatogr. A*, 2010, **1217**, 7074.
24. G. Guiochon and F. Gritti, *J. Chromatogr. A*, 2011, **1218**, 1915.
25. S. Fekete, K. Ganzler and J. Fekete, *J. Pharm. Biomed. Anal.*, 2011, **54**, 482.
26. N. Tanaka, H. Kobayashi, N. Ishizuka, H. Minakuchi, K. Nakanishi, K. Hosoy and T. Ikegami, *J. Chromatogr. A*, 2002, **965**, 35.
27. J. Bones, C. Duffy, M. Macka and B. Paull, *Analyst*, 2008, **133**, 180.
28. K. Miyamoto, T. Hara, H. Kobayashi, H. Morisaka, D. Tokuda, K. Horie, K. Koduki, S. Makino, O. Núñez, C. Yang, T. Kawabe, T. Ikegami, H. Takubo, Y. Ishihama and N. Tanaka, *Anal. Chem.*, 2008, **80**, 8741.
29. F. Svec and J. M. J. Fréchet, *Anal. Chem.*, 1992, **64**, 820.
30. C. Viklund, E. Pontén, B. Glad, K. Irgum, P. Hörstedt and F. Svec, *Chem. Mater.*, 1997, **9**, 463.
31. S. Xie, F. Svec and J. M. J. Fréchet, *J. Chromatogr. A*, 1997, **775**, 65.
32. M. Wang, J. Xu, X. Zhou and T. Tan, *J. Chromatogr. A*, 2007, **1147**, 24.
33. C. M. Zeng, J. L. Liao, K. Nakazato and S. Hjerten, *J. Chromatogr. A*, 1996, **753**, 227.
34. Q. C. Wang, F. Svec and J. M. J. Fréchet, *Anal. Chem.*, 1993, **65**, 2243.
35. S. T. Popovici and P. Schoenmakers, *J. Chromatogr. A*, 2004, **1060**, 237.
36. A. Premstaller, H. Oberacher and C. G. Huber, *Anal. Chem.*, 2000, **72**, 4386.
37. K. Nakanishi and N. Soga, *J. Am. Ceram. Soc.*, 1991, **74**, 2518.
38. G. Guiochon, *J. Chromatogr. A*, 2007, **1168**, 101.
39. L. Rieux, D. Lubda, H. A. G. Niederlander, E. Verpoorte and R. Bischoff, *J. Chromatogr. A*, 2006, **1120**, 165.
40. L. Xiong, R. Zhang and F. E. Regnier, *J. Chromatogr. A*, 2004, **1030**, 187.
41. N. Tanaka, H. Kobayashi, K. Nakanishi, H. Minakuchi and N. Ishizuka, *Anal. Chem.*, 2001, **73**, 420.
42. J. H. Knox, *J. Chromatogr. A*, 2002, **960**, 7.
43. J. H. Knox, *J. Chromatogr. A*, 1999, **831**, 3.

44. J. H. Knox, *J. Chromatogr. Sci.*, 1980, **18**, 454.
45. G. Desmet, D. Clicq and P. Gzil, *Anal. Chem.*, 2005, **77**, 4058.
46. G. Desmet, P. Gzil and D. Clicq, *LC-GC Eur.*, 2005, **18**, 403.
47. E. D. Katz, K. Ogan and R. P. W. Scott, *J. Chromatogr. Library*, 1985, **32**, 403.
48. M. Martin and G. Guiochon, *J. Chromatogr. A*, 2005, **1090**, 16.
49. C. Horvath and H-J. Lin, *J. Chromatogr.*, 1978, **149**, 43.
50. A. Jerkovich, J. Mellors and J. Jorgenson, *LC-GC Eur.*, 2003, **16**, 20.
51. J. Billen and G. Desmet, *J. Chromatogr. A*, 2007, **1168**, 73.
52. J. C. Giddings, *Anal. Chem.*, 1965, **37**, 60.
53. J. H. Knox and M. Saleem, *J. Chromatogr. Sci.*, 1969, **7**, 614.
54. G. Guiochon, *Anal. Chem.*, 1981, **53**, 1318.
55. H. J. Poppe, *J. Chromatogr. A*, 1997, **778**, 3.
56. X. Wang, D. R. Stoll, P. W. Carr and P. J. Schoenmakers, *J. Chromatogr. A*, 2006, **1125**, 177.
57. T. Hara, H. Kobayashi, T. Ikegami, K. Nakanishi and N. Tanaka, *Anal. Chem.*, 2006, **78**, 7632.
58. T. J. Causon, E. F. Hilder, R. A. Shellie and P. R. Haddad, *J. Chromatogr. A*, 2010, **1217**, 5057.
59. P. W. Carr, X. Wang and D. R. Stoll, *Anal. Chem.*, 2009, **81**, 5342.
60. A. de Villiers, F. Lestremau, R. Szucs, S. Gélébart, F. David and P. Sandra, *J. Chromatogr. A*, 2006, **1127**, 60.
61. U. D. Neue, *LC-GC Eur.*, 2009, **22**, 570.
62. D. Guillarme, J. Ruta, S. Rudaz and J-L. Veuthey, *Anal. Bioanal. Chem.*, 2010, **397**, 1069.
63. E. Oláh, S. Fekete, J. Fekete and K. Ganzler, *J. Chromatogr. A*, 2010, **1217**, 3642.
64. D. T-T. Nguyen, D. Guillarme, S. Rudaz and J-L. Veuthey, *J. Chromatogr. A*, 2006, **1128**, 105.
65. J. Ruta, D. Guillarme, S. Rudaz and J-L. Veuthey, *J. Sep. Sci.*, 2010, **33**, 2465.
66. S. Louw, F. Lynen, M. Hanna-Brown and P. Sandra, *J. Chromatogr. A*, 2010, **1217**, 514.
67. A. de Villiers, F. Lynen and P. Sandra, *J. Chromatogr. A*, 2009, **1216**, 3431.
68. A. Fanigliulo, D. Cabooter, G. Bellazzi, B. Allieri, A. Rottigni and G. Desmet, *J. Chromatogr. A*, 2011, **1218**, 3351.
69. K. Broeckhoven, D. Cabooter, F. Lynen, P. Sandra and G. Desmet, *J. Chromatogr. A*, 2010, **1217**, 2787.
70. G. Desmet and D. Cabooter, *LC-GC Eur.*, 2009, **22**, 70.
71. D. V. McCalley, *J. Chromatogr. A*, 2010, **1217**, 858.
72. U. D. Neue, D. H. Marchand and L. R. Snyder, *J. Chromatogr. A*, 2006, **1111**, 32.
73. U. D. Neue, H. B. Hewitson and T. E. Wheat, *J. Chromatogr. A*, 2010, **1217**, 2179.

74. S. Eeltink, G. Desmet, G. Vivó-Truyols, G. P. Rozing, P. J. Schoenmakers and W. Th. Kok, *J. Chromatogr. A*, 2006, **1104**, 256.
75. U. D. Neue, *J. Chromatogr. A*, 2005, **1079**, 153.
76. H. Poppe, J. Paanakker and M. Bronckhorst, *J. Chromatogr.*, 1981, **204**, 77.
77. L. R. Snyder, J. W. Dolan and J. R. Gant, *J. Chromatogr. A*, 1979, **165**, 3.
78. L. R. Snyder and D. L. Saunders, *J. Chromatogr. Sci.*, 1969, **7**, 195.
79. K. Broeckhoven, D. Cabooter and G. Desmet, *J. Chromatogr. A*, submitted.
80. A. Fanigliulo, D. Cabooter, G. Bellazzi, D. Tramarin, B. Allieri, A. Rottigni and G. Desmet, *J. Sep. Sci.*, 2010, **33**, 3655.
81. U. D. Neue, *J. Chromatogr. A*, 2008, **1184**, 107.
82. J. J. van Deemter, F. J. Zuiderweg and A. Klinkenberg, *Chem. Eng. Sci.*, 1956, **5**, 271.
83. S. Heinisch, G. Desmet, D. Clicq and J-L. Rocca, *J. Chromatogr. A*, 2008, **1203**, 124.
84. M. M. Fallas, U. D. Neue, M. R. Hadley and D. V. McCalley, *J. Chromatogr. A*, 2008, **1209**, 195.
85. U. D. Neue, C. J. Hudalla and P. C. Iraneta, *J. Sep. Sci.*, 2010, **33**, 838.
86. F. Lestremau, A. de Villiers, F. Lynen, A. Cooper, R. Szucs and P. Sandra, *J. Chromatogr. A*, 2007, **1138**, 120.
87. D. Cabooter, F. Lestremau, A. de Villiers, K. Broeckhoven, F. Lynen, P. Sandra and G. Desmet, *J. Chromatogr. A*, 2009, **1216**, 3895.
88. T. Hara, S. Makino, Y. Watanabe, T. Ikegami, K. Cabrera, B. Smarsly and N. Tanaka, *J. Chromatogr. A*, 2010, **1217**, 89.
89. O. Núñez, K. Nakanishi and N. Tanaka, *J. Chromatogr. A*, 2008, **1191**, 231.
90. S. Miyazaki, M. Takahashi, M. Ohira, H. Terashima, K. Morisato, K. Nakanishi, T. Ikegami, K. Miyabe and N. Tanaka, *J. Chromatogr. A*, 2011, **1218**, 1988.
91. J. Billen, D. Guillarme, S. Rudaz, J-L. Veuthey, H. Ritchie, B. Grady and G. Desmet, *J. Chromatogr. A*, 2007, **1161**, 224.
92. P. A. Bristow and J. H. Knox, *Chromatographia*, 1977, **10**, 279.
93. B. Chauve, D. Guillarme, P. Cléon and J-L. Veuthey, *J. Sep. Sci.*, 2010, **33**, 752.
94. D. Cabooter, A. de Villiers, D. Clicq, R. Szucs, P. Sandra and G. Desmet, *J. Chromatogr. A*, 2007, **1147**, 183.
95. D. Cabooter, S. Heinisch, J. L. Rocca, D. Clicq and G. Desmet, *J. Chromatogr. A*, 2007, **1143**, 121.
96. D. Guillarme, E. Grata, G. Glauser, J-L. Wolfender, J-L. Veuthey and S. Rudaz, *J. Chromatogr. A*, 2009, **1216**, 3232.
97. G. Desmet, D. Cabooter, P. Gzil, H. Verelst, D. Mangelings, Y. Vander Heyden and D. Clicq, *J. Chromatogr. A*, 2006, **1130**, 158.
98. G. Desmet, *LC-GC Eur.*, 2008, **21**, 310.
99. D. Cabooter, F. Lestremau, F. Lynen, P. Sandra and G. Desmet, *J. Chromatogr. A*, 2008, **1212**, 23.

CHAPTER 2

Instrumentation and Columns for UHPLC Separations

KENNETH J. FOUNTAIN* AND PAMELA C. IRANETA

Waters Corporation, 34 Maple Street, Milford, MA 01757, USA
*E-mail: kenneth_fountain@waters.com

2.1 UHPLC Instrumentation

2.1.1 Instrument Component Overview

The most common terms used for chromatography using pressures greater than 400 bar are ultra-performance liquid chromatography (UPLC®) and ultra-high-pressure liquid chromatography (UHPLC). The former term is trademarked by Waters Corporation. Waters Corporation introduced the first commercially available LC instrument capable of pressures up to 15 000 psi (1034 bar), and uses the term UPLC to describe separations performed with columns packed with sub-2 μm particles on low dispersion instruments specifically optimized to achieve faster, higher resolution, and more sensitive analyses. The term UHPLC is the more generic term that refers to separations performed at pressures above 400 bar. For the remainder of this Chapter, the term UHPLC will be used to encompass all of the columns and systems used for working at pressures above 400 bar, but special attention will be paid to the use of sub-2 μm particle columns and their influence on instrument and column design.

Whether the maximum operating pressure is 600 bar or 1000 bar, new instrumentation is required to properly operate columns packed with sub-2 μm particles. Decreasing the particle size by a factor of 2 (*i.e.*, from 3.5 μm to 1.7

RSC Chromatography Monographs No. 16
UHPLC in Life Sciences
Edited by Davy Guillarme and Jean-Luc Veuthey

Published by the Royal Society of Chemistry, www.rsc.org

μm) increases the backpressure by a factor of 4 at constant linear velocity.[1] In addition, the linear velocity at the minimum of the van Deemter curve increases as particle size decreases, which at optimum linear velocity causes the pressure to increase with a decrease in particle size cubed.[2] For a 2-fold reduction in particle size, the backpressure at optimal flow actually increases by a factor of 8. Thus, a UHPLC system must be capable of pumping liquid precisely at elevated pressure (well above 400 bar), as well as introducing the sample at high pressure.[3] Other requirements for achieving efficient separations with small particle columns include adequate detector sampling rates ($\geq$40 Hz) and fast detector time constants ($\leq$0.05 s for fast separations), fast injection cycle times to accommodate analyses under 1 min, and minimal extra-column dispersion and gradient delay volume. The difference between extra-column volume (*e.g.*, dispersion) and gradient delay volume will be explained later in the Chapter.

The next three Sections will provide more detailed information about the design of different components of UHPLC systems. While each module will be explained as a separate entity, it is important to keep in mind that virtually all manufacturers of UHPLC instruments design and market an entire system, whereby each component of the system is ideally suited to work in concert to achieve the best possible performance.

2.1.1.1 Solvent Delivery

The solvent delivery system, or pump, used for any chromatographic separation must be capable of delivering the correct flow rate and mobile phase composition at all times. Before discussing the requirements of UHPLC systems to properly operate sub-2 μm particle columns, we need to differentiate between low- and high-pressure mixing pumps. High-pressure mixing is performed by a multi-pump system and is typical of binary solvent delivery systems, whereas low-pressure mixing is performed by a single pump and is used in a quaternary solvent delivery system. The major operational difference between these two types of pumping systems, other than the number of different solvents that can be used simultaneously, is the location of solvent mixing. In a high-pressure (or binary) mixing system (Figure 2.1, top panel), the solvents are mixed after the pump, and usually undergo some expansion or contraction. The actual flow rate pumped may be different from what is programmed, which is compensated by the use of software algorithms. In contrast, in a low-pressure (or quaternary) mixing system (Figure 2.1, bottom panel), the solvents are mixed before the pump, usually using a device called a gradient proportioning valve (GPV). In this case, the flow accuracy is determined by the accuracy of the pump instead of the compressibility of the fluids.

Each mixing system has advantages and disadvantages. High-pressure mixing systems typically have lower gradient delay volumes (*e.g.*, volume between the pump and head of the column), which is important for high-throughput gradient analyses. This type of mixing is especially helpful at low

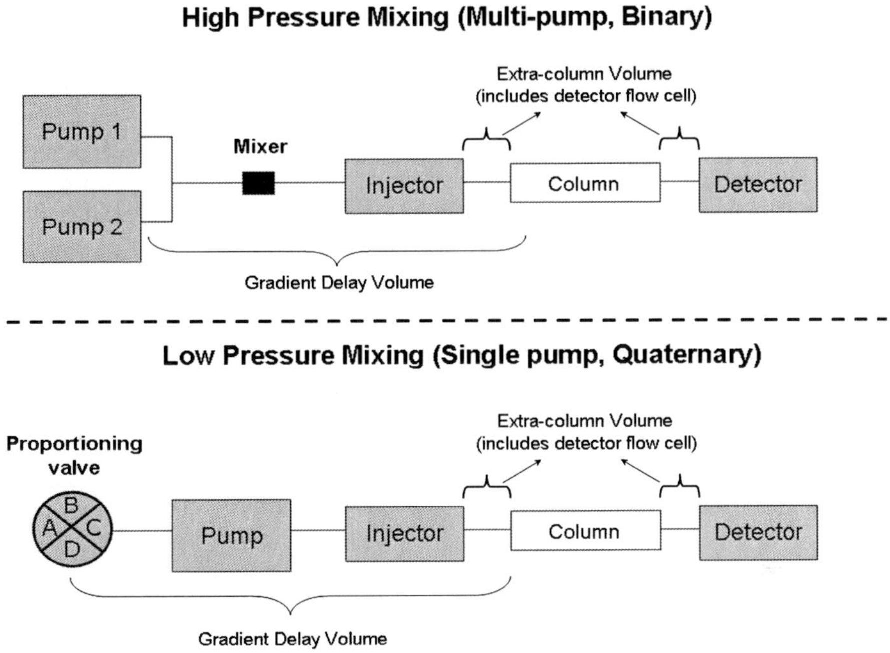

Figure 2.1 Schematics of a high-pressure and low-pressure mixing system. Gradient delay and extra-column volumes vary depending on the instrument.

flow rates, such as the ones used with small column diameters (≤2.1 mm) or for large molecule separations that require slower flow rates due to their lower optimum linear velocities.

While low-pressure mixing systems typically have larger gradient delay volumes than high-pressure mixing systems (3- to 4-fold), they are more flexible in that they allow the user to blend up to four solvents at a time, which is important when developing new methods, since each mobile phase does not have to be made in advance. At high flow rates, the larger gradient delay volume of low-pressure mixing systems causes a negligible increase in analysis time. Finally, low-pressure mixing systems can be less expensive than some high-pressure mixing systems due to the fact that there is only one pump instead of two.

For UHPLC separations, both of these mixing systems were re-engineered by manufacturers to achieve the expected performance benefits of small particle columns. All of the features of a conventional HPLC system were modified to meet throughput demands while routinely and robustly being able to withstand much higher operating pressures. As a result, many manufacturers incorporated changes into the materials and designs of plungers, seals, check valves, fittings, and ferrules. In addition, gradient delay volumes had to be minimized while still providing for adequate mixing of the mobile phases prior to delivery to the column. To achieve this, commercial UHPLC systems incorporate mixers after the pump, even with low-pressure mixing pumps which traditionally relied solely on the GPV. Depending on the application, these mixers can differ in volume from around 40 μL (routine small molecule separations) up to 500 μL. Larger mixing volumes are typically required for gradients that use ion-pairing agents such as TFA, because TFA absorbs UV light considerably stronger than water or acetonitrile (ACN), which results in significant baseline ripples.[4] This can compromise the signal-to-noise ratio of the peak, which significantly reduces the limit of detection (LOD) and limit of quantification (LOQ). Most separations done by UHPLC utilize mixers with volumes between 50 and 100 μL. Table 2.1 contains a listing of some commercially available UHPLC instruments with their associated specifications.

2.1.1.2 Injectors

The most challenging part of designing a UHPLC system is introducing the sample at very high pressures and in a small sample volume.[5] An additional consideration is the mechanical stability of UHPLC columns that are subjected to repeated injections at elevated pressure and gradient conditions where the pressure can cycle several thousand psi within a run. There are two main types of injectors: fixed loop style and variable flow through (sometimes referred to as a direct injector).

The first commercial UHPLC system contained a fixed loop style injector. In a fixed loop style injector, the sample needle is not part of the fluid path. Rather, the sample is drawn by the needle and then transferred to a loop that is brought into the fluid path during injection. Thus, only the loop volume, and not the sample needle, is flushed by the mobile phase. In a variable flow-through injector, the sample needle is part of the fluid path. Variable flow-through injectors can also have sample loops, however, both the sample needle and loop are brought into the fluid path during injection. As a result, the mobile phase flushes both the loop and needle during the run.

The benefits of using a fixed loop style injector are that they have lower dispersion, decreased gradient delay volume, and faster injection cycle times than variable flow-through injectors (see Table 2.1). This makes them ideal for use with smaller diameter columns since the injector contribution to the overall system dispersion is minor, and thus does not cause a major loss in the observed column efficiency. However, there are some drawbacks to using a fixed loop injector, which include the need for multiple wash solvents to reduce carryover, as well as increased dispersion with larger sample loops (≥20 μL) due to viscous flow or volume overload. By contrast, variable flow-through injectors keep the sample needle in the fluid path after injection and the entire path is flushed with the gradient, leading to a decrease in carryover.

Independent of the type of injector used for UHPLC separations, many of the fluidic components had to be redesigned to deliver robust operation at elevated pressure while decreasing dispersion and gradient delay volume. As with UHPLC pumps, this involved redesign of the needle, needle seat, injector

Table 2.1 Features of selected UHPLC instruments listed in order of increasing extra-column variance. The Alliance 2695 HPLC system is included for reference.

System name	*Vendor*	*Maximum pressure (psi)*	*Flow range (mL min^{-1})*	*Injector type*[a]	*Injection time (s)*	*Gradient delay volume (μL)*	*Extra-column variance*[b] *(μL^2)*	*%N*[c] *(2.1 mm × 50 mm, 1.7 μm)*
Acquity UPLC I-Class	Waters	18 000	0.01–2	FL or VTR	<15	<95	1.0	97
Acquity UPLC H-Class	Waters	15 000	0.01–2	VTR	<30	<400	3.2	81
Acquity UPLC	Waters	15 000	0.01–2	FL	15 to 30	<120	5.8	69
Ultimate 3000 RSLC	Dionex	15 000	0.1–8	FL or VTR	15	293	11.6	51
1290 Infinity LC	Agilent	17 400	0.001–5	VTR	19–25	<140	16.0	42
Accela High Speed LC	ThermoFisher	15 000	0.001–5	FL	<60	70 (pump only)	17.6	40
Alliance 2695 HPLC	Waters	5 000	Up to 10	VTR	40–45	<650	33.6[d]	25
Nexera	Shimadzu	18 800	0.0001–5	FL or VTR	10	<185	Unknown	Not calculated
Flexar FX-15	PerkinElmer	15 000	Unknown	Unknown	8	Unknown	Unknown	Not calculated
X-LC	Jasco	15 000	Up to 5	Unknown	30	Unknown	Unknown	Not calculated
PLATINblue	Knauer	15 000	Up to 5	Unknown	15–60	110	Unknown	Not calculated

[a]FL, fixed loop; VTR, variable flow-through. [b]Determined experimentally using protocol described in Section 2.1.2.2. [c]Assumes $h = 2$, $k' = 2.5$, $A = 0.212$, $B = 6$, $C = 0.108$, $\varepsilon_t = 0.67$, $D_m = 1.34\text{E-}05\ \text{cm}^2\ \text{s}^{-1}$, and $\tau^2 = 1.09\text{E-}6$, $u_0 = 0.783\ \text{cm s}^{-1}$. τ^2 is equal to the inverse of the sampling rate divided by 12.[1] [d]Configured for microbore use.

valves, fittings, and tubing. Today, there are commercial UHPLC systems that utilize both types of injectors to give optimum results depending on the application (Table 2.1).

Another component associated with the injector is the column thermostat compartment, which can also be a major source of extra-column dispersion and gradient delay volume. There are essentially two types of column heating/cooling available in all LC separations: passive and active. Passive preheating (cooling) is simply letting the column (preceded by an adequate length of tubing) sit in a thermostatically maintained oven while maintaining the temperature inside the oven compartment. With passive preheating, the mobile phase is brought to the oven temperature while it travels through a longer coiled section of tubing prior to the column.[6,7] Active preheating involves directly heating the mobile phase in a very short length of tubing prior to the column and maintaining the column in the oven at the same temperature, so that the column is more uniformly heated. Depending on the volume of the tubing required, the type of heating used can contribute to extra-column band spreading.

2.1.1.3 Detection

UHPLC separations produce very narrow peaks (1 to 3 s at the base, or 10 to 30 μL in volume for a 2.1 mm × 50 mm, 1.7 μm column). Therefore, detectors with fast acquisition times are needed to ensure this performance is maintained through the detection process. While on-column detection is the best approach in terms of eliminating post-column dispersion,[8–10] it suffers from low signal intensity,[5] and is not compatible with stainless steel columns. For ultraviolet (UV) absorbance detection, commercial UHPLC instruments utilize detector flow cells with low volumes (*e.g.*, 0.25 to 2 μL) to minimize the impact on extra-column dispersion. However, since the use of smaller volume flow cells negatively impacts the signal intensity due to a shorter path length, many UHPLC instrument manufacturers use light-guided flow cells.[5] Similar advances have been made with mass spectrometers (MS), where instruments are now capable of low dwell times (below 5 ms), as well as low inter-channel and inter-scan delay times. Other common detection techniques such as fluorescence (FL), evaporative light scattering (ELS), and refractive index (RI) detection are now available for use with UHPLC systems. A more detailed discussion on the influence of UV detector sampling rate and time constant on column performance will be presented in the next Section.

2.1.2 Extra-column Band Spreading

2.1.2.1 Sources of Extra-column Band Spreading

The term extra-column band spreading, or dispersion, is often confused or used interchangeably with the terms gradient delay volume and dwell volume. Gradient delay volume is the same as dwell volume, and is the volume between where the gradient is formed and the inlet of the column (Figure 2.1). The extent of the gradient delay volume depends on the type of mixing system, volume of the gradient mixer, and injector style. Extra-column band spreading is the unwanted broadening of a chromatographic peak from the point of injection to the point of detection.

In general, the broadening of a chromatographic peak is expressed as a variance, and is the sum of the broadening that occurs within the column as well as external to the column.[11] The lower the variance, the narrower the peak, and thus the higher efficiency separations that can be achieved. Eqn (2.1) below describes the performance of a column in the absence of extra-column band spreading:

$$\frac{\sigma^2{}_c}{L} = H = A \cdot d_p + \frac{B \cdot D_m}{u_0} + C \cdot \frac{d_p^2}{D_m} \cdot u_0 \tag{2.1}$$

In this equation, $\sigma^2{}_c$ is the length variance of the peak in the column, L is the column length, H is the theoretical plate height, u_0 is the linear velocity, D_m is the analyte diffusion coefficient, d_p is the particle size, and A, B, and C are the coefficients of the reduced van Deemter equation.[1] It is well known that minimizing the band spreading of a peak within the column is dependent on column parameters such as particle size, packing quality, *etc.*, as well as the operating conditions (*i.e.*, mobile phase, linear velocity). However, it is difficult to achieve 100% of the theoretical performance of a column due to the extra-column band spreading effects, of which there are two types. The first is volumetric in nature ($\sigma_v{}^2$) and results from the injection volume (and possibly injection solvent), the detector flow cell volume, the volume of connecting tubing between the injector and detector, and the quality of the connections.*

The second is time related, and results from parameters such as the detector sampling rate and filter time constant. When these two extra-column contributions are added to the van Deemter equation, it is possible to obtain the observed column performance taking into account total system band spreading:

$$H = A \cdot d_p + \frac{B \cdot D_m}{u_0} + C \cdot \frac{d_p^2}{D_m} \cdot u_0 + \frac{1}{L} \cdot \frac{\sigma_v^2}{\pi^2 \cdot r^4} \cdot \frac{1}{\varepsilon_t^2 \cdot (k'+1)^2} + \frac{1}{L} \cdot \frac{\tau^2 \cdot u_0^2}{(k'+1)^2} \tag{2.2}$$

L is the column length, $\sigma_v{}^2$ is the volumetric extra-column variance, r is the column radius, ε_t is the total porosity (mobile phase volume in the column divided by the empty column volume), k' is the retention factor, and τ is the standard deviation in time units.[1]

*While many UHPLC instruments have universal fittings for connecting a wide variety of columns with different end nut designs, it is advised to consult the UHPLC column care and use manual for the column being used to ensure the ferrule type and depth is correct. This will avoid unwanted leaking (especially at elevated pressures and flow rates), as well as ensure no extra-column band spreading is introduced, which can lead to decreases in observed column efficiency or peak capacity.

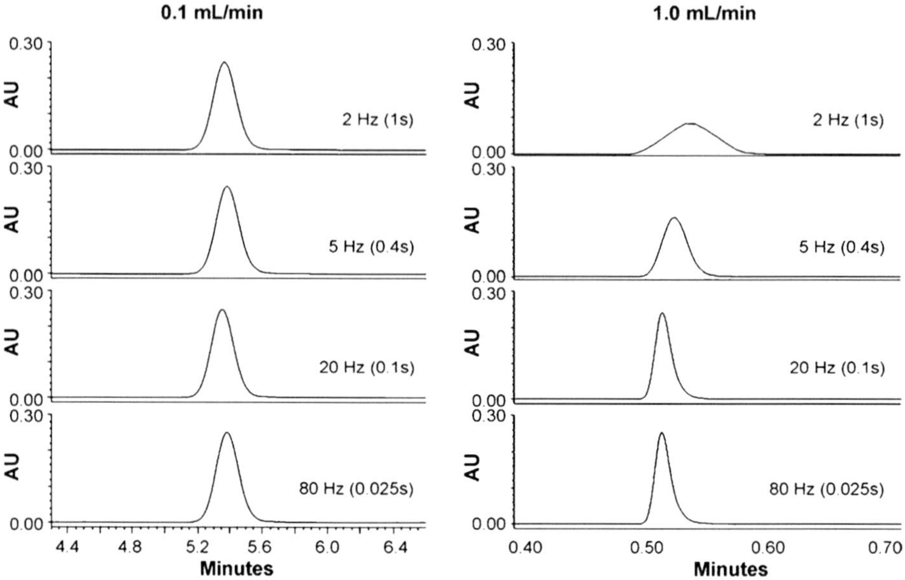

Figure 2.2 Influence of detector settings on peak distortion. The number in parenthesis is the filter time constant, in s. The test probe was acenaphthene (0.2 mg mL^{-1}); k' = 3.6. The column dimensions were 2.1 mm × 50 mm, 1.7 μm, ACQUITY UPLC® BEH C_{18}. Reproduced with permission from reference 1.

The extra-column variance resulting from detector time-based events is dependent on the sampling rate, and, more importantly, the filter time constant. The influence of the time constant is negligible at low linear velocities where the peaks are wide in the time domain, but becomes more important as the linear velocity increases. This is demonstrated in Figure 2.2, in which four different detector settings are compared at low and high flow rates for acenaphthene. At 0.1 mL min^{-1}, there is a minimal increase in the plate count for acenaphthene going from the 2 Hz, 1 s setting to the 80 Hz, 0.025 s setting. At 1.0 mL min^{-1}, the impact of the detector settings is much more apparent, so that there is an 8.5-fold improvement in the observed plate count and a 3-fold improvement in sensitivity for acenaphthene at the 80 Hz, 0.025 s setting over the 2 Hz, 1 s setting.[1] Fortunately, the extra-column band spreading due to time-related events is easily overcome by proper selection of the sampling rate and filter time constant for the particular application. Oversampling does not lead to band spreading effects but does contribute to increased background noise that can reduce signal-to-noise ratios and lead to the accumulation of large data sets. For quantitative analyses, 15–20 data points across the peak are more than adequate to capture the peak area. For routine UHPLC separations, a sampling rate of 20 to 40 Hz is recommended with a filter time constant of 0.05 s or less.

The extra volume required to ensure proper mixing and temperature control needs to be balanced with the need to keep extra-column volume minimized.

The higher the efficiency of the column (and smaller the diameter), the greater the need for minimizing all extra-column contributions to band spreading. From a practical perspective, the volumetric contributions to extra-column band spreading can occur anywhere between the injector and detector, including the loop volume, tubing diameter and length, type of fittings and ferrules, and detector flow cell volume.[1,12]

2.1.2.2 Measuring Extra-column Band Spreading

Extra-column band spreading varies with the flow rate and mobile phase conditions, thus it is important to use the same protocol when comparing the extra-column variance of two or more systems. Measuring extra-column variance, σ_V^2, is relatively straightforward and can give a prediction for how well a column of a certain length, radius, and particle size will perform for a given separation. In order to do this, the user can simply remove the LC column, replace it with a zero-volume union, and then inject a neutral marker under a specified mobile phase composition and flow rate to determine the peak variance in the absence of a column. Many of the extra-column variance measurements for the UHPLC instruments in Table 2.1 were measured using a 0.5 μL injection of 160 μg mL^{-1} caffeine into a mobile phase of 50:50 (v/v) ACN/water flowing at 0.5 mL min^{-1}. A detection wavelength of 273 nm was used along with a sampling rate of 40 Hz and no filter time constant. The peak width at 4.4% peak height (5σ) was then measured and multiplied by the flow rate to give the peak width in volume units (μL). This value was then divided by 5 to give the peak standard deviation (σ) and then squared to give the extra-column variance (σ^2) in μL^2. The lower the extra-column variance, the less contribution the system has on column performance, which allows higher efficiency separations to be realized with UHPLC columns. The data in Table 2.1 show that different UHPLC instruments give a range of extra-column variances, which will lead to differences in perceived column performance. Table 2.1 also shows the percentage of the theoretical plate count (*N*) that can be achieved on a 2.1 mm × 50 mm, 1.7 μm column based on the extra-column variance for each UHPLC system. As expected, UHPLC systems with lower extra-column band spreading are capable of delivering higher efficiency separations. The variability of the extra-column variances observed between different instruments is due to the differences in the designs of the injector, column heater, and detectors, as well as the parameters mentioned in Section 2.1.2.1.

2.1.2.3 Impact of Extra-column Band Spreading on Isocratic Separations

Once the extra-column variance is known, it can be used to predict the performance of columns having a certain dimension and particle size. Note that from eqn (2.2), the volumetric contribution to extra-column band spreading decreases with increasing column radius to the 4th power, increasing retention factor, and increasing column length. This means that for larger diameter columns of longer length, and analytes with moderate to high retention factors, extra-column variance has a negligible impact. However, for columns packed with sub-2 μm particles, which contribute much smaller values in the first three terms of eqn (2.2) compared to 5 μm particles, the resulting apparent column efficiency can decrease dramatically as the length, diameter, or retention factor (*k*) is decreased. For example, a well-performing HPLC system has an extra-column variance of approximately 52 μL^2.[1] Using eqn (2.2) and inputting this value for σ_v^2, we can calculate the plate count versus retention factor for columns having different diameters but the same length and particle size. Figure 2.3 shows that a 3.0 mm diameter column can be used on an HPLC system with minimal loss in efficiency above a *k*' value of 2.5, but suffers in performance for analytes with less retention. Using either a 2.1 mm

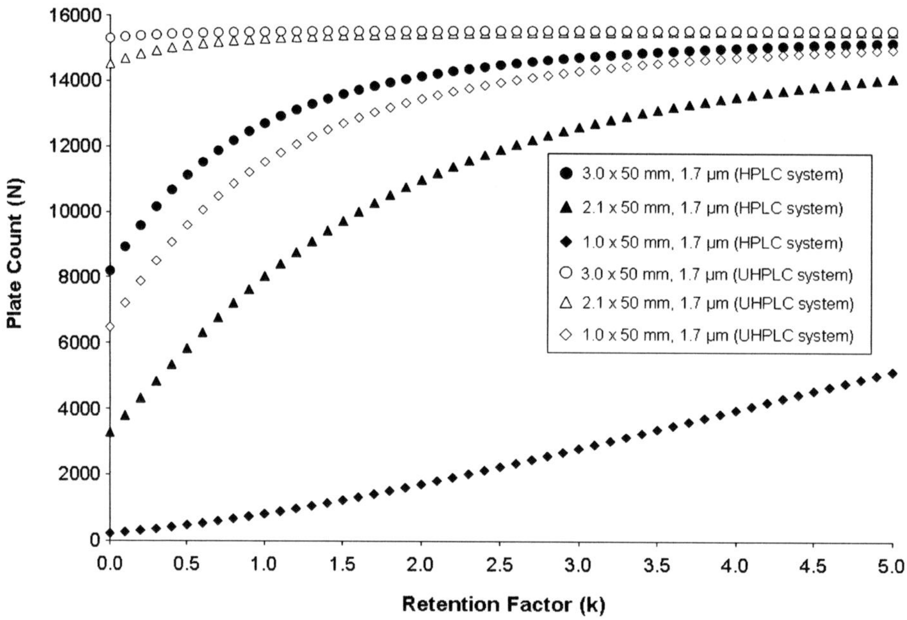

Figure 2.3 Calculated plate count (*N*) as a function of column diameter and retention factor. The extra-column band spreading for HPLC was measured on an Alliance® 2695 system (Waters); the variance is σ_v^2 = 52 μL^2. The extra-column band spreading for UPLC was measured on an Acquity UPLC® I-Class system (Waters) with a fixed loop injector and 5 μL loop; the variance is $\sigma_v^2 = 1$ μL^2. Sample was a 0.5 μL injection of 160 mg mL^{-1} caffeine in 50:50 ratio of ACN/water. The mobile phase was 50:50 ACN/water pumped at 0.5 mL min^{-1}. UV light was at 273 nm at the highest possible data rate (≥80 Hz) and lowest filter time constant setting (zero if available). Band spreading was measured at 5σ. A 10 cm length piece of 50 μm PEEKSil tubing was used in place of the analytical column.

or 1.0 mm diameter column packed with 1.7 μm particles on an HPLC system results in severe efficiency loss, even for analytes with large retention factors. Theoretically, a 4.6 mm diameter column packed with 1.7 μm particles can achieve the expected efficiency, even for analytes with low retention factors. In practice, however, the flow rates used to obtain the optimum plate count would cause a decrease in efficiency due to thermal effects. A solution to minimizing thermal effects is to use smaller diameter columns on instruments with low dispersion.

In contrast to the performance we see from the HPLC system in Figure 2.3, if a UHPLC system with an extra-column variance of 1.0 μL^2 is used, the plate counts for 2.1 and 1.0 mm diameter columns improve dramatically, to the point where little efficiency loss is seen for the 2.1 mm diameter column, even for analytes with extremely low retention factors. The 1.0 mm diameter column performance, while much improved from use on an HPLC system, still achieves only 90% of the theoretical efficiency for moderately retained analytes ($k' \sim 2.5$). This is the reason for using low dispersion instrumentation for UHPLC experiments.

It is interesting to consider the use of 4.6 mm diameter columns containing sub-2 μm particles on HPLC instrumentation in order to achieve UHPLC-like performance; however, there are some major barriers. The typical flow rate for a 2.1 mm × 50 mm, 1.7 μm particle column is approximately 0.5 to 0.6 mL min^{-1}. Geometrically scaled to a 4.6 mm diameter column, the flow rate is almost 3 mL min^{-1}. Not only does this higher flow rate cause the solvent usage to increase by a factor of almost 5, but it also requires a pressure of approximately 7000 to 8000 psi (assuming a 80:20 ratio of water/ACN at room temperature) in a typical gradient separation, which is not possible on a conventional HPLC system. Operating below this flow rate to stay under 5800 psi (400 bar) might compromise the efficiency and/or peak capacity that can be achieved with sub-2 μm particles, as well as throughput for small molecule separations. Typical optimal flow rates for peptides and larger molecules are lower, and separations using sub-2 μm particle columns can be successfully run on optimized HPLC systems using 4.6 mm diameter columns. However, these analyses require ~5-fold larger sample volumes (due to the increase in cross sectional area from a 2.1 mm to a 4.6 mm diameter column) in order to avoid a loss of sensitivity by the ~5-fold dilution in the higher flow rates required to maintain the same linear velocity used on a 2.1 mm diameter column. In addition, larger diameter columns are not as efficient at dissipating the heat generated by using small particle columns at higher linear velocities. This heat generated results in axial and radial thermal gradients in the column. The radial thermal gradients contribute to band spreading that can cause serious loss of efficiency for sub-2 μm columns under certain high flow rate conditions.[13]

Eqn (2.2) can also be used to calculate the minimum extra-column variance needed to achieve a certain plate count for a column of particular dimensions and particle size. Knowledge of the extra-column variance is useful for deciding which UHPLC system to purchase or use for a particular application. It can also be useful when designing the next generation of chromatographic instrumentation. Eqn (2.2) was solved for σ_v^2 to calculate the extra-column variance:

$$\sigma_v^2 = L\pi^2 r^4 \varepsilon_t^2 (k'+1)^2 \left[\frac{L}{N} - (A \cdot d_p) - \left(\frac{B \cdot D_m}{u_0}\right) - \left(C \cdot \frac{d_p^2}{D_m} \cdot u_0\right) - \left(\frac{1}{L} \cdot \frac{\tau^2 \cdot u_0^2}{(k'+1)^2}\right)\right] \quad (2.3)$$

Table 2.2 shows the extra-column variance needed to achieve 95% of the available plate count (N) for some commonly used HPLC and UHPLC column dimensions. Note that all of the chromatographic columns in the Table have the same resolving power, as indicated by their ratio of column length to particle size (L/d_p). Thus, they should have the same plate count. However, the actual plate count observed is highly dependent on the extra-column variance of the system that is being used to run each column. For all columns, the extra-column variance required to achieve the desired plate count increases with increasing retention factor. For the 5 μm particle column, the calculated extra-column variance needed to achieve 14 000 plates is high enough that any HPLC system is capable of delivering this efficiency. For 4.6 mm diameter columns packed with 2.5 and 3.5 μm particles, extra-column variance is only a concern for analytes with low retention factors. However, for smaller diameter columns packed with 1.7 μm particles, the extra-column variance needed to achieve 14 000 plates is very low, on the order of a few μL^2 for 2.1 mm × 50 mm columns. Cross-referencing the data in this Table with the extra-column variances shown for the UHPLC systems in Table 2.1 provides a benchmark for which system is needed to achieve the maximum plate count for some of the more common column dimensions in UHPLC separations.

2.1.2.4 *Impact of Extra-column Band Spreading on Gradient Separations*

A common misconception is that extra-column contributions to band broadening are negligible in gradient elution chromatography since the analytes are focused on the head of the column after injection and before the gradient starts. However, this is only partially true. All contributions coming before the column can be neglected,[14] but the post-column tubing as well as the flow cell can still contribute to extra-column dispersion in gradient separations. Lestremeau *et al.*[15] showed that a 1.0 mm diameter column run on a UHPLC system only produced approximately 70–85% of the peak capacity (and peak height) of a 2.1 mm diameter column.

Studies have also been performed in our laboratories to determine the effect of extra-column band spreading on peak capacity in gradient separations. Figure 2.4 shows the gradient separation of a mixture of six anaesthetics using the same column on two different UHPLC systems. The system used for the separation shown in Figure 2.4(A) employed a 0.5 μL detector flow cell

Table 2.2 Extra-column variance (in μL^2) needed to achieve 95% of the available plate count (N) for columns with the same L/d_p ratio. Columns are listed in order of decreasing inner diameter. Values for σ_v^2 were calculated using eqn (2.3). Assumptions: same as Table 2.1, except u_0 for 2.5, 3.5, and 5 μm particle columns was adjusted for particle size.

Diameter (mm)	*Length (mm)*	d_p *(μm)*	L/d_p	N_{ideal} *(h = 2)*	*95%* N_{ideal}	σ_v^2 *(μL^2)* *95%* N_{ideal} *(k' = 0.5)*	σ_v^2 *(μL^2)* *95%* N_{ideal} *(k' = 2.5)*	σ_v^2 *(μL^2)* *95%* N_{ideal} *(k' = 5.0)*
4.6	150	5	30 000	15 000	14 250	45.35	246.94	725.69
4.6	100	3.5	28 571	14 286	13 571	21.16	115.23	338.65
4.6	75	2.5	30 000	15 000	14 250	11.34	61.73	181.42
2.1	75	2.5	30 000	15 000	14 250	0.49	2.68	7.88
4.6	50	1.7	29 412	14 706	13 971	5.13	27.98	82.24
3.0	50	1.7	29 412	14 706	13 971	0.93	5.06	14.88
2.1	50	1.7	29 412	14 706	13 971	0.22	1.22	3.57
1.0	50	1.7	29 412	14 706	13 971	0.01	0.06	0.18

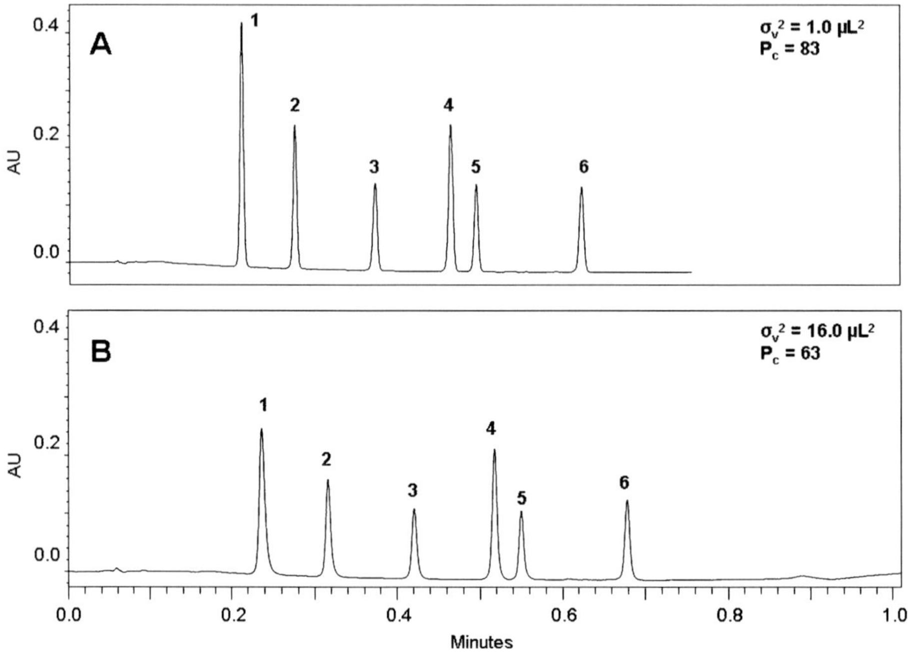

Figure 2.4 Gradient separation of six anaesthetics on two different UHPLC systems. The column was an Acquity UPLC® BEH C_{18}, 1.7 µm, 2.1 mm × 30 mm. Mobile phase A was 10 mM NH_4HCO_3, pH 10; mobile phase B was ACN. The gradient was 25 to 75% B in 1 min, and the total run time was 1.5 min. The injection volume was 2 µL, 50 °C, 1 mL min^{-1}, UV at 220 nm. Peaks: (1) benzocaine, (2) procaine, (3) prilocaine, (4) lidocaine, (5) tetracaine and (6) bupivacaine. Data courtesy of Patricia McConville and Tanya Jenkins of Waters Corporation.

volume, and the post-column tubing to the detector had an internal diameter of 63.5 µm. The system used for the separation shown in Figure 2.4(B) utilized a 1 µL flow cell volume and 127 µm diameter tubing. The smaller peak capacity observed for the UHPLC system in Figure 2.4(B) is a result of extra-column dispersion. Since the retention times for all of the peaks are similar to that of the separation shown in Figure 2.4(A), we can conclude that these systems have very similar gradient delay volumes. However, both the peak capacity and peak intensities on the system in Figure 2.4(B) are suffering from the increase in the diameter of the tubing connecting the column to the detector, as well as the increased volume of the flow cell compared with the system in Figure 2.4(A).

When choosing a UHPLC system and column combination, it is important to be aware of extra-column contributions to band spreading, even for gradient separations. Another important system parameter for gradients is the gradient delay (dwell) volume. The influence of this parameter on gradient separations will be covered further in the next Section.

2.1.3 Gradient Delay Volume

The gradient delay volume is also sometimes called the dwell volume, and it is not to be confused with the extra-column volume. As described in Section 2.1.2.1 and shown in Figure 2.1, the gradient delay volume is the volume between where the gradient is formed and the inlet of the column. This volume results in an unwanted isocratic hold at the beginning of the gradient that can differ from system to system (Figure 2.5). It is also a volume that needs to be flushed out at the end of the gradient in order to re-equilibrate the column prior to the next injection. High-pressure mixing systems typically have lower gradient delay volumes than low-pressure mixing systems. The gradient delay volume impacts the throughput of a gradient separation, as well as the transfer of a gradient method from one system to another. It does not impact the throughput and method transfer for isocratic separations.

Measuring the gradient delay volume of any LC system is essential to knowing when the sample that is injected will start to experience the real beginning of the gradient at the end of the isocratic hold. It is also important to know the gradient delay volume in order to properly design the gradient table. A general rule of thumb is that the gradient volume should be at least equal to the column void volume plus the gradient delay volume. While the column void volume can be estimated based on the dimensions, the gradient delay volume needs to be measured, as it can change even between systems of the same model depending on tubing, manufacturing tolerances, *etc.* Many manufacturers provide methods for measuring the gradient delay volume on their website or in product literature. There are many different methods

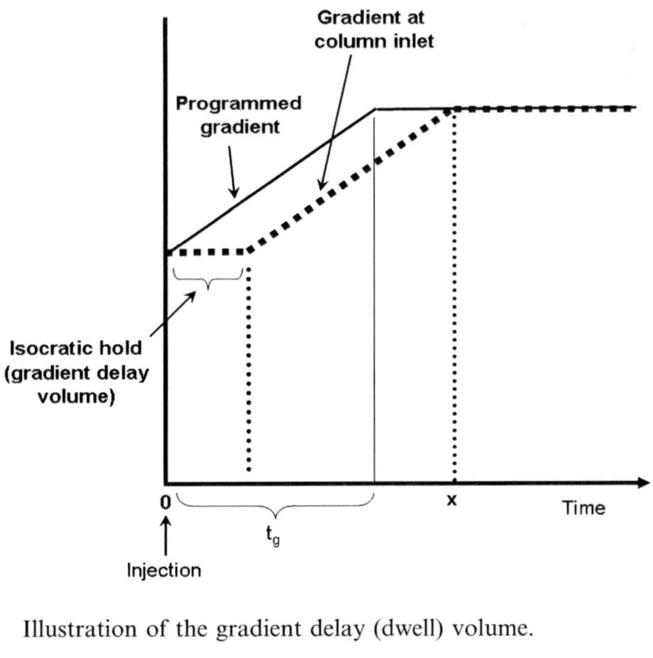

Figure 2.5 Illustration of the gradient delay (dwell) volume.

available, and, while one protocol is not particularly better than another, it is important to use the same method on every instrument whose gradient delay volume is being determined for the purpose of method transfer. One such method of determining the gradient delay volume is found below:

- Mobile phase A: ACN
- Mobile phase B: ACN with 7.5 mg L^{-1} propylparaben
- Flow rate: 0.75 mL min^{-1}
- UV at 254 nm
- Injection of ACN: 1 µL
- A 10 cm piece of 50 µm inner diameter fused silica capillary is installed in place of the column using two zero-volume unions.

Gradient table:

Time (min)	*Flow (mL min^{-1})*	*A (%)*	*B (%)*	*Curve*
Initial	0.75	100	0	–
5	0.75	100	0	6
15	0.75	0	100	6
25	0.75	0	100	1
35	0.75	100	0	1

Three replicate injections are recommended in order to properly determine the gradient delay volume of a system. A resulting chromatogram from this measurement method is shown in Figure 2.6, along with the steps to determine the gradient delay volume. Once the gradient delay volume is determined, it does not have to be measured again unless changes to the tubing between the pump (or point of gradient formation) and column inlet are made. Some UHPLC systems even have an input field in the instrument method to enter the gradient delay volume so that the injection is delayed to coincide with the arrival of the gradient at the head of the column. This makes it easier to transfer methods between systems and predict retention times for all peaks in the separation.

One of the major advantages for UHPLC separations is the ability to substantially decrease run times. This is due to the ability to run short columns (*i.e.*, 50 mm length) packed with sub-2 µm particles at higher linear velocities without compromising the separation. In order to achieve fast separation times in gradient elution, the UHPLC system must have a low gradient delay volume to properly perform the separation, high organic wash at the end of the gradient (if needed), and column re-equilibration prior to the next injection. Failure to achieve this can lead to poor peak shape, irreproducible retention times, late-eluting peaks (past the gradient window) and even loss of peaks, which can contribute to carryover or column deterioration.

Accounting for gradient delay volumes is especially important for successful method transfer from HPLC to UHPLC and even between different UHPLC systems. This ensures that the selectivity of the original separation is

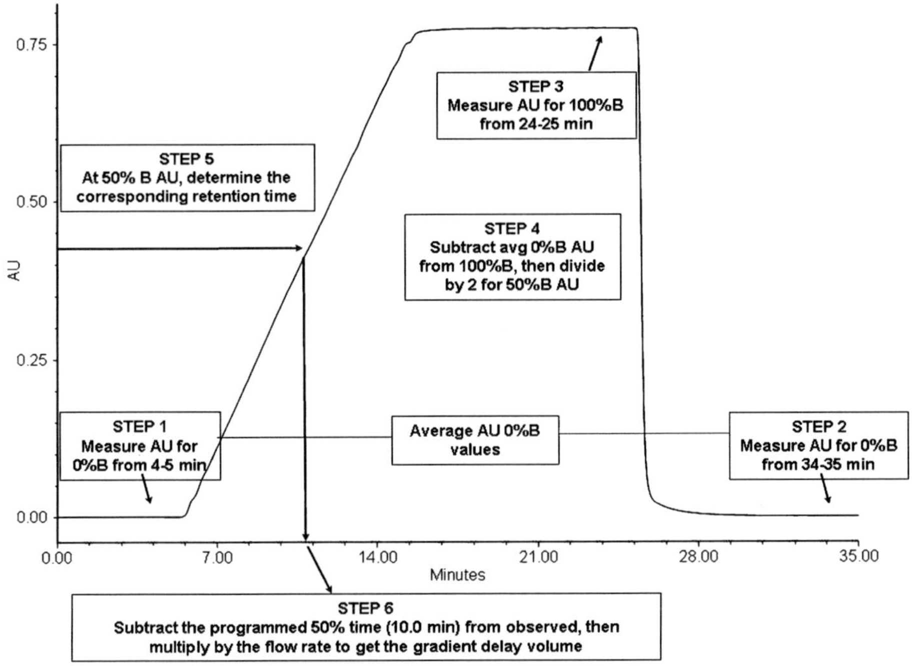

Figure 2.6 Step-by-step instructions for how to determine the gradient delay volume. avg, average; AU, absorbance units. Procedure and data courtesy of Hillary Hewitson of Waters Corporation.

maintained when transferring between systems and/or columns. The first step to successfully transferring HPLC separations to UHPLC is to select a column that has the same selectivity as the original method. A few column manufacturers provide the same stationary phase in both HPLC and UHPLC particle sizes to help facilitate method transfer. Next, the method can be transferred between HPLC and UHPLC using published calculations that account for the differences in column dimension, particle size, and gradient delay volume.[16] These calculations are similar to the ones used for transferring isocratic HPLC separations to UHPLC, but include compensation for the differences in the gradient delay volume in the form of an initial isocratic hold in the gradient table, or in some cases a pre-injector volume that is entered into the instrument method software. Many instrument manufacturers now have transfer calculators that automatically calculate the UHPLC conditions based on the original HPLC conditions (or *vice versa*). Before using one of these calculators, it is wise to ensure that there is a field in the calculator to input the gradient delay (dwell) volume so that the final separation has the same selectivity as the original method.

While using a calculator from an instrument manufacturer is the most convenient way to quickly transfer methods between different LC instruments, it is worth reviewing how the gradient delay volume results in either a pre-injector volume or an isocratic hold. We will consider the example of method transfer from HPLC to UHPLC, which allows one to realize the speed, resolution, and sensitivity benefits for older HPLC methods. Intuitively, one would think that because the HPLC system has a much larger gradient delay volume (up to 10 times in some cases) than the UHPLC system, an isocratic hold would not be required for the corresponding UHPLC method. For the most part this is true, but not strictly because of the differences in the measured gradient delay volumes. Rather than directly comparing the gradient delay volumes of the two systems, a proper method transfer calculator instead calculates the gradient delay volume as a fraction of the void volume for the column being used. For instance, a typical HPLC column dimension is 4.6 mm × 100 mm, 3.5 μm, and HPLC systems usually have a gradient delay volume of approximately 1 mL. The void volume in this case is 1.097 mL (assumes 34% of the column tube is occupied by packing material). Thus, the gradient delay volume is 0.912 void volumes (1 mL/1.097 mL). If an HPLC method developed on this column and system is then transferred to a UHPLC system, we must do the same calculation to obtain the ratio of the gradient delay volume to the void volume. The UHPLC column with the same length to particle size ratio (L/d_p) as the HPLC column above is a 2.1 mm × 50 mm, 1.7 μm column. If the UHPLC system being used has a gradient delay volume of 0.1 mL, the gradient delay volume is equivalent to 0.878 void volumes. Since the gradient delay volume is smaller on the UHPLC system (expressed as void volumes), a small isocratic hold needs to be applied to the beginning of the UHPLC gradient to ensure proper equilibration of the column after the analysis. However, this hold is on the order of seconds and most likely would have minimal impact on the separation, since the column will be properly equilibrated during the injection sequence. Conversely, if a 4.6 mm × 150 mm, 5 μm column is used for the original HPLC separation, a pre-injection volume is needed to ensure the sample is injected at the point where the gradient reaches the UHPLC column. This is because the gradient delay of the HPLC system is 0.608 void volumes and is now lower than the relative gradient delay volume of the UHPLC system. An example of an HPLC method that has been properly transferred to UHPLC using these calculations is shown in Figure 2.7.

Method transfer from UHPLC to HPLC is another common practice. While laboratories convert their instrumentation to UHPLC, they may need to utilize existing HPLC systems to run methods developed with UHPLC. Additionally, methods developed with UHPLC may need to be transferred globally to different laboratories that may not have UHPLC systems. In this case, the same calculations apply, and the use of the same stationary phase chemistry in both UHPLC and HPLC becomes even more critical. Figure 2.8(A) shows a separation developed for non-steroidal anti-inflammatory drugs (NSAIDs) on UHPLC. Baseline separation of all six peaks in the mixture is obtained in under 3 min, and the peak capacity is 29. In order to transfer this method to

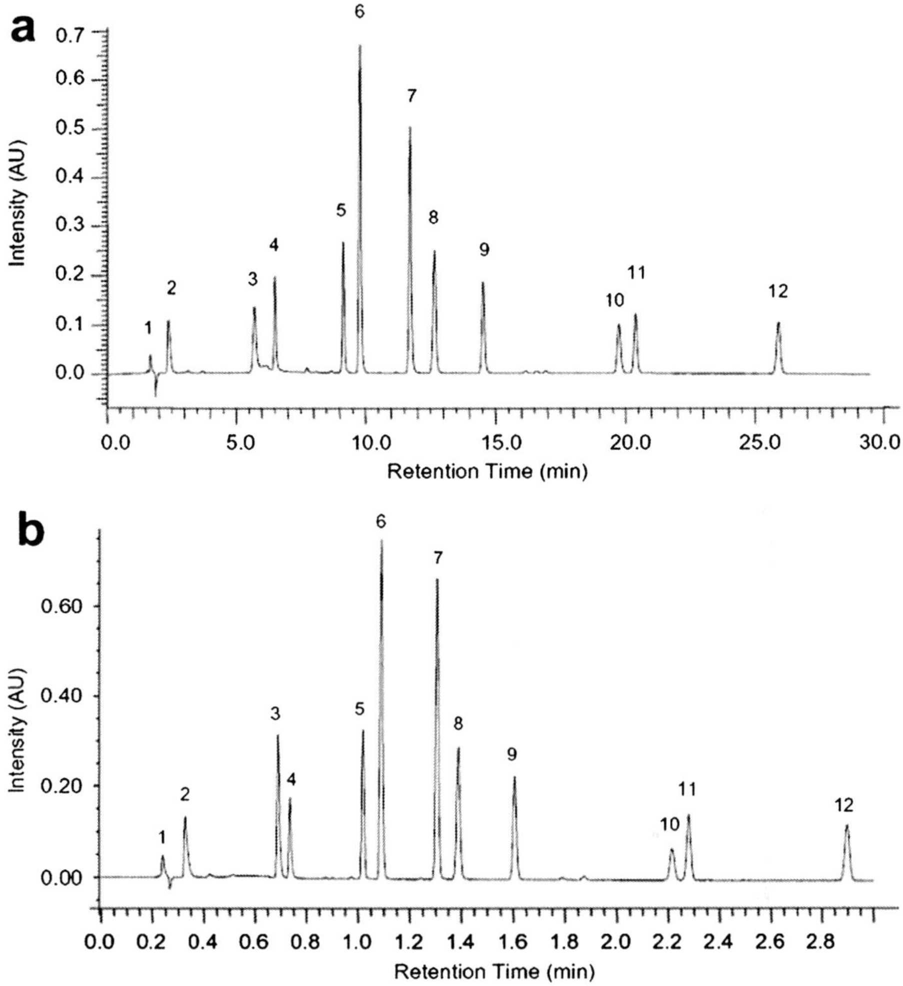

Figure 2.7 Separation of a pharmaceutical formulation containing the main product (6) and 11 impurities in gradient mode with a HPLC and UPLC system: (a) HPLC system; column, XBridge™ C_{18} 150 mm × 4.6 mm, 5 μm; flow rate, 1000 μL min^{-1}; injected volume, 20 μL; total gradient time, 45 min. (b) UPLC system; column, Acquity UPLC® BEH C_{18} 50 mm × 2.1 mm, 1.7 μm; flow rate, 610 μL min^{-1}; injected volume, 1.4 μL; total gradient time, 5.1 min. Reproduced with permission from reference 16.

HPLC, one could simply take the UHPLC column off of the UHPLC instrument and place it on the HPLC instrument and use the same gradient conditions. However, due to the pressure limitations of the HPLC instrument, the flow rate must be decreased, which causes all of the analytes to elute after the programmed gradient (Figure 2.8B). This is due to the larger gradient delay volume of the HPLC system compared to the UHPLC system (1.0 mL for HPLC *versus* 0.4 mL for UHPLC). As a result of the larger gradient delay

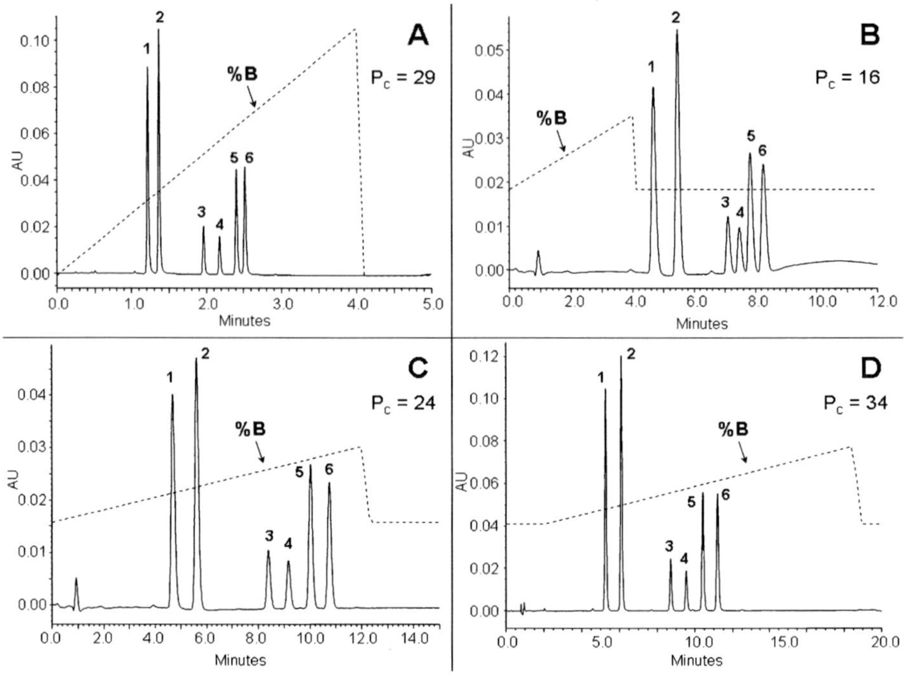

Figure 2.8 Separation of six non-steroidal anti-inflammatory drugs using (A) a UHPLC column on a UHPLC system, (B) a UHPLC column on an HPLC system with a 4 min gradient time, (C) a UHPLC column on an HPLC system with a 12 min gradient time, and (D) an HPLC column on an HPLC system with an 18.5 min gradient time (includes a 2 min initial hold). The UHPLC column was an Acquity UPLC® CSH™ Phenyl-Hexyl column, 2.1 mm × 50 mm, 1.7 μm. The HPLC column was an XSelect™ CSH™ Phenyl-Hexyl column, 4.6 mm × 100 mm, 3.5 μm. Mobile phase A was 0.1% formic acid in water; mobile phase B was 0.1% formic acid in ACN. The gradient was from 35 to 65% B. The flow rate for (A) was 0.6 mL min^{-1}, the flow rate for (B) and (C) was 0.2 mL min^{-1}, and the flow rate for (D) was 1.4 mL min^{-1}. UHPLC injection volume was 2 μL and HPLC was 19.2 μL. UV detection is performed at 270 nm. The UHPLC system was an Acquity UPLC® H-Class with an Acquity PDA. The HPLC system was an Alliance® 2695 with a 2998 PDA detector. Peak capacity was calculated as 1 + [($t_{R\,last}$ − $t_{R\,first}$)/(average peak width at 13.4% peak height)]. Data courtesy of Mia Summers of Waters Corporation.

volume, the analytes are subjected to an unprogrammed isocratic hold of almost 5 min prior to the gradient arriving at the head of the column, which causes peak distortion and loss of peak capacity. One solution is to adjust the gradient table in HPLC to compensate for the 3-fold decrease in flow rate (Figure 2.8C). Here, the gradient time was increased by a factor of 3 in order to maintain the same number of column volumes during the separation. We see that the analytes are now eluting within the gradient window, and the selectivity of the separation is similar to that of the UHPLC separation in

Figure 2.8(A), but the peak capacity is only ~80% of the UHPLC separation. In addition, the peak heights are 50% of the UHPLC separation and the run time is 3-fold longer. Figure 2.8(D) shows the HPLC separation on a 4.6 mm × 100 mm, 3.5 μm column that has the same stationary phase chemistry as the UHPLC column in Figure 2.8(A). The flow rate, gradient table, and injection volume were all scaled using the proper calculations,[16] and a 2 min hold was inserted at the beginning of gradient table to account for the gradient delay volume and ensure proper re-equilibration for the next injection (indicated by the %B trace). As a result, the HPLC separation on the 4.6 mm × 100 mm, 3.5 μm column provides the same selectivity, peak capacity, and peak heights as the UHPLC separation, but on an HPLC time scale.

2.2 UHPLC Columns

2.2.1 Properties of UHPLC Particles

Packing materials for UHPLC are prepared in a similar manner to those used in HPLC, and therefore have many of the same characteristics. The major differences are the particle diameter and particle strength. Fundamentally, UHPLC particles must be able to withstand the higher packing and operating pressures needed to produce rugged columns. The particles must also be able to deliver the expected plate counts and peak capacity predicted by theory, which means they must have the correct diameter and distribution, and be well packed into appropriate column hardware. The hardware and packing procedures used for UHPLC columns are proprietary to each column manufacturer and will not be covered in detail in this Chapter. General guidelines for producing well-packed columns can be found in references 17 and 18.

Perhaps the most critical property of any packing material is the particle size, which influences the plate count (thus, resolution and peak capacity) of the column. The particle size also determines the permeability of the column, which influences the back pressure.[18] Sizing of UHPLC particles is largely proprietary to each manufacturer, and there is little public information on the subject. However, classification techniques that may be used can be found in reference 19. Particle size measurements are typically performed using specific instruments such as Coulter counters,[20] and provide information on the mean particle diameter as well as the distribution. For LC, the particle diameter is reported either as the number-averaged particle size ($d_{p,n}$) or the volume-averaged particle size ($d_{p,v}$).[18] The volume-averaged particle size is more common, and is always larger than the number-averaged particle size, except in the case where particles are monodisperse.[18] The difference between these two averages depends on the particle size distribution, usually reported as the d_{90}/d_{10} particle diameter ratio, which is a measure of the width of the volume-averaged particle size distribution. For particles with a narrow size distribution, the difference between $d_{p,n}$ and $d_{p,v}$ is negligible. For particles with wider size distributions, $d_{p,n}$ tends to be a better indication of column back pressure, while $d_{p,v}$ provides a better fit to column efficiency and the C-term of the van Deemter equation.[18] The closer the d_{90}/d_{10} particle diameter ratio is to 1.0, the more uniform the particle size distribution around the mean value. The narrower the particle size distribution, the higher the plate count and lower the back pressure for a given mean particle size.[20]

The pore size, pore volume, and surface area of the packing material dictate its retention and loading.[21] Typical packing materials have an average pore size of approximately 100 Å (10 nm) and a specific surface area of 100 to 400 $m^2\ g^{-1}$. These materials are suitable for the separation of molecules with molecular weights < 500 Da. For large molecules (*i.e.*, peptides and proteins), larger pore diameters ($\sim$300 Å and above) are required. An increase in the pore size is accompanied by a proportionate decrease in the surface area,[20] resulting in a decrease in the loading capacity of the packing material (*i.e.*, less mass may be injected on the column). While the surface area of a particle influences retention, it is not the only determining factor. The phase ratio, which is the ratio of the stationary and mobile phase volumes in the column, is proportional to the surface area and directly proportional to the retention. As a result, the retention increases with a decrease in pore diameter and increase in the surface area for a given pore volume.[20] The pore volume affects the strength of a packing material. Packing materials that have a small pore volume tend to be stronger than those with larger pore volume.[21] For chromatographic techniques such as size exclusion chromatography (SEC), the pore volume also affects the separation.

2.2.2 Particle Preparation, Bonding and Endcapping

Packing materials for HPLC and UHPLC can be grouped into three categories: non-porous, superficially porous, and fully porous. In the late 1990s, Jim Jorgenson's group at the University of North Carolina built and tested an ultra-high-pressure LC apparatus using 30 μm diameter capillaries (50 to 70 cm in length) packed with commercially available, 1.5 μm non-porous silica C_{18} particles from Micra Scientific.[2] Capillaries were used to overcome the problem of heat generation using small particles at high flow rates.[2] As many as 480 000 plates per m were obtained at pressures of about 60 000 psi. The advantages of non-porous particles in general are that they eliminate pore diffusion and resistance to mass transfer (improved van Deemter C-term).[22] However, the major drawback to using non-porous particles is that they have extremely low surface area, which leads to considerably lower mass loadability compared with porous materials.[23] For this reason, a majority of the packings used in UHPLC are fully porous particles, a category which may be further split into silica-based and hybrid particles. There are also silica-based packing materials that are superficially porous. The following sections will briefly describe the process for the preparation, bonding, and endcapping of these particles, as well as provide examples of their utility for UHPLC.

2.2.2.1 Silica-based Particles

Fully porous silica particles can be made in several different ways, and the different options are described in reference 20, as well as in reference 21. An illustration of a high purity silica particle synthesis process is shown in Figure 2.9 (top panel). The chemical properties of silica-based stationary phases depend on the way the silica particles are prepared. The surface of silica is covered with silanols and siloxane bridges.[21] Silanols are weakly acidic and give the silica surface its polar adsorption characteristics.

Silanol groups can be modified with different surface groups to give different chromatographic selectivity. In reversed-phase (RP) chromatography, these surface groups are non-polar (*i.e.*, C_{18}, C_8), while normal phase (NP), ion exchange (IEX), and hydrophilic interaction chromatography (HILIC) phases contain polar surface groups.[21] The principles of bonding are almost identical irrespective of whether the particle is non-porous, fully porous (silica or hybrid), or superficially porous. Briefly, bonding of non-polar surface groups to a silica surface involves silanization using reactive silanes, the most common being chlorosilanes or alkoxysilanes. For polar functional groups, the silica surface can be bonded using alkoxysilanes or derivatized in an aqueous medium. The level to which the silica surface is derivatized depends on the way the silica particle is prepared, the type of

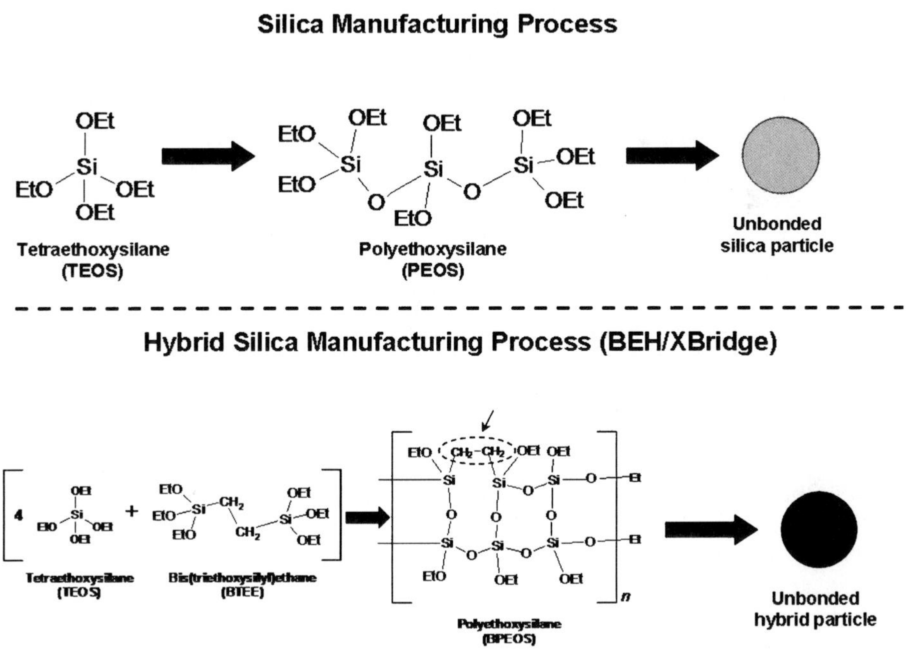

Figure 2.9 Synthesis processes for silica (top) and hybrid silica (bottom) particles. The bottom panel represents the process used to produce BEH and XBridge™ hybrid packing materials (Waters Corporation).

silane used for bonding, and the bonding conditions. Silanes can be monofunctional, difunctional, or trifunctional. Monofunctional silanes have one point of attachment to the surface, while difunctional and trifunctional silanes have two or three. Studies have shown that on average, trifunctional silanes only form two siloxane bonds with the particle surface.[21] Difunctional and trifunctional bonded phases have higher chemical stability in acidic mobile phases than monofunctional phases, and may also yield higher surface coverage (molar amount of surface group attached to the surface per unit surface area), which can impact the retention and resolution of some compounds.[24,25] The stability of monofunctional bonded phases can be increased using sterically-hindered silanes.[26]

After bonding, phases can be endcapped to remove residual surface silanols or silanols formed during bonding that may cause undesired chromatographic interactions.[21] This can be accomplished by using reactive di- or trimethylsilanes.[27–29] Endcapping not only reduces the deleterious effects of residual silanols, but also gives the particle a higher degree of protection from the mobile phase.

Sub-2 μm silica-based packings were available for use prior to the introduction of the first commercial UHPLC instrument, but they were not capable of being used above 400 bar. This is most likely due to some of the technical challenges associated with higher pressure operation, which include development of the proper column hardware rated to > 600 bar, packing optimization, and a process that produced particles capable of withstanding UHPLC pressures without breaking.[30] In fact, the first commercially available sub-2 μm particles capable of routine UHPLC operation up to 15 000 psi were hybrids (see Section 2.2.2.2). It was not until 2006 that the first sub-2 μm silica-based particle columns capable of withstanding pressures greater than 400 bar were introduced for use on UHPLC systems (Acquity UPLC® HSS columns from Waters Corporation and ZORBAX® Rapid Resolution High Throughput columns from Agilent). Soon after, refinements in small particle manufacturing and column packing resulted in the introduction of many more silica-based particle columns for use at pressures of 600 bar or 1000 bar. Table 2.3 lists a number of stationary phases currently available for UHPLC, along with their pH and pressure specifications.

An alternative to fully porous silica is superficially porous silica. The origins of superficially porous (also known as fused-core or shell) particles date back to the late 1960s when Horvath and Lipsky[31,32] proposed that these particles would have higher efficiency than fully porous particles due to the decreased diffusion distance. Separations using these types of particles were limited to biological macromolecules until the introduction of UHPLC.[33] Readers interested in the evolution of superficially porous particle technology are encouraged to read the review by Guiochon and Gritti.[34]

In 2007, superficially porous particles regained traction with the introduction of 2.7 μm stationary phases that consisted of a 1.7 μm non-porous silica core and a 0.5 μm shell.[35] Reduced plate heights for 4.6 mm diameter columns approached 1.5 and loading capacity was acceptable for most applications.[36] Several brands of sub-3-μm superficially porous particle columns were introduced between 2007 and 2010, with the first sub-2 μm superficially porous particle columns being commercialized in 2010 (Table 2.3). Modern superficially porous particles are prepared by a well-controlled growth process[37] that produces very uniform non-porous silica cores with d_{90}/d_{10} particle size ratios of approximately 1.1 (compared to 1.4 to 1.5 for fully porous particles).[34] These particles are then coated with several layers of silica sol nanoparticles to generate the final superficially porous packing with the desired pore size and layer thickness, resulting in a particle size distribution that is rather narrow.[21] These can then be functionalized with the same surface groups used with fully porous and non-porous particles. Superficially porous particles have shown improvements over fully porous particles in both the A- and B-terms of the van Deemter equation, as well as lower frictional heating effects compared to fully porous particles.[34] Improvements in the B-term are expected due to a higher fraction of the column that is occupied by non-porous silica, into which analytes cannot diffuse. Improvements in the A-term for 4.6 mm diameter columns are speculated to arise from either the narrower particle size distribution compared to fully porous particles or the external roughness of the particle. Due to the high surface roughness, commercially available columns packed with superficially porous particles typically have higher interstitial column porosity, which results in less efficiency loss due to frictional heating.[34]

It must be noted that since columns packed with superficially porous particles are also intended for UHPLC separations, they are susceptible to the same extra-column effects as fully porous particles, especially in small diameter columns. Many of the comparisons between fully porous particle columns and columns packed with superficially porous particles are performed on UHPLC instrumentation.[34] Superficially porous particles also suffer from similar chemical stability issues as silica-based packing materials, and still have lower loading capacity than fully porous materials. In addition, current processes used to produce superficially porous particles may not be easily scaled to produce columns for preparative analyses, compound isolation, and purification on a large scale.

2.2.2.2 Hybrid Particles

When the first commercial UHPLC instrument was introduced in 2004, the only column capable of routinely withstanding operating pressures up to 15 000 psi was packed with ethylene-bridged hybrid organic/inorganic particles (more commonly referred to as BEH). While silica was the dominant base particle for LC separations and had experienced 30 years of improvements to deliver high efficiency and mechanical stability, it had two major limitations: poor peak shape for bases due to interactions with residual surface silanols and poor chemical stability outside of pH 2 to 8.[38] Mobile phases with pH < 2 causes the bonded phase to hydrolyze.[39] while mobile phases with pH > 8

Table 2.3 Information for selected UHPLC columns (limited to sub-2 μm particle columns). Columns are listed by base particle support.

Name	*Base particle*	d_p *(μm)*	*Stationary phase types*	*Maximum pressure (psi)*	*pH range**	*Vendor*
Acquity UPLC BEH	Hybrid	1.7	C_{18}, C_{18} carbamate, C_8, C_4, phenyl, HILIC, amide, SEC	18 000	1–12 (C_{18}, C_8, phenyl), 2 to 11 (C_{18}-carbamate, amide), 1–10 (C_4), 1 to 9 (HILIC), 1–8 (SEC)	Waters
Acquity UPLC CSH	Hybrid	1.7	C_{18}, phenyl-hexyl, fluoro-phenyl	18 000	1–11 (C_{18} and phenyl-hexyl), 1–8 (fluoro-phenyl)	Waters
PolyRP	PS/DVB	1.0, 1.7	phenyl	10 000	1–14	Sepax
Proteomix	PS/DVB	1.7	SCX, WCX, SAX, WAX	10 000	2–12	Sepax
Kinetex	Superficially porous	1.7	C_{18}, C_8, PFP, HILIC	15 000	1.5–10 (C_{18}), 2–8 (PFP and HILIC)	Phenomenex
Acquity UPLC HSS	Silica	1.8	C_{18}, PFP, CN	18 000	2–8	Waters
VisionHT	Silica	1.5	C_{18}, HILIC, Silica	12 000	2–8	Grace
Sepak silica-based	Silica	1.8	C_{18}, C_8, C_4, phenyl, amino, CN	10 000	2–8.5	Sepax
Eclipse Plus RRHD	Silica	1.8	C_{18}, C_8, phenyl-hexyl, HILIC	18 000	2–9	Agilent
Extend RRHD	Silica	1.8	C_{18}	18 000	2–11.5	Agilent
StableBond RRHD	Silica	1.8	C_{18}, C_8, phenyl, CN	18 000	1–8	Agilent
Epic	Silica	1.8	C_{18}, C_8, phenyl, CN, SCX, phenyl-hexyl, PFP, diol, HILIC, silica	19 000	Not specified	ES Industries
Strategy	Silica	1.7	C_{18}, HILIC	Not specified	1–10 for C_{18}, 1.5–7 for HILIC	Interchim
BlueOrchid	Silica	1.8	C_{18}, C_8, C_4, phenyl, PFP, HILIC, CN, SAX, amido, silica	15 000	2–8 (2–9 for C_8; 1–11 for C_{18})	Knauer

Table 2.3 (*Continued*)

Name	*Base particle*	d_p (*μm*)	*Stationary phase types*	*Maximum pressure (psi)*	*pH range**	*Vendor*
Nucleodur	Silica	1.8	C_{18}, C_8, PFP, HILIC, CN, aminopropyl, silica	15 000	Varies, most 1 to 9 or 1 to 10	Macherey-Nagel
PinnacleDB and Ultra II	Silica	1.9	C_{18}, C_8, PFP, biphenyl, CN, silica	15 000	Need catalog	Restek
Hypersil GOLD	Silica	1.9	C_{18}, C_8, C_4, PFP, CN, phenyl, amino, AX, SAX, silica	15 000	Need catalog	ThermoFisher
Syncronis	Silica	1.7	C_{18}, C_8, phenyl, amino, HILIC, silica	Not specified	2–8	ThermoFisher
Pursuit UPS	Silica	1.9	C_{18}, diphenyl	Not specified	1.5–10 (C_{18}), 1.5–8 (diphenyl)	Varian
Brownlee Hres	Silica	1.9	C_{18}, Silica, CN, PFPP, biphenyl, IBD	Not specified	Not specified	PerkinElmer

*Data compiled from manufacturer websites, product literature, and reference 60.

dissolve the silica particle, resulting in loss of column efficiency and collapse of the chromatographic bed.[40]

Hybrid particles are produced by polymerizing two silane monomers to form an organic/inorganic structure (Figure 2.9, bottom panel).[20,21] The first generation of hybrid particles (commercial name XTerra®, Waters) incorporated a Si-CH_3 (methyl) moiety throughout the silica particle backbone. These hybrid particles not only had similar pore structures and bonding characteristics to silica, but also had similar efficiency and chromatographic performance.[41–44] In addition, due to the ~33% decrease in residual surface silanols, chromatographic performance for basic compounds was better than traditional silica-based packings. Finally, and perhaps most importantly, these hybrid materials had superior chemical stability at high pH, which allowed for use of a wider range of mobile phase pH values (pH 1 to 12) to achieve drastically different selectivity for the separation of charged analytes. In 2004, a second generation of hybrid materials was introduced under the commercial names Acquity BEH (1.7 μm) and XBridge™ (≥2.5 μm). These hybrids incorporate an ethylene bridge in the silica matrix.[38] and have improved tailing factors for basic compounds when compared with silica-based particles (Figure 2.10), higher chemical stability than the first generation hybrids, and give different selectivity compared to traditional silica C_{18} phases. Mellors and Jorgenson[45] utilized 1.5 μm BEH particles in 30 μm diameter capillaries at pressures up to 65 000 psi and found them to perform similarly to 1 μm non-porous silica particles, with the added advantage of a 35-fold improvement in loading capacity. Like silica, hybrid particles can be functionalized with a variety of different surface groups in order to alter the selectivity of separations (Table 2.3). Hybrid particles can also be functionalized with a low level of an ionizable group prior to bonding and endcapping to produce charged-surface hybrid (CSH™) particles. This type of particle formation was introduced by Waters Corporation in 2010, and overcomes some of the problems that many high purity silica-based packings have with slow equilibration and poor peak shape for basic compounds in low ionic strength mobile phases.[46] In addition, the application of a low level of positive surface charge causes shifts in the relative retention of charged analytes (Figure 2.11), providing a different set of selectivity choices for routine UHPLC method development and optimization.

2.2.3 Selectivity and Method Development in UHPLC

Decreasing the particle size is one way of improving the separation performance in LC. As mentioned previously, this also requires improvements to instrumentation since small particle columns cannot be used in larger diameter columns without detrimental effects due to heating. As the column diameter becomes smaller, there is a limit to the improvements that can be made to UHPLC instrumentation before there is no incremental gain in efficiency from small particles. Therefore, we must rely on techniques to alter the separation chemically rather than physically.

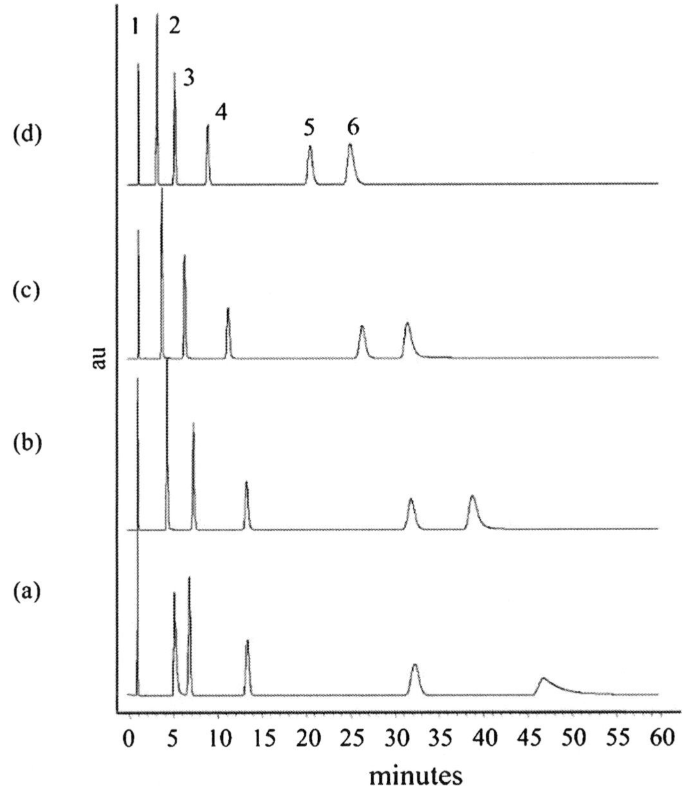

Figure 2.10 Chromatograms showing the separation of uracil (1), propranolol (2), butyl paraben (3), naphthalene (4), acenaphthene (5), and amitriptyline (6) on a silica-based C_{18} column (a) and three different BEH C_{18} columns: H1–C_{18} (b), H2–C_{18} (c), and H3–C_{18} (d). The three hybrid materials differ in their surface properties. Conditions: 4.6 mm × 150 mm columns, 1.4 mL min^{-1} of a 60:40 methanol/20 mM KH_2PO_4/K_2HPO_4 (pH 7.00) mobile phase. Reproduced with permission from reference 38.

The resolution of any chromatographic separation depends on the efficiency, retentivity, and selectivity of the column and conditions being used [eqn (2.4)].

$$R_s = \frac{\sqrt{N}}{4}\left(\frac{\alpha - 1}{\alpha}\right)\frac{k_2}{k_2+1} \tag{2.4}$$

R_s is the resolution, N is the plate count, α is the relative retention, and k_2 is the retention factor of the second peak of the two whose resolution is being calculated.[18] The plate count can be increased by using smaller particle columns and instrumentation with lower dispersion (as in UHPLC). Retentivity can be altered chemically through the use of particles having

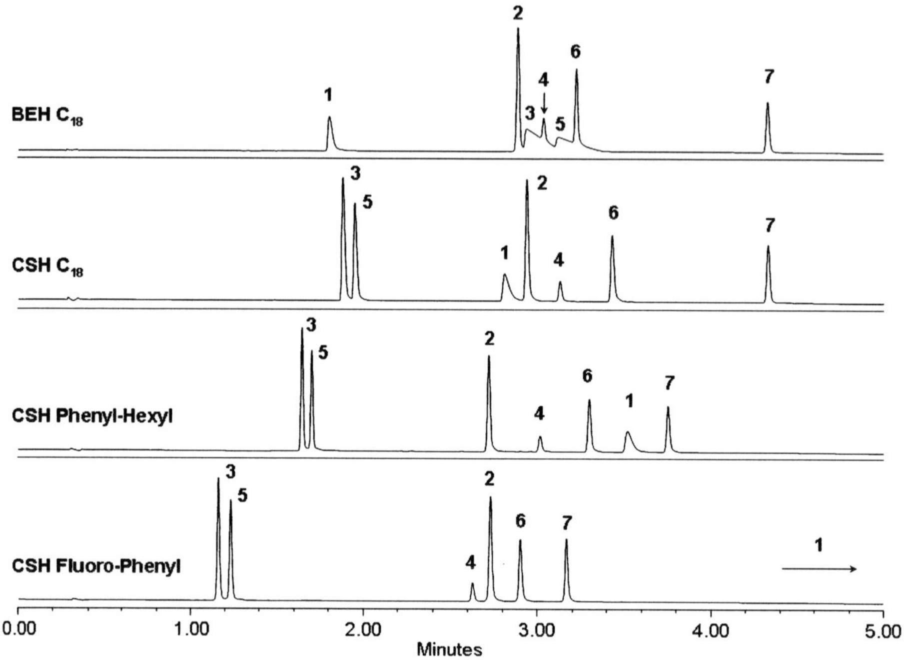

Figure 2.11 Analysis of a mixture of acids, bases and neutrals on a BEH column and three CSH™ columns under low pH conditions. The system was an Acquity UPLC® system equipped with a PDA detector and 4-position column manager. Mobile phase A was 0.1% formic acid in water, and mobile phase B was 100% ACN. The gradient was from 5 to 95% B in 5 min. The flow rate was 0.5 mL min^{-1}. UV detection was performed at 254 nm. The injection volume was 2 μL. The column temperature was 30 °C. The column dimensions were 2.1 mm × 50 mm, 1.7 μm. Peaks: (1) 1-pyrenesulfonic acid, (2) flavone, (3) imipramine, (4) fenoprofen, (5) amitriptyline, (6) diclofenac and (7) octanophenone. Data courtesy of Hillary Hewitson of Waters Corporation.

higher phase ratios and more retentive surface groups for the packing material, as well as by modifying the mobile phase. Selectivity is affected by the surface chemistry of the packing material and the mobile phase composition (*i.e.*, pH). Illustrating selectivity differences between stationary phases can present a challenge due to the strong dependency on the conditions and analytes being used. Specific analyte mixtures or mobile phases may not provide the diversity in interactions needed to demonstrate the differences in selectivity between the stationary phases, which may only be revealed in specific applications (*i.e.*, when a new impurity is discovered in a pharmaceutical active ingredient or formulation).

Fortunately, there are several established methods for selecting columns that provide very different selectivity. Published methods for determining the selectivity difference between two separations include the geometric method,[47] Pearson correlation coefficients,[48,49] selectivity coefficients,[50,51] and one of the most common – the Snyder–Dolan hydrophobic subtraction method (HSM)[52], which has been used to select different (or similar) columns from an interactive database (http://www.usp.org/USPNF/columnsPQRIapproach.html). The Snyder–Dolan scheme utilizes 16 different solutes under specific isocratic conditions to determine selectivity differences between different stationary phases.[53] The selectivity of these various solutes is representative of the five predominant solute–column interactions: hydrophobicity, steric resistance, hydrogen-bond acidity, hydrogen-bond basicity, and cation exchange. The advantage of this scheme is that a single selectivity function (F_s) can be calculated to compare two columns, which is useful in selecting a column with similar selectivity for method transfer, or for selecting a column with different selectivity during method development.

Other methods of determining selectivity such as selectivity coefficients can be used to determine the degree of difference between any two conditions (*i.e.*, column chemistry, mobile phase pH, elution solvent) for a particular separation. To determine the selectivity coefficient, the retention factor (k) for each analyte in the separation is plotted at two different conditions. A linear trend line is drawn and the correlation coefficient (r^2) is obtained. The selectivity difference, s^2, between the two conditions is then calculated according to eqn (2.5).

$$s^2 = 1 - r^2 \tag{2.5}$$

An s^2 value of 0 indicates that the two conditions (*e.g.*, columns, mobile phase conditions) do not give a selectivity difference. An s^2 value of 1 indicates that the two conditions being compared are completely orthogonal.

Independent of whether one is performing HPLC or UHPLC separations, selectivity is mainly affected by three factors: column chemistry, mobile phase buffer additive and pH, and the elution solvent. While temperature is known to affect selectivity for the separation of biological macromolecules and chiral compounds, it has much less impact on selectivity in small molecule reversed-phase separations[18]. In reversed-phase LC, the mobile phase pH has the largest impact on the selectivity, whereas in HILIC, the column chemistry has the largest influence.[54] While traditional C_{18} stationary phases can generally be used for 80 to 90% of all RP separations, there are advantages to exploring alternatives. When UHPLC was first commercially introduced in 2004, there was only one column chemistry available (C_{18}), which limited the use of UHPLC for certain separations. However, as shown by the listing of columns available in Table 2.3, there are now numerous selectivity options for UHPLC columns, which facilitate widespread adoption of UHPLC for all types of separations.

To demonstrate the impact of column chemistry for RP separations, a mixture of seven basic drugs was analysed on six different reversed-phase UHPLC stationary phases using the same gradient conditions (Figure 2.12). The selectivity coefficients relative to the endcapped BEH C_{18} column were

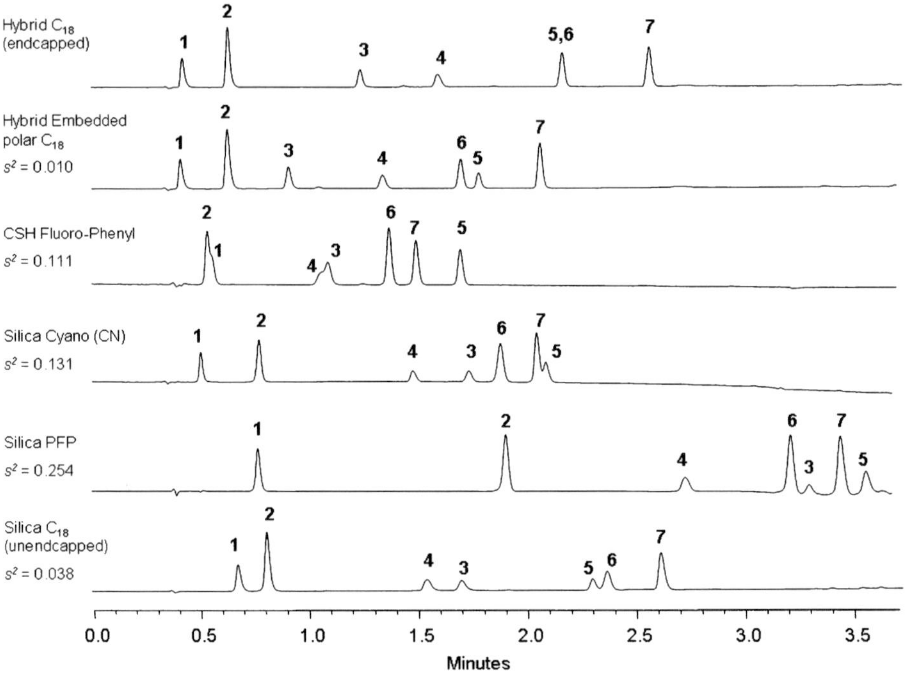

Figure 2.12 Gradient separation of basic drugs on six different UHPLC columns. All columns were 2.1 mm × 50 mm, 1.7 μm (hybrid) or 1.8 μm (silica). The system was an Acquity UPLC® H-Class system (Waters Corporation). Mobile phase A was 10 mM NH_4COOH, pH 3; mobile phase B was methanol. The gradient was 30 to 85% in 3 min, hold for 0.5 min at 85%, reset. The total run time was 4.5 min. The injection volume was 1 μL, 30 °C, 0.4 mL min^{-1}, UV detection at 260 nm. Peaks: (1) aminopyrazine, (2) pindolol, (3) quinine, (4) labetalol, (5) verapamil, (6) diltiazem and (7) amitriptyline. Data courtesy of Christopher Hudalla and Jane Xu of Waters Corporation.

also calculated using eqn (2.5). Notice that all of the columns, with the exception of the endcapped hybrid C_{18} phase, resolve verapamil from diltiazem, and only two of the columns give baseline separation of all seven peaks. In addition, the PFP stationary phase retains basic compounds much more than the other phases, which is beneficial for mixtures containing polar bases (*i.e.*, natural products, drug metabolites). Cyano stationary phases are well known for being able to separate closely related compounds such as steroids and are widely used in compendial methods. They can also be used for eluting non-polar compounds that do not easily elute from C_{18} columns. Unlike C_{18}, both PFP and CN can be used in RP, HILIC, and NP modes of chromatography.

As indicated by the selectivity coefficients in Figure 2.12, the CN and PFP stationary phases are the most different from the endcapped hybrid C_{18} phase, and thus it is beneficial to screen one of these types of columns during routine method development. It must be noted, however, that these types of phases are more difficult to produce with good batch-to-batch reproducibility and tend to be more variable and less stable than traditional alkyl phases. To compensate for this, resolutions of greater than 2.0 should be required in method development. These phases may also have more bleed than C_{18} columns, which can be detected with mass spectrometry because their bonded phase hydrolysis products are more easily ionized than those from C_{18} bonded phases.

While changing the column chemistry can impact the separation selectivity to a moderate degree, the mobile phase pH has the largest impact for RP separations of ionizable analytes. Until the introduction of hybrid particles in 1999, the pH range for silica-based packings was limited to approximately 8 or 9. This is due to the dissolution of the silica particle at elevated pH, which causes loss of column efficiency and column voiding.[38] With hybrid particles, one can increase the pH of the mobile phase above the pK_a of stronger bases (*i.e.*, tricyclic antidepressants such as amitriptyline) and neutralize the molecule, thus retaining it solely by hydrophobic interaction. As a result, basic compounds are more retained and give better peak shape, which is one way of overcoming the challenge of mass overloading with low pH, low ionic strength mobile phases (Figure 2.13), a common problem when analysing bases using MS-friendly mobile phases such as formic acid.[55] Also note in Figure 2.13 that selectivity coefficients much higher than those in Figure 2.12 can be achieved just by simply changing the pH of the mobile phase on the same hybrid C_{18} column. Since the retention times for basic compounds increase with increasing pH, these analytes elute in a higher percentage of organic solvent in RP, which can both enhance the signal intensity in MS and result in faster evaporation times for collected fractions from isolation and purification experiments.

Figure 2.13 also shows the influence of changing the organic elution solvent in RP separations. While changing the elution solvent will change the retention times of the analytes depending on the eluotropic strength, the retention times do not shift to the same degree for all the analytes. When switching to methanol from ACN, not only do the analytes increase retention (due to the weaker eluotropic strength of methanol), but changes in relative retention (selectivity) may cause some analytes to co-elute. While changing the elution solvent alone does not dramatically impact the selectivity coefficient for most separations, changing this parameter in combination with the column chemistry and/or mobile phase pH can lead to increases in the selectivity difference of the separation due to the additive effects on the selectivity coefficient.[56]

The selectivity of a separation can also be influenced by other modes of chromatography. A technique that is now popular for analysis of polar compounds is HILIC. Separations in HILIC mode are achieved using high organic (aprotic), low aqueous mobile phases in combination with a polar stationary phase (*i.e.*, bare silica or bare hybrid particles). HILIC uses the same mobile phases as RP chromatography, but in reverse order, and is a popular

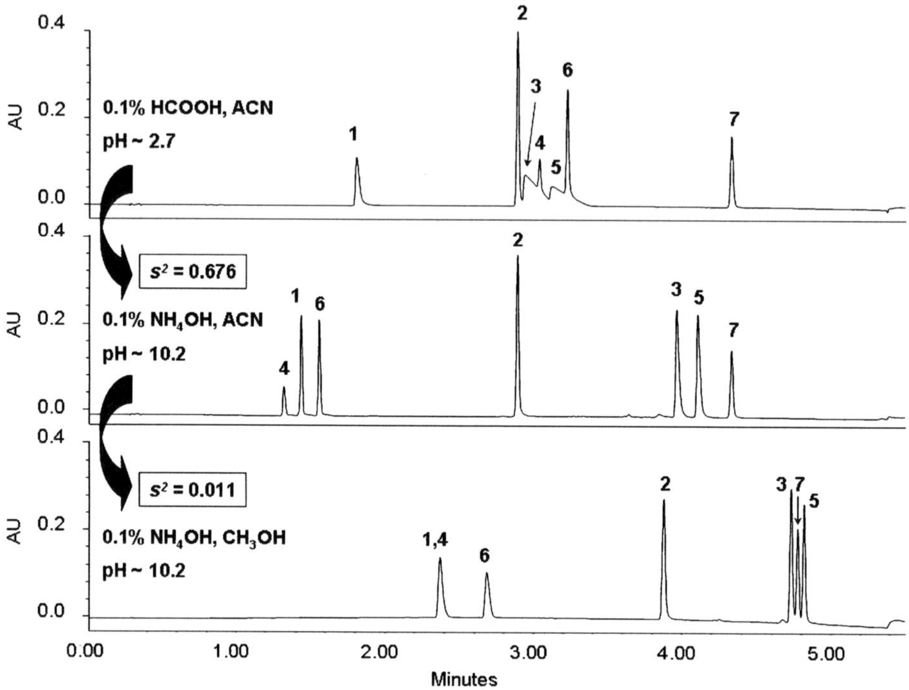

Figure 2.13 The influence of mobile phase pH and organic elution solvent on selectivity and retention in RP-UHPLC. The system was an Acquity UPLC® system with an Acquity PDA detector. The low pH mobile phase was 0.1% formic acid in water. The high pH mobile phase was 0.1% ammonium hydroxide in water. The organic solvent was ACN (top and middle) or methanol (bottom). The column was an Acquity UPLC® BEH C_{18} column, 2.1 mm × 50 mm, 1.7 μm. The gradient was 5–95% ACN in 5 min. The flow rate was 0.5 mLmin^{-1}. The column temperature was 30 °C and the UV detector was set to 254 nm. The injection volume was 2 μL. The sample diluent was 75:25 methanol (MeOH)/water. Peaks: (1) 1-pyrenesulfonic acid, (2) flavone, (3) imipramine, (4) fenoprofen, (5) amitriptyline, (6) diclofenac and (7) octanophenone. Data courtesy of Hillary Hewitson of Waters Corporation.

technique for retaining polar compounds that have little or no retention on traditional RP packings. Retention and selectivity in HILIC is driven by many different chemical interactions between the analyte, mobile phase, and stationary phase.[54] The major retention mechanism is thought to be partitioning of the analyte into a water-rich layer in the pores of the polar stationary phase.[57] As a result, HILIC separations have a high degree of orthogonality to RP separations (Figure 2.14), and provide more retention for polar compounds. Other published advantages of HILIC separations for polar compounds include increased signal in mass spectrometry and the ability to directly inject extracts prepared by ACN precipitation, solid phase extraction,

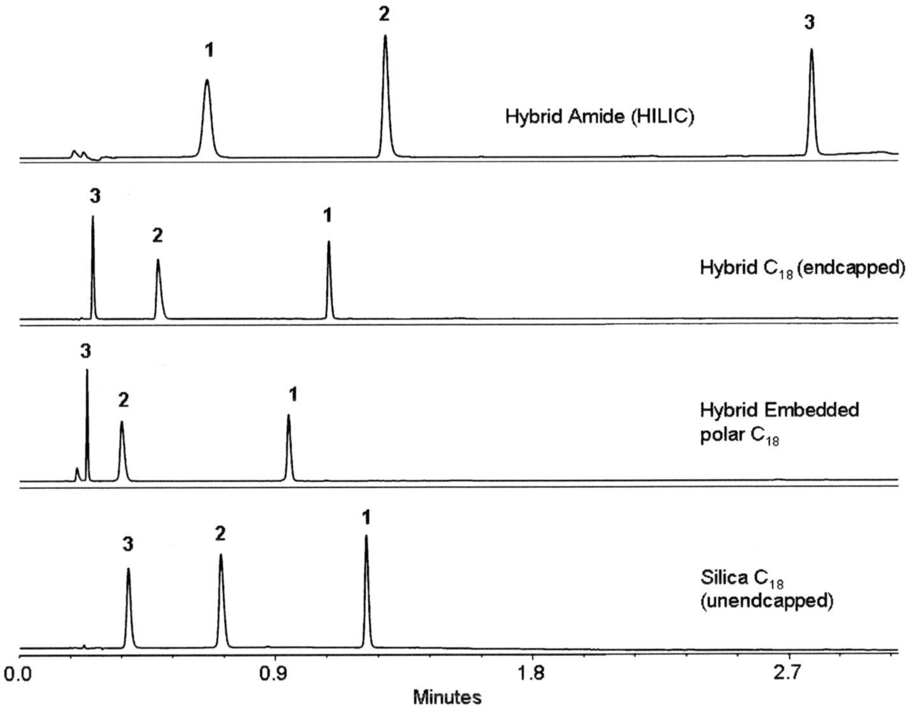

Figure 2.14 UHPLC separation of morphine-related compounds using HILIC and RP conditions. Separations were performed on an Acquity UPLC® H-Class system with an Acquity PDA. All columns were 2.1 mm × 50 mm, 1.7 μm. The mobile phase A was water the mobile phase B was ACN, and the mobile phase C was 200 mM ammonium formate, pH 3. The gradient for RP was from 5 to 95% B in 5 min with a constant addition of 5% C. The gradient for HILIC was from 90 to 50% B in 5 min with 5% C. The sample diluent for RP is 81:12:7 water/methanol (MeOH)/ACN/0.2% HCOOH. The sample diluent for HILIC is 75:25 ACN/MeOH. Each compound is present at a concentration of 25 μg mL^{-1}. UV detection was at 280 nm. Peaks: (1) 6-acetylmorphine, (2) morphine and (3) morphine-3β-D-glucuronide.

or liquid–liquid extraction.[54,58] Directly injecting these types of extracts can improve the throughput of certain assays by eliminating time-consuming evaporation and reconstitution steps.

Choosing columns with different selectivity is a critical step in method development. Based on column selection alone, a method meeting specific requirements can easily be developed 80 to 90% of the time, possibly from already established generic gradients. However, for the other 10 to 20% of separations, there are tools that are available that can aid in development of robust methods for challenging separations. Examples include software packages (*i.e.*, DryLab) that utilize statistical designs of experiments and quality-by-design (QbD) principles to establish a more well-defined separation space.[59] Another option is to explore the USP-PQRI database (http://www.usp.org/USPNF/columnsDB.html) to select a column for method development or method transfer.

Improvements in instrumentation, increases in column efficiency, and better quality and diversity of column chemistries have all been relatively recent developments. The breadth of published literature, software tools, and web-based information on chromatography reflects the level of development and sophistication the field has achieved. These advancements have made method development easier than it has ever been by leveraging the power of all the technical aspects needed to achieve robust separations.

Acknowledgements

We thank the following individuals for their contribution to this Chapter: Patricia McConville, Dr Kevin Wyndham, Hillary Hewitson, Tanya Jenkins, Chuck Phoebe, Dr Christopher Hudalla, Mia Summers, Jane Xu, Eric Grumbach, Dr Thomas Walter, Dr Martin Gilar Dr Andre Bourque, Dr Patrick McDonald, Joseph Arsenault, and Dr Tad Dourdeville . We dedicate this Chapter to the memories of Dr Marianna Kele and Dr Uwe D. Neue, whose contributions to the field and separations science, including UHPLC, will live on forever.

References

1. K. J. Fountain, U. D. Neue, E. S. Grumbach and D. M. Diehl, *J. Chromatogr. A*, 2009, **1216**, 5979.
2. J. E. MacNair, K. C. Lewis and J. W. Jorgenson, *Anal. Chem.*, 1997, **69**, 983.
3. D. T. Nguyen, D. Guillarme, S. Rudaz and J. L. Veuthey, *J. Sep. Sci.*, 2006, **29**, 1836.
4. G. Winkler, LC-GC, 1987, 5, 1044.
5. L. Novakova and H. Vlckova, *Anal. Chim. Acta*, 2009, **656**, 8.
6. U. D. Neue, *Chem. Eng. Process.*, 2010, **49**, 662.
7. H. Engelhardt and U. D. Neue, *Chromatographia*, 1982, **15**, 403.
8. J. E. MacNair, K. D. Patel and J. W. Jorgenson, *Anal. Chem.*, 1999, **71**, 700.
9. N. Wu, D. C. Collins, A. J. Lippert, Y. Xiang and M. L. Lee, *J. Microcolumn Sep.*, 2000, **12**, 462.
10. G. J. Dear, N. Patel, P. J. Kelly, L. Webber and M. Yung, *J. Chromatogr. B Analyt. Technol. Biomed. Life Sci.*, 2006, **844**, 96.
11. D. V. McCalley, *J. Chromatogr. A*, 2010, **1217**, 4561.
12. F. Gritti and G. Guiochon, *J. Chromatogr. A*, 2010, **1217**, 7677.
13. F. Gritti and G. Guiochon, *Anal. Chem.*, 2008, **80**, 5009.
14. J. W. Dolan, *LC-GC N. Am.*, 2005, **23**, 130.

15. F. Lestremeau, D. Wu and R. Szucs, *J. Chromatogr. A*, 2010, **1217**, 4925.
16. D. Guillarme, D. T. Nguyen, S. Rudaz and J. L. Veuthey, *Eur. J. Pharm. Biopharm.*, 2008, **68**, 430.
17. J. J. Kirkland and J. J. DeStefano, *J. Chromatogr. A*, 2006, 1126, 50.
18. U. D. Neue, *HPLC Columns: Theory, Technology, and Practice*, Wiley-VCH, New York, 1997.
19. F. Ullman, in *Ullmann's Encyclopedia of Industrial Chemistry*, Wiley-VCH Verlag GmbH, Weinheim, 1989.
20. L. R. Snyder, J. J. Kirkland and J. W. Dolan, in *Introduction to Modern Liquid Chromatography*, John Wiley and Sons, Hoboken, 2010, p. 200.
21. U. D. Neue, in *Encyclopedia of Analytical Chemistry*, ed. R. A. Meyers, John Wiley and Sons, Chichester, 2000, p. 11450.
22. T. J. Barder, P. J. Wohlman, C. Thrall and P. D. Dubios, *LC-GC*, 1997, **15**, 918.
23. M. Hanson and K. K. Unger, *LC-GC Int.*, 1996, **9**, 741.
24. L. C. Sander and S. A. Wise, *J. Chromatogr. A*, 1993, **656**, 335.
25. L. C. Sander and S. A. Wise, *Anal. Chem.*, 1995, **67**, 3284.
26. J. J. Kirkland, J. L. Glajch and R. D. Farlee, *Anal. Chem.*, 1989, **61**, 2.
27. A. Pryde, *J. Chromatogr. Sci.*, 1974, **12**, 486.
28. R. E. Majors and M. J. Hopper, *J. Chromatogr. Sci.*, 1974, **12**, 767.
29. K. Karch, I. Sebestian and I. Halasz, *J. Chromatogr.*, 1976, **122**, 171.
30. K. K. Unger, *J. Chromatogr. A*, 2004, **1060**, 1.
31. C. G. Horvath, B. A. Preiss and S. R. Lipsky, *Anal. Chem.*, 1967, **39**, 1422.
32. C. G. Horvath and S. R. Lipsky, *J. Chromatogr. Sci.*, 1969, **7**, 109.
33. J. J. Kirkland, F. A. Truszkowski, C. H. Dilks, Jr, and G. S. Engel, *J. Chromatogr. A*, 2000, **890**, 3.
34. G. Guiochon and F. Gritti, *J. Chromatogr. A*, 2011, **1218**, 1915.
35. J. J. DeStefano, T. J. Langlois and J. J. Kirkland, *J. Chromatogr. Sci.*, 2008, **46**, 254.
36. F. Gritti, A. Cavazzini, N. Marchetti and G. Guiochon, *J. Chromatogr. A*, 2007, **1157**, 289.
37. H. Geische, *J. Eur. Ceramic Soc.*, 1994, **14**, 205.
38. K. D. Wyndham, J. E. O'Gara, T. H. Walter, K. H. Glose, N. L. Lawrence, B. A. Alden, G. S. Izzo, C. J. Hudalla and P. C. Iraneta, *Anal. Chem.*, 2003, **75**, 6781.
39. J. N. Sagliano, T. R. Floyd, R. A. Hartwick, J. M. Dibussolo and N. T. Miller, *J. Chromatogr. A*, 1988, **443**, 155.
40. J. J. Kirkland, J. W. Henderson, J. J. DeStefano, M.A. van Straten and H. A. Claessens, *J. Chromatogr. A*, 1997, **762**, 97.
41. U. D. Neue, T. H. Walter, B. A. Alden, Z. Jiang, R. P. Fisk, J. T. Cook, K. H. Glose, J. L. Carmody, J. M. Grassi, Y-F. Cheng, Z. Lu and R. Crowley, *Am. Lab.*, 1999, **31**, 36.
42. Y-F. Cheng, T. H. Walter, Z. Lu, P. Iraneta, B. A. Alden, C. Gendreau, U. D. Neue, J. M. Grassi, J. L. Carmody, J. E. O'Gara and R. Fisk, *LC-GC*, 2000, **11**, 1162.

43. K. D. Wyndham, J. E. O'Gara, T. H. Walter, K. H. Glose, N. L. Lawrence, C. Hudalla, Y. Xu and G. Izzo, *PMSE Prepr.*, 2002, **87**, 274.
44. Z. Jiang, R. P. Fisk, J. O'Gara, T. H. Walter and K. D. Wyndham, *US Pat.* 6 686 035 B2, 2004.
45. J. S. Mellors and J. W. Jorgenson, *Anal. Chem.*, 2004, **76**, 5441.
46. P. C. Iraneta, K. D. Wyndham, D. R. McCabe and T. H. Walter, *Charged Surface Hybrid (CSH) Technology and Its Use in Liquid Chromatography*, Waters White Paper, literature code 720003929en, 2011.
47. M. Gilar, P. Olivova, A. E. Daly and J. C. Gebler, *Anal. Chem.*, 2005, **77**, 6426.
48. E. van Gyseghem, M. Jimidar, R. Sneyers, D. Redlich, E. Verhoeven, D. L. Massart and Y. Vander Heyden, *J. Chromatogr. A*, 2005, **1074**, 117.
49. S. Zhu, *J. Chromatogr. A*, 2009, **1216**, 3312.
50. U. D. Neue and A. Mendez, *J. Sep. Sci.*, 2007, **30**, 949.
51. U. D. Neue, J. E. O'Gara and A. Mendez, *J. Chromatogr. A*, 2006, **1127**, 161.
52. W. Fan, Y. Zhang, P. W. Carr, S. C. Rutan, M. Dumarey, A. P. Schelliger and W. Pritts, *J. Chromatogr. A*, 2009, **1216**, 6587.
53. J. J. Gilroy, J. W. Dolan and L. R. Snyder, *J. Chromatogr. A*, 2003, **1000**, 757.
54. K. J. Fountain, J. Xu, D. M. Diehl and D. Morrison, *J. Sep. Sci.*, 2010, **33**, 740.
55. D. V. McCalley, *Anal. Chem.*, 2006, **78**, 2532.
56. U. D. Neue, presented at the Pittcon Conference and Expo, Orlando, FL, 2010.
57. D. V. McCalley and U.D. Neue, *J. Chromatogr. A*, 2008, **1192**, 225.
58. E. S. Grumbach, D. M. Diehl and U. D. Neue, *J. Sep. Sci.*, 2008, **31**, 1511.
59. I. Molnar, *J. Chromatogr. A*, 2002, **965**, 175.
60. P. J. Eugster, D. Guillarme, S. Rudaz, J-L. Veuthey, P-A. Carrupt and J-L. Wolfender, *J. AOAC Intl.*, 2011, **94**, 51.

CHAPTER 3

Method Transfer Between Conventional HPLC and UHPLC

BENJAMIN DEBRUS[a], ERIC ROZET[b], PHILIPPE HUBERT[b], JEAN-LUC VEUTHEY[a], SERGE RUDAZ[a] AND DAVY GUILLARME*[a]

[a] School of Pharmaceutical Sciences, University of Geneva, University of Lausanne, Boulevard d'Yvoy 20, 1211 Geneva 4, Switzerland; [b] Laboratory of Analytical Chemistry, CIRM, Department of Pharmacy, University of Liège, Avenue de l'Hôpital 1, B36, B-4000 Liège, Belgium
*E-mail: davy.guillarme@unige.ch

3.1 Introduction

The goal of a chromatographic method is to sufficiently separate the components of a mixture within the shortest timeframe. To further increase throughput, ultra-high-pressure liquid chromatography (UHPLC), which involves the use of columns packed with sub-2 μm particles at pressures up to 1300 bar, became commercially available in 2004;[1–3] since then, UHPLC has quickly gained acceptance in numerous fields.[4–6] In addition to the throughput enhancement, an increase in the chromatographic resolution, a slight improvement in sensitivity and a significantly diminished solvent and sample consumption are possible with UHPLC, as demonstrated in numerous studies.[7–9] Because of these benefits, new UHPLC methods are often developed, whereas conventional high-pressure liquid chromatography (HPLC) methods

RSC Chromatography Monographs No. 16
UHPLC in Life Sciences
Edited by Davy Guillarme and Jean-Luc Veuthey

Published by the Royal Society of Chemistry, www.rsc.org

can be easily transferred to UHPLC using various approaches based on the analytical needs (throughput or resolution). When the column length is identical and only the particle size is reduced, an improvement in efficiency and peak capacity is expected.[10] On the contrary, when the particle size is simultaneously reduced with the column length, a reduction of analysis time is observed.[11] However, care must be taken in both cases to ensure that the operating flow rate, injection volume and gradient profile are appropriately scaled to obtain an equivalent or superior separation in UHPLC.[12,13]

In the present Chapter, the equations used for transferring an isocratic or a gradient method from conventional HPLC to UHPLC with limited influence on chromatographic resolution are presented and critically discussed, together with various examples that show the possibility of transferring methods toward UHPLC.

Except for cases that involve a qualitative transfer, it is often important to quantitatively evaluate the method and to validate it. After a conventional HPLC method to UHPLC has been transferred, there is a need to confirm that this new method still fits its intended purpose (*i.e.*, that is valid). A further requirement is to show that this new method will provide results that are equivalent to those obtained from the initial conventional HPLC method. Both of these requirements should be met to ensure that the results generated by the novel UHPLC assay will be reliable and that any critical decisions made (*e.g.*, batch release, non-clinical and clinical studies and patient status) will be at least as valid as those that would have been made with the conventional HPLC method.

3.2 Qualitative Transfer from HPLC to UHPLC

In general, the method transfer between HPLC and UHPLC is performed to increase the throughput or resolution of a conventional separative method. However, it is also possible to transfer methods from UHPLC to conventional HPLC. The second solution seems unnecessary but can be useful, for example, if UHPLC instruments are only available in R&D laboratories and the routine or quality control (QC) laboratories do not have UHPLC technology at their disposal. With the UHPLC approach, the time for method development can be drastically reduced, while the transfer to HPLC remains straightforward.

3.2.1 Theoretical Improvements with UHPLC

To maintain the same performance between HPLC and UHPLC, it is important that the ratio L/d_p be kept constant because the chromatographic efficiency of a column can be estimated with the following equation, in which h is the reduced height equivalent to a theoretical plate defined by Knox:

$$N = \frac{L}{h \times d_p} \tag{3.1}$$

As expected, this ratio is indeed constant between a conventional HPLC column of 150 mm × 4.6 mm, 5 μm (L/d_p = 30 000) and a UHPLC column of 50 mm × 2.1 mm, 1.7 μm (L/d_p = 29 400).

Because the analysis time of the transferred method (t_{ana2}) is directly proportional to the change in column dead time, it can be estimated according to the following equation:

$$t_{ana2} = t_{ana1} \cdot \frac{d_{p2}}{d_{p1}} \cdot \frac{L_2}{L_1} \tag{3.2}$$

The expected backpressure in UHPLC (ΔP_2) can be calculated from Darcy's law, which shows that ΔP is inversely proportional to $d_p{}^3$ when columns operate at the optimal linear velocity and that ΔP is strictly related to the column length:

$$\Delta P_2 = \Delta P_1 \cdot \frac{L_2}{L_1} \cdot \frac{d_{p1}^3}{d_{p2}^3} \tag{3.3}$$

Finally, the decrease in solvent consumption for the transferred method (V_2) can be calculated by taking into account the change in the internal diameter (I.D.) and length of the column:

$$V_2 = V_1 \cdot \frac{d_{c_2}{}^2}{d_{c_1}{}^2} \cdot \frac{L_2}{L_1} \tag{3.4}$$

Therefore, from a regular 150 mm × 4.6 mm I.D., 5 μm column to a UHPLC 50 mm × 2.1 mm I.D., 1.7 μm column, the theoretical reduction of analysis time is 9-fold, whereas the same efficiency, peak capacity and thus resolution are maintained. Another important benefit of UHPLC is the 14-fold decrease in solvent consumption, which was particularly beneficial during the acetonitrile shortage in 2009. However, the trade-off with UHPLC technology is the generated backpressure: an 8.5-fold increase is expected at the optimal mobile-phase flow rate, despite a 3-fold reduction in the column length. This increase confirms the need for dedicated instrumentation that withstands pressures greater than 400 bar. The changes in analysis time, backpressure and solvent consumption are reported in Table 3.1 for a number of different column geometries.

Finally, the increase in sensitivity between the two technologies is more difficult to assess because the column geometry changes simultaneously with the instrumentation (a UV detector can indeed be more sensitive on one system than on another). For this reason, it is even possible in some cases for UHPLC technology to exhibit a lower sensitivity. From a theoretical point of view, the increase in sensitivity should be related to the higher efficiency, which results in narrower peaks produced by UHPLC. However, when the injected volume is correctly scaled between HPLC and UHPLC and

Table 3.1 Changes in analysis time, backpressure and solvent consumption at optimal mobile-phase flow rate between a conventional HPLC column (first line of the Table) and six different column geometries providing a similar theoretical efficiency of 10 000 plates.

Column geometry	*Change in analysis time*	*Change in backpressure*	*Change in solvent consumption*
150 × 4.6 mm, 5 μm	–	–	–
100 × 4.6 mm, 3.5 μm	÷2.1	×1.9	÷1.5
50 × 4.6 mm, 1.7 μm	÷8.8	×8.5	÷3
150 × 2.1 mm, 5 μm	×1	×1	÷4.8
100 × 2.1 mm, 3.5 μm	÷2.1	×1.9	÷7.2
50 × 2.1 mm, 1.7 μm	÷8.8	×8.5	÷14.4

considering that an identical efficiency is obtained in both HPLC and UHPLC, there is no reason for the sensitivity to be increased, unless the UV technology has been improved, such as by the incorporation of the optofluidic waveguide technologies from Waters Corporation (Milford, MA, USA) or Agilent (Waldbronn, Germany).

3.2.2 Choice of UHPLC Column Dimensions and Chemistry for Method Transfer

In terms of column I.D., the standard in HPLC is 4.6 mm, whereas columns with diameters of 2.1 mm are commonly used in UHPLC. However, most of the column manufacturers have also commercialized 1, 3 and 4.6 mm I.D. columns for UHPLC. In our opinion, the 4.6 mm I.D. columns are not suitable because of the heat generated by small particles in the presence of a high mobile-phase flow rate, which generates a lack of repeatability for the retention times (see Section 3.2.5.2). In addition, the consumption of organic solvent is critical because the flow rate should be in the range 3–5 mL min^{-1}. Regarding the 1 mm I.D. columns, frictional heating is less problematic; however, the compatibility between a column with such a reduced volume and any UHPLC instrument is extremely critical (in particular, the tubing geometry is not adapted). Hence, the 2.1 mm I.D. column appears to be the best compromise, whereas the 3 mm I.D. UHPLC column should only be used with caution.

The column length should be selected based on the analytical needs (throughput or resolution) and the elution mode (isocratic or gradient).[8,9] Most of the providers propose a wide range of column lengths from 10 to 150 mm. In most of cases, 150 mm columns can be coupled in series with low-volume connection tubing to further extend the efficiency or peak capacity. In the isocratic mode, the chromatographic efficiency is known to be directly proportional to the column length. A column of 50 mm packed with sub-2 μm particles should generate approximately 10 000 plates, whereas the efficiency of

100 and 150 mm columns is equal to 20 000 and 30 000 plates, respectively. Therefore, the length should be selected according to the requirement of the separation and the trade-off between throughput and resolution.

In the gradient mode, the rules that govern column selection are different, and a recent paper has demonstrated that the longest column does not provide the highest peak capacity.[10] Indeed, the peak capacity depends on both the isocratic efficiency and the column dead time, each to a different extent. Therefore, to achieve the highest possible performance, a column of 50 mm should be selected in combination with a gradient time of less than 5 min, whereas the 100 mm column gives optimal performance for gradient times between 5 and 20 min, and the 150 mm column should only be used with gradient times longer than 20 min. As an example, the peak capacity is represented as a function of gradient time for different UHPLC columns in Figure 3.1.

The selection of suitable stationary-phase chemistry is the most critical step for efficient method transfer. Indeed, the selectivity of the chromatographic support should remain identical when methods from HPLC to UHPLC are transferred. For this purpose, it is possible to work with a stationary phase that is similar in nature but differs in column geometry. However, in most situations, HPLC methods are developed with previous-generation columns that are not available with sub-2 μm porous particles. In such cases, it is necessary to find a column that exhibits a retention and selectivity that is as close as possible to the original column. For this purpose, numerous tools can be found on the Internet or from column manufacturers. Most of these tools are based on the seminal works on the hydrophobic-subtraction model by

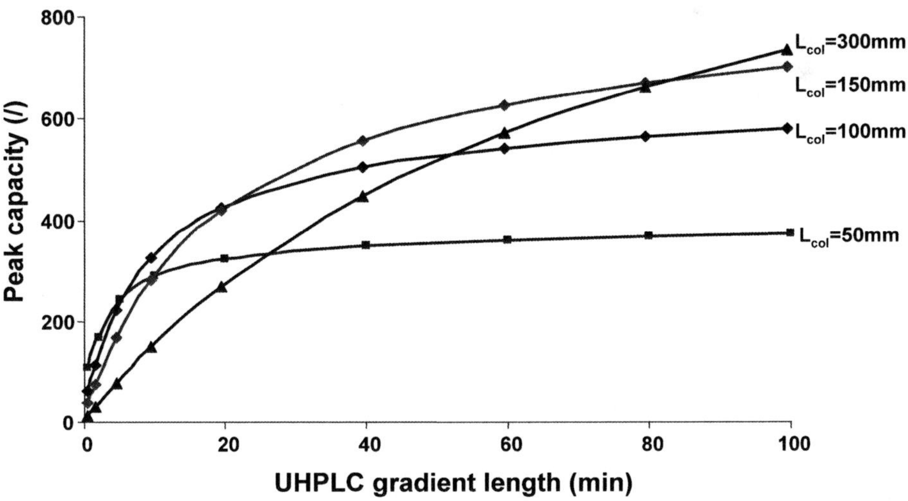

Figure 3.1 Evolution of peak capacity *vs.* gradient time, using different Waters Acquity C18 1.7 μm column lengths (L_{col}). Data are given for a 5–95% acetonitrile gradient. Peak capacities were systematically calculated for ΔP_{max} = 1000 bar. Adapted from Guillarme *et al.*[10] with permission.

Horvath, Kirkland, Snyder and Carr.[14–16] The application of this model to the retention data of various solutes and columns has provided new insights into the nature of solute–column interactions and allows for a complete characterization of columns based on hydrophobicity, steric hindrance, hydrogen-bonding acidity, hydrogen-bonding basicity and cation-exchange capacity.[17] Data have been published for more than 300 columns.[17] As shown in Table 3.2, a number of UHPLC columns are available from many different

Table 3.2 Summary of manufacturers for columns packed with sub-2 μm particles, with the number of available chemistries, the main types of chemistry, pressure tolerance and particle size. The information presented here was gathered from advertising and from the manufacturers' websites in January 2011.

Manufacturer	*Name*	*Number of chemistries*	*Main types of chemistry*	*Pressure tolerance (bar)*	*Particle size (μm)*
			500 bar < ΔP_{max} < 800 bar		
Agilent	RRHT	8	C_{18}, C_8, CN	600	1.8
Grace-Davison	Vision-HT	6	C_{18}, HILIC, silica	800	1.5
Sepax Technologies	Sepax UHPLC	11	C_{18}, C_8, C_4, phenyl, amino, CN, SCX, SAX, HILIC	600	1.8
VWR	LaChromUltra	1	C_{18}	600	2
YMC	UltraHT	2	C_{18}	500	2
Zirchrom Technologies	Zirchrom-PBD & -Phase	2	Zirconia-based material	700	2
			ΔP_{max} ≥ 1000 bar		
Agilent	RRHD	2	C_{18}	1200	1.8
ES Industries	Epic	7	C_{18}, PFP, HILIC, diol, silica plus 3 phases SFC	1000	1.8
Interchim	Strategy	2	C_{18}, HILIC	1000	1.7
Knauer	BlueOrchid	7	C_{18}, C_8, PFP, phenyl, CN, silica	1000	1.8
Macherey-Nagel	Nucleodur	5	C_{18}, C_8	1000	1.8
Restek	PinnacleDB and Ultra II	12	C_{18}, C_8, PFP, biphenyl, CN, silica	1000	1.9
Thermo	Hypersil GOLD	6	C_{18}, C_8, PFP, CN, phenyl	1000	1.9
Waters	Acquity BEH	6	C_{18}, C_8, phenyl, HILIC, amide	1000	1.7
Waters	Acquity HSS	3	C_{18}	1000	1.8

sources that can be employed for method transfer. However, when an equivalent column for UHPLC use cannot be found, the method cannot be simply transferred using basic geometric rules.

3.2.3 Rules for Method Transfer in the Isocratic Mode

In UHPLC, it is important to consider that the size of the frits contained within the column is much smaller than those for a regular HPLC packing. Generally, the inlet frit pore size of an HPLC column is of the order of 2 μm, whereas it can be reduced to only 0.2 μm in UHPLC columns from some column manufacturers. Therefore, small particles that are potentially present within the mobile phase and that do not affect HPLC materials can become critical under UHPLC conditions. To circumvent problems associated with UHPLC, several important rules have to be followed for mobile-phase preparation, including the use of high-grade organic solvents (ideally filtered through a 0.22 μm membrane), high-quality salts to prepare buffered mobile phases and ultra-pure water filtered through a 0.22 μm membrane (the Milli-Q® system or a similar high-quality water is recommended). In addition, it is mandatory to be careful with microbiological growth, which is particularly critical with phosphate buffers at pH levels of approximately 7, and to store columns with pure organic solvents (acetonitrile or methanol).[18]

Except for these common-sense rules for mobile-phase preparation in UHPLC and the need to select a column that exhibits a selectivity identical to that of the original one, the geometric rules to be applied are summarized below.

3.2.3.1 Geometrical Transfer Rules

In the isocratic mode, only two parameters have to be geometrically modified: the injection volume and the mobile-phase flow rate.[12]

To avoid a detrimental band broadening related to the instrument and to maintain equivalent levels of sensitivity in HPLC and UHPLC, the injection volume should be adapted in accordance with the column volume; the column volume is drastically reduced in UHPLC. The injected volume should represent only 1–5% of the column volume in HPLC. Because the HPLC column volume depends only on the column I.D. (d_c) and length (L), the injection volume is independent of the particle size. The new injected volume (V_{inj2}) can be simply calculated by holding the ratio of the column dead volume to the injected volume constant between regular HPLC and UHPLC. Thus, the UHPLC injected volume (V_{inj2}) can be calculated according to the following equation:

$$V_{inj_2} = V_{inj_1} \cdot \frac{d_{c2}^2}{d_{c1}^2} \cdot \frac{L_2}{L_1} \tag{3.5}$$

In this equation, subscripts 1 and 2 are related to HPLC and UHPLC column dimensions. For example, between a conventional 150 mm × 4.6 mm

I.D., 5 μm column and a UHPLC 50 mm × 2.1 mm I.D., 1.7 μm column, the injected volume should be decreased 14-fold. Injection volumes larger than those predicted can also be used to increase sensitivity. In this case, the sample should be dissolved in a solvent with an elution strength weaker than the initial mobile-phase composition to limit associated band broadening.[19] This approach, described as sample focusing (peak compression), allows for the enrichment of the analytes on the top of the column.[20,21]

The mobile-phase flow rate should be adapted to be as close as possible to the optimal linear velocity value of the Van Deemter curve (u_{opt}). In liquid chromatography (LC), the u_{opt} value is directly proportional to the square of the column diameter and depends on the particle size of the support. It is, however, completely independent of the column length.

For a successful method transfer, the product $u{\cdot}d_p$ must be maintained constantly to take into account simultaneous changes in the column diameter and particle size of the support. Therefore, for a geometrical transfer, the UHPLC flow rate (F_2) can be calculated with the following equation:

$$F_2 = F_1 \cdot \frac{{d_{c_2}}^2}{{d_{c_1}}^2} \cdot \frac{d_{p_1}}{d_{p2}} \tag{3.6}$$

In theory, between a regular 5 μm HPLC column and a UHPLC column of 1.7 μm particles, the mobile-phase flow rate should be increased by 3-fold according to the Van Deemter representation (Figure 3.2). However, because

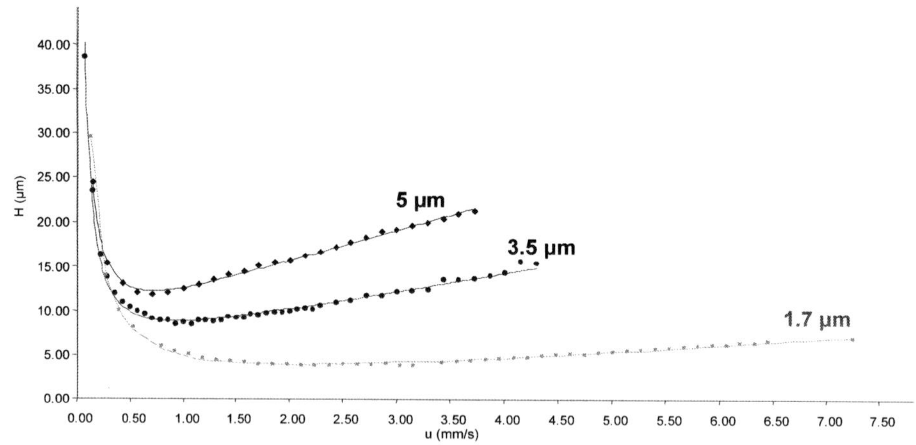

Figure 3.2 Impact of particle size reduction on the Van Deemter curves. These experimental curves were obtained with a mobile phase: H_2O/MeCN (60:40, v/v), with butylparaben 25 ppm in H_2O; UV detection: 254 nm; columns: XTerra RP18, 4.6 × 150 mm, 5 μm; XTerra RP18, 4.6 × 50 mm, 3.5 μm; Acquity BEH Shield RP18 2.1 × 50 mm, 1.7 μm. H, the height equivalent to a theoretical plate; u, linear velocity of the mobile phase. Adapted from Nguyen *et al.*[11] with permission.

the I.D. is also reduced between HPLC and UHPLC, the mobile-phase flow rate should be decreased by 1.6-fold between a regular 150 mm × 4.6 mm I.D., 5 μm column and a UHPLC 50 mm × 2.1 mm I.D., 1.7 μm column.

3.2.3.2 Importance of Extra-column Volume

As previously mentioned, a dedicated instrument that withstands pressures greater than 400 bar is mandatory to work with columns packed with sub-2 μm particles. In addition, because of the strong reduction of the column diameter and length, the extra-column volume (or system volume) should be proportionally reduced to limit the associated band broadening.[12,22] Indeed, the total band broadening is the sum of the column and the extra-column broadening, as shown with the following equation:

$$\sigma^2_{tot} = \sigma^2_{col} + \sigma^2_{ext} \tag{3.7}$$

In general, the ratio between extra-column variance, σ^2_{ext}, and total variance, σ^2_{tot}, should be less than 10% for a good separation.

The dispersion related to the chromatographic column itself (σ^2_{col}) is given by:

$$\sigma^2_{col} = \frac{V_R}{\sqrt{N}} = \frac{V_0 \cdot (1+k)}{\sqrt{N}} \tag{3.8}$$

where σ^2_{col} is the column variance (in μL^2), N is the number of theoretical plates and V_R is the retention volume, which is a function of the column dead volume V_0 and the retention factor k [$V_R = V_0 \cdot (1+k)$].

However, extra-column band broadening is expressed as the sum of three main dispersion sources:

$$\sigma^2_{ext} = \sigma^2_{inj} + \sigma^2_{det} + \sigma^2_{tubing} \tag{3.9}$$

where σ^2_{ext} is the extra-column variance, and σ^2_{inj}, σ^2_{det} and σ^2_{tubing} are the variances due to the injector, detector and tubing, respectively. As shown in eqn (3.10), σ^2_{ext} depends primarily on the injected volume V_{inj}, the tubing radius r and length l, the flow-cell volume V_{cell}, the detector time constant τ and the flow rate F according to the equation given below:

$$\sigma^2_{ext} = K_{inj} \cdot \frac{V^2_{inj}}{12} + K_{cell} \cdot \frac{V^2_{cell}}{12} + \tau^2 \cdot F^2 + \frac{r_c^4 \cdot l_c \cdot F}{7.6 \cdot D_m} \tag{3.10}$$

where K_{inj} and K_{cell} are constant (typically between 1 and 3)[19] and linked to the injection mode and the UV cell geometry, respectively. Based on this relationship, the following criteria have to be fulfilled to attain sharp peaks in UHPLC.

The tubing length should be as short as possible between the injector and the column inlet and between the column outlet and the UV detector. The diameter of the tubing should be selected as a compromise between an acceptable backpressure and a low volume. Thus, a system plumbed with 0.005 inch (1 inch = 2.54 cm) I.D. tubing and zero-dead-volume fittings seems optimal for UHPLC experiments.

As previously discussed, the injection volume should be selected in agreement with column geometry. Because most of the experiments conducted using UHPLC are performed with a 50 mm × 2.1 mm I.D. column ($V_0 = 120$ μL), the injected volume should be between 1 and 5 μL to limit band broadening. In addition, a fast injection cycle time is mandatory for analysis times shorter than 1 or 2 min.

Finally, the detector cell volume and time constant should also be adequately selected. The UHPLC detector should ideally possess a low cell volume (2 μL or less), whereas the sensitivity should not be altered. The detector time constant has to be sufficiently fast ($10 < \tau < 100$ ms) because peak widths are very small in UHPLC (only a few seconds).[9]

To evaluate the compatibility of a UHPLC instrument with a given column geometry, it is recommended to determine the extra-column volume. For optimal compatibility with fast separations, the extra-column volume should ideally be less than 20 μL.

3.2.3.3 *Examples from the Literature*

Numerous applications can be found in the literature to highlight the level of interest in transferring an HPLC method to UHPLC conditions. Many isocratic method transfers are involved with pharmaceutical applications.

For example, a simple pharmaceutical formulation "rapidocaïn", which comprises the active ingredient (lidocaine hydrochloride), two antimicrobial preservatives (methylparaben and propylparaben) and a degradation product (2,6-dimethylaniline) was analyzed for quality control purposes.[12] As reported in Figure 3.3, the original HPLC separation took approximately 8 min at a flow rate of 1 mL min^{-1} on a XBridge RP18 column, whereas the analysis time was reduced to only 1.2 min under UHPLC conditions at optimal mobile-phase flow rate ($F = 613$ μL min^{-1}) with a Acquity BEH Shield RP18; the resolution and selectivity between the HPLC and UHPLC methods were similar. Because it is possible to work beyond the optimal mobile-phase velocity in UHPLC without a significant loss in efficiency (due to the flat Van Deemter curves), the analysis time could be further reduced to 45 s at a pressure of 1000 bar and a flow rate of 1 mL min^{-1}. In this last case, the resolution was reduced by approximately 15%, whereas the analysis time was decreased 11-fold.

A screening analytical method for the determination of seven anxiolytics was originally developed using conventional HPLC with a Hypersil GOLD 150 mm × 4.6 mm I.D., 5 μm column. The analysis time was equal to 19 min. This

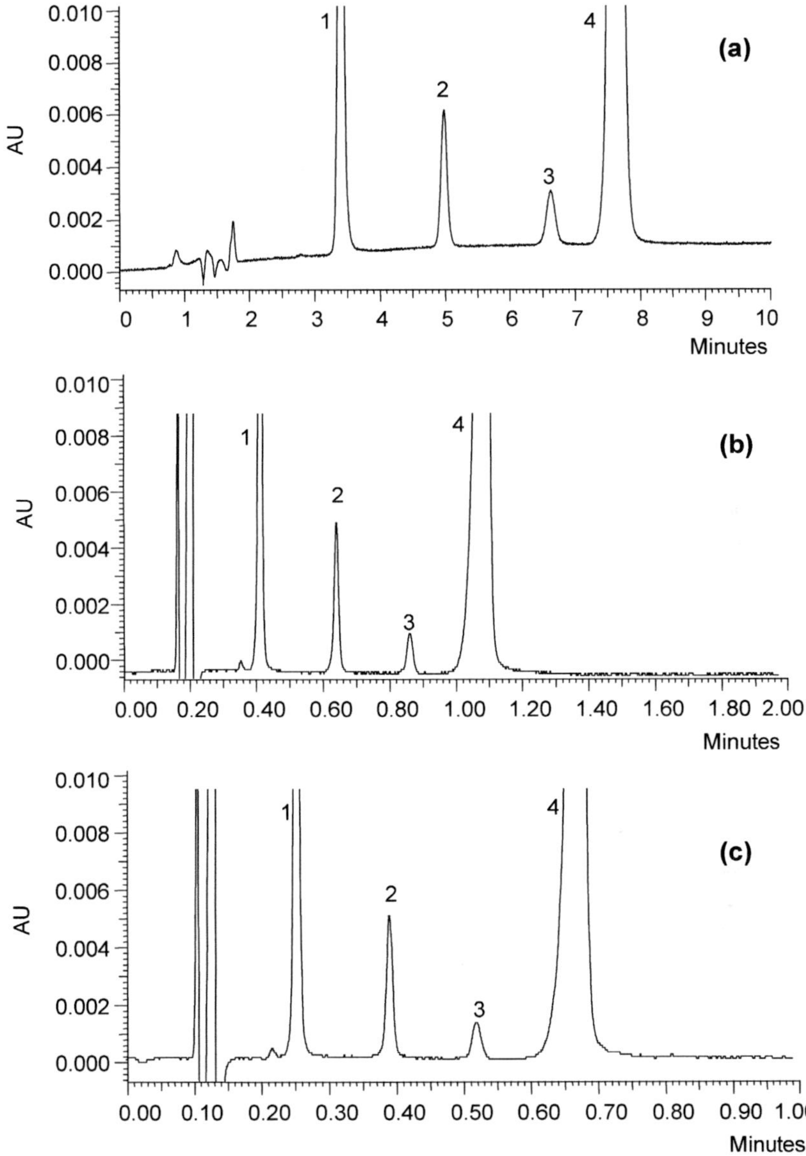

Figure 3.3 Isocratic method transfer from regular HPLC to UHPLC. Separation of a pharmaceutical formulation in isocratic mode with a mobile phase containing acetonitrile/phosphate buffer, pH 7.2 (40:60, v/v), T = 30 °C and λ = 230 nm. (a) Waters Xterra RP18, 150 × 4.6 mm, 5 μm, F = 1000 μL min^{-1}, V_{inj} = 20 μL. (b) Waters Acquity Shield RP18, 50 × 2.1 mm, 1.7 μm, F = 613 μL min^{-1}, V_{inj} = 1.4 μL. (c) Waters Acquity Shield RP18, 50 × 2.1 mm, 1.7 μm, F = 1000 μL min^{-1}, V_{inj} = 1.4 μL. 1, methylparaben; 2, 2,6-dimethylaniline; 3, propylparaben; 4, lidocaine. AU, absorbance units.

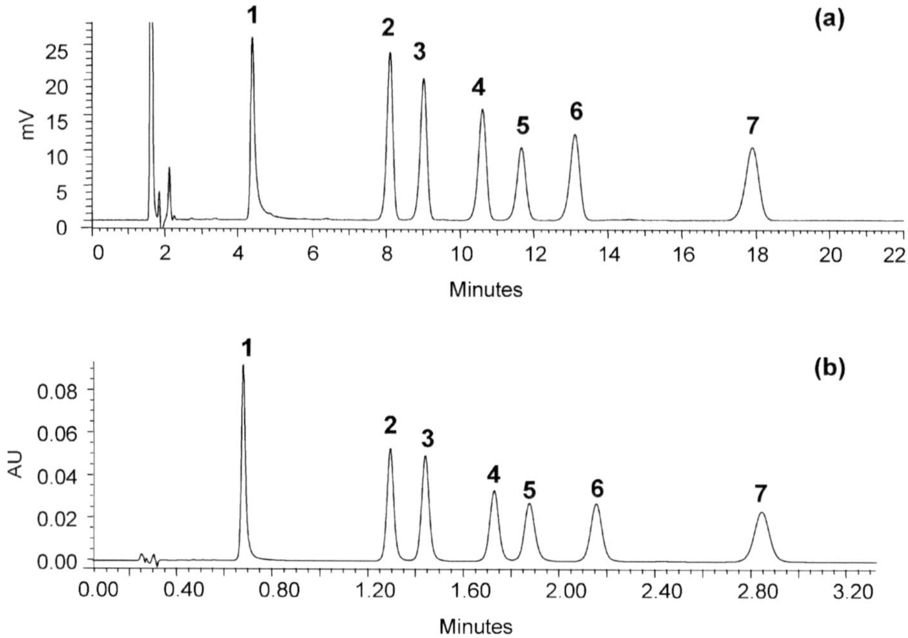

Figure 3.4 Isocratic method transfer from regular HPLC to UHPLC. Separation of a benzodiazepine mixture in isocratic mode with a mobile phase containing acetonitrile/water (31:69, v/v) with 0.1% formic acid, T = 30 °C and λ = 254 nm. (a) Hypersil GOLD ,150 × 4.6 mm, 5 µm, F = 1000 µL min^{-1}, V_{inj} = 20 µL. (b) Hypersil GOLD, 50 × 2.1 mm, 1.9 µm, F = 550 µL min^{-1}, V_{inj} = 1.4 µL. AU, absorbance units. Adapted from Russo et al.[23] with permission.

method was transferred to UHPLC conditions using a Hypersil GOLD 50 mm × 2.1 mm I.D., 1.9 µm column, and the analysis time was reduced to only 3 min at the optimal mobile-phase flow rate (i.e. F = 550 µL min^{-1}), as shown in Figure 3.4. For this separation, the minimal resolutions were equal to 2.09 and 1.93 in HPLC and UHPLC, respectively.[23]

These examples show that the isocratic method transfer is quite straightforward when the previously described rules are applied. However, the literature also contains numerous examples in which the authors mentioned a method transfer between HPLC and UHPLC for which the stationary-phase chemistry and mobile-phase composition are completely different.[24] In such cases, the denomination "method transfer" should be avoided.

3.2.4 Rules for Method Transfer in the Gradient Mode

Gradient elution is a very popular approach in LC for the analysis of complex samples, because it allows for a reduction of the analysis time, an improvement in peak capacity and an increase in sensitivity.[25] The conversion of a gradient method involves rules that are more complex than those for the isocratic mode, although the gradient rules are also based on basic principles of chromatography.

3.2.4.1 Geometrical Transfer Rules

First, the mobile-phase flow rate and injection volume should be scaled in a manner similar to that in the isocratic mode [eqns (3.5) and (3.6)].

Then, when linear and multilinear gradient elutions are involved, the gradient profile can be decomposed into a combination of isocratic and gradient steps.[26] For any isocratic step within the gradient (*i.e.*, the initial isocratic step, the isocratic step during a multilinear gradient and the re-equilibration time), the ratio between the isocratic step time (t_{iso}) and the column dead time (which depends on the mobile-phase flow rate, column I.D. and length) should be kept equivalent between conventional the HPLC and the UHPLC conditions. Therefore, the UHPLC isocratic step (t_{iso2}) can be determined with the following equation:

$$t_{iso_2} = t_{iso_1} \cdot \frac{F_1}{F_2} \cdot \frac{d_{c_2}^2}{d_{c_1}^2} \cdot \frac{L_2}{L_1} \tag{3.11}$$

As an example, for a method transfer from a regular 150 mm × 4.6 mm, 5 μm HPLC column to a UHPLC 50 mm × 2.1 mm I.D., 1.7 μm column, the isocratic steps that occur during the gradient process should be reduced 9-fold at the optimal mobile-phase linear velocity. The same rules have to be applied for the scaling of the re-equilibration time. Indeed, the usual 15–20 min required in HPLC experiments is reduced to approximately 2 min in UHPLC.

For slope segments, the rules described by Snyder and Dolan and recently updated by Carr *et al.* should be strictly followed.[27,28] To keep the relative position (*i.e.*, the selectivity) in the chromatogram unchanged, the retention factor must be kept constant. Because the elution strength increases during gradient elution, the retention factor k for each sample component decreases with time. Therefore, k, as defined in isocratic mode, is an unsuitable parameter for retention in gradient elution. For this reason, the linear solvent strength (LSS) theory developed by Snyder introduces the concept of k_e, which represents the retention factor of the solute in the eluted mobile-phase composition. The parameter k_e represents the "average" or "effective" value of k during gradient elution and should remain constant during method transfer. The definition of k_e is the following:

$$k_e = \frac{1}{2.3 \cdot b} \tag{3.12}$$

The b term is the gradient steepness, and can be obtained by:

$$b = \frac{t_0 \cdot \Delta\Phi \cdot S}{t_{\text{grad}}} \tag{3.13}$$

where, t_{grad} is the gradient time, t_0 is the column dead time, $\Delta\Phi = \%B_{\text{final}} - \%B_{\text{initial}}$ is the change in composition during the gradient, which ranges from 0 to 1, and S is a constant term for a given solute and organic modifier. The constant S corresponds to the elution strength of the organic modifier [slope of the logarithmic plot: d(log k)/dΦ]. Since S and $\Delta\Phi$ are constant, the gradient time should be scaled in proportion with the reduction in t_0 afforded by UHPLC.

In other words, it is important to scale the gradient volume in proportion to the number of column volumes to yield identical elution patterns,[13] whereas the initial and final gradient composition (%B) should be kept constant. Thus, the number of column volumes percolated during the gradient in the regular HPLC system should be equivalent to that of the UHPLC set-up. The new gradient time (t_{grad2}) can be expressed as:

$$t_{\text{grad}_2} = \frac{\left(\%B_{\text{final}_1} - \%B_{\text{initial}_1}\right)}{slope_2} \tag{3.14}$$

The gradient slope ($slope_2$) should be calculated to keep the product of the gradient slope and the column dead time constant. The new gradient slope ($slope_2$) can be expressed as:

$$slope_2 = slope_1 \cdot \frac{d_{c_1}^2}{d_{c_2}^2} \cdot \frac{L_1}{L_2} \cdot \frac{F_2}{F_1} \tag{3.15}$$

As an example, in the method transfer from a regular 150 mm × 4.6 mm I.D., 5 μm HPLC column to a UHPLC 50 mm × 2.1 mm I.D., 1.7 μm column, the gradient slope during the gradient process should also be increased 9-fold. As example, optimal gradient times for different types of columns are reported in Table 3.3.

Table 3.3 Calculated optimal gradient time according to column geometry.

Column geometry	*Flow rate (μL min⁻¹)*	*Gradient profile (%)*	*Optimal gradient time (min)*
150 × 4.6 mm, 5 μm	1000	5–95	30
50 × 2.1 mm, 1.7 μm	600	5–95	3.5
50 × 2.1 mm, 1.7 μm	1000	5–95	2.0
50 × 2.1 mm, 1.7 μm	600	10–60	2.0
100 × 2.1 mm, 1.7 μm	300	5–95	14
100 × 2.1 mm, 1.7 μm	300	10–60	8
150 × 2.1 mm, 1.7 μm	200	5–95	30
150 × 2.1 mm, 1.7 μm	200	10–60	17

3.2.4.2 Importance of the Dwell Volume/Gradient Delay Volume

When a gradient method is transferred from regular HPLC to UHPLC, some changes in selectivity may occur in the new gradient run because of differences in the dwell volumes between the instruments. The dwell volume (V_d) is also referred to as the gradient delay volume; it corresponds to the volume between the mixing point of the solvents and the inlet of the analytical column. After the gradient has been started, the selected proportion of solvent reaches the column some time later. The sample is thus subjected to an additional undesired isocratic migration in the initial mobile-phase composition.[29] Because the gradient dwell volume may differ from one system to another, this extra isocratic step also varies and could result in retention-time variations that affect the resolution of early eluting peaks when a method is transferred.[27]

To overcome this problem, the ratio of the system dwell time (t_d) to the column dead time (t_0) must be held constant while the column dimensions, particle size or mobile-phase flow rate are changed.[13,27]

Because the column dead time is reduced by approximately 9-fold between a regular 150 mm × 4.6 mm I.D., 5 μm column and a 50 mm × 2.1 mm I.D., 1.7 μm column, the system dwell time should be reduced by a similar factor. Therefore, it is mandatory to work with a UHPLC system that possesses a low dwell volume (no more than a few hundreds of μL). For this reason, it is recommended to work preferentially with high-pressure mixing systems in UHPLC because low-pressure mixing systems possess larger dwell volumes.

When the difference between t_d/t_0 ratios is excessive, the solution is either to add an isocratic hold at the beginning of the UHPLC gradient or to postpone the injection after the start of the gradient. If these solutions cannot be applied, the gradient method transfer could fail.

3.2.4.3 Examples from the Literature

Because gradient elution is more popular than the isocratic method, the literature contains an important number of applications that address the gradient method transfer between conventional and sub-2 μm column packings.

For example, an HPLC method was developed to separate an active pharmaceutical ingredient (API) and 11 different impurities (impurity profiling is a central type of analysis within the pharmaceutical industry).[13] The separation was achieved using a 150 × 4.6 mm I.D., 5 μm, C18 column and was further transferred to UHPLC with a 50 × 2.1 mm I.D., 1.7 μm C18 column with a strictly similar chemistry. For this application, the effect of dwell volume was negligible because the ratio t_d/t_0 was equivalent between the two sets of conditions. As shown in Figure 3.5, the original separation was performed in approximately 27 min; after being efficiently transferred to UHPLC, the separation was performed in less than 3 min (a reduction by a factor of 9). In addition, both separations were equivalent in terms of

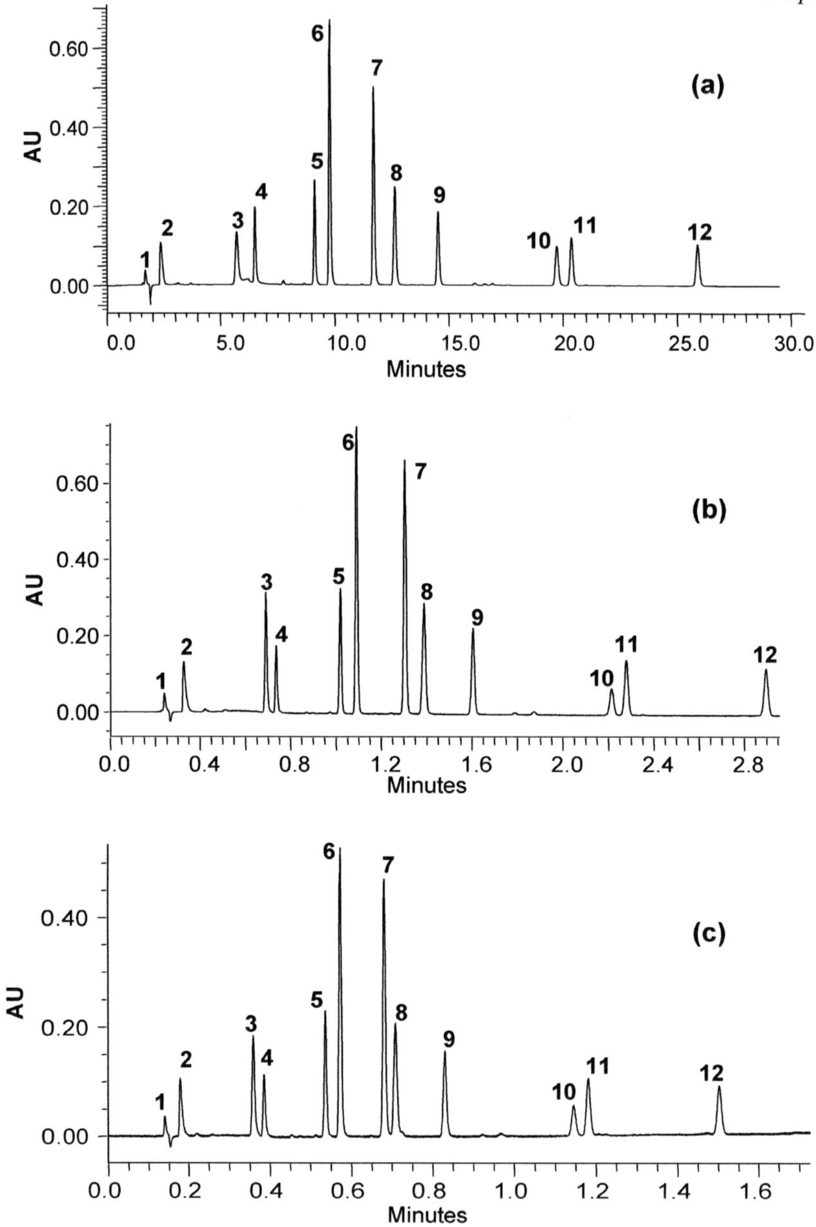

Figure 3.5 Gradient method transfer from regular HPLC to UHPLC. Separation of a pharmaceutical mixture containing the main product (6) and 11 impurities in gradient mode with HPLC and UHPLC systems. (a) Original HPLC method: column, XBridge C18, 150 × 4.6 mm, 5 μm; flow rate, 1000 μL min^{-1}; injected volume, 20 μL; gradient time, 30 min. (b) Transferred UHPLC method: column, ACQUITY BEH C18, 50 × 2.1 mm, 1.7 μm; flow rate, 613 μL min^{-1}; injected volume, 1.4 μL; gradient time, 3.4 min. (c) Transferred and optimized UHPLC method: column, Acquity BEH C18, 50 × 2.1 mm, 1.7 μm; flow rate, 1200 μL min^{-1}; injected volume, 1.4 μL; gradient time, 1.74 min. AU, absorbance units. Adapted from Guillarme *et al.*[13] with permission.

sensitivity, peak capacities and resolution. To further decrease the analysis time, the UHPLC mobile-phase flow rate was increased to 1200 μL min^{-1} (generating approximately 1000 bar), similar to what can be achieved in the isocratic mode. However, for an adequate transfer, it is necessary to adapt the gradient profile when the flow rate is increased, as illustrated by eqns (3.14) and (3.15). Therefore, when the mobile-phase flow rate is changed, the gradient slope and time of any isocratic step should be adapted to maintain equivalent selectivity. Figure 3.5 shows that the corresponding separation occurs in approximately 1.6 min for this complex mixture.

Another example of gradient transfer between HPLC and UHPLC conditions concerns the profiling of the standardized extract of a widely used phytomedicine: *Ginkgo biloba*. UHPLC is well established as allowing either a faster or a highly efficient separation in comparison with conventional HPLC. Using this complex biological matrix,[30] the authors have demonstrated these two possibilities by applying the basic rules for gradient method transfer. The corresponding chromatograms are presented in Figure 3.6. As expected from theory, a 9-fold reduction in the analysis time was obtained by transferring the original 60 min HPLC gradient on a 150 × 4.6 mm I.D., 5 μm column to a short UHPLC gradient on a 50 × 2.1 mm I.D., 1.7 μm column, with both operating at the optimal mobile-phase flow rate. However, a significant improvement in the resolution (*i.e.*, an approximately 3-fold improvement in the peak capacity) was observed with an equivalent column length (*i.e.*, 150 mm, 5 μm and 150 mm, 1.7 μm) and gradient time (60 min) in HPLC and UHPLC.

3.2.4.4 Method Translation Programs

Numerous method transfer calculators are available from both commercial and academic sources. The previously described equations have been integrated into these calculators. The user simply indicates the original and final column dimensions as well as the mobile-phase flow rate, injection volume and gradient profile to automatically determine the new conditions to be used in UHPLC. The locations of some of these programs are indicated in a recent paper by Majors,[31] and a freeware program for establishing method transfer conditions from our research group can also be downloaded.[32]

3.2.5 Problems Associated with Method Transfer

As shown in the previous examples, the method transfer from regular HPLC to UHPLC is straightforward provided that (i) the stationary-phase chemistry is identical in HPLC and UHPLC, (ii) the correct equations have been applied for the calculation of the new conditions based on column geometry and (iii) the UHPLC system is adequate in terms of extra-column and gradient delay volume. However, the method transfer between HPLC and UHPLC could fail in some cases, which can be attributed either to a problem of stationary-phase

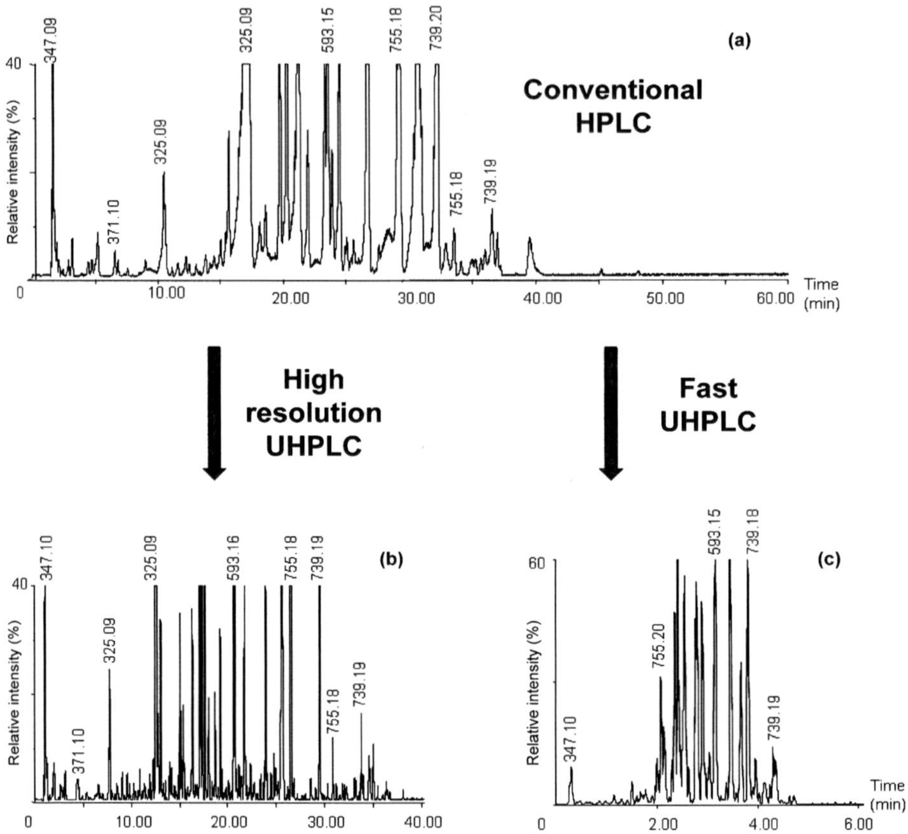

Figure 3.6 Comparison of chromatograms of a standardized *G. biloba* extract with method transfer. (a) Classical HPLC analysis carried out on a 150 × 4.6 mm I.D., 5 μm column with gradient of 5–40% acetonitrile (0.1% formic acid) in 60 min at 1 mL min^{-1}. (b) UHPLC method on a 150 × 2.1 mm I.D., 1.7 μm column, with the same run time, flow reduced to 0.35 mL min^{-1}. In this case, the resolution is notably increased. (c) UHPLC method on a 50 × 2.1 mm I.D., 1.7 μm column; the gradient time was reduced to 6.76 min (*i.e.*, 9-fold reduction), and the flow rate was 0.6 mL min^{-1}. Note that the performance should be the same, but it is slightly lower because of a relatively larger dead volume due to the electrospray ionization (ESI) probe. Detection: ESI–TOF-MS, *m*/*z* range 100–1000 in the negative mode. Adapted from Eugster *et al.*[30] with permission.

chemistry or to frictional heating effects that occur under very elevated pressures.

3.2.5.1 Packing Issue

In a recent study by Petersson and Euerby,[33] eight different sub-2 μm C18 stationary phases from five different vendors and their corresponding 3 and 5 μm columns were compared in terms of selectivity to highlight some potential problems during method transfer for sub-2 μm columns. To maintain a good relationship with the column vendors, the authors randomly coded the columns in the subsequent comparisons; consequently, the problematic columns packed with sub-2 μm particles were not clearly identified. However, the comparison of different packings from the same manufacturer is interesting because it emphasizes some potential problems for sub-2 μm columns. To compare the different materials, a modified version of the Tanaka protocol has been applied to determine the differences in selectivity and peak shape due to differences in the hydrophobicity, degree of surface coverage and silanol activity.[34,35]

The study by Petersson and Euerby[33] confirms that columns packed with sub-2, -3 and -5 μm versions of identical stationary-phase chemistries can generate potential method transferability problems for several of the columns because of selectivity differences. This fact is particularly true for columns that possess high silanol activity and low/intermediate surface coverage (two columns out of the eight tested). Because of the low surface coverage, the number of silanol groups is large and is responsible for secondary ionic interactions and poor peak shape for the analysis of protonated bases. Because these secondary interactions are not equivalent between the sub-2, -3 and -5 μm versions of the same packing, large differences in selectivity at intermediate pH were observed, as shown in Figure 3.7. This behavior is particularly critical for basic analytes because they strongly interact with negatively charged silanols.[36] At low pH, when the silanol groups are protonated, the selectivity differences are less pronounced. Thus, one possible transferability issue between HPLC and UHPLC may be related to a difference in the silanol activity of the stationary phases.

3.2.5.2 Frictional Heating

A frictional heating phenomenon is observed when columns packed with very fine particles are operated at high mobile-phase velocities, which generates a significant pressure drop.[37,38] Heat is induced by the friction of the mobile phase percolating through the column bed at very high pressure. There are two different ways to dissipate the induced temperature elevation: along and across the chromatographic column. This dissipation generates the formation of axial (longitudinal) and radial temperature gradients, which may not only influence column efficiency but also retention and selectivity.[39,40]

Radial temperature gradients are caused by a radial heterogeneity of the temperature distribution across the column, which means that the wall region of the column is cooler than the center. This heterogeneity may result in a significant loss of efficiency because of the influence of the temperature on mobile-phase viscosity. Such behavior has been mostly observed under isothermal conditions (*e.g.*, the column is placed in a forced-air oven or in a thermostatically controlled water bath).

In contrast, a longitudinal temperature gradient occurs when the column wall temperature is not kept constant (adiabatic case, *e.g.*, the column is placed

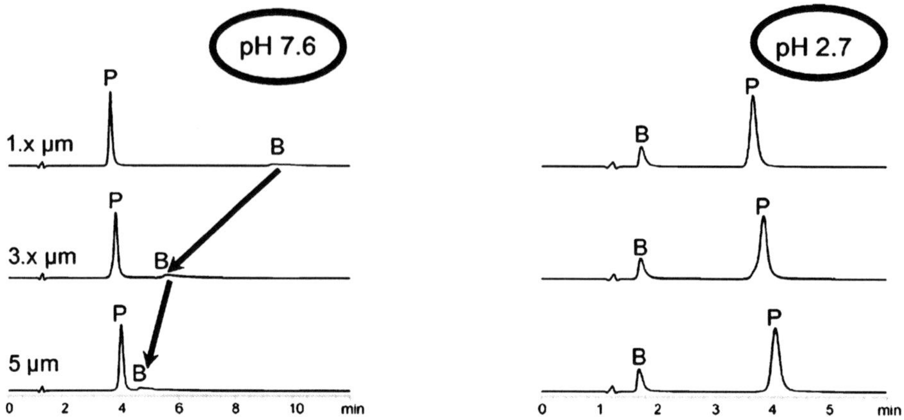

Figure 3.7 Separation of a basic (benzylamine, B) and a neutral (phenol, P) analyte on three columns that differ only in particle size. Two pH conditions were tested: pH 2.7 and 7.6. The black arrow highlights the change in retention between particle size, for the benzylamine when silanols are charged (pH 7.6). Adapted from Petersson *et al.*[33] with permission.

in a still-air oven) due to low heat transfer, which results in the temperature of the outlet being higher than that of the inlet. A longitudinal temperature gradient may alter the retention and selectivity through changes in the average column temperature, whereas column efficiency is not seriously affected.

For most UHPLC systems, the oven operates under adiabatic conditions (*i.e.*, the column is placed in a still-air oven); thus, the retention and selectivity can be changed according to the generated backpressure, which can be critical for method transfer because the temperature may be different between the original HPLC method and the UHPLC method.

As a measure of the frictional heat generated when a liquid is pumped through a packed LC column, the amount of power to be dissipated can be calculated with the following equation:[37,41]

$$Power = F{\cdot}\Delta P \tag{3.16}$$

where ΔP is the pressure drop over the column and F is the mobile-phase flow rate.

The effect of frictional heating is negligible with conventional 150 mm columns packed with 5 µm particles and operated at 100–200 bar.[38,42] However, frictional heating becomes significant with a 50–100 mm column packed with sub-2 µm particles and operated at pressure greater than 400 bar.[43]

In the literature, two different strategies have been applied to the experimental determination of the amplitude of temperature variation in UHPLC columns.

The first approach has been discussed in detail by Gritti and Guiochon, who have published numerous papers on the subject.[44–46] The applied methodology consists of measuring the temperature at the surface of the column and the temperature of the liquid exiting the column outlet, using several thermocouples with a precision better than ± 0.1 K. With this approach, the complete three-dimensional temperature profiles were calculated along and across various UHPLC columns of 30, 50, 100 and 150 mm with 1 and 2.1 mm I.D. at 1000 bar. This method allows for the accurate determination of the amplitude of the axial and radial temperature gradients in relation to the employed experimental conditions. The conclusions of these studies were that the temperature differences between the column outlet and the column inlet were in the range 13–20 °C and 3–16 °C for various lengths of 2.1 and 1 mm I.D. columns, respectively.[44–46] In all cases, the temperature difference was less pronounced for the longest columns.

An alternative strategy was proposed by Novakova *et al.*[47] who attempted to characterize the effect of frictional heating in UHPLC in real gradient-mode conditions, using a mixture that contained neutral, basic and acidic analytes in both acidic and basic pH conditions. The impact of frictional heating was evaluated, as well as the influence of pressure, temperature and pH, on UHPLC separations (up to 1000 bar). The effect of frictional heating in UHPLC was assessed by comparing chromatograms obtained for various mobile-phase temperature at low flow rates (*i.e.*, negligible frictional heating) and chromatograms obtained at various mobile-phase flow rates (*i.e.*, increasing backpressure) at 30 °C. As expected, some deflections, which were attributed to the effect of frictional heating, were observed from the initial separation performed at a low flow rate (0.1 mL min^{-1}, 100 bar) toward high flow velocity and pressure conditions. The importance of frictional heating was estimated at 300, 600 and 1000 bar, and a similar separation to that obtained at 100 bar was attained when the temperature difference between the original and final separations was +4 °C at 300 bar, +8 °C at 600 bar and +16 °C at 1000 bar on a 2.1 mm I.D. column, whereas the values were slightly lower on a 1.0 mm I.D. column (+3 °C, +6 °C and +12–13 °C, respectively). These results confirm those obtained by Guiochon and Gritti.[45]

In the case of method transfer, the effect of frictional heating should not be neglected, particularly if the less-resolved peaks are strongly influenced by the temperature of the mobile phase. A suitable solution to avoid changes in selectivity would be to work with a lower temperature in UHPLC. Indeed, if the original HPLC separation was performed at 30 °C, the UHPLC separation with a 2.1 mm I.D. column should be performed at 26, 22 and 14 °C for pressures of 300, 600 and 1000 bar, respectively. An example illustrating this approach is shown in Figure 3.8.

3.3 Normative Context for the HPLC to UHPLC Transfer

The transfer rules described in the previous sections can be freely applied to any LC geometrical transfer that needs to be scaled down (*e.g.*, HPLC to UHPLC) or scaled up (*e.g.*, HPLC to semi-preparative LC). Nevertheless, LC

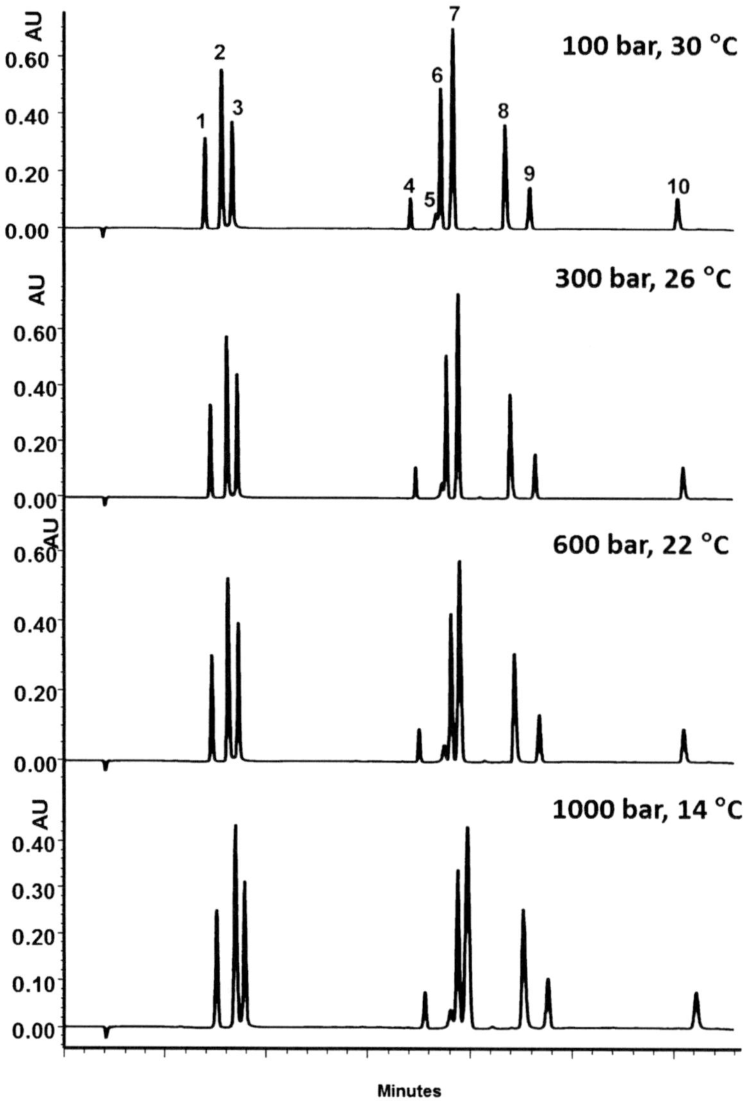

Figure 3.8 A potential solution for frictional heating: appropriate decrease of temperature in UHPLC at 300, 600 and 1000 bar. Acquity BEH C18 (50 × 2.1 mm, 1.7 μm) at 30, 26, 22 and 14 °C; mobile phase, ACN/ammonium formate (50 mM), pH 9.0; UV detection at 230 nm. Peaks were identified as: (1) paracetamol, (2) salicylic acid, (3) catechin, (4) ethacrynic acid, (5) oxycodone, (6) dexamethasone, (7) indapamide, (8) nortryptyline, (9) gestrinone, (10) thioridazine. Time scale is not indicated since chromatograms were aligned based on the column dead time for the different flow rates. AU, absorbance units. Adapted from Novakova *et al.*[47] with permission.

methods that are validated and that have to meet specific requirements (*e.g.*, requirements described in monographs) cannot be transferred without restrictions. In this case, the European Pharmacopeia (EP) and the United States Pharmacopeia (USP) provide some limitations to the modifications of the chromatographic parameters. Within the ranges of adjustments given in the EP or the USP, a transferred method does not have to be revalidated when it still meets the prescribed requirements and presents the same chromatographic performance.

3.3.1 Official Rules from the European Pharmacopeia and the United States Pharmacopeia for Method Transfer

The limitations described in the EP and the USP are summarized in Table 3.4.[48,49] As stated in the previous Sections, a transfer cannot be performed using the aforesaid rules or be named as such when the stationary-phase or the mobile-phase composition has been modified. For the case of a modified mobile-phase composition, the EP and the USP give some tolerance, and, in some cases, the composition of the mobile phase can be modified. However, such modifications should be avoided for geometrical method transfer. In Table 3.4, the other chromatographic parameters that can be modified are consistent with the rules of a geometrical transfer. Starting from a conventional HPLC column of 150 × 4.6 mm I.D., 5 μm, with a flow rate of 1 mL min^{-1}, the following parameters can be obtained for a transfer to UHPLC by complying with the EP and the USP rules. The column length and the I.D. can be reduced to 45 mm and 3.45 mm, respectively. The particles diameter can be decreased to 2.5 μm, and the flow rate can be adjusted between 0.5 mL min^{-1} and 1.5 mL min^{-1} using eqn (3.6). Therefore, the use of a UHPLC column of 50 × 2.1 mm I.D., 1.7 μm, would require a complete revalidation of the method. The transfer from an HPLC column of 100 × 2.1 mm I.D., 3.5 μm (F = 0.3 mL min^{-1}), to a UHPLC column of 50 × 2.1 mm I.D., 1.9 μm (F = 0.55 mL min^{-1}), however, can be performed without revalidation. Moreover, multiple transfers performed sequentially are prohibited, and only the initial and the final methods can be considered without intermediates. Analysts thus face a tricky problem caused by the lack of flexibility provided by the EP and USP in the framework of method transfer from HPLC to UHPLC.

3.3.2 Possible Solution for Transferring Methods Between HPLC and UHPLC without Revalidation

Rather than giving fixed limits for the modification of chromatographic parameters, a possible solution could be a performance-based approach. As mentioned in Section 3.2.1, the key factor defining the performance of a column is the ratio L/d_p. In the case of a transfer from an HPLC column of 150 × 4.6 mm I.D., 5 μm, to a UHPLC column 50 × 2.1 mm I.D., 1.7 μm, the ratio L/d_p is held constant (L/d_p = 30 000 and 29 400, respectively). The

Table 3.4 Limitations given by the EP and the USP for method transfer.

Chromatographic parameters	*Adjustments allowed by EP*	*Adjustments allowed by USP*
Composition of mobile phase	±30% relative or ±2% absolute of the minor solvent (no other component is altered by more than 10% absolute)	±30% relative of the minor solvent (no other component is altered by more than 10% absolute)
Column length	±70%	±70%
Column I.D.	±25%	±25%
Particles diameter	Reduction to 50%	Reduction to 50%
Mobile-phase flow rate	±50%	±50%
Column temperature	±10% (to a maximum of 60 °C)	±10 °C
pH of mobile phase	±0.2 (±1 when neutral compounds are to be examined)	±0.2
Concentration of buffer salts	±10%	±10%
Injection volume	May be decreased, provided detection and repeatability are satisfactory	Can be reduced as far as is consistent with accepted precision and detection limits

chromatographic performances of the systems are identical, although the analysis time using UHPLC is theoretically 9-fold shorter than that using HPLC. When the number of theoretical plates remains unchanged, the method should perform as before, and the basic chromatographic criteria (*e.g.*, resolution, tailing factor) should be identical. Furthermore, when some criteria do not reach their acceptance thresholds, the method can still be modified by following the rules given by the EP and USP. Importantly, an increase in the ratio L/d_p will result in a better performance. In a performance-based approach, a similar or better performance (identical or higher L/d_p ratio) should not require a revalidation of the method. Following this trend, Neue *et al.*[50] have submitted a "stimulus to the revision process" to the USP. Some regulatory agencies [*e.g.*, the US Food and Drugs Administration (FDA)] currently advocate a performance-based approach and could soon accept transfers from HPLC to UHPLC (*i.e.*, an improvement of method performance). We hope that this stimulus is a first step toward the ineluctable evolution of the view of pharmaceuticals authorities.

3.4 Validation of the UHPLC Methods and Equivalence of the HPLC–UHPLC Methods

Method validation is the documented confirmation that the method under study is fit for its intended purpose.[51] Concerning quantitative methods (*e.g.*, content assay methods, quantitative impurity assays and dissolution analytical methods), method validation must demonstrate that results obtained from the future routine use of the methods will, with a high guarantee, be within pre-specified acceptance limits.[52,53] The key validation criteria that should be assessed for quantitative assays are method trueness (systematic error, measured by a bias or recovery—sometimes called accuracy of the mean), method precision, repeatability and especially intermediate precision [random error, as measured by the relative standard deviation (RSD)] and accuracy of the results.

3.4.1 Validation Design

Validation standards (VSs) are, in general, used to assess the analytical method trueness, the precision and ultimately the results accuracy. These VS samples should match the matrix of the real samples and should therefore include all of the excipients, impurities, degradation products or bio-matrices components that will be present in the daily samples that will be studied in the routine application of the method. The extreme smallest and highest VS levels should be consistent with the aim of the method. According to most of the validation guidelines, at least three concentration levels of VSs should be used to cover the range over which the method is to be validated.

For each concentration level of the VS, at least three replicates should be analyzed and several runs or series should be performed to assess inter-run or inter-series precision and to allow for calculation of the method overall precision (*i.e.*, its intermediate precision). A minimum of three runs should be used for the calculation of the between-run standard deviation. It is important that each run or series involves different analysts, equipment, days or lots of reagents, when relevant. Failure to vary these parameters will provide an overly optimistic estimation of the method intermediate precision and, hence, an overestimation of accuracy of the results.

3.4.2 Decision Methodology

Several common methodologies are available to allow for decisions to be made about method validity that focus on the separate evaluation of method trueness and precision.[54,55] However, such methodologies fail to provide guarantees that the validated method will produce reliable results.[54] The decision about the validity of the method should be made after the accuracy of the results has been evaluated to guarantee that each future result will fall within pre-defined acceptance limits with high probability.[52,53] An adequate solution for achieving this objective is to compute statistical tolerance intervals for the results of the tested analytical method at each concentration level of the VSs that are used and to compare them to well-chosen acceptance limits.[52,56,57] This approach will allow an "accuracy profile", as shown in Figures 3.9(a) and 3.9(b), to be defined.[52] The tolerance intervals define a region where there is a high probability, as required by the user, of obtaining future results of the method.[58] If these intervals are within the pre-defined acceptance limits, then the method can be declared as valid; otherwise, the method is not valid. Full computational details are available in different publications to which the interested reader can refer.[59–62] Figure 3.9 illustrates accuracy profiles obtained for the validation of a conventional HPLC assay (Figure 3.9a) and the transfer to a UHPLC assay (Figure 3.9b). The conventional HPLC method used a Merck LaChrom system with an XTerra RP18, 150 × 4.6 mm I.D., 5 µm analytical column, whereas the UHPLC method used a Waters Acquity UPLC system with an Acquity Shield RP18, 50 × 2.1 mm I.D., 1.7 µm analytical column. The details of the analytical procedure have been reported elsewhere.[63] As evident in Figures 3.9(a) and 3.9(b), the accuracy profiles obtained for the conventional HPLC system and for the UHPLC system are similar. The tolerance interval limits computed with a probability β = 95% (dashed lines) are included within the pre-specified acceptance limits of ±5% of the reference concentration of the VS. These two accuracy profiles show that each future result, whether obtained using the HPLC or the UHPLC

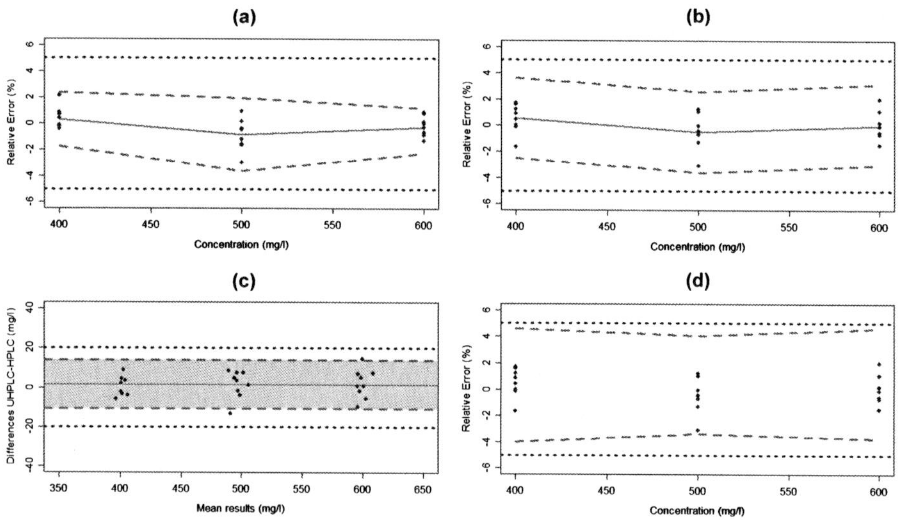

Figure 3.9 Illustration of the decision profiles used to assess methods validity and methods equivalence. Accuracy profiles for the validation of (a) the conventional LC method and (b) the UHPLC method. (c) Bland–Altman plot to assess the equivalence of the conventional LC and UHPLC methods in the paired case. (d) Decision plot to assess the LC and UHPLC equivalency in the unpaired case. The dotted lines are the acceptance limits to decide either about method validity (a, b) or methods equivalence (c, d); the dashed lines are the tolerance intervals (a, b, d) or the agreement limits (c). The continuous lines represent the systematic error or bias. The dots are the results obtained by the chromatographic methods.

method, will fall within the settled acceptance limits with at least 95% probability.

3.4.3 Methods Comparison and Equivalence

Having shown that the UHPLC method is valid is not always sufficient because another question still remains: do the two methods provide equivalent results?

This question can be addressed by a method comparison or a method equivalence study. Method comparison studies are used to assess the relative agreement between two analytical methods that measure the same chemical substance, primarily to assess the performance of a newly released or in-house-developed method. The final objective of method comparison is the determination of whether the results obtained by the test method or the reference method can be interchanged without leading to any inconsistency in the interpretation of the measured results. The statistical methods used for assessing agreement depend on the way the data are collected.

Method comparison studies are extremely frequent in clinical and medical laboratories.[64–66] Indeed, the Clinical and Laboratory Standards Institute (CLSI, formerly the NCCLS) has a specific guideline relative to method comparison.[67] Method comparison can also be useful in the general analytical chemistry sector, such as to compare classical slower methods with new faster methods.[68] Also, method comparisons are recognized as a way to validate new analytical methods. The FDA (2001), for example, considers method comparison studies as a type of method validation and titles these types of studies under the heading of cross-validations.[69] According to an International Conference on Harmonization (ICH) document, one way to assess accuracy is to compare "the results of the proposed analytical procedure with those of a second well-characterized procedure, the accuracy of which is stated and/or defined."[51] This statement may refer to method comparison studies. In the USP, several references are made to method comparisons and assessments of methods equivalency, such as statements in the General Notice and in Chapter <1223>.[70] However, none of these documents propose a clear guideline as to how method comparison studies should be performed and interpreted. Several approaches have thus been proposed.

Usually, the first step when comparing results from two methods is to compute the Pearson correlation coefficient from the linear regression analysis obtained between the conventional HPLC method and the UHPLC method to assess the method equivalence. For instance, a correlation coefficient greater than 0.99 (or even greater than 0.999) is sometimes believed to imply that the two methods are equivalent. However, the correlation coefficient is well known to be inappropriate for this purpose.[64,65] Indeed, the results obtained by the two assay methods could exhibit a high correlation coefficient but with a strong systematic difference (*i.e.*, a bias) between them. In addition, the range of the results also influences this coefficient: the larger this range is, the higher the correlation coefficient will be. Similarly, statistical hypothesis tests to evaluate whether the slope of the regression is different from 1 and the intercept is different from 0 do not allow for conclusions to be drawn about the agreement between the methods.[53,71–73] Therefore, other approaches have to be used to meet the aim of the method comparison study. Two types of equivalence can be assessed: average equivalence and individual equivalence.

3.4.4 Study Design

Whatever the type of equivalence (*i.e.*, average equivalence or individual equivalence), the study design is similar. The samples used in method comparison studies are typically real study samples (*e.g.*, incurred samples and samples from industrial batches). These samples should be randomly chosen, and they should cover, to the largest possible extent, the concentration range over which the two methods are intended to work. The inclusion in the study design of as many sources of variation as possible [*e.g.*, operators, equipment, time (days)] is important to obtain results that are as close as possible to the routine application of the methods.[52,53]

In addition, the same sets of samples could be used in both methods, thus corresponding to a paired case, or the samples could represent two sets of samples coming from the same initial population, thus corresponding to an unpaired case. The data analysis for paired or unpaired cases is different and these study designs should be included when a decision methodology is selected.

3.4.5 Average Equivalence

Average equivalence aims to show that, on average, both methods, the conventional HPLC assay and the UHPLC assay, provide equivalent results.[71,74–76] To realize average equivalence, the acceptance limits must be defined before the method comparison test is conducted. The acceptance limits are represented by $[L^- \; ; \; L^+]$, where (L^-) and (L^+) define the minimum and maximum differences, respectively, between the mean values of the conventional HPLC and UHPLC methods that can be considered as acceptable for the routine application of the analytical method. The average equivalence then involves two one-sided Student's *t*-tests.[76] The first *t*-test tests the hypothesis $H_{01}: \mu_{LC} - \mu_{UHPLC} \leq L^-$ *vs.* $H_{A1}: \mu_{LC} - \mu_{UHPLC} > L^-$, and the second one tests the hypothesis $H_{02}: \mu_{LC} - \mu_{UHPLC} \geq L^+$vs. $H_{A2}: \mu_{LC} - \mu_{UHPLC} >< L^+$. Nonetheless, in practice, the (1–α)% confidence interval of the difference between the means (*i.e.*, the methods bias) is computed and compared to the interval $[L^- \; ; \; L^+]$. This confidence interval is given by $(\hat{\mu}_{LC} - \hat{\mu}_{UHPLC}) \pm t_{v;1-\alpha}\, \hat{\sigma}_{\mu_{LC}-\mu_{UHPLC}}$, where $\hat{\mu}_{LC}$ and $\hat{\mu}_{UHPLC}$ are the estimates of the mean results of the conventional HPLC method and of the UHPLC method, respectively. The term $\hat{\sigma}_{\mu_{LC}-\mu_{UHPLC}}$ is the standard error of the difference between the means, and ν is the degrees of freedom associated with

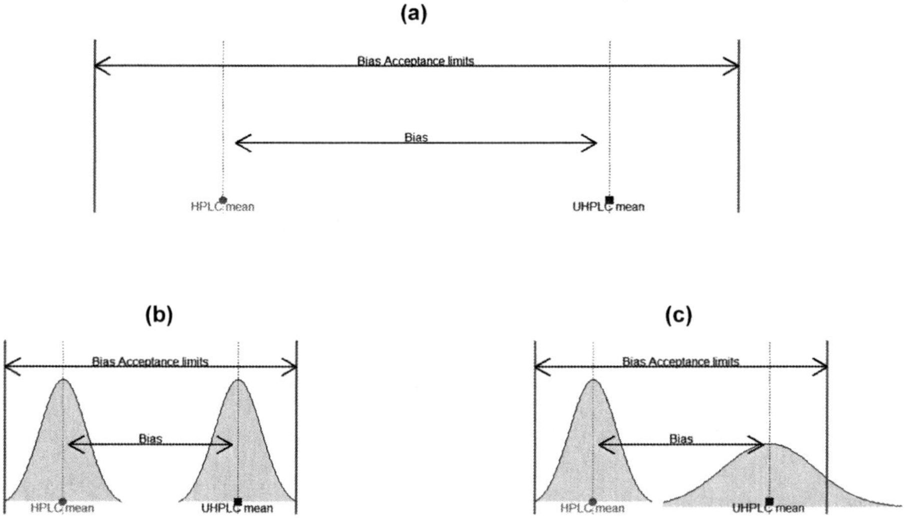

Figure 3.10 The flaws of the average equivalence approach. An acceptable bias between two methods (a) and two potential situations: (b) a small variability of the UHPLC method where results will be reliable and (c) a higher variability of the UHPLC method where results are too frequently out of specifications.

the standard error. The computational details to obtain these statistics for the unpaired case,[62,76] as well as for the paired case are available elsewhere.[74]

Figure 3.10 illustrates a case of average equivalence and its potential weakness. In Figure 3.10(a), the difference of the means of the two methods (*i.e.*, the methods bias) is included within the vertical acceptance limits, which indicates that the methods are equivalent on average. However, two situations are possible that are not assessed with the average equivalence study. The first situation, as depicted in Figure 3.10(b), shows that the variability of the UHPLC method is not excessive, which ensures that most of the results using UHPLC will be within the acceptance limits. Figure 3.10(c) depicts the case in which the variability of the UHPLC method is so large that, although both methods are equivalent on average, the results generated using the UHPLC method will too frequently fall outside of the acceptance limits. Therefore, care should be exercised when average equivalence is assessed because the variability of the test method (the UHPLC in this example) is not controlled. This variability could lead to the two methods being declared equivalent (on average), although the results obtained by these two methods are significantly different.

3.4.6 Individual Equivalence

To demonstrate that the two methods are equivalent on an individual basis (*i.e.*, for the individual results and not only for the means of the results), it is thus recommended that the study be designed to obtain paired results. However, it is also possible to assess individual equivalence on the unpaired case, as will be demonstrated in the following sections.

3.4.6.1 Paired Results

When the method comparison design has been performed by analysis of the same set of samples by the conventional HPLC method and the UHPLC method, the Bland–Altman approach could be used.[64,65,77,78] The advantages of this approach are that the statistics to be performed are not tedious, and it provides a visual tool to assess the method comparison and equivalence. Agreement limits are computed based on the results obtained using both analytical methods and are compared to pre-defined acceptance limits: $[L_{BA}^{-}\ ;\ L_{BA}^{+}]$. The agreement limits represent a region where there is a 95% probability that differences between the results of the two methods will be obtained: $\hat{\bar{\delta}} - 1.96\hat{\sigma}_{\delta}$ and $\hat{\bar{\delta}} + 1.96\hat{\sigma}_{\delta}$, where $\hat{\bar{\delta}}$ is the estimated mean of all the individual differences between the conventional HPLC and UHPLC results and $\hat{\sigma}_{\delta}$ is the estimated standard deviation of these differences.[64,65,77,78] The acceptance limits define the minimum (L_{BA}^{-}) and maximum (L_{BA}^{+}) differences between the individual results of the conventional HPLC and UHPLC methods that can be considered to be acceptable for the routine application of the analytical method. This approach is usually summarized in a so-called Bland–Altman plot or difference plot that is used as a visual tool to assess the methods comparison, as shown in Figure 3.9(c).

This plot represents the individual differences on the *y*-axis *versus* the mean results on the *x*-axis. In addition, limits of agreement that provide information about the distribution of the individual differences are computed and drawn on the plots together with the acceptance limits $[L_{BA}^{-}\ ;\ L_{BA}^{+}]$. Both methods are defined as equivalent when the agreement limits are included within the acceptance limits over the entire investigated concentration range. Figure 3.9(c) shows a Bland–Altman plot based on the same two methods described previously. In this Figure, the *y*-axis represents the differences of the results obtained using the UHPLC method minus those obtained using the conventional HPLC method. The *x*-axis is the average of the results obtained by both methods. The dashed lines are the agreement limits, which define a region where 95% of the differences between the results of the UHPLC and HPLC methods are expected to fall (grey area). As evident in the Figure, these agreements limits are fully included into the acceptance limits set at ± 20 mg L^{-1}, which confirms that the two methods provide equivalent results.

3.4.6.2 Unpaired Results

In the case of unpaired results, a similar methodology to that proposed for method validation can be performed, as reported by Dewé *et al.*[79] From the test method results represented by the UHPLC method, tolerance intervals that contain each future result with a user-defined probability (*e.g.*, 90, 95 or 99%) are computed for each sample, and these intervals are compared to acceptance limits $\left[L_{\text{Adj}}^{-} \,;\, L_{\text{Adj}}^{+}\right]$ that have been adjusted around the mean results of each sample obtained using the reference method (*i.e.*, conventional HPLC). These acceptance limits are adjusted in view of the fact that the mean results provided by conventional HPLC are only estimations of the real mean values and are thus estimated with uncertainty. Therefore, the acceptance limits are adjusted based on the lower and upper limits of the confidence interval of the mean of the results obtained using the conventional HPLC method.[79] A complete worked-out example of the computations of this approach is available in the work of Dewé *et al.*[79] for the comparison of two laboratories' results. However, the methodology is fully transferable to the comparisons of two methods. The two LC methods can be considered to be equivalent as long as the tolerance intervals computed with the UHPLC method results are included within the acceptance limits that are adjusted with the confidence interval of the mean results obtained from the conventional HPLC results. Figure 3.9(d) illustrates the comparison of the same two methods used in the previous examples. As evident in this graph, the tolerance intervals (dashed lines) based on the results obtained using the UHPLC method computed with a probability $\beta = 95\%$ are fully included within the acceptance limits $\left[L_{\text{Adj}}^{-} \,;\, L_{\text{Adj}}^{+}\right]$ (dotted lines). This result highlights the fact that each future result obtained with the UHPLC method has a high probability (at least a 95% chance in this example) of being within 5% of the results obtained using the conventional HPLC method. Hence, methods equivalence is again demonstrated.

3.5 Conclusions

This Chapter describes the rules for the successful transfer of conventional HPLC methods to UHPLC. Two methodologies were depicted for the isocratic and gradient modes. The reduction of particle size without modification of the column length led to an improvement in the efficiency and the peak capacity. However, when the column length was reduced concurrently with particle size, the analysis time decreased for an identical efficiency or peak capacity. Various examples were presented to support the feasibility of both approaches. As additional advantages, the transfer of a method to UHPLC could also improve the sensitivity and significantly decrease the solvent and sample consumption, thereby contributing to the current trend toward green chemistry.

Nevertheless, these geometric transfer rules do not take into account the quantitative performance of chromatographic methods. The normative situation currently limits the modifications of method parameters and narrows the transferability of numerous HPLC methods. Pending a change in the rules of formal transfer, a possible solution has been proposed that involves maintaining (or increasing) the L/d_p ratio between HPLC and UHPLC.

Indeed, if chromatographic performance is unchanged or improved, quantitative performance should be ensured.

After a conventional HPLC method has been transferred to UHPLC, the results that will be generated by this new method must be demonstrated to remain reliable. The first step is to demonstrate that the UHPLC method is still fit for its intended purpose (*i.e.*, a method validation should be performed). In this Chapter, we have focused on the methodologies to demonstrate that the quantitative performance of the UHPLC method remains acceptable. We have further stressed that the core requirement is to show that results of adequate quality will be generated. When one method is transferred to a new one, there is also a question concerning the comparability and equivalence of the methods. Two types of method equivalence studies have been proposed to address this question. Furthermore, the validity of both methods and their equivalence strengthen the fact that, when geometrical rules of transfer are followed and the L/d_p ratio is held constant (or increases), an original method can be validated after the transfer, independent of the ranges currently mentioned by the EP and USP.

References

1. D. Guillarme, D. T. T. Nguyen, S. Rudaz and J. L. Veuthey, *J. Chromatogr. A*, 2007, **1149**, 20.
2. L. Novakova, D. Solichova and P. Solich, *J. Sep. Sci.*, 2006, **29**, 2433.
3. E. S. Grumbach, T. E. Wheat, M. Kele and J. R. Mazzeo, *LC-GC Eur.*, 2005, **12**, 40.
4. S. A. C. Wren and P. Tchelitcheff, *J. Chromatogr. A*, 2006, **1119**, 140.
5. M. Petrovic, M. Gros and D. Barcelo, *J. Chromatogr. A*, 2006, **1124**, 68.
6. E. Grata, J. Boccard, D. Guillarme, G. Glauser, P. A. Carrupt, E. E. Farmer, J. L. Wolfender and S. Rudaz, *J. Chromatogr. B*, 2008, **871**, 261.
7. A. D. Jerkovich, J. S. Mellors and J. W. Jorgenson, *LC-GC Eur.*, 2003, **16**, 20.
8. N. J. Wu and A. M. Clausen, *J. Sep. Sci.*, 2007, **30**, 1667.
9. D. T. T. Nguyen, D. Guillarme, S. Rudaz and J. L. Veuthey, *J. Sep. Sci.*, 2006, **29**, 1836.
10. D. Guillarme, E. Grata, G. Glauser, J. L. Wolfender, J. L. Veuthey and S. Rudaz, *J. Chromatogr. A*, 2009, **1216**, 3232.
11. D. T. T. Nguyen, D. Guillarme, S. Rudaz and J. L. Veuthey, *J. Chromatogr. A*, 2006, **1128**, 105.
12. D. Guillarme, D. T. T. Nguyen, S. Rudaz and J. L. Veuthey, *Eur. J. Pharm. Biopharm.*, 2007, **66**, 475.
13. D. Guillarme, D. T. T. Nguyen, S. Rudaz and J. L. Veuthey, *Eur. J. Pharm. Biopharm.*, 2008, **68**, 430.
14. C. Horvath, W. Melander and I. Molnar, *J. Chromatogr.*, 1976, **125**, 129.
15. N. S. Wilson, M. D. Nelson and J. W. Dolan, *J. Chromatogr. A*, 2002, **961**, 171.

16. J. Kohler, D. B. Chase and R. D. Farlee, *J. Chromatogr.*, 1986, **352**, 275.
17. L. R. Snyder, J. W. Dolan and P. W. Carr, *J. Chromatogr. A*, 2004, **1060**, 77.
18. http://shop.perkinelmer.com/content/applicationnotes/wht_guidelinesfor-uhplcinstruments.pdf, consulted in September 2011.
19. A. Pruss, C. Kempter, J. Gysler and T. Jira, *J. Chromatogr. A*, 2003, **1016**, 129.
20. J. P. Chervet, M. Ursem and J. B. Salzmann, *Anal. Chem.*, 1996, **68**, 1507.
21. J. P. C. Vissers, A. H. de Ru, M. Ursem and J. P. Chervet, *J. Chromatogr. A*, 1996, **746**, 1.
22. K. J. Fountain, U. D. Neue, E. S. Grumbach and D. M. Diehl, *J. Chromatogr. A*, 2009, **1216**, 5979.
23. R. Russo, D. Guillarme, D. T. T. Nguyen, C. Bicchi, S. Rudaz and J. L. Veuthey, *J. Chromatogr. Sci.*, 2008, **46**, 199.
24. J. B. Wang, H. F. Li, C. Jin, Y. Qu and X. H. Xiao, *J. Pharm. Biomed. Anal.*, 2008, **47**, 765.
25. A. P. Schellinger, D. R. Stoll and P. W. Carr, *J. Chromatogr. A*, 2005, **1064**, 143.
26. D. R. Stoll, C. Paek and P. W. Carr, *J. Chromatogr. A*, 2006, **1137**, 153.
27. J. W. Dolan and L. R. Snyder, *J. Chromatogr. A*, 1998, **799**, 21.
28. A. P. Schellinger and P. W. Carr, *J. Chromatogr. A*, 2005, **1077**, 110.
29. J. W. Dolan, *LC-GC N. Amer.*, 2006, **24**, 458.
30. P. J. Eugster, D. Guillarme, S. Rudaz, J. L. Veuthey, P. A. Carrupt and J. L. Wolfender, *J. AOAC Int.*, 2011, **94**, 51.
31. R. E. Majors, *LC-GC N. Amer.*, 2011, **29**, 476.
32. http://www.unige.ch/sciences/pharm/fanal/lcap/telechargement.htm, consulted in September 2011
33. P. Petersson and M. R. Euerby, *J. Sep. Sci.*, 2007, **30**, 2012.
34. K. Kimata, K. Iwaguchi, S. Onishi, K. Jinno, R. Eksteen, K. Hosoya, M. Araki and N. Tanaka, *J. Chromatogr. Sci.*, 1989, **27**, 721.
35. M. R. Euerby and P. Petersson, *J. Chromatogr. A*, 2005, **1088**, 1.
36. D. V. McCalley, *J. Chromatogr. A*, 2010, **1217**, 858.
37. I. Halasz, R. Endele and J. Asshauer, *J. Chromatogr.*, 1975, **112**, 37.
38. F. Gritti and G. Guiochon, *J. Chromatogr. A*, 2006, **1131**, 151.
39. G. Desmet, *J. Chromatogr. A*, 2006, **1116**, 89.
40. A. De Villiers, H. Lauer, R. Szucs, S. Goodall and P. Sandra, *J. Chromatogr. A*, 2006, **1113**, 84.
41. H. Poppe, J. C. Kraak, J. F. K. Huber and H. M. Van Der Berg, *Chromatographia*, 1981, **14**, 515
42. F. Gritti and G. Guiochon, *J. Chromatogr. A*, 2006, **1138**, 141.
43. F. Gritti, M. Martin and G. Guiochon, *Anal. Chem.*, 2009, **81**, 3365.
44. F. Gritti and G. Guiochon, *J. Chromatogr. A*, 2007, **1166,** 47.
45. F. Gritti and G. Guiochon, *Anal.Chem.*, 2008, **80**, 5009.
46. F. Gritti and G. Guiochon, *J. Chromatogr. A*, 2008, **1187**, 165.
47. L. Novakova, J. L. Veuthey and D. Guillarme, *J. Chromatogr. A*, in press.

48. European Pharmacopeia, *EP 5.0, 2.2.46. Chromatographic Separation Techniques*, 2005, p. 69.
49. United States Pharmacopeia, *USP 32–NF 27, Chromatography <621>*, Rockville, MD, USP, 2009, p. 227.
50. U. D. Neue, D. McCabe, V. Ramesh, H. Pappa and J. DeMuth, *Pharm. Forum*, 2009, **35**, 1622.
51. International Conference on Harmonization (ICH) of Technical Requirements for Registration of Pharmaceuticals for Human Use, Topic Q2 (R1): Validation of Analytical Procedures: Text and Methodology, Geneva, 2005.
52. Ph. Hubert, J-J. Nguyen-Huu, B. Boulanger, E. Chapuzet, P. Chiap, N. Cohen, P-A. Compagnon, W. Dewé, M. Feinberg, M. Lallier, M. Laurentie, N. Mercier, G. Muzard, C. Nivet and L. Valat, *J. Pharm. Biomed. Anal.*, 2004, **36**, 579.
53. E. Rozet, A. Ceccato, C. Hubert, E. Ziemons, R. Oprean, S. Rudaz, B. Boulanger and Ph. Hubert, *J. Chromatogr. A*, 2007, **1158**, 111.
54. A. Bouabidi, E. Rozet, M. Fillet, E. Ziemons, E. Chapuzet, B. Mertens, R. Klinkenberg, A. Ceccato, M. Talbi, B. Streel, A. Bouklouze, B. Boulanger and Ph. Hubert, *J. Chromatogr. A*, 2010, **1217**, 3180.
55. R. Kringle, R. Khan-Malek, F. Snikeris, P. Munden, C. Agut and M. Bauer, *Drug Inf. J.*, 2001, **35**, 1271.
56. B. Boulanger, E. Rozet, F. Moonen, S. Rudaz and Ph. Hubert, *J. Chromatogr. B*, 2009, **877,** 2235.
57. D. Hoffman and R. Kringle, *Pharm. Res.*, 2007, **24,** 1157.
58. G.J. Hahn and W.Q. Meeker, *Statistical Intervals: A Guide for Practitioners*, Wiley, 1st edn, 1991.
59. D. Hoffman and R. Kringle, *J. Biopharm. Stat.*, 2005, **15**, 283.
60. E. Rozet, V. Wascotte, N. Lecouturier, V. Préat, W. Dewé, B. Boulanger and Ph. Hubert, *Anal. Chim. Acta*, 2007, **591,** 239.
61. Ph. Hubert, J-J. Nguyen-Huu, B. Boulanger, E. Chapuzet, N. Cohen, P-A. Compagnon, W. Dewé, M. Feinberg, M. Laurentie, N. Mercier, G. Muzard, L. Valat and E. Rozet, *J. Pharm. Biomed. Anal.*, 2007, **45**, 82.
62. E. Rozet, W. Dewé, E. Ziemons, A. Bouklouze, B. Boulanger and Ph. Hubert, *J. Chromatogr. B*, 2009, **877**, 2214.
63. D. Guillarme, D. T-T. Nguyen, S. Rudaz and J-L. Veuthey, *J. Chromatogr. A*, 2007, **1149** 20.
64. D. G. Bland and J. M. Altman, *Statistician*, 1983, **32**, 307.
65. J. M. Bland and D. G. Altman, *Lancet*, 1986, **1**, 307.
66. P. H. Petersen, D. Stöckl, O. Blaabjerg, B. Pedersen, E. Birkemose, L. Thienpont, J. F. Lassen and J. Kjeldsen, *Clin. Chem.*, 1997, **43**, 2039.
67. CLIA, Clinical and Laboratory Standards Institute, (formerly NCCLS), Method comparison and bias estimation using patient samples; approved guidelines. Document EP9-A2, Wayne, PA, 2002.

68. A. Nguyen Minh Nguyet, A. M. van Nederkassel, L. Tallieu, S. Kuttatharmmakul, E. Hund, Y. Hua, J. Smeyers-Verbeke and Y. Vander Heyden, *Anal. Chim. Acta*, 2004, **516**, 87.
69. Guidance for Industry: Bioanalytical Method Validation, US Department of Health and Human Services, US Food and Drug Administration (FDA), Center for Drug Evaluation and Research (CDER), Center for Biologics Evaluation and Research (CBER), Rockville, May 2001.
70. United States Pharmacopeia, *USP 33-NF 28*, Rockville, MD, USPC, 2010.
71. W. W. Hauck, A. J. DeStefano, T. L. Cecil, D. R. Abernethy, W. F. Koch and R. L. Williams, *Pharm. Forum.*, 2009, **35**, 772.
72. R. F. Martin, *Clin. Chem.*, 2000, **46**, 100.
73. D. Stöckl, K. Dewitte and L. M. Thienpont, *Clin. Chem.*, 1998, **44**, 2340.
74. W. Dewé, *J. Chromatogr. B*, 2009, **877**, 2208.
75. D. Chambers, G. Kelly, G. Limentani, A. Lister, K. R. Lung and E. Warner, *Pharm. Technol.*, 2005, **29**, 64.
76. D. J. Schuirmann, *J. Pharmacol. Biopharm.*, 1987, **16**, 657.
77. J. M. Bland and D. G. Altman, *Lancet*, 1995, **346**, 1085.
78. J. M. Bland and D. G. Altman, *Stat. Methods Med. Res.*, 1999, **8**, 135.
79. W. Dewé, B. Govaerts, B. Boulanger, E. Rozet, P. Chiap and Ph. Hubert, *Chemometr. Intell. Lab. Syst.*, 2007, **85**, 262.

CHAPTER 4

Using Elevated Temperature in UHPLC: Interest and Limitations

SABINE HEINISCH

Institut des Sciences Analytiques, UMR CNRS 5280, Université de Lyon, 43 Boulevard du 11 Novembre 1918, 69622 Villeurbanne Cedex, France
*E-mail: sabine.heinisch@univ-lyon1.fr

4.1 Introduction

Although the interest in the use of high temperatures in liquid chromatography (HTLC) was first stimulated in 1980[1] and even before,[2] it has grown considerably over the last decade among the scientific community and many reviews have been published recently, providing either a general overview of HTLC[3–6] or more specific features such as detection in HTLC,[7] high temperatures in hydrophilic interaction liquid chromatography (HILIC),[8] potential use of high temperature as a second dimension of a two-dimensional LC system (2DLC)[9] or the use of superheated water as mobile phase.[10,11] However, the pharmaceutical industry has still not considered this approach in routine analysis.

Both kinetic and thermodynamic behaviour of solutes are affected by increasing temperature. This explains why column temperature plays an important role in HPLC and also why higher temperatures result in a complex variation of the chromatographic process, which is sometimes not easy understand.

RSC Chromatography Monographs No. 16
UHPLC in Life Sciences
Edited by Davy Guillarme and Jean-Luc Veuthey

Published by the Royal Society of Chemistry, www.rsc.org

HTLC refers to any separation performed at temperature higher than 40 °C with a mobile phase maintained in a liquid state. This separation technique is attractive for two major reasons: first, the reduction of mobile phase viscosity which leads to faster separations, and, second, the significant improvement in peak shapes of polar and charged compounds which is of prime interest in life science analysis. Both features will be discussed in this Chapter under the view of combining high temperature and ultra-high pressure. The interest of combining UHPLC and HTLC for 2DLC separations will be also discussed in the last Section and illustrated by 2D separation of biological samples.

4.2 High Throughput and High Resolution in HT-UHPLC

4.2.1 Knox Curves

A major goal of using higher temperature is usually to shorten analysis time in order to obtain higher analysis throughput. As early as 1993,[12] an ultra-fast gradient separation of four proteins at 120 °C was presented. This was carried out in less than 10 s on a 2 µm pellicular C18 silica column. Antia and Horvath[13] emphasized the importance of increasing temperature to reduce viscosities and hence to increase diffusivity, thereby leading to faster analysis. Several empirical relationships between solvent viscosity (of acetonitrile/water and methanol/water) and both temperature and solvent composition can be found in the literature;[12,14–16] some of them are valid up to 200°C.[6] It is noticeable that these curves are smoother at high temperature, thereby greatly reducing the variation of viscosity with the organic content in the mobile phase.

The reduced plate height ($h = H/d_p$, where d_p is the particle size and H is the column plate height) can be plotted against the reduced linear velocity [$\nu = (u \times d_p)/D_m$, where u is the linear velocity and D_m is the molecular diffusion coefficient of the solute in the mobile phase) and usually well fitted using the Knox equation.[17,18] For neutral compounds, with a given reduced linear velocity, the independence of h upon temperature was shown,[19] suggesting that there was no change in packing quality with temperature. For a given stationary phase and for neutral solutes, it can be assumed that the Knox equation coefficients only depend on both the packing quality and the solute retention factor, thereby leading to a unique curve as highlighted by the experimental h–v plots obtained at three different temperatures on a sub-2 µm silica-based stationary phase (Figure 4.1).[20]

The relationship between the diffusion coefficient, D_m, the solvent viscosity, η, and the absolute temperature, T, can be approximated by the Wilke–Chang equation:[21]

$$D_m = 7.4 \times 10^{-8} \frac{(\Phi M)^{1/2} T}{\eta V_a^{0.6}} \quad (4.1)$$

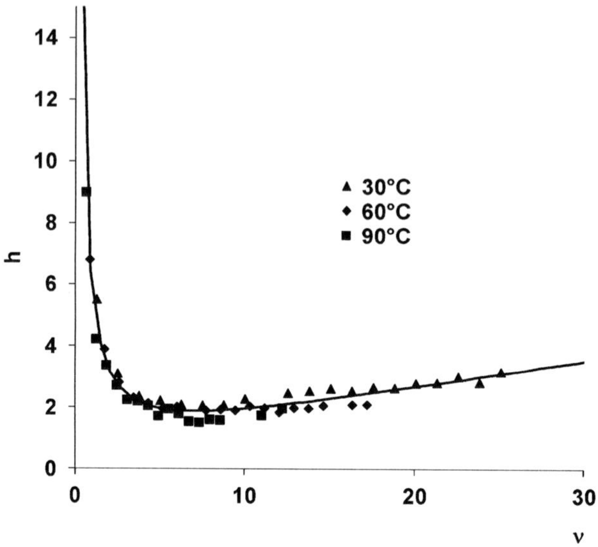

Figure 4.1 *h versus v* for experimental data obtained on a 50 × 2.1 mm, 1.9 μm Hypersil Gold-C18 column at (▲) 30°C with 68% acetonitrile (ACN), (◆) 60°C with 53% ACN and (■) 90°C with 58% ACN. Solute, butylbenzene (k = 3). The data points were fitted using the Knox equation ($h = 0.28v^{1/3} + 5.44/v + 0.09v$) (from reference 20 with permission).

where D_m is the solute diffusion coefficient (cm^2 s^{-1}), M is the molecular mass (g mol^{-1}), T is the absolute temperature (in K), η is the solvent viscosity (cP), V_a is the solute molar volume (cm^3 g^{-1} mol^{-1}) and Φ is a solvent association factor (dimensionless). From this equation, it clearly appears that the product $D_m(\eta/T)$ is constant for a given solute in a given mobile phase. As a result, once the diffusion coefficient is known in given mobile phase composition and temperature, it can be further calculated in any mobile phase conditions by using appropriate correlations for viscosity calculation. For large molecules such as peptides, numerous empirical relationships between D_m in water at 25 °C and the molecular mass can be found in the literature, such as the following one which was found to mimic the Wilke–Chang equation for a molecular mass of less than 1000 Da:[22]

$$D_{m,water,25^\circ C} = 10^{-9}\left(2.2M^{-0.32} + \frac{62}{M}\right) \tag{4.2}$$

4.2.2 Obtaining Faster Separations or Higher Resolutions

As highlighted in a previous paper,[6] the gain in analysis speed should not be balanced by a loss of overall separation power. The concept of peak capacity is a valuable tool for assessing the separation power. The peak capacity was defined by Giddings[23] as the number of peaks that can be ideally placed between the first (often represented by the column dead time) and the last peak of interest with a resolution unity between all peaks. In isocratic analysis, it is given by:[2]

$$n_c = 1 + \frac{\sqrt{N}}{4} \times \ln(1 + k_n) \tag{4.3}$$

where k_n is the retention factor of the more retained compound.

According to Neue,[24] a general relationship for peak capacity in reversed phase linear gradient elution is given by:

$$n_c = 1 + \frac{\sqrt{N}}{4} \times \frac{1}{1+2.3b} \ln\left(\frac{1+2.3b}{2.3b} e^{2.3.S.\Delta C} - \frac{1}{2.3b}\right) \tag{4.4}$$

where ΔC is the range of composition covered by the gradient, S is the slope of the relationship between log(k) and the organic solvent concentration, which is not expected to vary much with temperature, and b is the linear-solvent-strength (LSS) gradient steepness [$b = S \times \Delta C(t_0/t_G)$, where t_0 is the column dead time and t_G is the gradient time].[25] S usually increases with molecular size and thus is higher for peptides than for small pharmaceuticals. Eqn (4.4) is valid provided that: (1) the gradient elution is of the so-called LSS form[26] and (2) the contribution of the dwell volume to the retention is not significant.

According to eqns (4.3) and (4.4), it appears that the determination of the time saved by increasing temperature is correct provided that: (1) the comparison is made with the same plate number and (2) k_n (for isocratic separation) or ΔC and t_0/t_G (for gradient elution separation) are kept constant in order to maintain the same separation power. Alternatively, the benefits of high temperatures in the field of high throughput analysis might be overestimated.

For fixed column length and variable column pressure, the maximum plate number that can be achieved, N_{max}, is obtained at the optimum reduced linear velocity, ν_{opt}, corresponding to the minimum reduced plate height, h_{min}. In these conditions, the column dead time is given by:

$$t_0 = N_{max} \times \frac{h_{min}}{\nu_{opt}} \times \frac{d_p^2}{D_m} \tag{4.5}$$

As a result, the gain in analysis time when increasing temperature from 20 °C to T corresponds to the ratio of the diffusion coefficients:

$$\frac{t_{0.20\,^\circ C}}{t_0 T} = \frac{D_m T}{D_{m20\,^\circ C}} = \frac{\eta_{20\,^\circ C} \times T}{\eta_T \times 293} \tag{4.6}$$

More generally, for any (h, v) coordinates, the gain in analysis time to achieve N [$N = L/(h \times d_p)$] is given by eqn (4.6). The ratio of the diffusion coefficients is strongly dependent on both the type and the volume fraction of organic modifier. It is much higher with mixtures of water and methanol than with water and acetonitrile. At 200 °C, it can reach 25 with methanol, whereas it only reaches 9 with acetonitrile. At 100 °C, the gain is in the range 2 to 3.5 with acetonitrile and 4 to 6 with methanol.[6] It should be noted that these values assume a constant solvent composition while the temperature is increased. Actually, it should not be the case. The content of organic solvent has to be decreased to maintain the same solute retention (see Section 4.3.4). As a result, the variation of solvent viscosities and hence gain values may be slightly different, especially in the range of high concentrations of the organic modifier.

For a given plate number with a given column (given h and v coordinates), the higher temperature involves a higher column pressure. The resulting increase in column pressure drop is given by:

$$\frac{\Delta PT}{\Delta P_{20\,^\circ C}} = \frac{T}{293} \tag{4.7}$$

As shown by eqn (4.7), the pressure increase is approximately 30% from 20 °C to 100 °C and 60% from 20 °C to 200 °C. Moreover, at a higher temperature, additional tubing is required as a heat exchanger for preheating and cooling the mobile phase. This tubing leads to further increase in pressure. This issue is much misunderstood as it is sometimes claimed that elevating temperature leads to decrease in pressure. This may be true when keeping the same linear velocity and hence accepting the change in column plate number. However, in this latter case, the comparison of both techniques is no longer valid as highlighted above.

The comparison of performance limits between two temperatures is more correctly assessed by using either kinetic performance curves, as proposed by Giddings[27] and Poppe,[28] or the more recent and convenient kinetic plot concept developed by Desmet and co-workers.[29] Kinetic plots take into account the instrument constraints and, in particular, the maximum affordable column or instrument pressure. The method consists in plotting either the minimum column dead time (t_0) required to reach a given column plate number (N) *versus* N (isocratic separation) or the minimum gradient time (t_G) required to reach a given peak capacity (n_c) *versus* n_c (gradient separation).[30] In both cases, the column length is allowed to vary while maintaining a fixed column pressure. In addition to the acceptable pressure, other constraints can be taken into account such as the maximum flow-rate that can be delivered by the instrument, the maximum column length that can be contained into the oven and also the contribution of extra-column dispersion to band broadening.[20] Simulated kinetic curves based upon the Knox coefficients given in Figure 4.1 and current UHPLC instrument limitations are shown in Figure 4.2. The continuous lines represent theoretical (*i.e.* unconstrained) kinetic curves at 20 °C and 90 °C. As indicated by the arrow, the gain in analysis speed is approximately 3 for a desired plate number of 10 000. However, the gain decreases when the desired plate number increases. The two curves intersect at a critical plate number close to 80 000, suggesting that 90 °C is beneficial for lower efficiencies and detrimental for higher values. It should be pointed out that these curves do not take into account any loss in performance originating from extra-column band broadening, extra-column pressure drop and/or instrument limitations in both flow rate and column length. Theoretical kinetic plots can be constrained by applying instrument limitations.[20] The contribution of the different constraints prevents the achievement of the theoretically expected performance as shown by the dotted lines in Figure 4.2. While theoretical curves only depend on column parameters (permeability, packing quality), solvent parameters (viscosity) and solute parameters (molecular mass, retention factor), the constrained kinetic curves are also dependent on the instrument.

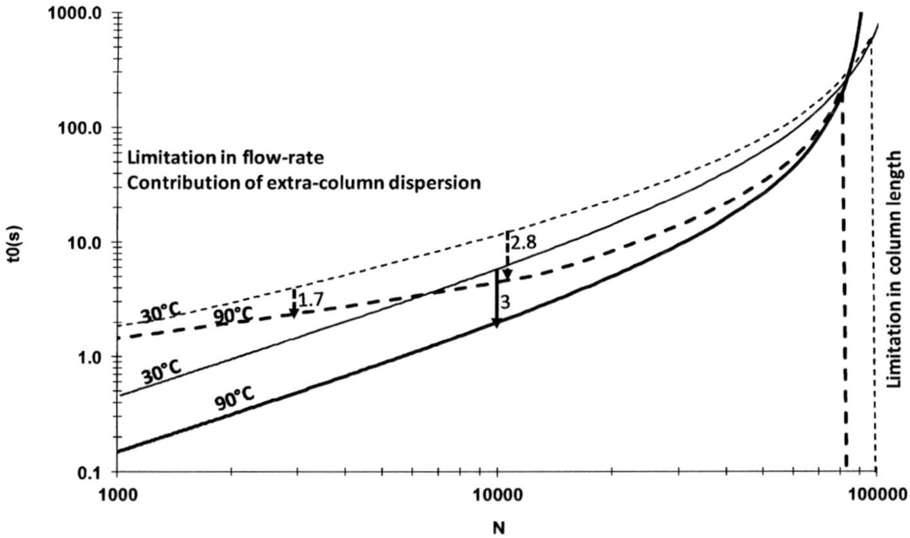

Figure 4.2 Simulated kinetic plots in isocratic elution, log(t_0) against log(N) at 20 °C and 90 °C in UHPLC conditions for a neutral solute (k = 3). The curves were obtained by assuming the following parameters: u_0-based flow resistance = 500; $D_{m(20\ °C\ in\ water)} = 10^{-9}\ m^2\ s^{-1}$; d_p = 1.9 µm; column i.d. = 2.1 mm; Knox equation as in Figure 4.1; mobile phase composition = 20% acetonitrile. The continuous lines represent unconstrained kinetic curves at maximum pressure of 800 bar. Dashed lines represent kinetic curves with additional instrumental constraints: maximum flow-rate, 2000 µL min^{-1}; maximum column length, 600 mm; tubing, 500 × 0.127 mm. The extra-column variance was calculated for an Acquity UPLC instrument (Waters) from empirical equations given in Figure 4.5.

Theoretically, the maximum plate number that can be achieved is finite and can be calculated from the kinetic plot equations according to:

$$N_{max} = \Delta P_{max} \times \frac{K_o}{BD_m \eta} \tag{4.8}$$

where K_0 is the u_0-based column permeability, DP_{max} is the maximum allowable pressure and B is the coefficient of the contribution to longitudinal diffusion in the Knox equation. As previously underlined, $D_m\eta$ is proportional to the temperature and hence N_{max} decreases with temperature as suggested by the continuous lines at 20 °C and 90 °C in the high plate range. It should be noted, however, that this theoretical maximum plate number should be reached after an infinite time. The limitation in column length will fix the real maximum plate number that can be reached. It is indicated by a straight vertical dotted line in Figure 4.2. It is close to 100 000 and 80 000 at 20 °C and 90 °C respectively with the present instrument. As can be observed, a higher efficiency and hence a higher resolution is achieved at 20 °C compared to 90 °C. In the range of low efficiencies (<10 000), the gain in analysis time between 20 °C and 90 °C is lower than that predicted by theoretical curves (*e.g.* 1.7 instead of 3 for 3000 plates). This is mainly due to the limitation in flow rate (here F_{max} = 2000 μL min^{-1}), which is more detrimental at high temperatures. However, this is also due to contribution of extra-column band broadening, which is more significant at high temperatures due to a higher flow rate (see Section 4.3.3). In gradient elution, extra-column band broadening mainly occurs in the detector cell and in the connecting tubing between the column outlet and the detector. As a result, extra-column band broadening is less a concern. Similarly, kinetic curves can be simulated in gradient elution. They are given in Figure 4.3 for the same small solute as in Figure 4.2 and for a larger analyte with a lower diffusion coefficient. The use of larger analytes such as peptides usually leads to both lower diffusion coefficient and higher S value. Kinetic plots were obtained from eqn (4.4) for different values of the gradient steepness, b, ranging from 0.03 to 3. Instrument limitations were taken into account and the contribution of extra-column band broadening was assumed to be negligible. The curves are restricted in the low and high plate range because of limitations in the flow rate and the column length respectively. As observed in Figure 4.3(a), the gain in analysis time for a given peak capacity is the same as in isocratic elution (Figure 4.2). Furthermore, similar to isocratic elution, this gain decreases when the desired peak capacity increases, resulting, for small compounds, in a critical peak capacity which depends on the gradient steepness value. For larger compounds (Figure 4.3b), the critical peak capacity is never reached because of the present limitation in column length. As a result, increasing temperature up to 90 °C is expected to be beneficial within the whole range of achievable peak capacities. In the range of low peak capacities (very fast gradient, *i.e.* $t_G \leq 2$ min), the use of higher temperature is beneficial for both small and large solutes. This

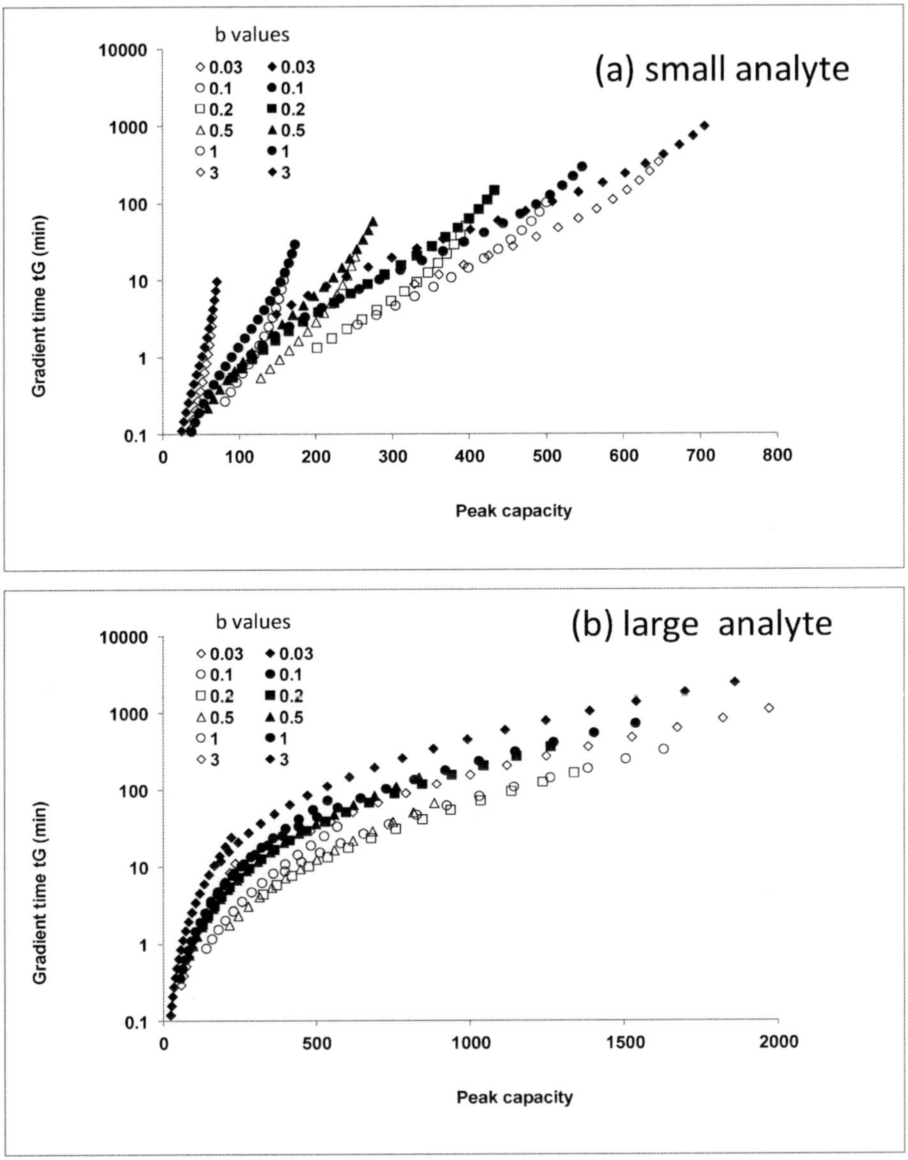

Figure 4.3 Simulated kinetic plots in gradient elution, log(t_G) against peak capacity at 20 °C (closed symbols) and 90 °C (open symbols) in UHPLC conditions (maximum pressure of 800 bar) for different values of the gradient steepness ranging from 0.03 to 3. (a) For a small neutral compound; composition range ΔC = 100%; S = 0.03; and (b) for a large compound [$D_{m(20\ °C\ in\ water)} = 2 \times 10^{-10}\ m^2\ s^{-1}$] assuming the same Knox equation as in Figure 4.1; composition range ΔC = 40%; S = 0.2; same other parameters as in Figure 4.2.

observation is of prime interest for the second dimension of comprehensive on-line two-dimensional separations (2DLC). This will be further developed in Section 4.5.

The illustration of two very fast gradient separations is given in Figure 4.4. Both separations were performed on a sub-2 μm stationary phase by combining the potential of high temperature (90 °C) and ultra-high pressure (800 bar). Figure 4.4(a) shows a separation of eight β blockers (small analytes) in approximately 25 s with methanol as organic solvent. Figure 4.4(b) shows a separation of a tryptic digest of three proteins (large analytes) in 100 s.[31]

4.2.3 Concluding Remarks

- At high temperatures, the theoretical gain in speed compared to low temperature is dependent on the desired plate number. It is at a maximum (close to the D_m ratio) for low efficiencies (short columns), whereas it decreases (and can even be reversed) in the range of high efficiencies.
- For a desired analysis time (desired t_0 or t_G), a higher efficiency should be attained at higher temperatures provided that t_0 (or t_G) is below the critical point (*e.g.* $t_{0,\ critical}$ = 200 s in Figure 4.2).
- UHPLC and HTLC are not competing techniques but fully complementary ones. High temperatures can provide faster separations for a given efficiency, whereas high pressure can extend the range of attainable plate number.
- Finally, it is strongly recommended to take instrument features into account to have a true assessment of the high temperature potential in the field of high throughput or high resolution.

4.3 Limitations of HTLC and HT-UHPLC

4.3.1 Thermal stability of Stationary Phases

The thermal stability of stationary phases is a major issue which currently limits the use of very high temperatures. Column manufacturers usually recommend a maximum temperature of 60 °C for silica-based stationary phases; yet, some of them can be used at higher temperature. However, presently, no silica-based columns are stable enough to withstand temperatures exceeding 90 °C for a long time. Furthermore, the thermal stability of the columns strongly depends on the mobile phase conditions (pH, buffer type, buffer concentration ...).[32] More generally, silica-based columns which are stable at both high pH values and ambient temperatures are expected to be stable at both neutral pH (or low pH) and moderate temperature (up to 90 °C). Non-silica-based stationary phases can usually withstand much higher temperatures. These are based on either metal oxide (zirconia or titania) porous graphitic carbon or organic polymers. Table 4.1 summarizes the

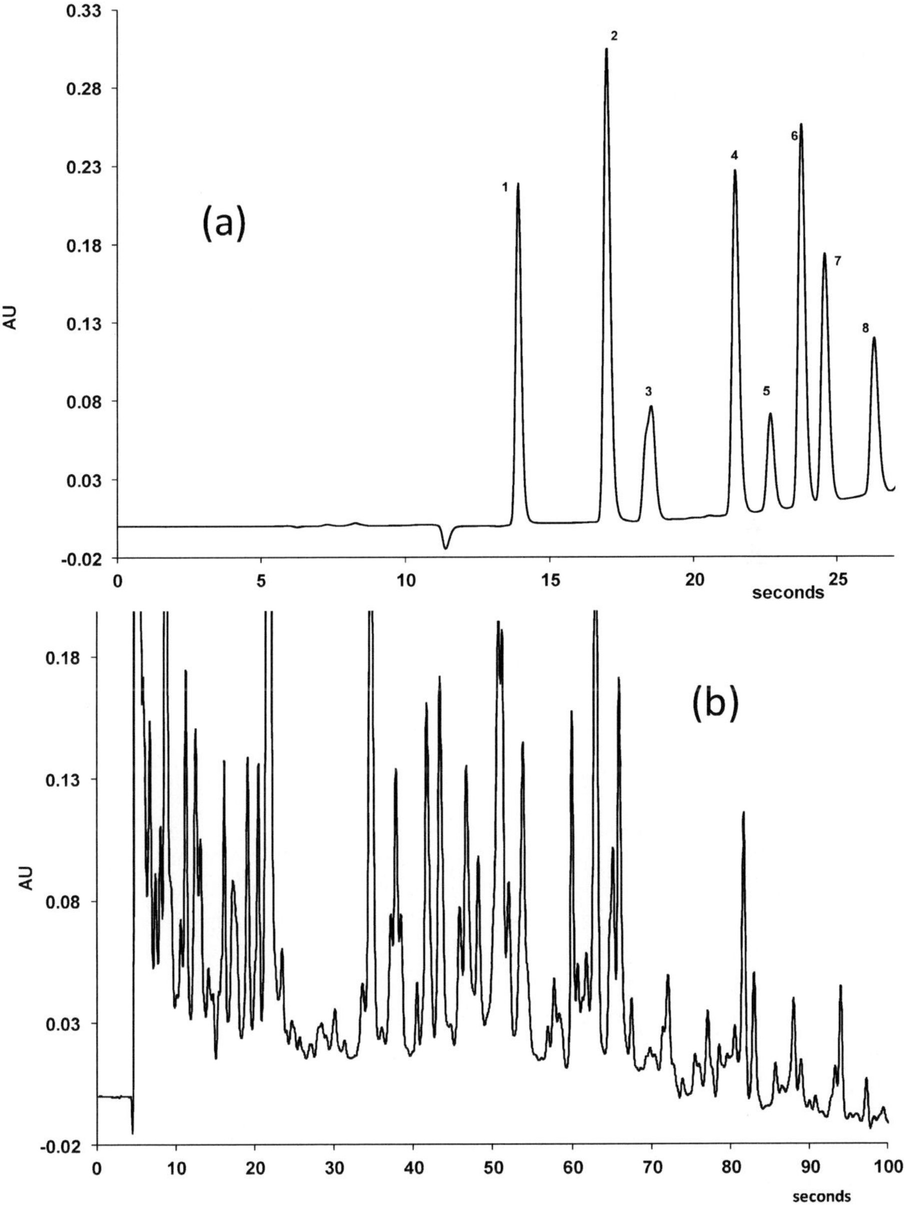

Figure 4.4 Ultra-fast gradient separations in UHPLC at 90 °C. (a) Eight β-blockers; gradient conditions: 10 mM ammonium acetate/methanol, 95:5 (v/v) to 40:60 (v/v) in 20 s. Flow-rate: 1200 μL min^{-1}; initial pressure: 800 bar; solutes: atenolol (1), pindolol (2), nadolol (3), acebutolol (4), metoprolol (6), oxprenolol (7) and metipranol (8). (b) Tryptic digest of three proteins: BSA, lysozyme and myoglobin; gradient conditions: 10 mM ammonium acetate/ acetonitrile, 99:1 (v/v) to 68:32 (v/v) in 120 s. Flow-rate: 1600 μL min^{-1}; initial pressure: 900 bar. Column: Acquity BEH C18, 50 × 2.1 mm, 1.7 μm; instrument: Acquity UPLC (Waters) (from reference 31 with permission).

Table 4.1 Commercially available columns for high-temperature RP separations and their maximum permitted temperature.

Stationary phase	*Manufacturer*	*Base material*	*Stationary phase*	T_{max} *(°C)*
Zorbax Stablebond	Agilent Technologies	Silica, steric protection	C18, C8, CN, phenyl	90 80
Zorbax extend	Agilent Technologies	Silica, bidentate	C18, C8, C3, CN, phenyl	80
Nucleodur Gravity	Macherey Nagel	Silica, densely bonded	C18	80
Blaze 200	Selerity Technologies	Silica, polydentate	C18	150
Xbridge-Acquity	Waters	Hybrid silica	C18, C8, C6, phenyl	90
Discovery-Zirconia	Supelco	Zirconium oxyde	Polybutadiene, Polystyrene Carbon clad C18	150 200
PLRP-S	Polymer Laboratories	polystyrene-divinylbenzene		200
PRP-1	Hamilton	polystyrene-divinylbenzene		200
Hypercarb	Thermo Fisher	Porous graphitized carbon		>200

reverse-phase (RP)-stationary phases that can be used above 80 °C. Unfortunately, there is currently a lack of information about the long-term stability of stationary phases and, above all, there is no universal test which could provide an objective comparison. As a result, the analyst has to acquire his own experience and hence to assess the number of runs which can be performed in his own mobile phase conditions before damaging the separation. Finally, one should be aware that the bleeding of the stationary phase from the column is an additional problem of thermal instability which can be very detrimental with mass spectrometry detection.[33,34]

4.3.2 Thermal Stability of Analytes

While the thermal instability of some analytes can limit the use of very high temperatures, it is usually less of a concern in case of ultra-fast HTLC or HT-UHPLC separation. Numerous compounds of pharmaceutical interest are prone to degradation depending on solvent conditions. However, their analysis at high temperatures may be possible provided that an on-column reaction does not happen. The occurrence of an on-column reaction depends on both the residence time in the column and the reaction rate. As suggested by Antia and Horvath,[35] an on-column reaction may be insignificant if the increase in the reaction rate is well compensated for by a decrease in column residence time due to the necessary increase in flow rate at high temperatures.

It has also been reported that analyte degradation might be enhanced by the stationary phase itself because of a catalytic activity, particularly with zirconium-based columns.[34] Nevertheless, for unstable analytes, it is recommended to make sure of the reliability of the HTLC analysis.

4.3.3 Specific Equipment

HTLC (and HT-UHPLC) instruments have to be designed to work at high temperatures. No UHPLC instrument is currently designed for working at very high temperatures (up to 200 °C) but external ovens, equipped with preheater and cooling devices, are commercially available and can be used at very high temperatures combined with UHPLC (VHT-UHTLC).[36] However, very high temperatures (> 90 °C) are still not widely used in UHPLC. This is due to the fact that, presently: (1) silica-based columns cannot withstand very high temperatures and (2) non-silica-based columns cannot withstand very high pressures (> 350 bar). As a result, two different ranges of temperature have to be distinguished: the first one ranging from ambient to 90 °C and the second one from 100 °C to 200 °C, which is not yet adapted to UHPLC separations. HTLC separations require suitable ovens to heat the column, as well as efficient preheating and cooling devices. Unfortunately, mobile phase preheating has to be carried out after the injector valve in order to prevent rotor degradation. The preheating device leads therefore to additional extra-column band broadening. The difference in temperature between the incoming eluent and the column walls is recommended to be within ±5 °C in order to keep optimal column performance.[37–39] Consequently, whatever the design of the preheating device (long-coiled tubing, heater block or active preheater), preheating always requires an additional length of tubing between the injector and the column inlet. Moreover, a cooling device is often required, in particular with UV detection to avoid baseline mismatch. The cooling device is also necessary to maintain the mobile phase in a liquid state in the case of very high temperatures. For both preheating and cooling, the required tubing length is strongly related to column temperature and mobile phase flow rate. As a result, the problem of additional extra-column band broadening and additional extra-column pressure is more important at very high temperatures (> 100 °C) and/or with large column diameters and/or in UHPLC with higher flow rates. For both efficient preheating and efficient cooling, the tube length is decisive, whereas the tube internal diameter (i.d.) is not essential. Consequently, to minimize extra-column band broadening, the tube i.d. should be reduced as much as possible. HT-UHPLC instruments are currently equipped with stainless steel tubing (0.127 mm i.d.) as the preheater. The variation of measured extra-column variance as a function of flow rate is shown in Figure 4.5 for two HT-UHPLC instruments equipped with a suitable preheating device. As can be observed, variance values increase with flow rate according to an exponential behavior.[20] For a given flow rate, they are slightly lower at 90 °C than at 30 °C. Since the optimum flow rate can be 3- to 6-fold

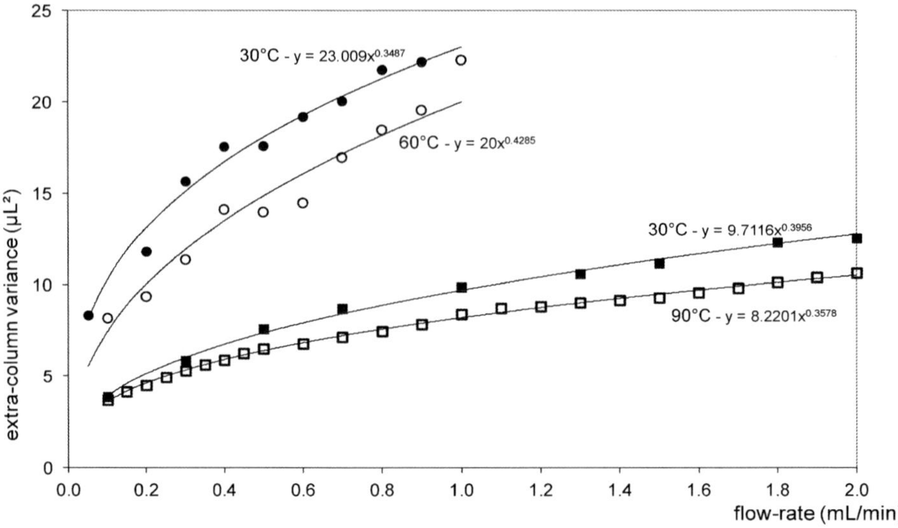

Figure 4.5 Plots of extra-column variance versus flow-rate at 30 °C (closed symbols) and 90 °C (open symbols) with (■, □) an Acquity UPLC instrument (Waters) equipped with a 10 µL loop and (●, ○) an Accela instrument (Thermo Fisher) equipped with a 20 µL loop. Variances were calculated by the statistical second-order moment. Data are fitted with exponential curves (from reference 20 with permission).

higher at 90 °C depending on the organic solvent, the extra-column variance and hence peak band broadening is usually higher at 90°C than at 30 °C.

This issue restricts the use of current HT-UHPLC systems to gradient elution. However, isocratic elution can be suitable if retention factors are high enough (> 3–4). With micro-bore columns (1 mm i.d.), neither isocratic nor gradient separations are recommended as highlighted by the gradient separation of aromatic compounds at 90 °C on the same 1.7 µm stationary phase with two different column i.d., 2.1 mm (Figure 4.6a) and 1 mm (Figure 4.6b)). Transfer rules were applied to both gradient time and injected volume. Thus, both separations should have been similar if the contribution of extra-column band broadening was the same. As can be observed in Figure 4.6(b), this contribution is too much important with micro-bore columns. It can be noted that the slight difference in selectivity for the less retained compounds arises from a difference in dwell time due to a lower flow rate with the micro-bore column. Very recently, much narrower tubes with 0.065 mm i.d. have appeared. Based on nickel and cobalt, they are suitable for preheating and hence expected to extend the use of HT-UHPLC systems to micro-bore columns. However, it is important to note that instead of additional band broadening, this small i.d. tubing is expected to generate significant additional extra-column pressure.

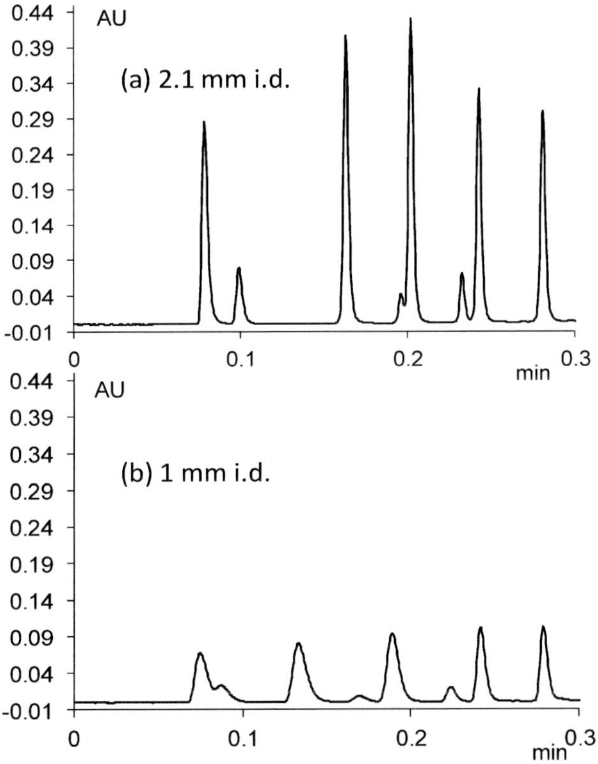

Figure 4.6 Extra-column band broadening depending on column i.d. Gradient separation of aromatic compounds at 90 °C. (a) Acquity BEH C18 column, 50 × 2.1 mm, 1.7 μm, 25 to 75% acetonitrile in water in 0.29 min; 1800 μL min^{-1}; 0.5 μL injected. (b) Acquity BEH C18 column, 50 × 1 mm, 1.7 μm, 25 to 75% acetonitrile in 0.2 min; 600 μL min^{-1}; 0.1 μL injected. Solutes: uracil, caffeine, methylparabene, *o*-cresol, ethylparabene, β-naphtol, propylparabene and butylparabene. UV detection is at 254 nm. Instrument: Acquity UPLC (Waters). AU, arbitrary units.

4.3.4 Reduction of Eluent Strength

When increasing the temperature, eluent strength, and hence retention factors, have to be kept constant (isoeluotropic conditions). It is therefore necessary to reduce the content of organic solvent used at room temperature. This reduction is attractive for "green chemistry", as the consumption of the organic solvent decreases. In some particular cases, it is even possible to use pure water as mobile phase which may be also attractive with some universal detectors that are sensitive to the presence of organic solvents (*e.g.* flame ionization detector).[7] Conversely, less organic solvent in the mobile phase may be detrimental to electrospray-ionization mass spectrometry (ESI-MS) detection. In case of neutral compounds, an increase in temperature most usually leads to a decrease in retention similar to an increase in the percentage of organic modifier. It was shown in a study[40] on different stationary phases that a variation of 40% acetonitrile in water is similar, in terms of solute retention, to a variation of 160 °C. That means that every increase of 40 °C should involve a decrease of approximately 10% acetonitrile in the mobile phase in order to maintain the same eluent strength. Similar conclusions were drawn with methanol as the organic solvent.[41] The reduction in eluent strength with temperature is beneficial to large hydrophobic molecules which are strongly retained on RP-stationary phases at ambient temperature, even with a high percentage of organic solvent in the mobile phase. However, very high temperatures can be detrimental to many compounds of pharmaceutical or biological interest such as very polar metabolites or any charged compounds that are poorly retained even at ambient temperature.

4.4 Advantage of High Temperature in Life Science Analysis

4.4.1 Great Improvement of the Kinetic Performance of Charged Compounds at High Temperatures

In life science analysis, the molecules are very often ionized or partially ionized. A major advantage in elevating temperature is the significant improvement of peak shapes. Poor peak shapes can be due to different factors including slow kinetic behaviour, dual retention mechanism and/or overloading effects. The beneficial effect of temperature on peak shape has been extensively reported for charged compounds[42–44] and even for neutral polar ones.[45,46] Peak tailing could be due to kinetic phenomena with slow kinetics of interaction with some strong sites, such as silanol groups, together with fast kinetics of interaction with weak sites, such as alkyl chains. Therefore, increasing the temperature might result in increasing kinetics of mass transfer in such a way that both kinetics (strong and weak sites) tend to be similar thereby improving the peak shape for charged compounds. Moreover, the dissociation rate of solutes and silanols changes with temperature, thereby probably contributing also to the improvement in the peak shapes. Very poor peak shapes are usually observed for drugs with non-silica-based columns (Hypercarb, organic polymer- or zirconia-based columns) at temperatures below 90 °C as shown, for example, in Figure 4.7 for the peak of diphenhydramine on a polystyrene–divinylbenzene column. The necessity of elevating temperature is indicated by the regular decrease in the observed peak asymmetry with temperature. In fact, non-silica-based columns being thermally stable, it is worth working at very high temperatures for the separation of charged compounds. Unfortunately, none of them are designed for UHPLC conditions so far.

Silica-based columns perform much better at ambient temperature, especially those dedicated to the separation of basic compounds. However, peak asymmetries are usually higher than 1.2 depending on various parameters including mobile phase pH value, ionic strength and type of buffer. In such

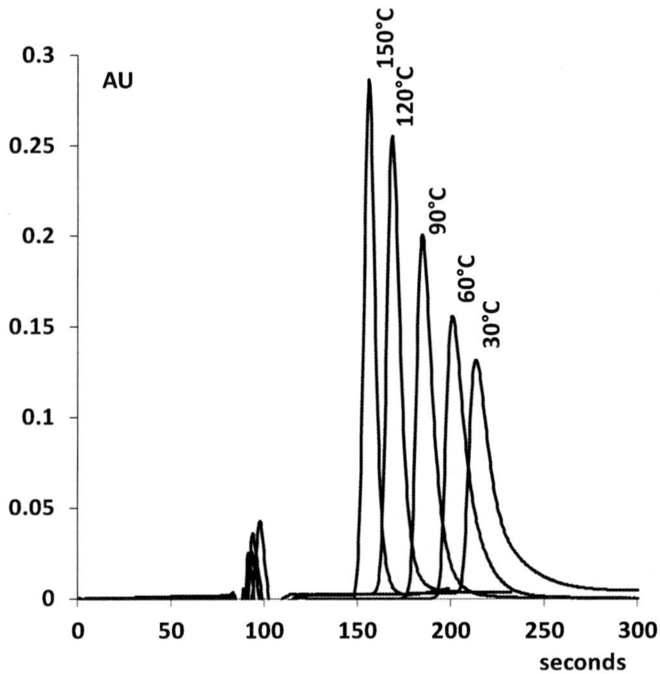

Figure 4.7 Improvement of peak shape with temperature. Temperatures ranging from 30 °C to 150 °C on a PLRP-S 150 × 4.6 mm, 5 μm column. Solute: diphenhydramine. Mobile phase : ammonium acetate (pH 3.8)/acetonitrile, 40:60 (v/v). Flow-rate: 1000 μL min^{-1}. AU, arbitrary units.

conditions, the measurement of peak variance at half peak height is hazardous. It may result in a significant underestimation of the reduced plate height. Accordingly, with the objective of obtaining reliable *h–v* plots, measurements of peak variance have to be made either numerically by statistical second order moment or directly by using the empirical Dorsey–Foley equation[47] which is known to be well correlated to the true peak variance for small peak asymmetry (< 2.5):

$$\sigma^2_{peak} = \frac{\left({}^{b}\!/_{a} + 1.25 \right) \times w_{0.1}}{41.7} \tag{4.9}$$

where $w_{0.1}$ is the peak width and *b*/*a* is the ratio of the right to the left half width, both measured at 10% of the peak height. In a recent study,[31] the kinetic behaviour of small pharmaceuticals and peptides was compared to that of neutral ones on sub-2 μm silica-based C18 columns using various volatile buffers (MS-compatible) at 30 °C, 60 °C and 90 °C. The obtained results with charged compounds were quite different from those with neutral compounds, showing much higher reduced plate height values in case of charged compounds. By elevating the temperature, *h–v* plots were significantly shifted towards the bottom in all studied conditions, except at acidic pH with formic acid 0.1%. Results obtained with 10 mM ammonium acetate are shown in Figure 4.8. As can be seen, better results were obtained at 60 °C with *h*–*v* plots of the three studied peptides, becoming closer to those of the neutral solute (caffeine, continuous line). Unlike non-silica-based columns, a small elevation of temperature (30 °C to 60 °C in Figure 4.8) is sufficient to dramatically enhance kinetic performance.

For a given sample, the concept of "sample peak capacity", as derived by Dolan *et al.*,[48] can provide a reliable comparison of separation quality in different gradient conditions. The sample peak capacity can be determined by:

$$n_c = \frac{V_n - V_1}{W_{average}} \tag{4.10}$$

where V_n and V_1 are the gradient retention volumes of the most and the least retained solutes and $w_{average}$ is the average 4σ peak width in volume units. In addition to column efficiency, the sample peak capacity takes into account the retention strength. To have an objective comparison of the separation quality between two different temperatures, both the range of composition covered by the gradient, ΔC, and the gradient volume, V_G, have to be kept constant. An example of the effect of temperature on the quality of separation is shown in Figure 4.9 with the gradient separation of seven peptides on a sub-2 μm RP-column at 30 °C, 60 °C and 90 °C, with caffeine being the reference neutral

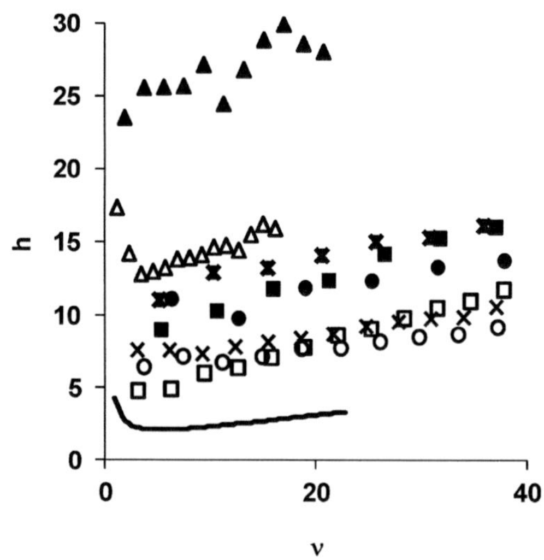

Figure 4.8 Plots of *h versus v* for different charged compounds and caffeine. Mobile phase: water/acetonitrile with 10 mM ammonium acetate. Column: Kinetex-C18, 50 × 2.1 mm; 1.7 μm at 30 °C (closed symbols) and 60 °C (open symbols). Solutes: influenza haemagglutinin (■, □), bombesin (●, ○), [Ile]-angiotensin (×, ×), diphenhydramine (▲, △) and caffeine (—). *k* is approximately 3 for all solutes (adapted from data presented in reference 31).

solute. Sample peak capacity values are indicated at the top left in Figure 4.9. As previously mentioned, a major issue with the use of high temperatures is the decrease in retention which may result in a reduction of the retention window in gradient elution which, in turn, can affect the peak capacity. A small reduction of the retention window can, indeed, be observed between 30 °C and

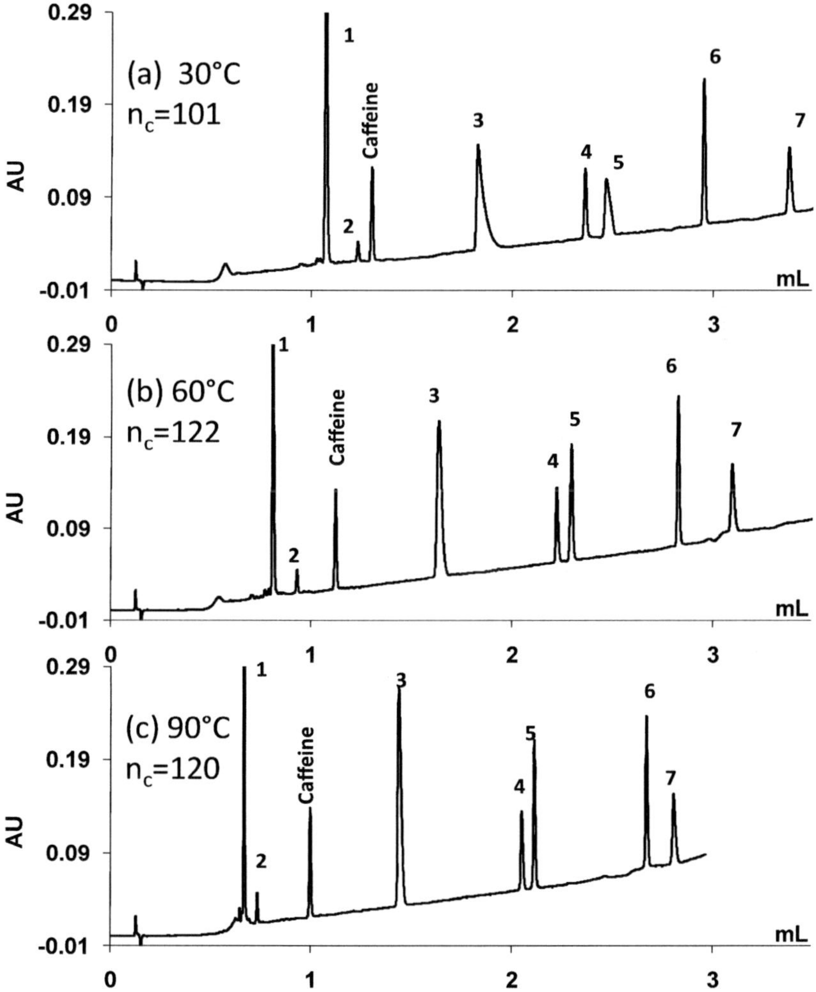

Figure 4.9 Effect of temperature on the gradient separation of seven peptides and caffeine. Column: Acquity BEH C18, 50 × 2.1 mm; 1.7 µm. Conditions: gradient from 1 to 35% ACN with a gradient volume of 4 mL; (a) 30 °C and 500 µL min^{-1} flow-rate; (b) 60 °C and 700 µL min^{-1} flow-rate; (c) at 90 °C and 1000 µL min^{-1} flow-rate; mobile phrase: 10 mM ammonium acetate; 1.0 µL injected. Solutes: WDDHH (1), FLAG (2), caffeine, bradykinin fragment 1–5 (3), [Ile]-angiotensin (4), influenza haemagglutinin (5), substance P (6) and bombesin (7). Sample peak capacities are indicated in the Figure (from reference 31 with permission). AU, arbitrary units.

90 °C. However, it seems to be well compensated for by the dramatic enhancement of peak efficiency, thereby increasing, on the contrary, the sample peak capacity (indicated in Figure 4.9) between 30 °C and 60 °C and maintaining a quite constant value between 60 °C and 90 °C.

4.4.2 Temperature as a Useful Parameter for Method Development

A great majority of molecules of pharmaceutical interest contain basic functional groups. Temperature plays an important role in the ionization process of these molecules. As a result, changing the temperature allows one to modify their dissociation rates depending on their p*K*a and hence to control selectivity.

The dissociation rate, τ, is given by:

$$\tau = \frac{1}{1 + 10^{{}_{S}^{S}pH_{T} - {}_{S}^{S}pK_{a,solute,T}}} \tag{4.11}$$

where ${}_{s}^{s}pH_T$ and ${}_{s}^{s}pK_{a,solute,T}$ refer to the pH of the mobile phase and pK_a of the solute respectively, both measured at the same temperature, T, in the same medium using a pH meter calibrated with standards in the same medium. From eqn (4.11), it appears that the variation of the dissociation rate with temperature is linked to the variation of $({}_{\downarrow}s^{\uparrow}s)pH - [[({}_{\downarrow}s^{\uparrow}s)pK]]_{\downarrow}$ with temperature. It should be emphasised that pH is usually measured at room temperature in aqueous medium. This pH can be written ${}_{w}^{w}pH_{rt}$ according to IUPAC recommendations, in which the subscript "rt" refers to room temperature. The difference between the two pH values represent the difference between the dissociation constants of buffer in both medium[49] and hence:

$${}_{s}^{s}pH_{T} = {}_{w}^{w}pH_{rt} + \left({}_{s}^{s}pK_{a,buffer,T} - {}_{w}^{w}pK_{a,buffer,rt}\right) \tag{4.12}$$

For a given mobile phase, the variation of the acid dissociation constant with the absolute temperature is given by:

$$\frac{d\left({}_{s}^{s}pk_{a}\right)}{d(1/T)} = \frac{\Delta H_{a}^{o}}{2.3R} \tag{4.13}$$

where ΔH_a^o is the standard enthalpy of ionization and R is the gas constant. Thus, according to eqns (4.12) and (4.13), the following equation can be written:

$$\frac{d\left({}_{S}^{S}pH - {}_{S}^{S}pK_{a,solute}\right)}{d(1/T)} = \frac{d\left({}_{S}^{S}pK_{a,buffer}\right)}{d(1/T)} - \frac{d\left({}_{S}^{S}pK_{a,solute}\right)}{d(1/T)} = \frac{\left(\Delta H_{a,buffer}^{o} - \Delta H_{a,solute}^{o}\right)}{2.3R} \tag{4.14}$$

Thus, according to the preceding equations, the variation of the dissociation rate with temperature is directly dependent on the difference in standard enthalpies of ionization of both buffer and solute. It is therefore expected to be significant when the difference is large. Since enthalpies of ionization are usually large for basic compounds (cationic acids) and small for acidic ones (neutral or anionic acids),[50–52] large variations can be expected with basic solutes in acidic buffers.[53,54] Using a method based on chromatographic data and described elsewhere,[55] Heinisch *et al.*[44] studied the variation of the dissociation rate with temperature for various ionizable compounds (acid and basic) in different buffers. It was shown that the dissociation of a basic compound in acidic buffer, such as phosphate buffer, is quite identical when the pH value is varied by 2 units at a given temperature compared with when the column temperature is varied by 60 °C at a given pH value. The resulting van't Hoff plots are not linear and, furthermore, the retention may increase with temperature, as highlighted in Figure 4.10 for different pharmaceuticals with potassium phosphate (Figure 4.10a) and ammonium acetate (Figure 4.10b) as buffers.[56] Consequently, temperature may be as efficient as pH to vary the dissociation rate of basic compounds and hence to control selectivity. Using temperature rather than pH in method development is attractive for many reasons: (1) the relationship between retention and temperature is smooth, as can be observed in Figure 4.10, and hence more convenient for modelling; (2) temperature is a more readily adjustable parameter; and (3) optimizing temperature can give rise to more rugged analysis. As previously stated, a 1.5 °C change in temperature can lead to the same change in solute

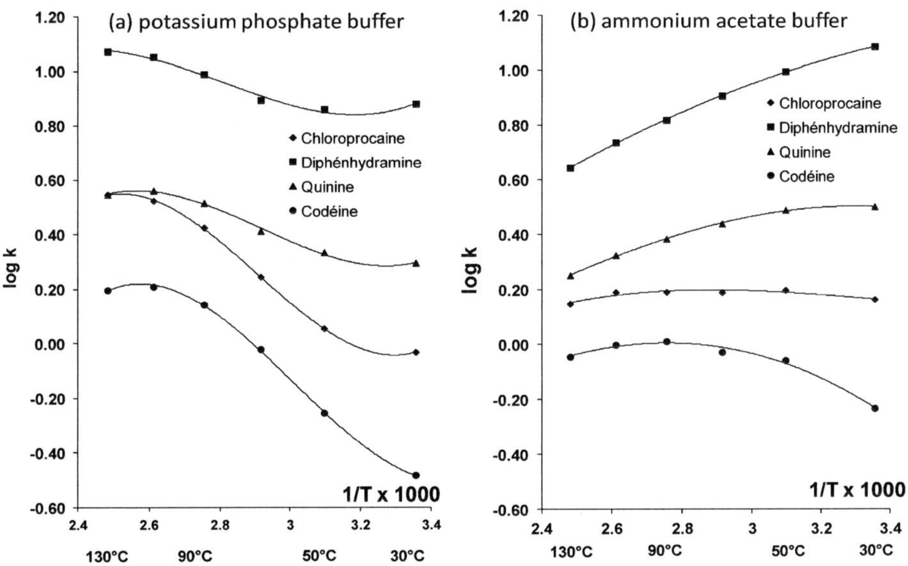

Figure 4.10 van't Hoff plot for different pharmaceuticals on a Nucleodur Gravity C18 column with (a) 15 mM potassium phosphate/ACN, 70:30 (v/v), and (b) 15 mM ammonium acetate/ACN, 60:30 (v/v), both at ${}^{w}_{w}pH_{rt}$ = 6.0 (adapted from reference 56).

dissociation rate as a 0.05 pH unit change. Actually, a temperature deviation of less than 1.5 °C is much easier to control with currently available ovens than pH deviation of 0.05 pH units.

Method development can be supported by powerful optimization software able to achieve retention modelling, simulation of chromatograms and response function calculation over the entire parameter space. An example is given in Figure 4.11, with the optimization of both temperature and mobile phase composition for the separation of a pharmaceuticals sample.[44] The response function is a multi-criteria function, taking into account the

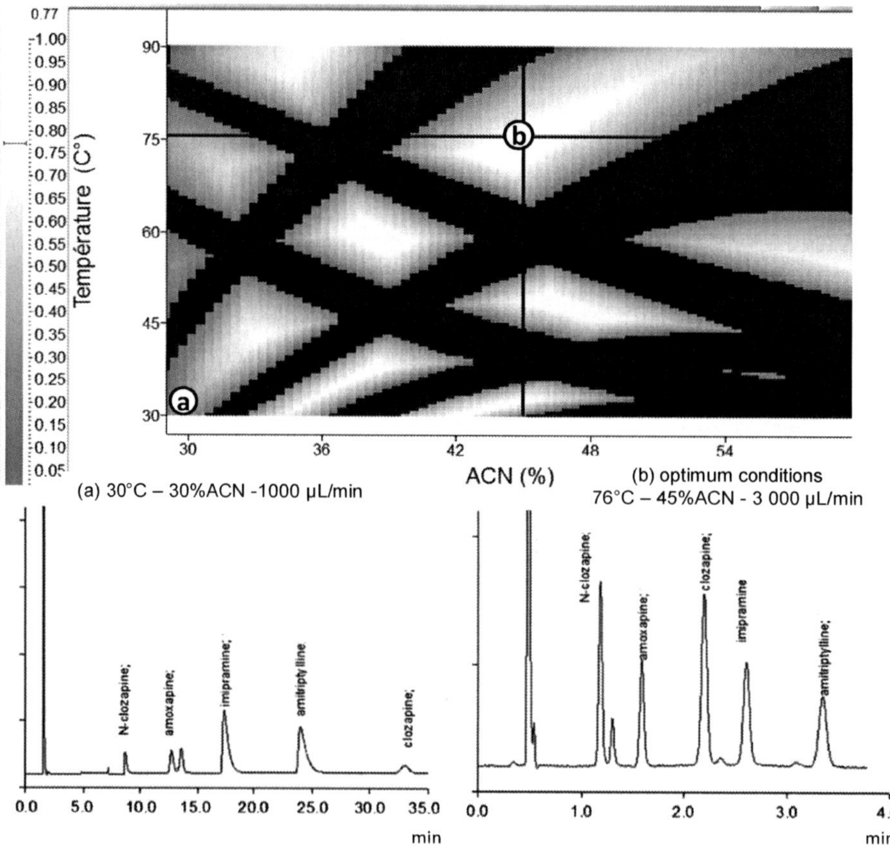

Figure 4.11 Example of optimization of both temperature and mobile phase composition for the separation of pharmaceuticals. The response surface and the response scale on the left are produced by OSIRIS software (Datalys, Grenoble, France). The presented experimental chromatograms correspond to two different set of robust conditions located in the response surface by two circles: (a) 30 °C; 30% ACN; 1000 µL min^{-1} and (b) optimum conditions 76 °C; 45% ACN; 3000 µL min^{-1}. Column: RPXterra C18 150 × 4.6mm; 5 µm (from reference 44 with permission).

resolution between each pair of peaks, the analysis time and the robustness of the method. It varies from 0 (black areas) to 1 (white areas) with a zero value when one at least among the three criteria has not reached the threshold value set by the chromatographer. The response surface displays many light areas indicating that different possibilities of robust conditions exist and suggesting many retention order reversals. Two experimental chromatograms corresponding to two sets of conditions are given in Figure 4.11. In addition to a significant reduction of peak tailing, the optimum conditions led to a gain in analysis time of approximately 10 compared to the best conditions at 30 °C.

4.5 HT-UHPLC in Comprehensive On-Line Two-Dimensional Liquid Chromatography (LC × LC)

4.5.1 Limits of One-Dimensional Liquid Chromatography: Advantage of Comprehensive Two-Dimensional Liquid Chromatography

In the field of pharmaceutical, biological or environmental analysis, the available samples are sometimes very complex with an impressive number of components to be separated. As a result, such separations require very high peak capacities that cannot be attained in one-dimensional liquid chromatography, even if one accepts a very long analysis time. Considering eqn (4.4), a rapid calculation shows that, with a very large gradient time $t_G = 50t_0$, the theoretical peak capacity that can be attained is close to $2\sqrt{N}$ for small molecules (typically with $S \times \Delta C = 3$) and to $5\sqrt{N}$ for larger ones (typically with $S \times \Delta C = 10$). Actually, higher peak capacities could theoretically be attained but it should be achieved at the expense of a dramatic increase in the gradient time. This is also pointed out in Figure 4.3 in case of UHPLC or HT-UHPLC conditions. The gradient kinetic plots show that 1000 min (17 h) are necessary to attain peak capacities of approximately 700 for small and 1700 for larger compounds by using a 60 cm long column packed with sub-2 μm particles, thereby leading to approximately 100 000 plates (see Figure 4.2). The main parameters that limit the peak capacity are the affordable column backpressure, the maximum column length and the acceptable analysis time. One million plates have been obtained by using 12 m capillary silica-based monolithic columns.[57] However, the column dead time was close to 2.5 h and, even with this huge plate number, the theoretical peak capacity could just reach 2000 (*i.e.* $2\sqrt{N}$) with a separation lasting 5 days.

Comprehensive 2D liquid chromatography (LC × LC) is a powerful technique for the separation of complex mixtures. In proteomic research, LC × LC can be coupled to MS detection to further enhance the resolving power, resulting in a very large number of identified peptides.[58,59] In addition to the analysis of complex samples, tracking impurities in pharmaceuticals is also an attractive possible use of LC × LC.

In LC × LC, the total peak capacity is theoretically the product of peak capacities in each dimension ($n_{c,total} = n_{c1} \times n_{c2}$).[60] This multiplicative rule is true as long as two conditions are fulfilled: (1) the peaks are eluted over the entire space defined by the two ranges of mobile phase compositions (first and second dimensions), which involves sufficient degree of orthogonality between the two chromatographic systems, and (2) the sampling rate in the first dimension is sufficient so that the resolution obtained in the first dimension is not lost in the second. In such favourable conditions, with a rather small peak capacity of 50 in each dimension, the total peak capacity should be 2500 in LC × LC, which represents a much higher value than the best expected one in one-dimensional LC.

4.5.2 Requirement for HT-UHPLC in the Second Dimension

In comprehensive on-line LC × LC, fractions from the first dimension are collected in a sample loop and continuously introduce in the second dimension *via* a ten-port switching valve equipped with two identical loops. The on-line approach is more attractive than the off-line one because it avoids sample contamination as well as possible loss or degradation during sample handling. Furthermore, it reduces the total analysis time and hence leads to faster 2D separations. It has been shown that there should be at least three sampling per first-dimensional peak to obtain the highest two-dimensional resolution.[61] Accordingly, the analysis time in the first dimension is given by:

$$t_1 = 3 \times n_{c,1} \times t_2 \tag{4.15}$$

where t_1 and t_2 are the first- and second-dimensional analysis times respectively and $n_{c,1}$ is the peak capacity in the first dimension. Eqn (4.15) clearly shows that ultra-fast separations in the second dimension (short t_2) are of prime importance to reduce t_1, which represents the total analysis time of the 2D separation.

The peak capacity in the second dimension becomes the limited parameter, and it is of prime importance to keep it as high as possible and hence to maintain the gradient steepness low enough. For example, as can be observed in Figure 4.5, b should be lower than 3 to expect a peak capacity higher than 50. As a result, considering a typical value of 3 for the product $S\Delta C$, the ratio of the gradient time to the column dead time should be higher than unity ($t_{G,2}/t_{0,2} > 1$). This ratio is given by:

$$\frac{t_{G,2}}{t_{0.2}} = \frac{t_2}{t_{0.2}} - \left[\frac{V_{D,2}}{V_{0.2}} + (x + 1)\right] \tag{4.16}$$

where $V_{D,2}$ is the second instrument dwell volume (typically close to the column dead volume) and x is the number of column volumes used for column re-equilibration, which has to be as small as possible. It was shown that two column volumes ($x = 2$) are usually sufficient with charged compounds to have a good run-to-run repeatability,[62] which means that $t_2/t_{0,2}$ should be higher than 5. t_2 also corresponds to the time required to fill the loop ($t_2 = V_{i,2}/F_1$ where $V_{i,2}$ is the injection volume in the second dimension and F_1 is the flow rate in the first dimension). $V_{i,2}$ is a fraction, f_2, of the column dead volume ($V_{i,2} = f_2 \times V_{0,2}$). Thus eqn (4.16) can also be written as:

$$\frac{t_{G,2}}{t_{0.2}} = f_2 \times \frac{F_2}{F_1} - \left[\frac{V_{D,2}}{V_{0.2}} + (x+1)\right] \tag{4.17}$$

It was recently shown that, for RPLC × RPLC separations,[63] f_2 should not be higher than 0.1 in order to keep more than 70% of the column plates with the injection process. For RPLC × HILIC, f_2 should be much lower due to great difference in eluent strength between the two mobile phases.[64,65] Thus, if the goal is $t_{G,2}/t_{0,2} > 1$, F_2/F_1 has to be as high as possible with a minimum value of 50 in RPLC × RPLC. Assuming an equal porosity between the two columns, the ratio of the flow-rates can be expressed as:

$$\frac{\mathrm{F}_2}{\mathrm{F}_1} = \left(\frac{\mathrm{d}_{i,2}}{\mathrm{d}_{i,1}}\right)^2 \times \frac{\mathrm{u}_2}{\mathrm{u}_1} \tag{4.18}$$

The i.d. of the second dimension must be larger than that of the first one. However, the difference should not be too high because of solute dilution. Typically, a ratio of 2 between the two column i.d. is a good compromise (*e.g.* 1 mm and 2 mm in the first and second dimensions respectively). Accordingly, for $t_{G,2}/t_{0,2} > 1$, the minimum value for the ratio of the linear velocities, u_2/u_1 is 12 in RPLC × RPLC and probably much higher in RPLC × HILIC. To achieve that, u_1 and u_2 are usually located in the B-term and in the C-term dominated region of the Van Deemter curves respectively. Plots of simulated peak capacity in the second dimension *versus* flow rate are shown in Figure 4.12 for conventional HPLC, HTLC, UHPLC and HT-UHPLC using 1 mm and 2.1 mm column i.d. in the first and the second dimensions respectively. The calculations were carried out with a flow rate of 10 μL min^{-1} in the first dimension, 50 mm column length in the second dimension and $f_2 = 0.1$. Such conditions lead to an analysis time of approximately 30s. Negligible dispersion caused by the injection process was assumed. As shown, peak capacity increases with F_2 up to a maximum value, depending on the maximum pressure. The maximum values that can be provided by the four techniques are 45, 70, 100 and 170 respectively, thereby showing the great advantage of HT-UHPLC in the second dimension.[66] Unfortunately, the flow rate is often limited to 2 mL min^{-1} on current UHPLC instruments, and, thus, the maximum peak capacity is lower than expected (155 instead of 170 as shown in Figure 4.12). Despite this current instrument limitation, HT-UHPLC is strongly recommended in the second dimension in order to limit the total analysis time and increase peak capacity.

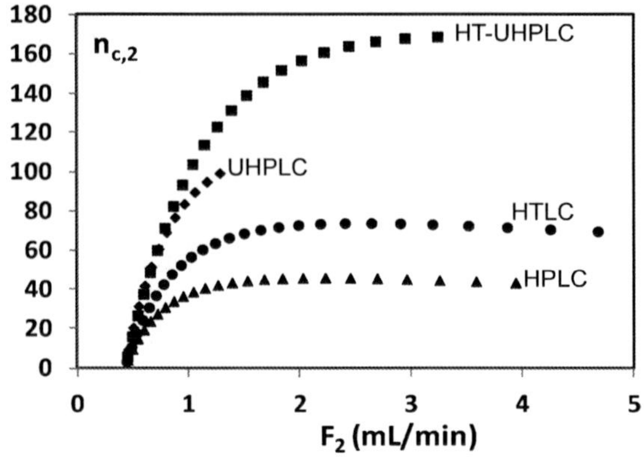

Figure 4.12 Interest of HT-UHPLC in the second dimension. Simulated peak capacity as a function of flow-rate in the second dimension with 50 × 2.1mm column in (▲) conventional HPLC (20 °C, d_p = 5 μm, ΔP_{max} = 350 bar); (●) HTLC (90 °C, d_p = 5 μm, ΔP_{max} = 350 bar); (♦) UHPLC (20 °C, d_p = 1.9 μm, ΔP_{max} = 800 bar) and (■) HT-UHPLC (90 °C, d_p = 1.9 μm, ΔP_{max} = 800 bar). F1= 10 μL min^{-1}; dwell volume = 120 μL; two column volumes for re-equilibration (× = 2); other parameters are the same as in Figure 4.2.

4.5.3 Application to the Separation of a Tryptic Digest of Proteins

HT-UHPLC in the second dimension was applied to the on-line LC × LC separation of a tryptic digest of three proteins[66] with RPLC as first dimension, and RPLC and HILIC as second dimension. RPLC × RPLC is attractive for charged compounds because: (1) the mobile phases are highly compatible; (2) peak shapes are usually correct; and (3) the degree of orthogonality is satisfactory provided that the mobile phase pH is different between the two separations. Although mobile phases are less compatible in RPLC × HILIC, the use of HILIC in the second dimension may be attractive because of the high degree of orthogonality between RPLC and HILIC. Furthermore, the ESI-MS detection of peptides is expected to be more sensitive in HILIC conditions due to the high percentage of acetonitrile in the mobile phase (> 70%) compared to RPLC conditions (usually <40% for peptides). Finally, a high percentage of acetonitrile allows higher flow rate in the second dimension and hence higher F_2/F_1 values. The two 2D separations are shown in Figure 4.13. In RPLC × RPLC (Figure 4.13a), peak volumes are very small with measured average peak width (4σ) of approximately 0.5 s in the second dimension. Consequently, more than 100 resolved peaks can be counted in the RPLC × RPLC chromatogram, whereas the expected number of peptides after the tryptic digestion was in the order of 130. As shown by the very short analysis time in the second dimension (30 s), the use of HT-UHPLC made it possible to obtain fast RPLC × RPLC separations (200 min) while keeping an

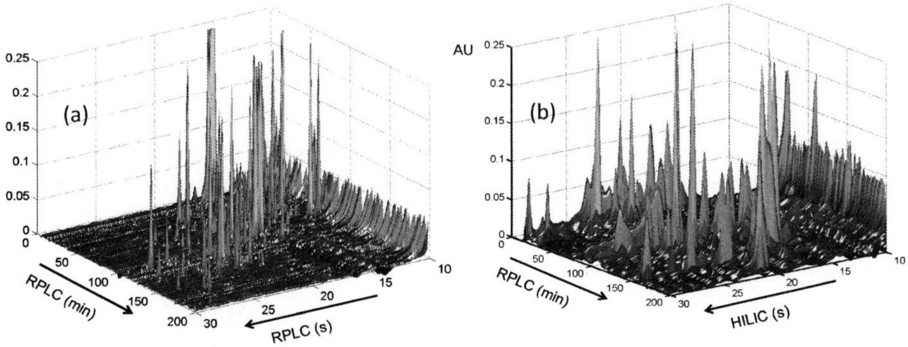

Figure 4.13 LC × LC separations of a tryptic digest of three proteins (BSA, lysozyme and myoglobin) with HT-UHPLC in the second dimension. First dimension RPLC: Hypersil Gold 50 × 1 mm, 3 μm; flow-rate, 10 μL min^{-1}; buffer A, 10 mM ammonium acetate; solvent B, acetonitrile; 2–30% B in 200 min; 30 °C; 5 μL injected. Second dimension: (a) RPLC Kinetex XB-C18 50 × 2.1 mm, 2.6 μm; 1800 μL min^{-1}; solvent A, water with 0.1% formic acid; solvent B, acetonitrile with 0.1% formic acid; 1–35% B in 30 s: 60 °C; 750 bar at 20% acetonitrile; (b) HILIC Acquity-BEH 50 × 2.1 mm, 1.7 μm; 1800 μL min^{-1}; solvent A, acetonitrile; buffer B, 10 mM ammonium acetate; 1–31% B in 24 s; 80 °C; 700 bar at 70% acetonitrile; detection at 210 nm (adapted from reference 65).

impressive peak capacity. In RPLC × HILIC (Figure 4.13b), for similar retention times in both dimensions, the number of resolved peaks is less impressive (<50). Large peak widths can be observed in the second dimension, some peaks exhibiting severe tailing. As previously mentioned, the worse results in RPLC × HILIC can be explained by the significant difference in eluent strength between injection solvent and solvent at elution. However, it is noteworthy that the peaks cover the entire separation space, suggesting that the two separations are quite orthogonal. As a result, this two-dimensional system is potentially very interesting but it requires additional studies to improve peak shapes and hence to increase peak capacity.

4.6 Conclusions

The basics of HTLC and its practical aspects have been covered in this Chapter. Medium temperatures (up to 100 °C) combined with UHPLC conditions are very attractive for the separation of pharmaceutical and biological compounds. Both techniques are fully complementary. Higher temperature offer faster separations and peak shape improvement for charged and polar compounds. The use of temperature as a convenient parameter to vary band spacing in pharmaceutical analysis has also been highlighted. Efficient equipment, including UHPLC instruments and stable silica-based stationary phases, are currently commercially available for HT-UHPLC separations at temperature below 100 °C. Future development of HT-UHPLC instruments should be achieved to limit extra-column volumes and hence to reduce the contribution of extra-column band broadening with micro-bore columns. The development of stable sub-2 μm stationary phases, as well as advances in instrumentation, are necessary before working at very high temperatures (>100 °C) in UHPLC. However, one should be aware that when very high temperatures are used, a balance must be struck between the benefits and the risks that include column life reduction, decrease in retention, increase in pressure and possible analyte degradation.

Finally, HT-UHPLC is of great interest to speed up the second separation in on-line 2DLC. The two LC × LC separations of peptides, presented in this Chapter, clearly show that LC × LC is a powerful technique for biological samples and that HT-UHPLC in the second dimension can lead to impressive peak capacities in a rather short analysis time.

References

1. W. R. Melander and Cs. Horvath, in *High-Performance Liquid Chromatography, Advances and Perspectives*, ed. Cs Horvath, vol. 2, Academic Press, New York, 1980, p. 113.
2. E. Grushka, *Anal. Chem.*, 1970, **42**, 1142.
3. J. W. Dolan, *J. Chromatogr. A*, 2002, **965**, 195.
4. T. Greibrokk and T. Andersen, *J. Chromatogr. A*, 2003, **1000**, 743.
5. G. Vanhoenacker and P. Sandra, *J. Sep. Sci.*, 2006, **29**, 1822.
6. S. Heinisch and J-L. Rocca, *J. Chromatogr. A*, 2009, **1216**, 642.
7. D. Guillarme and S. Heinisch, *Separ. Purif. Rev.*, 2005, **34**(2), 181.
8. Z. Hao, B. Xiao and N. Weng, *J. Sep. Sci.* 2008, **31**, 1449.
9. D. R. Stoll, X. Li, X. Wang, P. W. Carr, S. E. G. Porter and S. C. Rutan, *J. Chromatogr. A*, 2007, **1168**, 3.
10. Y. Yang, *J. Sep. Sci.*, 2007, **30**, 1131.
11. R. M. Smith, *J. Chromatogr. A*, 2007, **1184,** 441.
12. H. Chen and Cs. Horvath, *Anal. Meth. Instrument.*, 1993, **1**, 213.
13. F. D. Antia and Cs. Horvath, *J. Chromatogr.*, 1988, **435**, 1.
14. H. Colin, J. D. Diez-Masa, T. Czajkowska, I. Miedziak and G. Guiochon, *J. Chromatogr.*, 1978, **167**, 41.
15. J. D. Thompson and P. W. Carr, *Anal. Chem.*, 2002, **74**, 4150.
16. D. Guillarme, S. Heinisch and J. L. Rocca, *J. Chromatogr. A*, 2004, **1052**, 39.
17. J. H. Knox and M. Saleem, *J. Chromatogr. Sci.*, 1969, **7**, 745.
18. G. J. Kennedy and J. H. Knox, *J. Chromatogr. Sci.*, 1972, **10**, 549.
19. J. H. Knox and G. Vasvari, *J. Chromatogr.*, 1973, **83**, 181.
20. S. Heinisch, G. Desmet, D. Clicq and J-L. Rocca, *J. Chromatogr. A*, 2008, **1203**, 124.
21. C. R. Wilke and P. Chang, *Am. Inst. Chem. Eng. J.*, 1955, **1**, 264.
22. M. A. Stadalius, H. S. Gold and L. R. Snyder, *J. Chromatogr.*, 1984, **296**, 31.

23. J. C. Giddings, *Anal. Chem.,* 1967, **39,** 1027.
24. U. W. Neue, *J. Chromatogr. A*, 2005, **1079**, 153.
25. L. R. Snyder and J. W. Dolan, *High-Performance Gradient Elution*, Wiley-Interscience, Hoboken, 2007.
26. L. R. Snyder, J. W. Dolan and J. R. Gant, *J. Chromatogr.*, 1979, **165**, 3.
27. J. C. Giddings, *Anal. Chem.*, 1965, **37,** 60.
28. H. Poppe, *J. Chromatogr. A*, 1997, **778**, 3.
29. G. Desmet, D. Clicq and P. Gzil, *Anal. Chem.*, 2005, **77**, 4058.
30. K. Broeckhoven, D. Cabooter, F. Lynen, P. Sandra and G. Desmet, *J. Chromatogr. A*, 2010, **1217**, 2787.
31. S. Heinisch, A. D'Attoma and C. Grivel, *J. Chromatogr. A*, 2012, **1228**, 135–147.
32. H. A. Claessens and M. A. van Straten, *J. Chromatogr. A*, 2004, **1060**, 23.
33. E. Grata, D. Guillarme, G. Glauser, J. Boccard, P-A. Carrupt, J-L. Veuthey, S. Rudaz and J-L. Wolfender, *J. Chromatogr. A*, 2009, **1216**, 5660.
34. G. Vanhoenacker, A. Dos Santos Pereira, T. Kotsukab, D. Cabooter, G. Desmet and P. Sandra, *J. Chromatogr. A*, 2010, **1217**, 3217.
35. F. D. Antia and Cs. Horvath, *J. Chromatogr.*, 1988, **435**, 1 .
36. H. G. Gika, G. Theodoridis, J. Extance, A. M. Edge and I. D.Wilson, *J. Chromatogr. B*, 2008, **871**, 279.
37. H. Poppe, J. C. Kraak, J. F. K. Huber and J. H. M. van den Berg, *Chromatographia*, 1981, **14,** 515.
38. R. G. Wolcott, J. W. Dolan, L R. Snyder, S. R.Bakalyar and M. A. Arnold, *J. Chromatogr. A*, 2000, **869,** 211.
39. J. D. Thompson, J. S. Brown and P. W. Carr, *Anal. Chem.*, 2001, **73**, 3340.
40. S. Heinisch and J-L Rocca, *Advantages and Limits of HTLC*, published online at Chromedia.nl, June 2008.
41. Y. Yang, *J. Sep. Sci.*, 2007, **30**, 1131.
42. D. V. McCalley, *J. Chromatogr. A*, 2000, **902**, 311.
43. T. Greibrokk and T. Andersen, *J. Sep. Sci.*, 2001, **24**, 899.
44. S. Heinisch, G. Puy, M-P. Barrioulet and J-L. Rocca, *J. Chromatogr. A*, 2006, **1118**, 234.
45. Y. Mao and P. W. Carr, *Anal. Chem.*, 2000, **72**, 2788 .
46. D. Guillarme, S. Heinisch, *Spectra Anal.*, 2010, **274**, 24.
47. J. P. Foley and J. G. Dorsey, *Anal. Chem.,* 1983, **55**, 730.
48. J. W. Dolan, L. R. Snyder, N. M. Djordjevic, D. W. Hill and T. J. Waeghe, *J. Chromatogr. A*, 1999, **857**, 1.
49. S. Heinisch and J-L. Rocca, *J. Chromatogr. A,* 2004, **1048**, 183.
50. R. G. Bates, *Determination of pH Theory and Practice*, Wiley, New York, 2nd edn, 1973.
51. W. R. Melander, J. Stoveken and Cs. Horvath, *J. Chromatogr.*, 1979, **185**, 111.
52. C. B. Castells, C. Rafols, M. Roses and E. Bosch, *J. Chromatogr. A*, 2004, **1042**, 23.

53. C. B. Castells, C. Rafols, M. Roses and E. Bosch, *J. Chromatogr. A*, 2003, **1002**, 41.
54. L. G. Gagliardi, C. B. Castells, C. Rafols, M. Roses and E. Bosch, *J. Chromatogr. A*, 2005, **1077**, 159.
55. S. Goga, S. Heinisch and J-L. Rocca, *Chromatographia,* 1998, **48**, 237.
56. M. P. Barrioulet, S. Heinisch and J-L. Rocca, *Spectra Anal.*, 2007, **36**, 38.
57. K. Miyamoto, T. Hara, H. Kobayashi, H. Morisaka, D. Tokuda, K. Horie, K. Koduki, S. Makino, O. Nunez, C. Yang, T. Kawabe, T. Ikegami, H. Takubo, Y. Ishihama and N. Tanaka, *Anal. Chem.*, 2008, **80**, 8741.
58. P. Donato, F. Cacciola, E. Sommella, C. Fanali, L. Dugo, M. Dacha, P. Campiglia, E. Novellino, P. Dugo and L. Mondello, *Anal. Chem.*, 2011, **83**, 2485.
59. S. Di Palma, P. J. Boersema, A. J. R. Heck and S. Mohammed, *Anal. Chem.*, 2011, **83**, 3440.
60. A. Van der Horst and P. J. Schoenmakers, *J. Chromatogr. A*, 2003, **1000**, 693.
61. R. E. Murphy, M. Schure and J. P. Foley, *Anal. Chem.*, 1998, **70**, 1585.
62. C. Grivel, J-L. Rocca, D. Guillarme, J-L. Veuthey and S. Heinisch, *J. Chromatogr. A*, 2010, **1217**, 459.
63. M. Martin, C. Grivel and S. Heinisch, presented at the International Symposium on Chromatography (ISC2010), Valencia, September 12–16, 2010.
64. J. Ruta, S. Rudaz, D. V. McCalley, J-L. Veuthey and D. Guillarme, *J. Chromatogr. A*, 2010, **1217**, 2230.
65. A. D'Attoma and S. Heinisch, *J. Chromatogr. A,* submitted for publication.
66. C. Grivel, A. Dechenaux and S. Heinisch, *Chromatography Today*, 2010, **Aug/Sep,** 8–11.

CHAPTER 5

Comparison of the Performance of Totally Porous and Core-shell Particles

SZABOLCS FEKETE* AND JENO¥ FEKETE

Budapest University of Technology and Economics, Department of Inorganic and Analytical Chemistry, 1111, Szt. Gellért tér 4., Budapest, Hungary
*E-mail: Szabolcs.Fekete@unige.ch

5.1 Introduction

Very fast liquid chromatographic separations are of great interest to increase the analysis throughput. The first priority for performing a fast separation is the selection of the appropriate stationary phase. Mobile phase optimization is also crucial to achieve the required selectivity and promote the short separation time, but in this Chapter we are focusing on column selection. The speed of a chromatographic separation can be increased with different approaches. One opportunity to obtain higher efficiency is to reduce the particle size of the stationary phase. The particle size is a beneficial factor as it improves the rate of mass transfer and reduces the eddy dispersion effect, resulting in smaller plate height and higher optimum linear velocity as the particle size is reduced.[1,2]

Increasing of temperature in HPLC also offers a chance to decrease the analysis time. High temperature reduces the viscosity of mobile phase and increases the mass transfer. Analysis time can be shortened without the loss of resolution through column heating.[3–6]

Another possibility to increase the speed of the separation is the reduction of the intrinsic flow resistance by increasing the external porosity and the

RSC Chromatography Monographs No. 16
UHPLC in Life Sciences
Edited by Davy Guillarme and Jean-Luc Veuthey

Published by the Royal Society of Chemistry, www.rsc.org

flow-through pore size of the packing. The monolith approach, originally initiated by the work of Hjerten *et al.*,[7] Svec and Frechet,[8] Horvath and co-workers,[12] Tanaka and co-workers,[10,11] and Huber and co-workers,[12] had already led to a number of well performing, commercially available polymeric and silica monolith columns.

In the recent development of particle technology targeted for liquid chromatography, the use of core-shell (or superficially porous, fused-core, shell) particles has received considerable attention. Core-shell particles manifest the advantages of porous and non-porous particles. The concept of superficial or shell stationary phases was introduced by Horváth and co-workers[13,14] in the late 1960s. Later, Kirkland[15] showed that 30–40 μm diameter superficially porous packing provide much faster separations compared with the large porous particles used earlier in liquid chromatography. Later on, the core diameter was reduced and the thickness of active layer was decreased to 0.5 μm and was used for fast separation of peptides and proteins.[16] Now core-shell packing materials are commercially available in various diameters (5, 2.7, 2.6 and 1.7 μm), with different shell thickness (0.5, 0.35, 0.25, 0.23 and 0.15 μm) (Table 5.1).[17–20] The thickness of the porous layer plays a major role in governing the porosity of the particles.[20] While this benefit of core-shell packing is hardly visible with small molecules, the improvement is most significant with peptide and protein molecules.

The kinetic performance of different columns is generally compared using plate height plots. There are several plate height models but the model described by van Deemter is generally the accepted one.[21] Alternative approaches are mostly based on the kinetic principles that were first expounded by Giddings.[22] Later, Poppe[23] proposed a representation in which the plate time (t_0/N) was plotted against the plate number (N). Peak capacity, a concept first described by Giddings,[24] and was put to good use by Horvath and Lipsky[25] for gradient chromatography. It is a measure of the separation power that includes the entire chromatographic space together with the variability of the peak width over the chromatogram.

This Chapter is a critical comparison of the practical possibilities and limitations of commercially available columns, packed with sub-2 μm particles, and the more recently developed sub-3 μm core-shell particles, both under isocratic and gradient conditions. Theoretical and practical efficiency of core-shell particles is discussed in details. The physical properties of sub-3 μm core-shell particles, such as particle size distribution (PSD), surface roughness, loading capacity and column packing difficulties, are also presented.

5.2 Column Performance

5.2.1 Column Performance in Isocratic Elution Mode

In isocratic elution mode, the peak dispersion is generally characterized by the theoretical plate height (H) and the number of theoretical plates (N). A more

Table 5.1 Particle structure of core-shell packings.

Vendor	*Column/product name*	*Average particle diameter (μm)*	*Shell thickness (μm)*	*Stationary phase chemistry*
Agilent	Poroshell 300	5	0.25	C18, C8, C3
Agilent	Poroshell 120	2.7	0.50	EC-C18, SB-C18
Advanced Material Technology	Halo	2.7	0.50	C18, C8, HILIC, RP-amide, phenylhexyl, pentafluorophenyl
Advanced Material Technology	Halo Peptide-ES 160 Å	2.7	0.50	C18
Sigma–Aldrich	Ascentis Express	2.7	0.50	C18, C8, HILIC, RP-amide, phenylhexyl, pentafluorophenyl
Sigma–Aldrich	Ascentis Express Peptide-ES 160 Å	2.7	0.50	C18
Phenomenex	Kinetex	2.6 1.7	0.35 0.23	C18, XB-C18, C8, HILIC, pentafluorophenyl
Thermo Scientific	Accucore	2.6	0.50	C18, aQ, RP-MS, HILIC, phenylhexyl, pentafluorophenyl
Macherey–Nagel	Nucleoshell	2.7	0.50	RP-18, HILIC
Sunniest	SunShell	2.6	0.50	C18
PerkinElmer	BrownLee SPP	2.7	0.50	C18, C8, HILIC, RP-amide, phenylhexyl, pentafluorophenyl, Peptide ES C18
Commercially not available	Eiroshell	1.7 1.7 1.7	0.35 0.25 0.15	C18

detailed diagnosis of the kinetic performance is based on the dependency of the plate height (H) as the function of the linear velocity (u) of the mobile phase.[26] The treatment of the mass transfer processes and the distribution equilibrium between the mobile and stationary phase in a column led to equations which link the theoretical plate height as the decisive column performance parameter to the properties of the LC systems such as the linear velocity of eluent, the diffusion coefficient of analyte, the retention coefficient of analyte, column porosity, *etc.*

First van Deemter proposed an equation, which described the column performance as a function of the linear velocity for a packed column in gas chromatography coated with a stationary liquid layer.[27] Since then, several plate height equations were derived for LC by numerous researchers. The most accepted equations were introduced by Giddings,[28] Snyder and Kirkland,[29] Huber and Hulsman,[30] Kennedy and Knox,[31] Horvath and Lin,[32] and Yang *et al.*;[33] however, the H *versus* u plots are generally referred as van Deemter plots.

Knox suggested a three-term equation to describe the dependency of the theoretical plate height of a column as a function of linear velocity:

$$H = \frac{L}{N} = Au^{1/3} + \frac{B}{u} + Cu \tag{5.1}$$

where A, B and C are constants, determined by the magnitude of band broadening due to eddy dispersion, longitudinal diffusion and resistance to mass transfer, respectively.[34,35] The constant A depends on the quality of the column packing. The B and C terms of the plate height equation depend on the analyte retention. The B term increases with analyte retention as more time is available for diffusion to take place in the stationary phase (surface diffusion). The C term expresses the resistance to mass transfer.

The application of reduced parameters is common in chemical engineering to compare the performance of columns in unit operations. It is useful to convert H and u into dimensionless parameters according to the following simple formulae:

$$h = \frac{H}{d_p} \tag{5.2}$$

$$\nu = \frac{ud_p}{D_M} \tag{5.3}$$

where h is the reduced plate height, d_p is the particle size of the column packing material, ν is the reduced linear velocity and D_M is the analyte diffusion coefficient. The particular advantage of this approach is the ability to compare the performance of columns packed with particles of different sizes or structures.

van Deemter and h–ν plots lack permeability considerations. Alternative approaches such as kinetic plots are useful tools for visualizing the compromise between separation speed and efficiency.[23] It is very straightforward to map the kinetic performance potential of a given chromatographic support type by taking a representative set of van Deemter curve data and re-plotting them as H^2/K_{v0} *versus* $K_{v0}/(uH)$ instead of as H *versus* u (K_{v0} is the unretained component-based column permeability). Multiplying both quantities with the same proportionality constant (being the ratio of the available pressure drop, ΔP, and the mobile phase viscosity, η), the obtained values correspond directly to the minimal t_0 time needed in a column taken exactly long enough to yield a given number of theoretical plates (N) at the available pressure drop. N and t_0 can be calculated according to the following equations, which have been introduced by Desmet *et al.*:[36]

$$N = \frac{P}{\eta}\left(\frac{K_{V0}}{u \cdot H}\right) \tag{5.4}$$

$$t_0 = \frac{P}{\eta}\left(\frac{K_{V0}}{u^2}\right) \tag{5.5}$$

For the construction of kinetic plots, certain defining experimental parameters are used, including the maximum operating pressure ($\Delta P_{\max}$), the column reference length and flow resistance or permeability (K_{v0}), temperature, mobile phase viscosity (η) and the diffusion coefficient of the analyte in the mobile phase (D_M). Column permeability can be determined experimentally using the following relation:

$$K_{V,0} = \frac{u\eta L}{P} \tag{5.6}$$

in which ΔP is the pressure drop over the column with length (L) and u is the linear velocity.

In the thus obtained plot, each data point corresponds to a column with a different length, *i.e.*, the length that yields the maximally allowable pressure drop $\Delta P_{\max}$ for the value of u under consideration. This is automatically the fastest way to achieve the corresponding number of plates.

The hydrodynamic properties expressed by the column pressure *versus* flow dependency provide an insight into the flow behaviour. From these data, the column permeability can be calculated. The Darcy law (for laminar flow) expresses that the back pressure of a column increases in inverse proportion to the square of the particle diameter:

$$P = \frac{\phi \eta L u}{d_p^2} \tag{5.7}$$

in which ϕ is the flow resistance of the packed bed.

Viscosity values can be calculated using equations derived by Chen and Horváth,[37] and analyte diffusion coefficients can be obtained generally by the Wilke–Chang equation:[38]

$$D_m = 7.4 \cdot 10^{-8} \frac{(M_S)^{0.5} \cdot T}{\eta \cdot V_A^{0.6}} \tag{5.8}$$

where Ψ is the solvent association factor, M_S is the molecular mass of the mobile phase (g mol^{-1}), η is its viscosity, T is the temperature and V_A is the molar volume of the solute at its boiling point (cm^3 mol^{-1}).

5.2.2 Efficiency in Gradient Elution Mode

There are a number of objective measures of chromatographic performance, for example, peak efficiency, resolution and peak capacity. Of these, resolution and peak capacity are relevant to gradient systems. Peak capacity was first introduced by Giddings and later Horvath applied this concept for gradient chromatographic separations.[25,39] It is a measure of the separation power that includes the entire chromatographic space together with the variability of the peak width over the chromatogram. For samples, which contain many components, peak capacity is a useful measure of the comparative separating power of different columns. In gradient separation, peak capacity is a function of column efficiency, gradient time, flow-rate and analyte/mobile phase characteristics.

The general expression of the theoretical peak capacity P_c in liquid chromatography, assuming a resolution of unity between the successively eluted peaks can be written as:[40]

$$P_C = 1 + \int_{t_I}^{t_F} \frac{1}{4\sigma} dt \tag{5.9}$$

where t_I is the retention time of the first eluted peak, t_F is the retention time of the last eluted peak, dt is a dummy time variable and σ is the time standard deviation of a peak. For the practical comparison of the efficiency of different columns in gradient elution mode, two experimental formulas are often used. The conditional peak capacity is directly related to the average peak resolution and is computed from experimental data as:[41]

$$n_C = \frac{t_{R,n} - t_{R,1}}{w} \tag{5.10}$$

where $t_{R,n}$ and $t_{R,1}$ are the retention times of the last and the first eluting peaks, and w is the average 4σ peak width. Here n_c is called the conditional peak capacity, because it depends strongly on all of the experimental conditions of the gradient elution program including temperature, flow-rate, initial and final mobile phase compositions, as well as the column parameters and the properties of the sample.[42] If the peak width pattern over a chromatogram is similar, as it is in most reversed-phase gradient separations, one can use the following popular formula of experimental peak capacity:[43–45]

$$P_C = 1 + \frac{t_g}{\bar{w}} \tag{5.11}$$

where $\bar{w}$ is the average peak width (at 4σ) and t_g is the gradient run time. The procedure of eqn (5.11) is also applicable for implementation in generic methods, where the gradient run time is a predetermined value.[46] By analogy to Knox's separation impedance concept, similar representation of efficiency can be constructed by calculating the peak capacity per unit time and per unit pressure values, according to the following formula:[47,48]

$$PTP = \frac{P_c}{t_g \cdot P} \quad (5.12)$$

where *PTP* is the peak capacity per unit time and per unit pressure value, t_g is the gradient run time and ΔP is the column pressure drop. Therefore *PTP* is proportional to the achievable separation speed.

5.3 Possibilities of Recent Core-shell Technology

The initial intend of applying pellicular particles was the separation of macromolecules. The rationale behind this concept was to improve column efficiency by shortening the pathways that analyte molecules must travel and, so doing, to improve their mass transfer kinetics.[49] Recently, the pressing needs to improve analytical throughputs forced particle manufacturers to find a better compromise between the demands for higher column efficiency that require short diffusion paths of analyte molecules in columns and the need for columns that can be operated with the conventional instruments for liquid chromatography, which operate with moderate column back-pressures.[49]

Core-shell particles are made of a solid, non-porous core surrounded by a shell of a porous material that has properties similar to those of the fully porous materials conventionally used in HPLC. It is expected that the axial and eddy dispersion contributions to the efficiency of columns packed with these particles would correspond to the external diameter of the particle, but the internal mass transfer resistances would correspond to the thickness of the porous layer.

Different plate height models are written as the sum of four different contributions such as (1) longitudinal diffusion, (2) eddy dispersion, (3) the external film mass transfer and (4) the trans-particle mass transfer resistance. The trans-particle mass transfer resistance for shell particles was derived by Kaczmarski and Guiochon.[50] According to this theory, the intra-particle diffusivity depends on the ratio (ρ) of the diameter of the solid core to that of the particle in a core-shell particle (Figure 5.1). As this ratio increases, the mass transfer kinetics become faster through the shell particles compared to through totally porous particles.

The reduced longitudinal diffusion term (h_{long}) can be written by the following equation:

$$h_{long} = 2\frac{\gamma_e + (1 - \varepsilon_e)\dfrac{1 - \rho^3}{\varepsilon_e}}{\nu} \quad (5.13)$$

where γ_e is the obstruction factor for diffusion in the inter-particle volume, ε_e is the interstitial porosity, Ω is the ratio of the intra-particle diffusivity of the

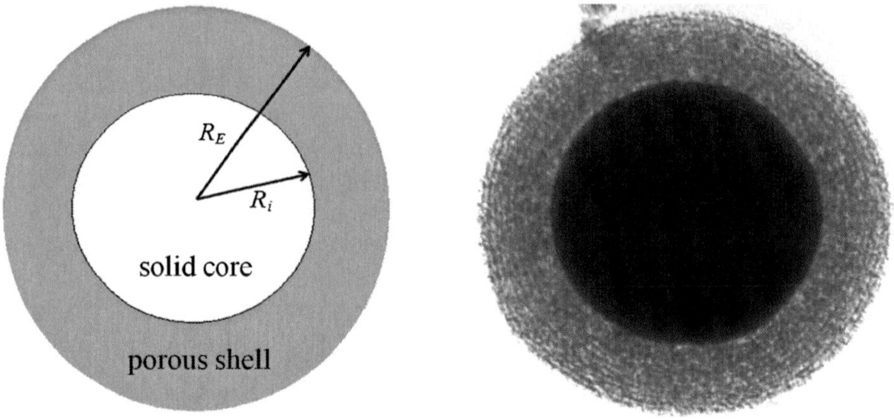

Figure 5.1 Schematic structure of a core-shell particle (left-hand side) and a SEM image (right-hand side) (from reference 20 with permission).

sample through the porous shell (D_{shell}) to the bulk diffusion coefficient, and $\rho = R_i/R_e$ is the ratio of the diameter of the solid core (R_i) to that of the particle (R_e).

The eddy dispersion term (h_{eddy}) comprises sources of four different origins, differing in the length scale considered, *e.g.*, the trans-channel ($i = 1$), the short-range inter-channel ($i = 2$), the long-range inter-channel ($i = 3$) and the trans-column flow heterogeneities ($i = 4$). For the eddy dispersion a general expression is given by:[51,52]

$$h_{eddy} = \sum_{i=1}^{i=4} \frac{1}{\frac{1}{2\lambda_i} + \frac{1}{\omega_i v}} \cong 2\sum_{i=1}^{i=4} \lambda_i \tag{5.14}$$

The values of λ_1–λ_3 were estimated by Giddings.[51] The value of λ_4 can be derived from the flow distribution across the column diameter. For quadratic flow profile distributions, the following expression was derived as:[52]

$$\lambda_4 = \frac{2}{45}\frac{L}{d_p}\omega_{\beta,c}^2 \tag{5.15}$$

Where $\omega_{\beta,c}$ is the relative flow velocity difference between the centre and the wall of the column and L is the column length.

The external film mass transfer term (h_{film}) was derived from the Laplace transform of the general rate model equations.[53] It can be expressed as:

$$h_{film} = \frac{\varepsilon_e}{1 - \varepsilon_e}\frac{k_1^2}{(1 + k_1)^2}\frac{1}{3Sh}v \tag{5.16}$$

where $Sh = (k_f\, d_p/D_m)$ is the Sherwood number, k_f is the film mass transfer coefficient and k_1 is given for superficially porous particles by the next formula:[50]

$$k_1 = \frac{1 - \varepsilon_e}{\varepsilon_e}\left(\varepsilon_p + \frac{1 - \varepsilon_p(1 - \rho^3)}{1 - \rho^3} K_a\right)(1 - \rho^3) \tag{5.17}$$

where ε_p is the porosity of the porous shell of the particle and K_a is the Henry's constant of adsorption on the walls of the porous shells.

The trans-particle mass transfer resistance ($h_{particle}$) is given by the following equation:[50]

$$h_{particle} = \frac{\varepsilon_e}{1 - \varepsilon_e}\frac{k_1^2}{(1 + k_1)^2}\frac{1}{30}\frac{1 + 2\rho + 3\rho^2 - \rho^3 - 5\rho^4}{(1 + \rho + \rho^2)^2}\nu \tag{5.18}$$

This equation is consistent with the one applied for totally porous particles when $\rho = 0$. As ρ increases, the apparent intra-particle diffusivity of the probe studied increases and the mass transfer kinetics become faster through the shell particles compared with through totally porous particles.

According to this theory, approximately 2.3- and 1.7-times faster intra-particle diffusivity can be expected for commercially available sub-3 μm shell packing (Kinetex and Halo/Ascentis Express/Poroshell-120, where $\rho = 0.73$ and $\rho = 0.63$ respectively), than for fully porous particles (if the particle size is assumed to be the same).

Figure 5.2 illustrates a comparison of theoretical *h*–*v* curves of porous and core-shell particles. The only difference is the solid core to core-shell ratio; all other column properties are assumed to be the same (same particle size, same quality of packing, *etc.*) when constructing these plots. It can be seen that both plate height and the slope of the *C* term are more favourable for a core-shell material than for a fully porous packing. For a well-packed column a minimum reduced plate height value of $h_{min} = 2$ can be expected, whereas core-shell columns can perform much lower plate heights. A reduced plate height minimum of $h_{min} = 1.4$–1.5 can be predicted for shell particles when $\rho = 0.63$, whereas particles with $\rho = 0.73$ should perform at a $h_{min} = 1.1$–1.2 plate height minimum.

In contrast to the theory, several different reduced plate height minimum values were reported in the last 3–4 years obtained with core-shell packing. The 5 cm long columns packed with core-shell particles can achieve a maximum plate count of N = 9000–19 000 when small molecular mass compounds are separated.[17,54–59] The reported maximum plate number of a 5 cm long core-shell column (Kinetex 1.7 μm) is $N \sim$ 19 200.[58] The 10 cm long columns provide plate numbers in the range of N = 14 000–32 000, whereas the 15 cm long columns can achieve N = 30 000–44 000.[18,19,54,60–62] The standard bore [4.6 mm internal diameter (ID)] Poroshell-120 columns provide an approximate $h_{min} = 1.4$–1.6, it is in good agreement with the theory, but the narrow bore

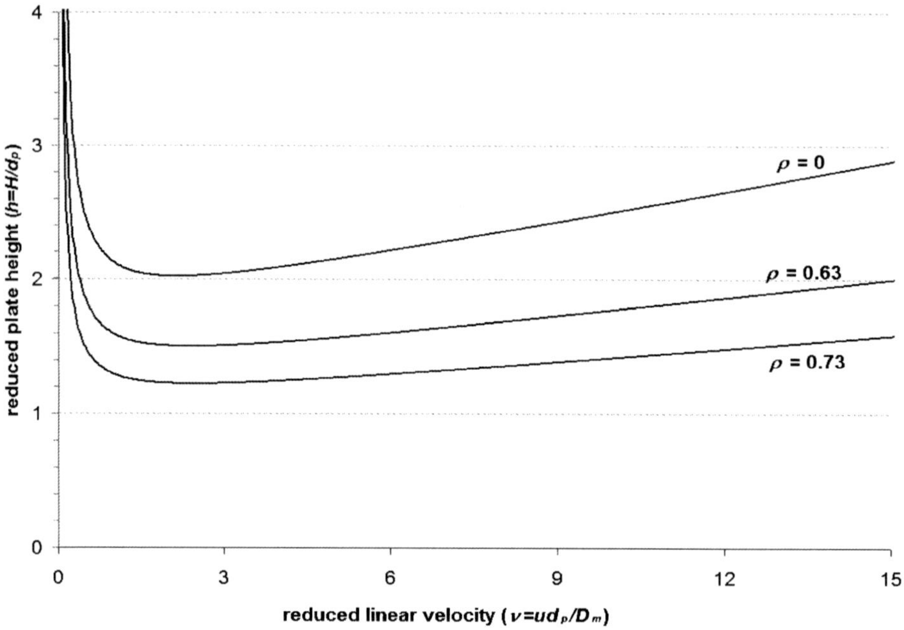

Figure 5.2 Theoretical h–v curves of fully porous and core-shell packing ($\rho = 0$, $\rho = 0.63$ and $\rho = 0.73$).

Poroshell-120 columns perform at $h_{min} \geq 2.0$. The 4.6 mm Halo columns provide $h_{min} = 1.5$–1.7 for small molecules, and 2.1 mm columns packed with this particles perform very similar reduced plate height values ($h_{min} = 1.6$–1.8). The 2.6 μm Kinetex particles achieve an $h_{min} = 1.2$–1.4 when packed in standard bore column, and an $h_{min} = 1.5$–1.9 when packed in narrow bore column.

A recent study has shown that columns packed with shell particles having the same overall particle diameter, but varied in the ratio of solid core to shell particle, strongly influences the mass transfer kinetic of the column.[20] Omamogho and Glennon[20] presented experimental results on the kinetic efficiency of core-shell particles ($d_p = 1.7$ μm). The Eiroshell™ particles were studied with different solid core to core-shell ratio: (a) 1.0 μm solid core, 0.35 μm shell thickness, (b) 1.2 μm solid core, 0.25 μm shell thickness, and (c) 1.4 μm solid core, 0.15 μm shell thickness. There was an obvious correlation between the h_{min} and shell thickness. Particles with a 0.35 μm shell thickness gave an h_{min} value of 2.5 and it was 2.2 for a 0.25 μm shell thickness, whereas the particles with the thinnest porous layer (0.15 μm shell) produced a minimum reduced plate height value of 1.9.[20]

5.4 PSD and Roughness of Core-shell Particles

Some recent studies, focused on particles with a different design such as the superficially porous particles, have suggested that particles displaying a very narrow PSD can lead to unprecedented low minimal plate heights.[63–65] It is, however, unclear whether this finding can be purely related, because there are also other factors that might influence the packing quality. Superficially porous particles have a higher density and some of them are rougher than fully porous particles.[64,66] This might also have had an influence on the achieved packing quality, apart from the PSD.

A strong (nearly linear) correlation has been observed between the width of the PSD of several commercially available HPLC particle types (both fully porous and superficially porous) and some commonly used parameters that reflect the quality of a packing, namely the minimum reduced plate height, the *A* term and the minimum-reduced separation impedance.[60] These observations have been made despite the fact that the studied particles have a number of other differences in addition to PSD, such as particle porosity, pore size, pore structure and bonding conditions. Covering a wide group of fully porous as well as porous-shell particles, these observations confirm the most recent views in the field, stating that there is a strong relation between the PSD of the packing and the quality of the packing. The observed relationship also suggests that the performance of the current generation of fully porous particle columns could be significantly improved if the PSD of these particles could be reduced. However, the original purpose of developing core-shell particles was to separate high molecular mass compounds such as proteins or polymers, which have low diffusivities, and these shell particles also show significant advantages for small molecule separations. Interestingly, it was recently shown that the enhanced performance of columns packed with shell particles in the separation of small molecules was due to the combination of lower longitudinal diffusion (*B* coefficient) and eddy diffusion (*A* coefficient) terms of the general van Deemter equation.[65]

Scanning electron microscopic (SEM) images were used in a recent study to show the roughness of the particles and to determine the particle size distribution of the core-shell type Kinetex, Ascentis Express and the totally porous Waters BEH material (Figure 5.3).[19,58] It can be evidently seen that the Kinetex and totally porous particles have smooth surface; they are much softer than those of the Ascentis Express or Halo. According to Gritti and Guiochon[57] the external roughness of the particles probably affects the column efficiency. The nature of the particle surface, its smoothness or degree of roughness considerably affects the film mass transfer kinetics, hence the column efficiency, while the film mass transfer is nearly negligible with ultra-smooth particles. If the particles have rugose surface, this may cause the formation of large stagnant pools of liquid phase surrounding the particles. The elevated *C* term of the Ascentis Express or Halo column probably can be explained by the rough surface. The Kinetex shell particles have a much softer surface than the Ascentis Express (or Halo) particles and, in accordance with this fact, the Kinetex particles provide much higher efficiency in the range of high linear velocities.

The particle size distribution is presented as the histograms (Fig.5.4.). The Kinetex particles have similar narrow particle size distribution, similar to the

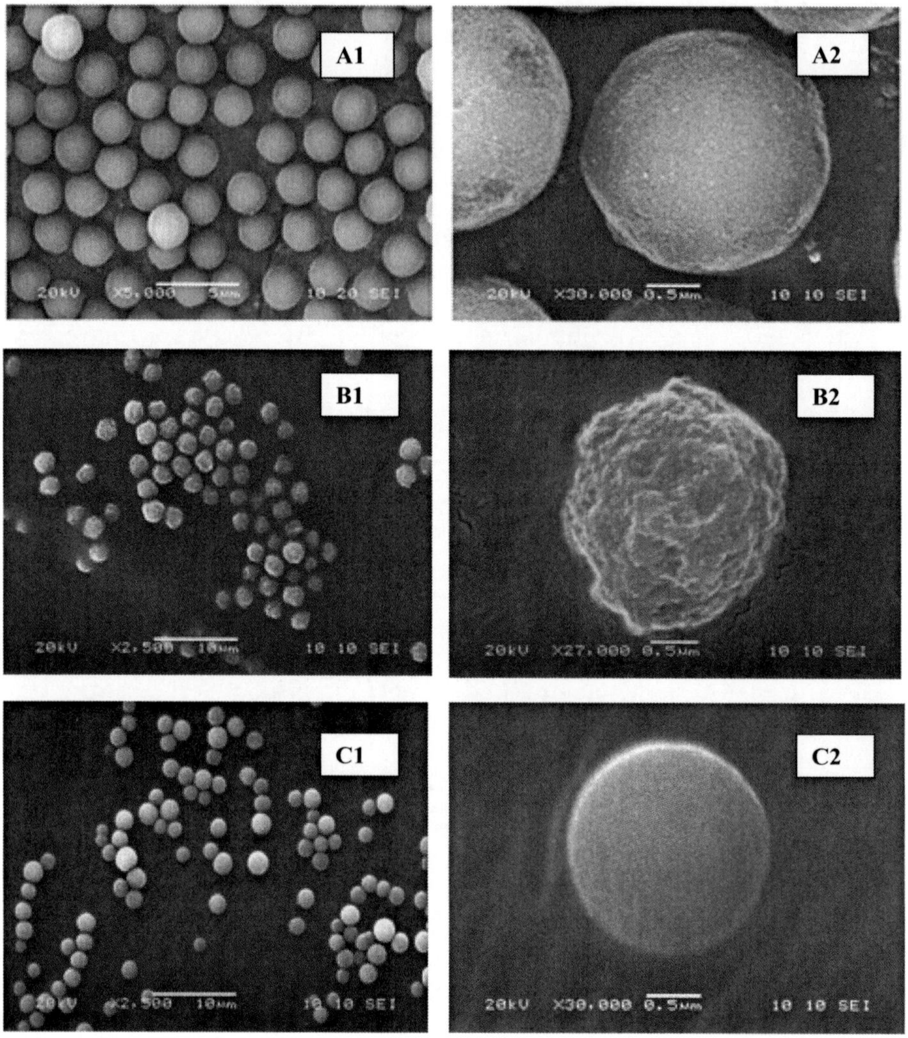

Figure 5.3 SEM pictures of Kinetex 2.6 μm core-shell particles (A1, A2), Ascentis Express 2.7 μm core-shell particles (B1, B2) and Waters BEH 1.7 μm fully porous particles (C1, C2) (from references 19 and 58 with permission).

Ascentis Express particles (Table 5.2). The totally porous sub-2 μm particles show much wider distribution (RSD > 15%).

5.5 Loading Capacity of Core-shell Particles

Core-shell particles can provide lower mass transfer resistance and higher efficiency than totally porous particles, but porous particles have greater surface areas and can provide a much higher sample loading capacity. Another

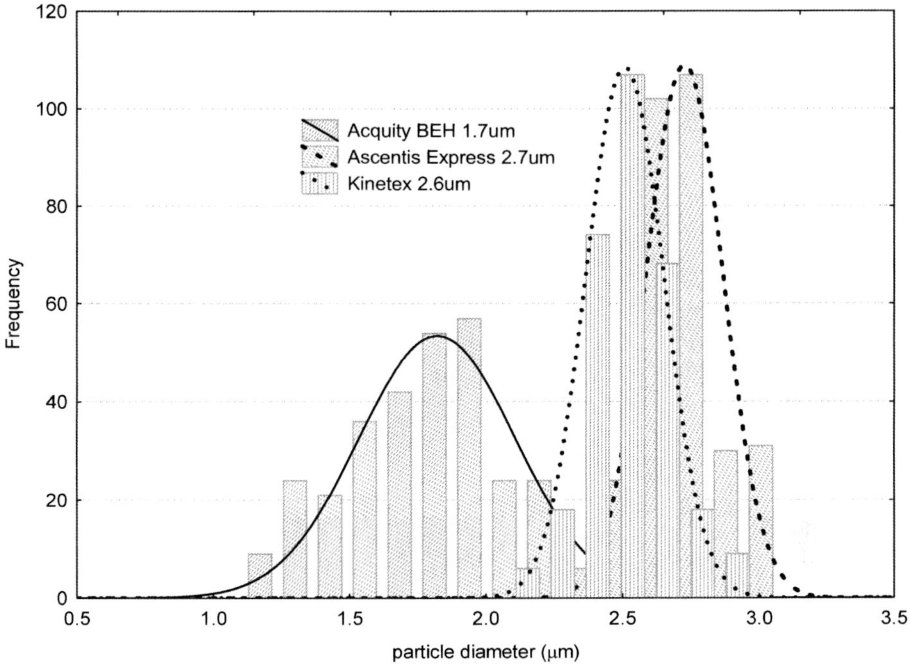

Figure 5.4 Particle size distribution of Waters UPLC BEH 1.7 μm porous particles, Ascentis Express 2.7 μm shell particles and Kinetex 2.6 μm shell particles (for reconstructing the plots, the data were taken from references 19 and 58).

Table 5.2 Particle size distribution of fully porous and core-shell particles (from ref. 58 with permission).

Material	*Type*	d_{10}	d_{90}	$d_{90/10}$	*Mean (μm)*	*RSD (%)*
2.6 μm Kinetex	Core-shell	2.32	2.68	1.15	2.50	5.94
2.7 μm Ascentis Express	Core-shell	2.56	2.96	1.16	2.72	5.30
1.7 μm Waters Acquity BEH	Porous	1.43	2.19	1.53	1.81	15.79
1.9 μm Hypersil Gold	Porous	1.50	2.26	1.51	1.95	15.18

issue is the lower retention on core-shell particles compared to totally porous ones.

Generally, the acceptable loading capacity for a packed analytical column can be estimated as:[66]

$$M_{\max}^{10\%} = \frac{2}{N}\left[\frac{k_{\infty} + 1}{k_{\infty}}\right]^2 \beta V_S \tag{5.19}$$

where $M_{max}^{10\%}$ is the maximum loading capacity for a solute for a 10% plate height increase, N is the theoretical plate number for the column, k_∞ is the retention factor at infinite dilution, β is a constant dependent of the phase ratio and type of solute, and V_s is the volume of the stationary phase for the packings. The equation shows that, for a given packed column, the sample loading capacity is proportional to the stationary phase volume. The stationary phase volume is, in turn, proportional to the carbon loading percentage and density of the packing.

The separation power of superficially porous particles increases with decreasing shell thickness if the strength of the eluent is decreased to compensate for the retention change caused by the decreased surface area of the stationary phase.[67] On the other hand, if the mobile phase remains the same, the resolution of low molecular sized compounds may decrease. The smaller the diffusivity of the solutes, then the larger the increase of the separation power compared with that of fully porous particles. However, the theory suggests that the thickness of the porous layer should be decreased drastically (down to ρ = 0.90–0.95) in order to increase the separation efficiency of the columns for large molecular sized compounds; this is not the proper option because decreasing the shell thickness and selecting particles with large ρ values decreases markedly the loadability of the column, making column overload easier, broadening the bands and decreasing the separation efficiency.[67]

A major drawback of reducing shell thickness is the reduced porosity and thus the surface area, which strongly affects sample loading capacity and analyte retention. Therefore, the optimum shell thickness in reality is likely to be a compromise between efficiency, sample loading capacity and analyte retention, and is strongly sample dependent.

The porous volume of the Halo or Ascentis Express and Poroshell-120 particles represents 75% of the total volume of the particle; in the case of Kinetex particles it gives a smaller fraction of porous material in the particle, which is approximately 61%. Overload problems are likely to be more severe for both sub-2 μm porous, as well as core-shell particles, due to the very high efficiencies produced by both types of column.[68]

Overload is a serious problem when analysing ionised acids and bases with both types of column. A recent study showed that when using volatile 20 mM ammonium formate buffers, the column efficiency was halved by the injection of sample concentrations only 85 mg L^{-1} for an ionised base and 141 mg L^{-1} for an ionised acid with a sub-2 μm column, even when using small sample volumes (1 μL for a 0.21 mm ID column).[68] For the shell phase, the values were even lower at 48 and 88 mg L^{-1} respectively when using a scaled sample volume of ~5 μL for a 0.46 mm ID column. The lower capacity of shell particles was found to be in line with the proportion of the particle that was porous (~61%).[68]

5.6 Limited Efficiency when Core-shell Particles are Packed in Narrow Bore Columns

The brands of core-shell packing materials made of fine particles are available in both conventional (4.6 mm and 3 mm ID) and narrow bore (2.1 mm ID) columns. It is a general observation that the efficiency of the former tends to be markedly higher than that of the latter. It was shown that the landmark performance of columns packed with the Kinetex 2.6 μm particles is only limited to the standard bore column (4.6 mm ID); however, when packed in a narrow bore column (2.1 mm ID), a reduced plate height of 1.9 was the minimum achieved.[19] This suggests that the packing of narrow bore columns does not provide comparable packed bed homogeneity to that of the standard bore columns. Gritti and Guiochon[55] studied the mass transfer kinetics of the Kinetex 1.7 μm C18 packed in a 2.1 mm ID column, and a minimum reduced plate height above 2.0 was obtained. This provides further suggestion that the problematic situation of packing narrow bore columns is compounded when the packing materials are finer such as the sub-2 μm particles. According to Gritti and Guiochon,[63] the difference in efficiency is accounted for by a contribution to the column HETP of the long-range eddy diffusion term that is larger in the 2.1 mm than in the 4.6 mm ID columns. While the associated relative velocity biases are of comparable magnitude in both types of columns, the characteristic radial diffusion lengths are of the order of 100 and 40 μm in the wall regions of narrow bore and conventional columns, respectively.[63]

Another observation is that the 4.6 mm ID beds packed with 2.6–2.7 μm superficially particles are more homogeneous than those of the 2.1 mm ID narrow bore beds packed with 1.7 μm fully porous particles.[63] The external roughness of the core-shell particles might explain the origin of this advantageous property, because the shear stress that takes place during the slurry packing process is stronger between rugged particles than between smooth ones. Therefore, particles move less with respect to each other and the amount of strain occurring through the bed is smaller. Thus, the distribution of the external porosity throughout the bed of rugged particles is more homogeneous from the centre (low packing stress) to the wall of the column (high packing stress) than through beds of smooth particles.[69] Figure 5.5 clearly demonstrates the lower efficiency of narrow bore core-shell columns.

5.7 Extra-column Effects – Contribution to Band Broadening

Every serious progress in column technology requires important progress in instrument design and manufacturing.[70] Extra-column band spreading affects the measured performance of columns packed with small particles, especially for columns with an ID smaller than the classical standard of 4.6 mm.[71]

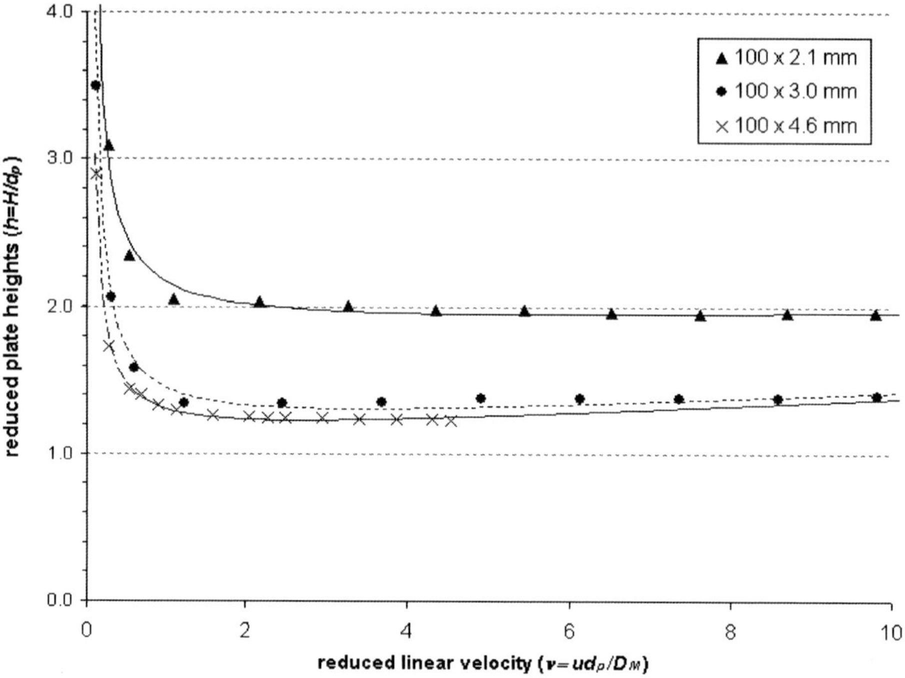

Figure 5.5 Experimental *h*–*ν* plots of 2.6 μm core-shell type Kinetex (100 mm × 2.1 mm, 100 mm × 3.0 mm and 100 mm × 4.6 mm) columns (peak widths were corrected for the extra-column broadening). Mobile phase: 48:52, acetonitrile/water; temperature: 35 °C; injection: 0.5 μL; D_M = 1.15×10^{-5} cm^2 s^{-1}; test analyte, oestradiol (from reference 19 with permission).

Recently, several papers focused on the extra-column effect as a major factor that negatively impacts the apparent performance of columns packed with core-shell or sub-2 μm particles.[19,58,70–72] Conventional HPLC systems contribute to the measured peak variance of approximately 40–200 μL^2, whereas standard ultra-performance or ultra-high pressure chromatographic systems have a contribution typically in the range of 4–9 μL^2.[19,58,70–72] In the case of very efficient columns, the extra-column variance of the commercially available LC systems with very low dispersion (<10 μL^2) is not negligible. The extra-column peak dispersion of the UPLC (Waters) system causes an efficiency loss of approximately 25–50% for the Kinetex 1.7 μm (5 cm × 2.1 mm) column when low molecular mass analytes are separated.[58] Further optimization of UHPLC systems, such as using smaller volume needle seat capillary, narrower and shorter connector capillary tubes and a smaller volume detector cell, can provide a significant decrease in extra-column contribution down to approximately 1–5 μL^2.[20,70] With these improvements, the efficiency loss can be significantly reduced.

5.8 Performance of Core-shell and Totally Porous Particles in Isocratic Elution Mode

5.8.1 Separation of Small Compounds and Pharmaceuticals

Several studies compared the efficiency of columns packed with sub-3 μm core-shell and sub-2 μm porous particles and monolithic columns in isocratic elution mode.[17,19,20,58] According to these results, currently the columns packed with sub-3 μm core-shell particles provide the most favourable plate time (t_0/N) values and therefore offer the shortest analysis time when small test compounds or pharmaceutical test analytes are separated.[19]

In everyday practice, chromatographers usually work above the optimum linear velocity. Therefore, it is necessary to emphasize that in the range of high linear velocities, the Kinetex column performs higher plate numbers as other core-shell packings (Ascentis Express or Halo) or columns packed with totally porous sub-2 μm particles, and offers more efficient separation when fast analysis is required. The particle size of Kinetex is 2.6 μm therefore the pressure drop is much lower compared to the sub-2 μm columns, thus higher linear velocity can be applied during the analysis.

In contrast with the Ascentis Express or Halo particles (0.5 μm porous silica layer) the Kinetex particles (0.35 μm porous silica layer) produce much lower C terms for small molecules.

The intra-particle diffusivity depends on the ratio (ρ) of the diameter of the solid core to that of the particle in a core-shell particle. As this ratio increases the mass transfer kinetic becomes faster through the shell particles than it is through totally porous particles. The ratio of solid core and particle diameter for the Kinetex material is $\rho = 0.73$ and for the Ascentis Express is $\rho = 0.63$. Figure 5.6 shows experimental isocratic Poppe plots of a steroid (~300 Da) obtained on five different columns at 400 bar (HPLC limitation) and at the maximum applicable pressure for each column to represent the utilization of maximum performance (UHPLC application).

On Figure 5.6, the curves, one for each column, demonstrate the maximum speed obtainable at a given required plate number (N) and also demonstrate the effect of the choice of column (stationary phase type, totally porous particles, shell particles, monolith column). Please note that the plate times depend on the maximum allowable pressure drop, which is different for the tested columns. A column can offer faster separation if it has a stationary phase with stronger mechanical stability (higher ΔP_{max}) than the column which has lower mechanical stability.

In the case of this compound (~300 Da), when the maximum available pressure is limited to 400 bar (Figure 5.6A) the Ascentis Express column provides the most favourable plate time values and therefore offers the shortest analysis time if the separation requires 10 000–90 000 plate counts. The second choice in this plate number range is the Kinetex column, which presents similar plate time values to the monolith column if the required plate number is higher

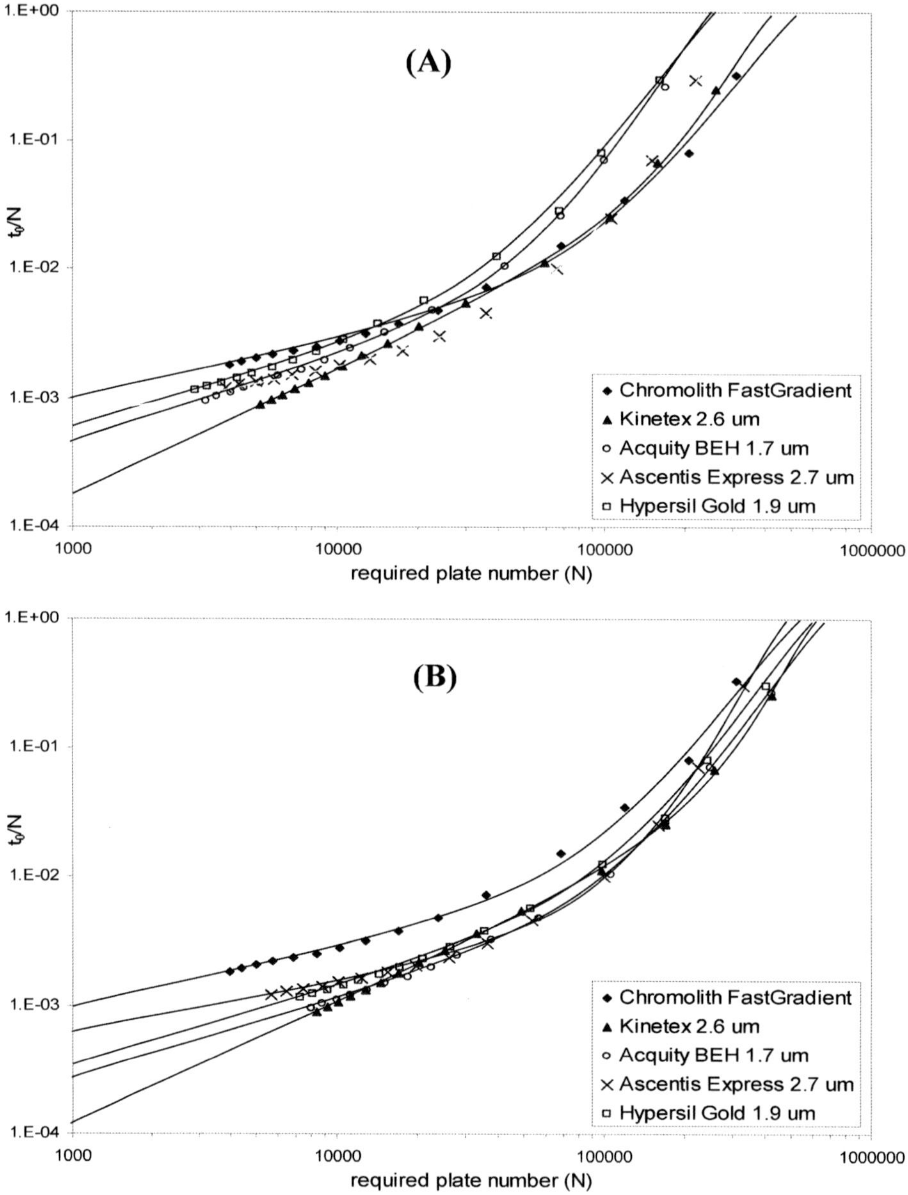

Figure 5.6 Isocratic Poppe plots of oestradiol. Experiments were conducted on 5 cm long narrow bore columns in 48:52, acetonitrile/water, η = 0.85 cPoise, at 35 °C. Available maximum pressure: (A) 200 bar for Chromolith column and 400 bar for the packed columns, (B) 1000 bar for Waters BEH and Hypersil Gold columns, 600 bar for Ascentis Express and Kinetex columns and 200 bar for the Chromolith column (from reference 19 with permission).

than 30 000. The Kinetex column outperforms the other totally porous sub-2 μm and 2.7 μm shell columns when low plate numbers ($N < 10\ 000$) and also when very high plate numbers ($N > 90\ 000$) are required for the separation. The first case can be beneficial when very short columns (L = 2–5 cm) are applied and only a small number of analytes are planned to be separated, whereas the second case is practical when long columns are used (or columns are coupled).

If the maximum performance of an UHPLC system is utilized (Figure 5.6B), the column packed with 1.7 μm totally porous particles and the column packed with 2.7 μm shell particles (0.5 μm porous shell) could give a good chance for the shortest analysis when the separation requires a plate count value in the range of N = 20 000–120 000. The Kinetex column is preferred when low plate number is enough ($N < 11\ 000$) for a given separation.

Figure 5.7 shows the experimental isocratic Poppe plots of a compound with a molecular mass of ~900 Da. Under HPLC conditions (ΔP_{max} = 400 bar), the Kinetex column gives the fastest separation when less than 80 000 plate numbers is required (Figure 5.7A). This 80 000 plates count can be achieved with an approximately 43 cm long Kinetex column. So in every day practice, when 5–25 cm long columns are used, the Kinetex column provides the most favourable plate time values and offers the shortest analysis time.

Under UHPLC conditions the column packed with 1.7 μm totally porous particles outperforms the Ascentis Express column if $N < 200\ 000$ is required (Figure 5.7B). If the analysis demands a plate count $N < 70\ 000$, the Kinetex column performs the fastest separation. The 70 000 plates count can be achieved with an approximately 48 cm long Kinetex column.

Figure 5.8. demonstrates the very high separation power of core-shell columns when applied for the separation of pharmaceuticals under isocratic elution mode.

5.8.2 Separation of Peptides and Proteins

The efficiency of recently introduced sub-2 μm core-shell particles (1.7 μm Kinetex)— applied for the separation of peptides (a 4.1 kDa polypeptide)— was studied in detail.[58] Figure 5.9. shows the *H–u* plots obtained with the 1.7 μm core-shell (Kinetex), the 2.7 μm core-shell (Ascentis Express) and the 1.7 μm porous (Acquity BEH) packing.

The minimum plate height obtained with 1.7 μm core-shell particles $HETP_{min}$ = 6.3 μm is approximately two times lower than the $HETP_{min}$ value obtained with the column packed with the same sized fully porous particles ($HETP_{min}$ = 12.1 μm). The Ascentis Express column performed a $HETP_{min}$ = 9.2 μm in this comparison. The optimum in linear velocity was measured as 0.15 cm s^{-1} in the case of Kinetex 1.7 μm column, 0.12 cm s^{-1} for the Ascentis Express column and 0.11 cm s^{-1} for the Waters BEH column. The 1.7 μm Kinetex column performed at an approximately 2 times lower C value

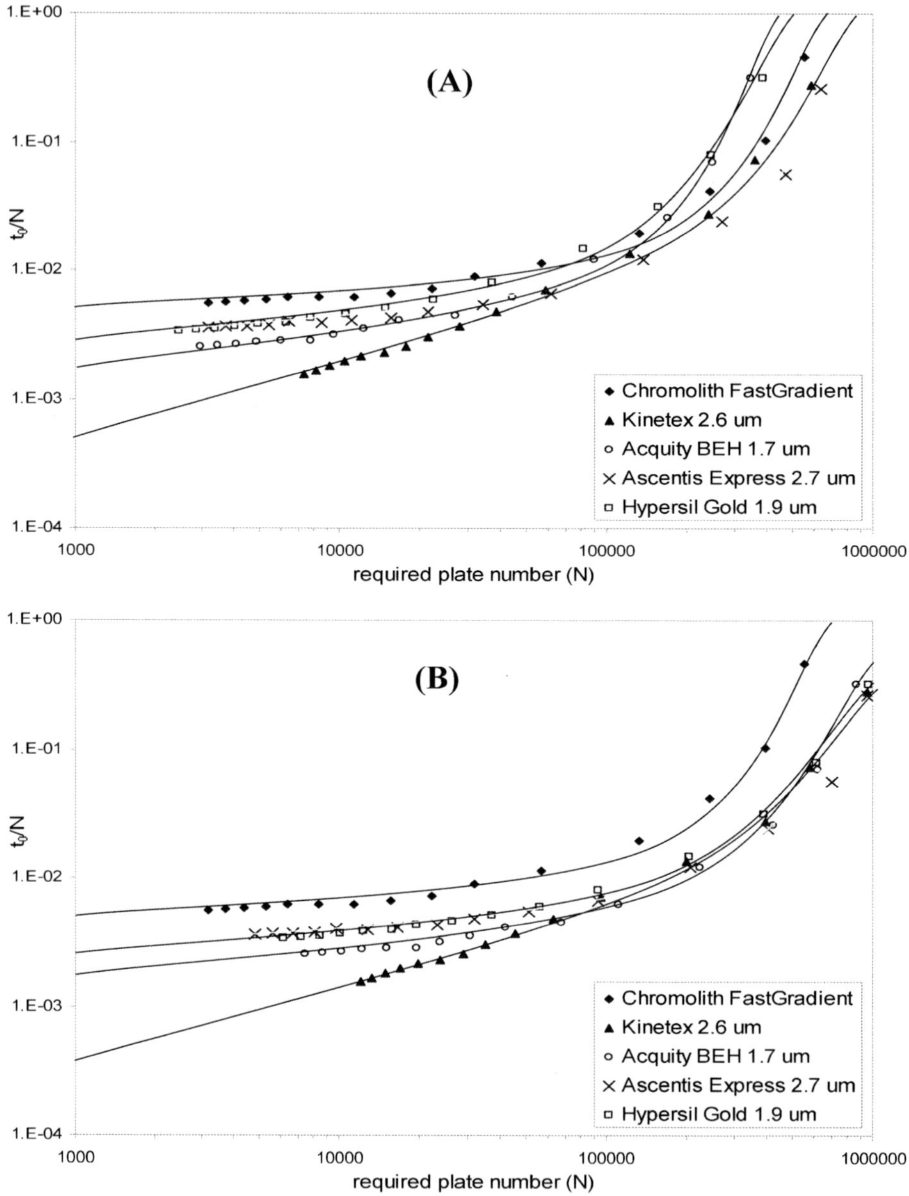

Figure 5.7 Poppe plots of ivermectin. Experiments were conducted on 5 cm long narrow bore columns in 95:5, acetonitrile/water, η = 0.33 cPoise, at 35 °C. Available maximum pressure: (A) 200 bar for Chromolith column and 400 bar for the packed columns, (B) 1000 bar for Waters BEH and Hypersil Gold columns, 600 bar for Ascentis Express and Kinetex columns and 200 bar for the Chromolith column (from reference 19 with permission).

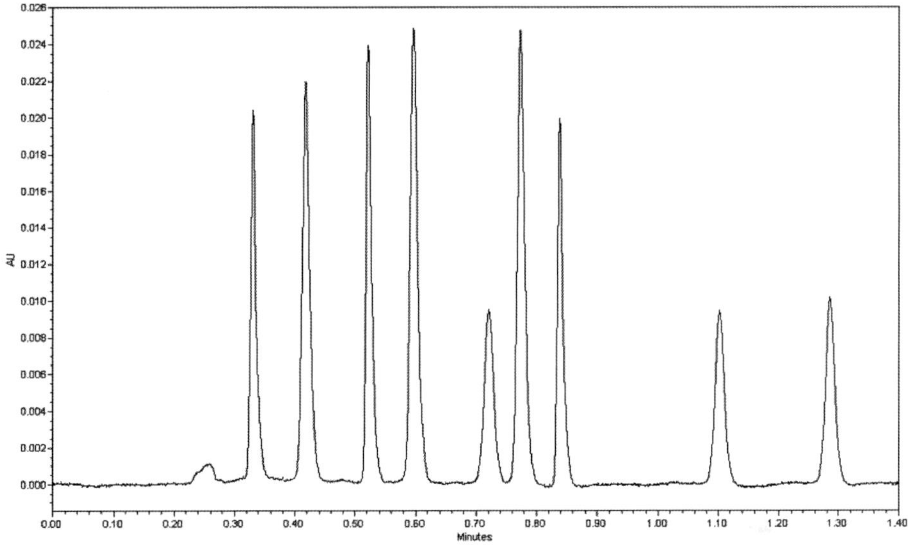

Figure 5.8 Column: Kinetex C18 2.6 μm (15 cm × 3.0 mm), mobile phase: 47:53 (v/v), acetonitrile/water; flow-rate: 1.8 ml min^{-1} (p = 516 bar); column temperature: 60 °C; injection volume: 1 μl; detection: 215 nm; analytes: steroids (neutral polar API) (from reference 19 with permission).

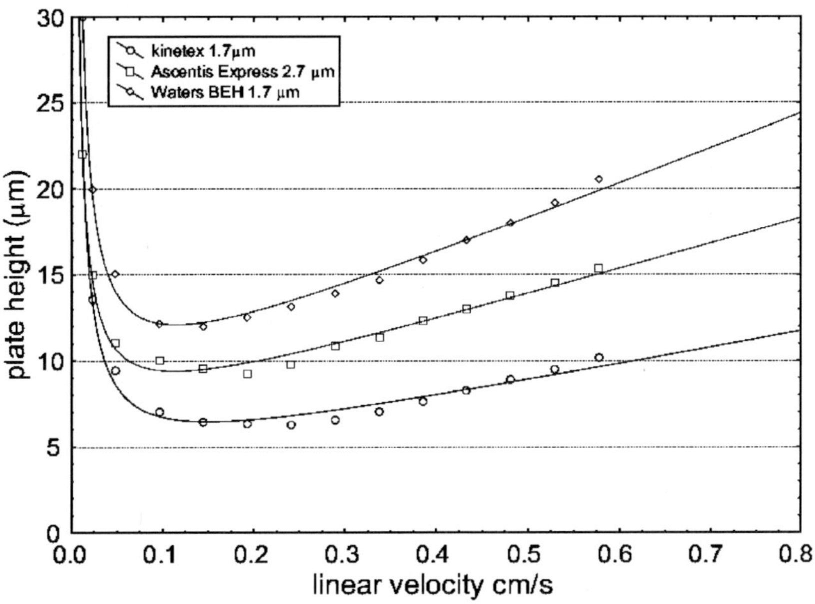

Figure 5.9 Experimental van Deemter plots of 1.7 μm core-shell (Kinetex C18, 5 cm × 2.1 mm), 2.7 μm fused-core (Ascentis Express C18, 5 cm × 2.1 mm) and 1.7 μm fully porous (Waters BEH C18, 5 cm × 2.1 mm) columns. Mobile phase: 140:860:1, acetonitrile/water/TFA; temperature: 35 °C; injection: 0.5 μl; test analyte: 4.1 kDa polypeptide (from reference 58 with permission).

than the column packed with same sized fully porous particles. It is in good agreement with the theory of Kaczmarski and Guiochon.[50]

The major advance of 1.7 μm core-shell particles is the approximately 50% decrease in plate heights compared to the fully porous 1.7 μm particles. The permeability of the Kinetex 1.7 μm and Waters 1.7 μm column is approximately the same, thus faster separation can be expected on the core-shell column when peptides are separated. Figure 5.10 shows the plate time *versus* required plate number plots (isocratic Poppe plot) to present that significantly shorter analysis time can be expected with the Kinetex 1.7 μm column than with the Waters BEH or the Ascentis Express column in the case of peptide separation.

The advantage of core-shell particles when applied to the separation of proteins (18.8 kDa, 38.9 kDa and 66.3 kDa) is more obvious. Figure 5.11 shows the experimental *H–u* curves of the Kinetex 1.7 μm and Waters BEH columns.

As the weight (and size) of the protein is larger, the slope of the *H–u* curves becomes steeper. Considering that the particle size of the two packing is similar and the Kinetex column gives approximately 2 times higher plate counts, a considerably faster analysis can be expected with the Kinetex column than with the Waters BEH column in the case of large proteins.

It is necessary to mention that the nominal pore size of the two packing is approximately 10 nm and 13 nm, thus these columns are not completely suitable for large protein separation and some size exclusion effect is inevitable. Producing sub-2 μm core-shell particles with larger pore size (*e.g.* 300 Å—which is widespread for general protein separation) should result in the full advantage of core-shell sub-2 μm particles for protein separations.

Recent development and commercialization of the latest generation of core-shell columns have not included media with pore sizes of 300 Å, which are required for macromolecule separations. However, a 160 Å packing was recently introduced by Advanced Material Technology (AMT) and Sigma-Aldrich under the brand names of Halo Peptide ES-C18 and Ascentis Express Peptide ES-C18, respectively.[73]

While the benefit of core-shell particles is evident with small molecules, the improvement is more significant with peptide and protein molecules.

5.9 Performance of Core-shell and Totally Porous Particles in Gradient Elution Mode

5.9.1 Separation of Small Compounds and Pharmaceuticals

A systematic study compared the performance of three modern core-shell (Kinetex, Poroshell 120 and Ascentis Express) and one totally porous (Acquity BEH) materials under gradient elution conditions, applied for the separation of small pharmaceutical (molecular wieght: 270–430 g mol^{-1}) compounds.[74] The two experimental variables of flow-rate and gradient duration were investigated. The former has a direct influence on the column efficiency,

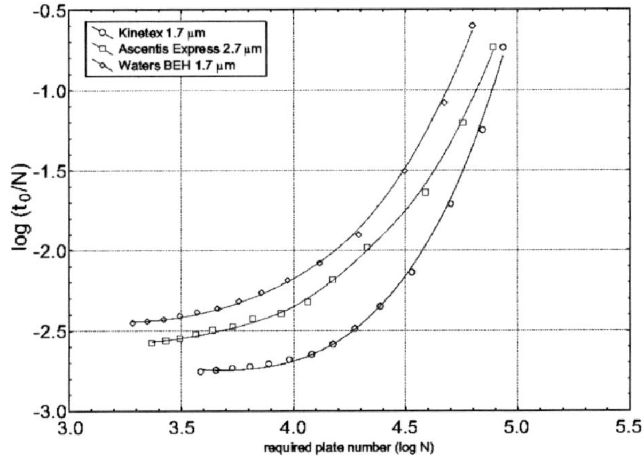

Figure 5.10 Poppe plots of 4.1 kDa polypeptide. Experiments were conducted on 5 cm long narrow bore columns in 140:860:1, acetonitrile/water/TFA, η = 0.99 cPoise, at 35 °C. Available maximum pressures: 600 bar for the Ascentis Express column and 1000 bar for the Waters BEH and Kinetex 1.7 µm columns (from reference 58 with permission).

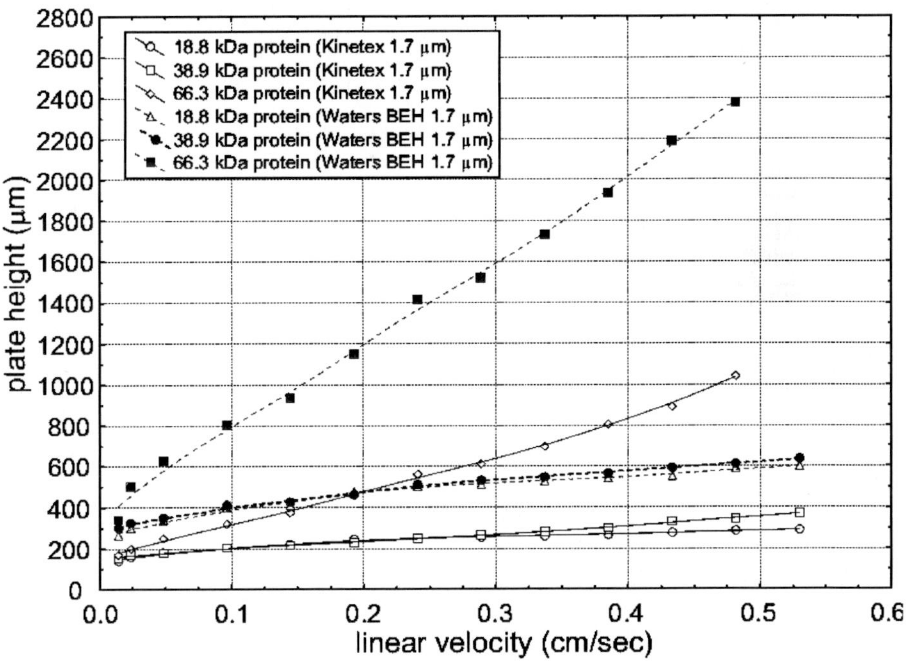

Figure 5.11 Experimental *H–u* curves of 1.7 µm core-shell (Kinetex C18, 5 cm × 2.1 mm) and 1.7 µm fully porous (Waters BEH C18, 5 cm × 2.1 mm) columns. Mobile phases: 440:560:1 (for the 18.8 kDa protein), 470:530:1 (for the 38.9 kDa protein) and 610:390:1 (for BSA, 66.3 kDa) ratio of acetonitrile/water/TFA; temperature: 60 °C; injection: 0.5 µl; test analytes: proteins (from reference 58 with permission).

whereas gradient duration plays an important role in the resolution under gradient conditions. Figure 5.12 illustrates how considerably the experimental conditions affect the peak capacity of core-shell packing (Kinetex 5 cm $\times$ 2.1 mm, 2.6 μm).

According to the theory, the efficiency of core-shell columns depends on the ratio (ρ) of the diameter of the solid core to that of the particle in a core–shell particle. This ratio is $\rho = 0.73$ for the Kinetex column and $\rho = 0.63$ for both the Ascentis Express and Poroshell columns. The nominal diameters of the particles are very similar ($d_p = 2.6$ μm for the Kinetex and $d_p = 2.7$ μm for the Ascentis Express and Poroshell columns), thus similar efficiency can be expected for the Ascentis Express and Poroshell columns, whereas the Kinetex column probably provides higher separation power.

When a very fast gradient separation was achieved (3 min gradient time), the Kinetex column outperformed all the other three tested columns at all of the applied flow-rates (Figure 5.13A). At 0.3 ml min^{-1} the Kinetex column produced a peak capacity value of $P_c = 35$, whereas the Ascentis Express, Poroshell and Waters BEH columns achieved a $P_c \sim 28$ (5 cm long narrow bore columns). In the range of a 0.3–0.5 ml min^{-1} flow-rate, the Ascentis Express, Poroshell and Waters BEH columns provided practically the same efficiency. If the flow-rate was higher than 0.5 ml min^{-1}, the column packed with fully porous sub-2 μm particles performed at a lower peak capacity than those obtained with core-shell particles. At high flow-rate (0.8 ml min^{-1}), the Kinetex column yielded an efficiency of $P_c = 55$. The Ascentis Express, Poroshell and Waters BEH columns gave approximately 10, 20 and 25% lower peak capacity than the Kinetex column when 0.8 ml min^{-1} flow-rate was set. The results showed that in the case of a 3 min gradient span, the Ascentis Express column gradually outperforms the Poroshell column when the flow-rate is increased ($F > 0.5$ ml min^{-1}), in spite of the same shell thickness and particle diameter. Another finding is that the slope of the peak capacity curve of the Waters BEH column is flatter than it is for the columns packed with larger partially porous particles. A possible explanation of this phenomenon could be due to the heat friction released at high flow-rates. The heat power friction released in the BEH column per volume unit is typically twice as large than that released in the columns packed with 2.6–2.7 μm particles.[55]

When a relatively slow gradient separation was executed with these 5 cm long columns (flat gradient program with a gradient time of 21 min), significantly different results were obtained than in the case of very fast (steep) gradient separation (Figure 5.13B). The first interesting result is that the core-shell-type Ascentis Express packing produced a lower peak capacity than the Waters BEH column which is packed with totally porous particles. It can probably be explained by the rough surface of the Ascentis Express particles in which the mass transfer rate is reduced through the outer stagnant liquid.[17,75]

The other surprising result is that as the flow-rate increases, the peak capacity of the Poroshell column ($\rho = 0.63$) becomes closer to that obtained with the Kinetex column ($\rho = 0.73$). The Poroshell column provides 15% less

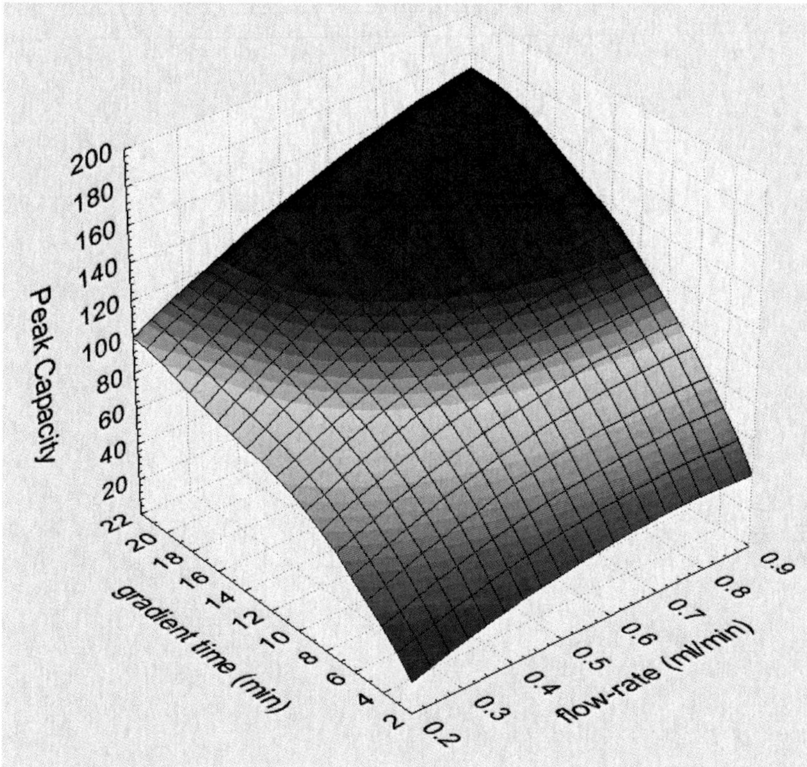

Figure 5.12 Peak capacity surface plot as function of flow-rate and gradient time. Column: Kinetex 5 cm × 2.1 mm, 2.6 μm. Conditions: the volume fractions of acetonitrile at the beginning and at the end of the gradient were set constant at 40 and 90%, respectively ($\Delta\phi$ = 0.50), the columns were thermo-stated at 30 °C and the injected volume was 0.5 μl (from reference 58 with permission).

peak capacity at 0.3 ml min^{-1}, whereas at an enhanced flow-rate (0.8 ml min^{-1}) it performs only at 2% less peak capacity than in the case of the Kinetex column. A probable reason for this result can be the larger average pore size of Poroshell material which could have some benefit when longer separations are performed. The average pore size of the new Poroshell material is 11.9 nm, whereas in the case of the Kinetex packing it was measured as 9.5 nm.[74] The advantage of the Kinetex column is most significant at a 0.3–0.4 ml min^{-1} flow-rate. At a 0.3 ml min^{-1} flow-rate, the Kinetex column produces a 15–20% higher peak capacity than the other tested columns. The 5 cm long narrow bore Kinetex and Poroshell columns can generate a peak capacity of 160–170% within a 21 min gradient separation when high flow-rates are applied.

These results confirm the excellent performance of the new generation columns packed with 2.6–2.7 μm core-shell particles under gradient elution mode when very fast separations are achieved. For the separation of

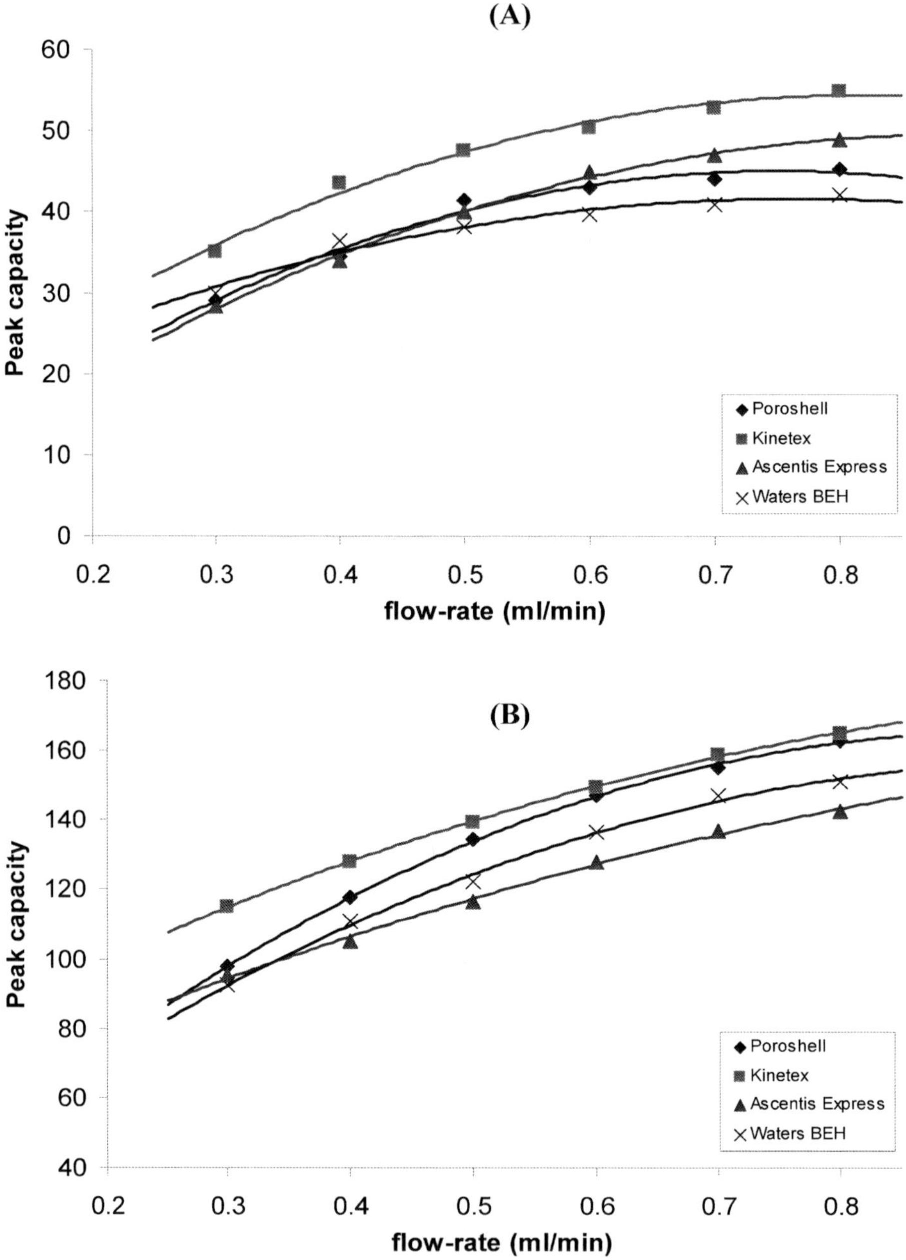

Figure 5.13 Peak capacity plots of 5 cm × 2.1 mm columns as a function of flow rate at 3 min gradient time (A) and at 21 min gradient time (B). Conditions: the volume fractions of acetonitrile at the beginning and at the end of the gradient were set constant at 40 and 90%, respectively ($\Delta\phi = 0.50$), the columns were thermo-stated at 30 °C and the injected volume was 0.5 μl (from reference 74 with permission).

pharmaceutical compounds, the peak capacity of these 5 cm long narrow bore columns is similar or higher to those of the 1.7 μm totally porous particles. The highest peak capacity was obtained with the Kinetex column in all cases. It is necessary to emphasize that the efficiency of Poroshell-120 column is very similar to the Kinetex when relatively high flow-rate and long gradient separation are performed. The efficiency of the new generation of sub-3 μm partially porous packing is comparable to those obtained with totally porous sub-2 μm particles in fast gradient separations of pharmaceuticals in the case of short narrow bore columns.

5.9.2 Separation of Peptides and Proteins

A critical comparison of column efficiency under gradient conditions of two core-shell (Kinetex and Halo) and one sub-2 μm porous packing (Acquity BEH) were investigated in depth by Gritti and Guiochon.[76]

For the separation of the peptides in protein digests (β-lactoglobulin digest), the peak capacity of the Kinetex column was similar to those of the 2.7 μm Halo shell particles and the 1.7 μm Acquity BEH totally porous particles at a small reduced linear velocity ($v < 2$) and a small intrinsic gradient steepness ($G < 1$). However, the Kinetex column maintained 100% of its performance at increasing flow velocities of the eluent. It provides peak capacities of approximately 170 within 4 min (15 cm long columns), for a gradient steepness $G \sim 1$, whereas peak capacities of only 150 and 125 were generated by the Halo and the Acquity BEH columns, respectively. For the separation of authentic proteins by gradient elution, the peak capacities of all three columns significantly decreased with increasing the gradient slope at a constant gradient steepness. This observation is due to the slow mass transfer kinetics of proteins through the porous shell and the porous particles. The performance of the Kinetex column is slightly better than that of the Acquity BEH column when the peak capacity is derived from the peakwidth of insulin, but it is comparable when it is derived from the peakwidth of lyzozyme. These observations show the advantage of using shell particles to separate large biomolecules, because their intra-particle diffusivity is less than a few percent of their bulk diffusivity under non-retained conditions. The Kinetex column can provide a peak capacity of 85 with proteins of molecular mass of approximatley 15 kDa within 3.5 min at a gradient steepness $G \sim 6$, whereas the Acquity BEH and Halo columns can only generate peak capacities of 75 and 30 respectively.[76] Figure 5.14 shows the obtained chromatograms of β-lactoglobulin digest.

Kirkland and co-workers[77] compared the efficiency performance of the 160 Å Halo Peptide ES C18 column to the original 90 Å Halo C18 column for mixtures of peptides and small proteins. The smaller peptides showed equivalent efficiencies on both columns. However, the small proteins (ribonuclease, insulin, cytochrome C and lysozyme) showed broadened peaks on the 90 Å Halo C18 column, indicating restricted diffusion, but narrow peaks on the 160 Å Halo Peptide ES C18 column, indicating unrestricted diffusion.[77]

Gritti and Guiochon[73] also investigated the possibilities of the new 160 Å Halo Peptide ES packing. Their results also demonstrated that the Halo-ES-peptide column, designed to resolve mixtures of large molecules, provides markedly better kinetic performance than the first generation of Halo particles (Halo 90 Å).[73]

The physico-chemical explanations for these observations might not necessarily be those anticipated during the design of this new packing material. The initial incentive was to increase the mesopore size in order to reduce the hindrance to diffusion through the porous shells of the particles. Successfully enough, the sample diffusivity in the porous shells was indeed increased. For all that, however, their results proved that the trans-particle

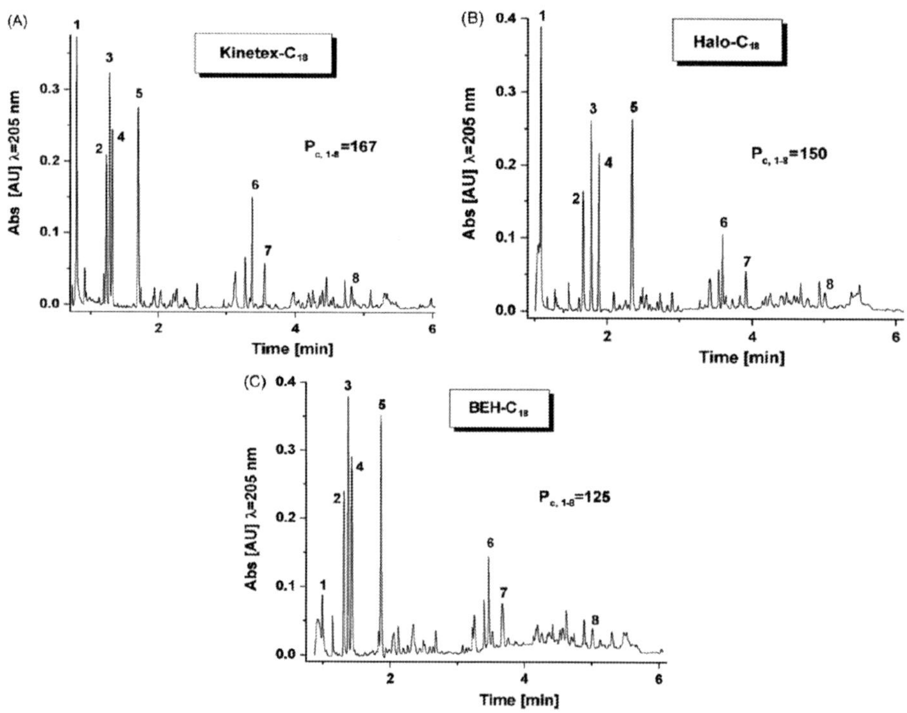

Figure 5.14 Chromatograms obtained after the injection of 5 μL of β-lactoglobulin digest in gradient elution. The concentration of acetonitrile at the beginning and at the end of the gradient is equal to 5 and 45%, respectively. The linear velocity was constant for all three columns at 0.37 cm s^{-1}. T = 295 K. (A) 10 cm × 4.6 mm, 2.6 μm Kinetex-C18 column, (B) 5 cm × 4.6 mm, 2.7 μm Halo-C18 column, and (C) 10 cm × 3.0 mm, 1.7 μm BEH-C18 column (from reference 76 with permission).

mass transfer resistance term is not the limiting kinetic factor that controls the solid–liquid mass transfer resistance in Halo particles. Actually, the van Deemter *C* term for large molecules is mostly accounted for by a slow external film mass transfer.[73]

They concluded that the improvement of the column efficiency of large molecules such as peptides and proteins (500 < molecular mass < 20 000 Da) observed with the Halo-ES-peptide column is related to the easier access of their molecules to the internal volume. The probability for a molecule to penetrate the shell of the particles increases when its internal porosity increases. As for small molecules, the improvement in column performance is also due to the eddy diffusion term of the Halo-ES-peptide 160 Å column being 25% smaller than that of the first generation of Halo 90 Å column. This diminution of the *A* term emphasizes the important role of sample diffusivity through porous particles in the relaxation of radial concentration gradients caused by short-range inter-channel and trans-column velocity biases.[73] Figure 5.15 shows the obtained chromatograms of β-lactoglobulin digest on the Halo-ES-peptide 160 Å column (15 cm × 4.6 mm, 2.7 μm). A peak capacity value of P_c = 410 was observed during a 90 min gradient span.

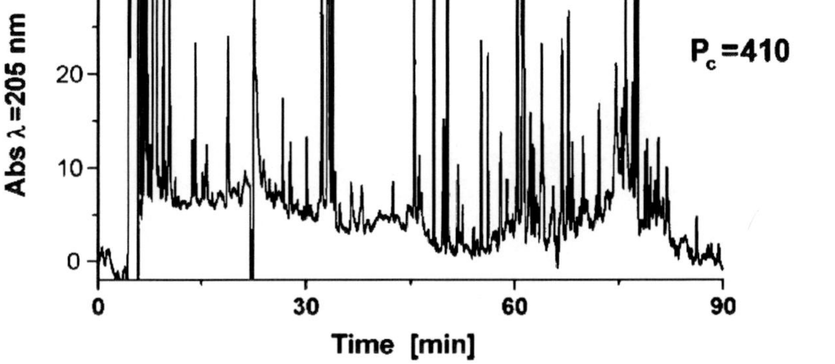

Figure 5.15 Resolution of a peptide mixture (tryptic digest of the protein β-lactoglobulin) on the Halo-ES-peptide 160 Å column. Flow rate: 0.3 mL min^{-1}. Gradient: 5 to 40% acetonitrile in 90 min. T = 295 K. The average peak width was approximately 13 s (from reference 73 with permission).

5.10 Conclusions

This Chapter demonstrates why the new generation of sub-3 μm core-shell particles are applied with great success and are becoming more and more popular in the field of both pharmaceutical and biomedical separations.

As a final conclusion, we summarize the obvious advantages of the recently developed core-shell particles against modern fully porous sub-2 μm particles. These advantages are listed below:

- The short diffusion path of shell particles allows superior mass transfer kinetics, and leads to better kinetic performance at high mobile phase flow-rates.
- The narrow particle size distribution and surface roughness of core-shell particles probably contribute to a better column packing quality. This is possibly the main reason for their great success in separating small analytes.
- Sub-3 μm core-shell columns (2.6 μm and 2.7 μm) generate only approximately half the back pressure of the sub-2 μm fully porous particles under their own optimal flow rates.

Recent core-shell packings can be applied with a very high efficiency for the separation of both small and macro-molecules. However, recently it is not possible to utilize the real efficiency—separation power—of these very efficient columns because of the ineligible extra-column effects of the commercially available UHPLC systems.

References

1. M. E. Swartz and B. Murphy, *Am. Lab.*, 2005, **37**, 22.
2. M. E. Swartz, *J. Liq. Chromatgr.*, 2005, **28**, 1253.
3. J. A. Blackwell and P. W. Carr, *J. Liq. Chrom.*, 1991, **14**, 2875.
4. C. Zhu, D. M. Goodall and S. A. C.Wren, *LC-GC North Am.*, 2005, **23**, 1.
5. H. A. Claessens and M. A. van Straten, *Reduction of Analysis Times in HPLC at Elevated Column Temperatures*, Eindhoven University of Technology, 2004.
6. P. T. Jackson and P. W. Carr, *Chem.-Tech.*, 1988, **28**, 29.
7. S. Hjerten, J. L. Liao and R. Zhang, *J. Chromatogr.*, 1989, **473**, 273.
8. F. Svec and J. M. Frechet, *J. Anal. Chem.*, 1992, **64**, 820.
9. I. Gusev, X. Huang and C. Horvath, *J. Chromatogr. A*, 1999, **885**, 273.
10. H. Minakuchi, H. Nagayama, N. Soga, N. Ishizuka and N. Tanaka, *J. Chromatogr. A*, **797**, 121.
11. T. Ikegami, E. Dicks, H. Kobayashi, H. Morisaka, D. Tokuda, K. Cabrera and N. Tanaka, *J. Sep. Sci.*, 2004, **27**, 1292.
12. H. Oberacher, A. Premstaller and C. G. Huber, *J. Chromatogr. A*, 2004, **1030**, 201.
13. C. Horvath, B. A. Preiss and S. R. Lipsky, *Anal. Chem.*, 1967, **39**, 1422.
14. C. Horvath and S. R. Lipsky, *J. Chromatogr. Sci.*, 1969, **7**, 109.

15. J. J. Kirkland, *Anal. Chem.*, 1969, **41**, 218.
16. J. J. Kirkland, F. A. Truszkowski, C. H. Dilks Jr and G. S. Engel, *J. Chromatogr. A*, 2000, **890**, 3.
17. S. Fekete, J. Fekete and K. Ganzler, *J. Pharm. Biomed. Anal.*, 2009, **49**, 64.
18. F. Gritti and G. Guiochon, *J. Chromatogr. A*, 2011, **1218**, 907.
19. E. Olah, S. Fekete, J. Fekete and K. Ganzler, *J. Chromatogr. A*, 2010, **1217**, 3642.
20. J. O. Omamogho, J. P. Hanrahan, J. Tobin and J. D. Glennon, *J. Chromatogr. A*, 2011, **1218**, 1942.
21. U. D. Neue, *HPLC Columns*, Wiley-VCH, New York, 1997.
22. J. C. Giddings, *Anal. Chem.*, 1965, **37**, 60.
23. H. Poppe, *J. Chromatogr. A*, 1997, **778**, 3.
24. J. C. Giddings, *Anal. Chem.*, 1967, **39**, 1027.
25. C. Horvath and S. R. Lipsky, *Anal. Chem.*, 1967, **39**, 1893.
26. K. K. Unger, R. Skudas and M. M. Schulte, *J. Chromatogr. A*, 2008, **1184**, 393.
27. J. J. van Deemter, F. J. Zuiderweg and A. Klinkenberg, *Chem. Eng. Sci.*, 1956, **5**, 271.
28. J. C. Giddings, *J. Chromatogr.*, 1961, **5**, 61.
29. L. R. Snyder and J. J. Kirkland, *Introduction to Modern Liquid Chromatography*, Wiley, New York, 2nd edn, 1979.
30. J. F. K. Huber, J. A. R. J. Hulsman, *Anal Chim. Acta*, 1967, **38**, 305.
31. G. J. Kennedy and J. H. Knox, *J. Chromatogr. Sci.*, 1972, **10**, 549.
32. C. Horvath and H-J. Lin, *J. Chromatogr. A*, 1978, **149**, 43.
33. Y. Yang, A. Velayudhan, C. M. Ladish and M. R. Ladish, *J. Chromatogr. A*, 1992, **598**, 169.
34. A. D. Jerkovich, R. LoBrutto and R. V. Vivilecchia, *LC-GC N. Am.*, 2005, **5**, 15.
35. J. H. Knox, *J. Chromatogr. A*, 2002, **960**, 7.
36. G. Desmet, D. Clicq and P. Gzil, *Anal. Chem.*, 2005, **77**, 4058.
37. H. Chen and C. Horváth, *J. Chromatogr. A*, 1995, **705**, 3.
38. C. R. Wilke and P. Chang, *AIChE J.*, 1955, 264.
39. J. C. Giddings, *Anal. Chem.*, 1967, **39**, 1027.
40. U. D. Neue, *J. Chromatogr. A*, 2005, **1079**, 153.
41. X. Wang, D. R. Stoll, A. P. Schellinger and P. W. Carr, *Anal. Chem.*, 2006, **78**, 3406.
42. J. W. Dolan, L. R. Snyder, N. M. Djordjevic, D. W. Hill and T. J. Waeghe, *J. Chromatogr. A*, 1999, **857**, 1.
43. U. D. Neue and J. R. Mazzeo, *J. Sep. Sci.*, 2001, **24**, 921.
44. U. D. Neue, J. L. Carmody, Y-F. Cheng, Z. Lu, C. H. Phoebe and T. E. Wheat, *Adv. Chromatogr.*, 2001, **41**, 93.
45. U. D. Neue, Y-F. Cheng and Z. Lu, in *HPLC Made to Measure: A Practical Handbook for Optimization*, ed. S. Kromidas, Wiley-VCH, Weinheim, 2006, p. 47.
46. U. D. Neue, *J. Chromatogr. A*, 2008, **1184**, 107.

47. P. A. Bristow and J. H. Knox, *Chromatographia*, 1977, **10**, 279.
48. X. Wang, W. E. Barber and P. W. Carr, *J. Chromatogr. A*, 2006, **1107**, 139.
49. G. Guiochon and F. Gritti, *J. Chromatogr. A*, doi:10.1016/j.chroma.2011.01.080
50. K. Kaczmarski and G. Guiochon, *Anal. Chem.*, 2007, **79**, 4648.
51. J. Giddings, *Dynamics of Chromatography*, Marcel Dekker, New York, 1965.
52. F. Gritti and G. Guiochon, *Chem. Eng. Sci.*, 2011, **66**, 3773.
53. G. Guiochon, A. Felinger, A. Katti and D. Shirazi, *Fundamentals of Preparative and Nonlinear Chromatography*, 2nd edn, Academic Press, Boston, 2006.
54. D. A. Robinson and S. P. Friedman, *Physica A*, 2005, **358**, 447.
55. F. Gritti and G. Guiochon, *J. Chromatogr. A*, 2010, **1217**, 5069.
56. F. Gritti and G. Guiochon, *J. Chromatogr. A* 2010, **1217**, 5137.
57. F. Gritti and G. Guiochon, *J. Chromatogr. A*, 2007, **1138**, 141.
58. S. Fekete, K. Ganzler and J. Fekete, *J. Pharm. Biomed. Anal.*, 2011, **54**, 482.
59. G. Desmet, K. Broeckhoven, J. De Smet, S. Deridder, G. V. Baron and P. Gzil, *J. Chromatogr. A*, 2008, **1188**, 171.
60. D. Cabooter, A. Fanigliulo, G. Bellazzi, B. Allieri, A. Rottigni and G. Desmet, *J. Chromatogr. A*, 2010, **1217**, 7074.
61. S. Deridder and G. Desmet *J. Chromatogr. A*, 2011, **1218**, 46.
62. B. G. Yew, J. Ureta, R. A. Shalliker, E. C. Drumm and G. Guiochon, *AIChE J.*, 2003, **49**, 642.
63. F. Gritti andG. Guiochon, *J. Chromatogr. A*, 2011, **1218**, 1592.
64. J. J. DeStefano, T. J. Langlois and J. J. Kirkland, *J. Chromatogr. Sci.*, 2008, **46**, 254.
65. F. Gritti, I. Leonardis, J. Abia and G. Guiochon, *J. Chromatogr. A*, 2010, **1217** 3819.
66. F. Gritti, A. Cavazzini, N. Marchetti and G. Guiochon, *J. Chromatogr. A*, 2007, **1157**, 289.
67. H. Poppe and J. C. Kraak, *J. Chromatogr.*, 1983, **255**, 395.
68. K. Horváth, F. Gritti, J.N. Fairchild and G. Guiochon, *J. Chromatogr. A*, 2010, **1217**, 6373.
69. D. V. McCalley, *J. Chromatogr. A*, 2011, **1218**, 2887.
70. B. G. Yew, J. Ureta, R. A. Shalliker, E. C. Drumm and G. Guiochon, *AIChE J.*, 2003, **49**, 642.
71. F. Gritti, C. A. Sanchez, T. Farkas and G. Guiochon, *J. Chromatogr. A*, 2010, **1217**, 3000.
72. K. J. Fountain, U. D. Neue, E. S. Grumbach and D. M. Diehl, *J. Chromatogr. A*, 2009, **1216**, 5979.
73. F. Gritti and G. Guiochon, *J. Chromatogr. A*, 2010, **1217**, 7677.
74. F. Gritti and G. Guiochon, *J. Chromatogr. A*, 2011, **1218**, 907.
75. S. Fekete and J. Fekete, *Talanta*, 2011, **84**, 416.

76. F. Gritti and G. Guiochon, *J. Chromatogr. A*, 2007, **1166**, 30.
77. F. Gritti and G. Guiochon, *J. Chromatogr. A*, 2010, **1217**, 1604.
78. S. A. Schuster, B. M. Wagner, B. E. Boyes and J. J. Kirkland, *J. Chromatogr. Sci.*, 2010, **48**, 566.

CHAPTER 6

Shell Particle and UHPLC Technologies for Fast Analysis of Polar Compounds in the HILIC Mode

DAVID V. McCALLEY

Department of Applied Sciences, University of the West of England, Frenchay, Bristol BS16 1QY, UK
E-mail: David.Mccalley@uwe.ac.uk

6.1 Introduction

Hydrophilic interaction chromatography (HILIC) is a liquid chromatographic (LC) separation technique that it is ideally suited for the separation of polar and hydrophilic species, which may be poorly retained in reversed-phase (RP)-LC. It uses a polar stationary phase (for example, bare silica or a polar ligand bonded to a silica or organic polymer substrate) in conjunction with a mobile phase containing an appreciable concentration of water together with a high proportion of an organic solvent [very often acetonitrile (ACN)]. The water content is usually in the range ~2.5–40% (v/v), and thus HILIC differs from normal phase (NP, or "adsorption chromatography") in which efforts are often made to exclude the presence of water due to its disruptive effects on the reproducibility of the separation.[1] However, the order of elution of solutes is somewhat similar in HILIC to that in NP, with retention generally increasing with an increase in solute polarity, hence the technique is sometimes referred to

RSC Chromatography Monographs No. 16
UHPLC in Life Sciences
Edited by Davy Guillarme and Jean-Luc Veuthey

Published by the Royal Society of Chemistry, www.rsc.org

as "aqueous normal phase chromatography" (ANP). Indeed, it is possible that there are some common features in the retention mechanisms of NP and ANP separations. The presence of a much higher water concentration in HILIC compared with NP separations gives rise to much more stable retention times and good reproducibility of the separation.

The term HILIC was coined by Alpert in 1990,[2] who not only showed some excellent separations of amino acids, peptides and proteins but also made some important contributions to understanding the mechanism of these separations. He demonstrated that, typically, the order of solute elution was the opposite to that obtained in RP chromatography, and that partitioning between the more hydrophobic mobile phase and a layer of mobile phase enriched with water and partially immobilised on the stationary phase was a significant contributor to the separation mechanism. However, as pointed out by Alpert, the conditions of HILIC had already been used for the analysis of sugars and oligosaccharides.[3] Indeed, it could be argued that the origins of HILIC date back even to the earliest days of column LC, when Martin and Synge[4] separated amino acids on a silica column using water-saturated chloroform as the mobile phase, claiming that the separation mechanism was partitioning between the water layer on the silica surface and the chloroform. However, in this case, the two solvents (chloroform and water) are immiscible, whereas ACN and water are fully miscible, and the polarity of chloroform is different from that of ACN/water, so this proposal is debatable. As expected for a separation mechanism to which partition is contributory, the magnitude of solute retention in HILIC shows some correlation with log D values, where D is the distribution coefficient of the solute (i.e. unionised and ionised) between octanol and water. However, it will be seen that the mechanism of HILIC is complex, involving processes additional to partition, thus providing a rationale for the deviations often observed for solute retention compared with that expected from log D values.

In addition to giving good retention of hydrophilic species, HILIC has a number of other advantages compared with RP-LC. Due to the high concentrations of organic solvent used, typical mobile phases have a much lower viscosity than those used in RP-LC. This factor leads to enhanced solute diffusion coefficients in the mobile phase, improved mass transfer and thus limited increase in the C term in the van Deemter equation at high flow rates.[5,6] Columns can also be operated at much lower pressures, prolonging their lifetime or, alternatively, allowing much longer columns to be operated at reasonable backpressures. Higher sensitivities can be obtained using electrospray mass spectrometry due to more efficient spraying and desolvation of the mobile phase.[7] Further advantages are the good peak shapes that can often be obtained for compounds difficult to analyse by RP-LC, such as ionised bases.[5]

While the first applications of UHPLC were in RP separations, and only a handful of HILIC separations using sub-2 μm particles were published prior to 2008, there has been a rapid increase in interest, with over 30 publications referring to the use of these small particles in HILIC in 2009–2010. Diverse applications include the determination of animal growth promoters,[8] metabolomic analysis,[9] anti-viral agents,[10] drugs[11] and glycans.[12] Sub-3 μm superficially porous or "shell particles" have also been shown to give excellent results in the HILIC mode[13] The driving force for the application of UHPLC has been the desire to obtain fast separations of the target compounds while maintaining high efficiency. Particularly in the pharmaceutical industry, there is interest in reducing the time and therefore the cost of the analysis of large numbers of samples.[14] Due to the lower pressures experienced with HILIC mobile phases, it may also be possible to use sub 2 μm columns on conventional HPLC equipment, at least those modified to reduce the high bandspreading typical of these instruments.[15]

In this Chapter, we will consider the complex separation mechanism of HILIC, which contrasts with its simple practical implementation, the kinetic performance of small particle UHPLC HILIC columns and survey some of the different applications of HILIC that have been published using UHPLC techniques.

6.2 Analytical Conditions for Performing HILIC

6.2.1 Choice of Mobile Phase

In HILIC, the mobile phase strength increases in line with increasing the mobile phase polarity in the order:

acetone < acetonitrile < isopropanol < ethanol < methanol < water
increasing strength →

Water is typically used as the strong solvent; in the first instance, the aprotic solvent ACN is normally selected as the weak solvent. An appropriate blend of ACN and water to give a reasonable k value (*e.g.* $2 < k < 10$) can be established either by trial and error or by running a gradient of increasing water concentration. Acetone can be substituted for ACN in applications which do not involve UV detection. Alcohols such as methanol or ethanol can be used in place of water as the strong solvent. As these alcohols are weaker solvents than water, their use gives increased retention, as shown for peptide separations by Yoshida.[16] Bicker *et al.*[17] carried out a detailed study of the analysis of nucleobases and nucleotides on silica materials modified in-house with short alkyl chains carrying hydroxyl functionalities, substituting the water component of the mobile phase with various alcohols. The increase in retention followed the order:

ethanol > methanol > 1,2-ethanediol

Some useful selectivity effects also resulted. Nevertheless, the vast majority of applications continue to use ACN/water combinations.

While simple mixtures of ACN and water can be used for the separation of neutral molecules, a buffer is necessary for the analysis of ionised solutes. The buffer stabilises the ionisation both of the solute and the column surface. Due to the limited solubility of many inorganic buffer constituents in the high concentrations of ACN typically used in HILIC, ammonium formate, ammonium acetate and formic or acetic acids have been used. Figure 6.1(c) shows the separation of 3-phenylpropanol, caffeine and phenol in ACN/water, 85:15, v/v, containing 0.2% formic acid, under conditions where all solutes are uncharged and give excellent peak shape.

Figure 6.1(a) shows the separation of 2-naphthalenesulfonic acid, nortriptyline and propranolol (two strong bases) in the same mobile phase at the same concentration (100 mg L^{-1}) in which all three compounds are fully charged. This resulted in very poor peak result. An improvement was obtained by considerable dilution of the injected solution to a concentration of 1 mg L^{-1} (Figure 6.1b). However, use of 15 mM ammonium formate buffer at approximately the same pH as that of 0.2% formic acid ($_{w}^{w}$pH 3.0) gave excellent peak shapes for the acidic and basic compounds similar to that for the neutrals (results not shown in Figure). Formic acid becomes a very weak acid

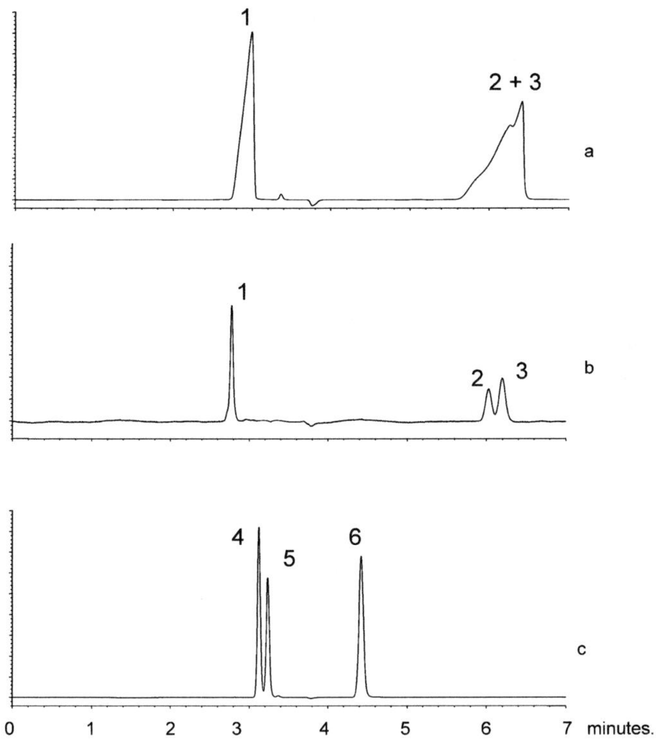

Figure 6.1 (a) Analysis of (1) 2-naphthalenesulfonic acid (ionised acid), (2) nortriptyline (ionised based) and (3) propranolol (ionised base), at a concentration of 100 mg L^{-1}; (b) same as (a) but the concentration was 1 mg/L; (c) (4) phenol, (5) 3-phenylpropanol and (6) caffeine (neutral compounds), at a concentration of 100 mg L^{-1}. Column, Atlantis silica 25 × 0.46 cm. Mobile phase, ACN/water, 85:15 (v/v), containing 0.2% formic acid (v/v). Reproduced from reference 5 with permission.

in 85% ACN such that the ionic strength of the mobile phase is extremely low, which seems to give rise to this effect. In contrast, the salt ammonium formate should be completely ionised in the mobile phase, giving much improved results.[5] An alternative explanation for this behaviour is that the positively charged ammonium ions contained in ammonium formate could block the residual silanols, thus decreasing the peak tailing. However, in a separate experiment in which the effect of buffer concentration on the peak shape of basic compounds was investigated, peak tailing did not seem particularly sensitive to buffer concentration (at least above a threshold value), casting some doubt on this reasoning (see below). A difficulty in HILIC procedures is the effect of high concentrations of ACN on the pH of the mobile phase. pH is conventionally measured in HPLC before addition of the organic solvent (${}_{w}^{w}pH$). However, the pH after addition of the organic component (${}_{w}^{s}pH$, electrode calibrated in aqueous buffers) is markedly different, and the changes are much greater than typically found in RP-LC due to the higher organic concentrations typically used.[5] For example, 15 mM ammonium formate (${}_{w}^{w}$ pH 3.0) in 85% ACN gives ${}_{w}^{s}pH > 5$. A further difficulty is the estimation of the extent of ionisation of any ionogenic groups on the stationary phase under HILIC conditions. These considerations can make the prediction and interpretation of the retention of ionised compounds difficult. As a general rule, mobile phase conditions that result in the ionisation of solutes act to increase retention, as the hydrophilicity of the solute is increased. However, this general rule can be strongly influenced by ionic interactions with charged groups on the stationary phase (see below).

6.2.2 Choice of Injection Solvent

It is important that the injection solvent in HILIC is not stronger than the eluting solvent, otherwise poor peak shapes will be obtained. Guillarme and co-workers[18] recommended that the lowest concentration of water (strong solvent) should be used for injection, regardless of the nature of the compounds. The best peak shapes were usually obtained when the injection solvent was pure ACN. Figure 6.2 shows the effect of the proportion of water in the dissolution solvent on the peak shapes of some small molecules using a gradient of decreasing ACN concentration (starting at 95% ACN, v/v) in ammonium formate buffer, pH 3.1, and a UHPLC 1.7 μm bridged ethyl hybrid silica column. When the volume fraction of the aqueous component in the injection solvent was 20% or greater, a serious deterioration in peak shape occurred, with the effect being greatest for the early peaks. Strongly aqueous dissolution solvents have the greatest difference in elution strength compared with the mobile phase during the early part of the gradient. Nevertheless, the solubility of some compounds in high concentrations of ACN can be problematic. In this case, ACN/isopropanol (50:50, v/v) could be a viable alternative.[18] The negative effects of the strength of the injection solvent can be reduced by using the minimum sample volume possible.

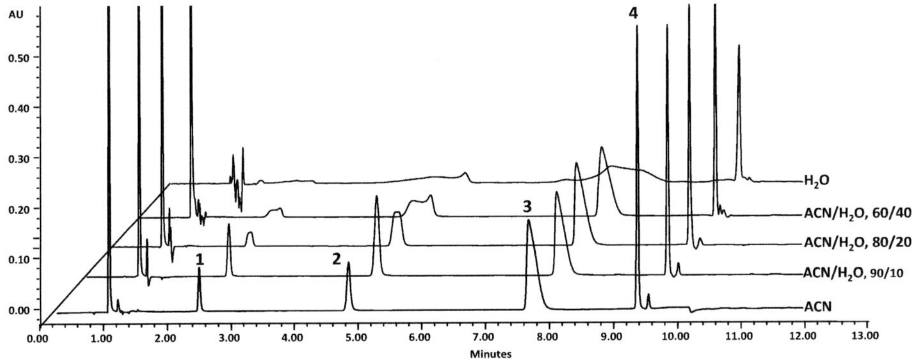

Figure 6.2 Effect of the proportion of water in the dissolution solvent on peak shape of: (1) hypoxanthine, (2) cytosine, (3) nicotinic acid and (4) procainamide. Column, Acquity BEH (15 × 0.21 cm, 1.7 μm). Mobile phase, 50 mM ammonium formate, pH 3.1, modified with ACN. Gradient, 95–75% ACN in 5 min with a 6 min initial hold. Flow rate, 0.5 mL min^{-1}. Injection volume, 5 μL. Reproduced from reference 18 with permission.

6.2.3 Stationary Phases for HILIC

A variety of different stationary phases have been used in HILIC. Most bonded HILIC phases are based on a silica matrix, although polymer-based phases are also commercially available. The most widely used phase is still bare silica, although the commercial availability of increasing numbers of different phases (at least in conventional particle size format) with polar bonded ligands has the potential to challenge the use of bare silica. However, the number of these phases available in sub-2 μm particle size or shell particle format is still rather restricted, and, unfortunately, small particle versions of only the bare silica and amide silica versions were commercially available at the time of writing. Nevertheless, a brief review of column types available at least in conventional particle sizes will be given in the expectation that some at least are likely to be available soon in these new small particle formats.

6.2.3.1 Bare Silica Phases

Bare silica columns are the simplest variety of a HILIC stationary phase. They have the advantage that there is no bonded ligand to cleave off, which may be useful in combination with sensitive detection techniques such as mass spectrometry. Although few formal studies appear to have been published, the stability of a bare silica surface in contact with appreciable amounts of water is likely to be reduced at neutral–high pH values compared with bonded silicas. Silanol groups on bare silica phases are likely to be ionised over the whole range of typical HILIC operation (${}_{w}^{w}$pH 2.5 to ${}_{w}^{w}$pH 7.5), giving the potential for interaction with ionised solutes. Nevertheless, at least with modern "Type B" silica materials (based on highly pure silica), good peak shapes can be obtained with many basic compounds that can give tailing peaks in RP separations.[5] A possible problem with bare silica phases is the difference in chromatographic behaviour (selectivity and efficiency) that might exist between columns from different manufacturers. However, comparison of three bare silica columns (Atlantis, Supelco), Luna (Phenomenex) and Atlantis (Waters) using a test mix of acidic, basic and neutral compounds and an ACN/ammonium formate buffer, pH 3, showed very similar separations.[5] Nevertheless, these columns all contain modern Type B high purity silica, and greater differences would seem likely if older Type A silica columns, which can contain high and variable metal contents, had been included in the comparison.

6.2.3.2 Columns with Neutral Ligands

This variety of phase (*e.g.* diol, amide) has also been popular in HILIC. The group is typically linked to the column surface *via* a spacer such as a propyl group. These phases have been used for a number of applications; for instance, a review of peptide separations on amide silica columns has been published.[16] While the bonded ligand remains uncharged over the whole pH range of HILIC separations, there is still the possibility of charge resulting from the presence of the base material itself (*e.g.* silica), and thus ionic interactions occurring with solutes.

6.2.3.3 Columns with Ionised Ligands

Other types of column have ionogenic bonded groups, for example, aminopropyl silica has been used extensively for the separation of carbohydrates. An important feature of aminopropyl phases is their ability to generate an alkaline environment in the columns when used for these separations with ACN/water mixtures. Sugars can exist in two anomeric forms that can slowly interconvert at low pH. However, at alkaline pH or at high temperature, the anomers rapidly interconvert, giving rise to a single peak which is much easier to quantitate. A problem can be the instability of silica-based phases of this type. However, the stability of the phases can be increased by polymeric bonding of the ligand using trifunctional silanes, or use of a polymer as the column base material. Polysulfoethyl A is a hydrophilic poly(2-sulfoethyl aspartamide) silica which acts as a strong cation exchange material. The selectivity for peptides changes markedly when the mobile phase contains high levels of ACN, and can be used either in the RP or HILIC mode.[2]

6.2.3.4 Zwitterionic Phases

This type of phase has been commercialised by the company SeQuant (now Merck) and contain a polymeric layer with sulfobetaine groups $-CH_2$ $N^+(CH_3)_2$ $(CH_2)_3$–SO_3^- attached to either silica or a polymeric base material.

The presence of a positively charged quaternary ammonium group and a sulfonic acid group in a 1:1 ratio in close proximity gives rise to a zero net charge (but see also below), and a strong ability to bond water to the surface.[19]

6.2.3.5 Mixed Mode Phases

A variety of such phases is now available such as the Dionex Acclaim HILIC-1 which incorporates a diol group in a long bonded carbon chain. The column can be used in the RP mode when the mobile phase contains relatively low concentrations of ACN, but in the HILIC mode with high ACN concentrations.[20] This phase was shown to have particular application for the separation of certain types of surfactant.

6.2.4 Separation Mechanisms in HILIC

Partition into a water layer on the column surface or adsorption on to polar groups deactivated by water has long been considered to contribute to retention in HILIC. Evidence for the existence of the water layer has been provided by both experimental and theoretical studies. An experimental investigation of the retention of sugars on an amino bonded phase showed that water from the mobile phase was extracted on to the surface of the stationary phase.[21] A different experimental approach (see Figure 6.3) investigated the variation in the retention time of benzene and toluene as a function of the water concentration in an ACN/water mobile phase on a superficially porous bare silica phase (2.7 μm particles). Benzene and toluene were used as unretained void volume markers in HILIC. The decrease in retention, which occurs up to approximately 30% water (v/v), can be attributed to the establishment of a mobile phase layer rich in water on the silica surface.[22] Due to the limited solubility of these hydrophobic compounds in water, it can be

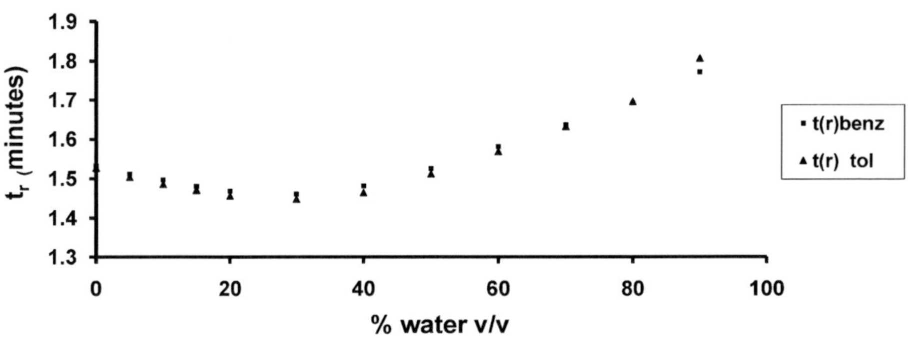

Figure 6.3 Retention time of benzene and toluene (min) as a function of water content of aqueous ACN mobile phase. Flow rate, 1.0 mL min^{-1}. UV detection was carried out at 254 nm. Injection volume, 5 μL. Column, Halo bare silica (2.7 μm shell particles, 15 × 0.46 cm). Reproduced from reference 22 with permission.

assumed that they partition almost entirely into the bulk mobile phase and cannot penetrate the water layer. The difference between the retention volume of the probe in pure ACN and in mobile phases containing up to approximately 30% ACN can be used to estimate the proportion of the pore volume occupied by an impenetrable layer of water, giving values of up to approximately 15%. The existence of a water layer in silica particles under HILIC conditions has also been indicated by molecular simulation dynamics studies performed by Tallarek and co-workers.[23] They found that water was enriched and acetonitrile depleted inside cylindrical silica pores of 3 nm diameter compared with the bulk solution. The simulations were performed with water/acetonitrile mixtures of molar ratios 1:3, 1:1 and 3:1, corresponding to approximate volumetric ratios of 10:90, 25:75 and 50:50 (v/v). The computations indicated the molar ratios of water/acetonitrile in the silica pores (as opposed to the bulk mobile phase) were considerably higher at 1.5, 3.2 and 7.0. Thus the *relative* water fraction in the pores increases with decreasing water content of the bulk mobile phase. The results suggested a layer close to the column surface (< 0.45 nm) where water hydrogen bonds preferentially to silanol groups, with only scarce silanol–acetonitrile bonds. In this region, the water molecules appeared to be nearly immobilised to the silanol groups. Beyond the immediate surface region, water–water hydrogen bonding is preferred, although there are some acetonitrile–water hydrogen bonds.

It is of interest in the consideration of the mechanism to compare the separation of the same mixture of compounds on different stationary phases under the same mobile phase conditions. Figure 6.4 shows the separation of a mixture of two neutral compounds (phenol and caffeine), two strong acids [*p*-xylene-2-sulfonic acid (p-XSA) and naphthalene-2-sulfonic acid (2-NSA)] and four basic compounds (nortriptyline, diphenhydramine, benzylamine and procainamide) on five different HILIC phases using a mobile phases of ACN/water (95:5, v/v) containing 5 mM ammonium formate, ${}_{w}^{w}$pH 3.0. This group of stationary phases contains examples of all the different types discussed in Section 6.2.3, including a mixed-mode phase containing a diol moiety attached to a long carbon chain that can be used either as a HILIC or RP column, in the latter case if low concentrations of ACN are utilised. Clearly, the different columns show widely different retention and selectivity for the probe compounds. This result confirms that the mechanism of separation of HILIC is complex, and that the stationary phase cannot function merely as an inert support for a water layer into which the solutes partition from the bulk mobile phase.[6] The retention of the bases is considerably greater on the bare silica column than the rest (note the time axis scale is a factor of 2 different for this column). One reason may be the greater surface area of this column compared with the others, which should increase the volume of water associated with the stationary phase. A further reason is considerable ionic retention of bases on the dissociated silanol groups of this phase. However, the ionised acidic solutes are repelled from the surface of the bare silica and elute at low retention times, although ionised species would otherwise be expected to

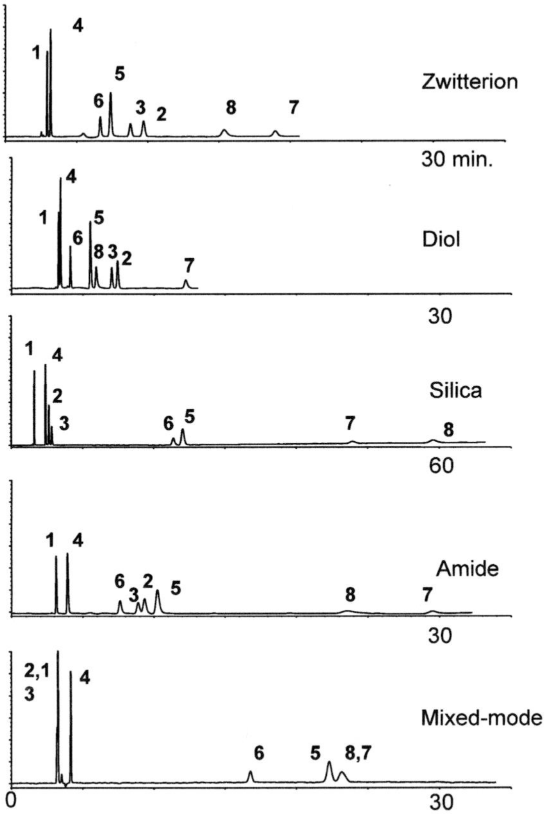

Figure 6.4 Analysis of (1) phenol, (2) 2-naphthalenesulfonic acid, (3) p-xylenesulfonic acid, (4) caffeine, (5) nortriptyline, (6) diphenhydramine, (7) benzylamine and (8) procainamide on different HILIC phases. Mobile phase, ACN/water (95:5, v/v) containing 5 mM ammonium formate, pH 3.0; flow rate, 1 mL min^{-1}. Reproduced from reference 6 with permission.

partition preferentially into the aqueous stationary phase layer (see above). In contrast, better retention of the acid compounds was obtained on the zwitterionic and diol columns where the effect of the column substrate material is likely to be screened by the polymeric ligand bonding of the particular column types used in this study.[6]

Clearly, ion exchange is an important contributor to retention of ionised compounds. Studies of the retention factor of the same set of bases as shown in Figure 6.4 as a function of buffer cation concentration at constant pH indicated that ion exchange contributed to the retention of the basic compounds on all of the columns studied, which were silica-based. The smallest contribution of ion exchange to the retention of the bases was obtained on the diol column (3–17% dependent on the solute) when using 10 mM ammonium formate, ${}_{w}^{w}$pH 3.0, in 90% ACN. The greatest contribution was found on the silica column (25–43%) and for the zwitterionic column (32–55%) when using the same mobile phase. Cationic retention might arise on the zwitterionic phase due to the sulfonic acid groups which, being on the end of the bonded chain, are more accessible to solutes.

Attempts have been made to assess whether partition or adsorption is the major contributor to the (non-ionic) portion of the retention mechanism.[6,19] Retention in adsorption chromatography can be described by the Snyder–Soczewinski equation:

$$\log k = \log k_{\mathrm{B}} - n\log X_{\mathrm{B}} \tag{6.1}$$

where X_{B} is the mole fraction of the strong solvent B (in this case water) in the mobile phase, k_{B} is the retention factor with pure B as the eluent, n is the number of B solvent molecules displaced by the solute. Alternatively, for a partition mechanism, as is thought to contribute substantially to retention in RP chromatography, the following empirical equation gives an approximate description of retention. Here, φ is the volume fraction of solvent B in the mobile phase (at least for neutral solutes), the empirical equation:

$$\log k = \log k_{\mathrm{w}} - S\varphi \tag{6.2}$$

describes retention approximately, where φ is the volume fraction of solvent B in the mobile phase and k_{w} is the hypothetical retention factor when the mobile phase contains no B solvent, *i.e.* contains only the weak solvent. However, the results of these investigations have been inconclusive, with, in some cases, the linearity of the plot indicating a more adsorption-like mechanism and in other cases a more partition-like mechanism.

At low concentrations of ACN (typically 10% ACN or less), hydrophobic type retention can even be demonstrated on bare silica columns.[5] The order of elution of the solutes can be different from that under HILIC conditions. Further decreasing the concentration of ACN at these low organic solvent concentrations gave increased retention of all solutes in this study, the opposite to that using HILIC conditions. It appears that the retention mechanism is a mixture of RP retention on siloxane bonds combined with ionic retention. Nevertheless, the contribution of RP retention is likely to be small for most HILIC applications, considering the high concentrations of ACN generally employed.

In summary, retention in HILIC can be envisaged as a complex, multi-parametric process, consisting of partitioning of solutes between a surface water layer and the bulk mobile phase, adsorption *via* processes such as hydrogen bonding and other dipole–dipole interactions, and electrostatic forces between charged column ligands (or ionised silanols on silica materials) and the ionised solutes.[17,24,25]

6.2.5 Kinetic Performance of HILIC Columns

A study of the efficiency of a variety of conventional (5 μm particle size) HILIC columns has demonstrated the generally flat van Deemter curves that can be obtained with such columns.[6] They show generally low values of the *C* term, thus preserving the high efficiency of the columns at elevated flow velocity. An exception was found to be some types of column that may have polymerised stationary phase layers, which can be detrimental to mass transfer. The good results obtained with many phases can be explained in terms of the higher solute diffusion coefficients in mobile phases of high organic solvent content (thus low viscosity) leading to improved mass transfer. The higher solute diffusion also leads to somewhat larger *B* terms for HILIC columns, which is detrimental to performance at low flow rate, although this observation is likely to be of only academic interest, as most practical separations are performed at or above the optimum flow rate. Nevertheless, increased *B* terms are detrimental to the performance of very long or serially coupled columns intended to generate maximum efficiency within the pressure constraints of a particular system. Such columns may be required to operate under conditions of low flow in order to avoid exceeding the maximum operating pressure of the system.

While most studies published on the sub-2 μm or sub-3 μm superficially porous (shell) HILIC columns have presented applications, a few more fundamental studies have looked at the kinetic performance of these columns. Veuthey and co-workers[26] studied the performance of these two types of small particle HILIC column using a set of 15 model compounds of diverse physicochemical properties. Figure 6.5(a) compares the plate height for cytosine on two sub-2 μm columns (Waters Acquity, Grace Davison Vision HT) together with a shell particle (Ascentis Express) column (all with dimensions 5 × 0.21 cm) under conditions where the retention factor $k > 4.5$ in order to limit the effects of instrumental band broadening on the results. The optimum flow velocity for all phases was approximately 3 mm s^{-1}, which corresponded to a flow rate of 0.45 mL min^{-1}. The minimum plate height *H* of the Acquity column at approximately 4.0 μm was about 30% lower than that of the Ascentis column (6.0 μm), whereas the lower performance of the Vision HT column was attributed to poorer packing. Nevertheless, the backpressure of the Ascentis column was about a factor of 2 lower than the Acquity column, allowing the possible use of longer shell particle columns. Figure 6.5(b) shows alternatively plots of the reduced plate height h (H/d_p) against the reduced velocity v (ud_p/D_m) using the same data. These reduced plots take into account the differences in particle diameter of the phases and thus also indirectly, the different back pressures of operation. In this case, it appears that the performance of the Acquity and Ascentis phases is roughly equivalent, with h_{min} in both cases being ~2.3 μm, although somewhat flatter plots were obtained with the shell particle column at higher velocity. The good performance of the shell particle column (Ascentis) is, however, tempered somewhat by the lower pressure limitation of 600 bar recommended by the manufacturer (compared with 1000 bar for the Acquity column).

Other authors have investigated the performance of wider-bore (0.46 cm i.d.) shell particle silica columns used in the HILIC mode.[13] It seems that there may

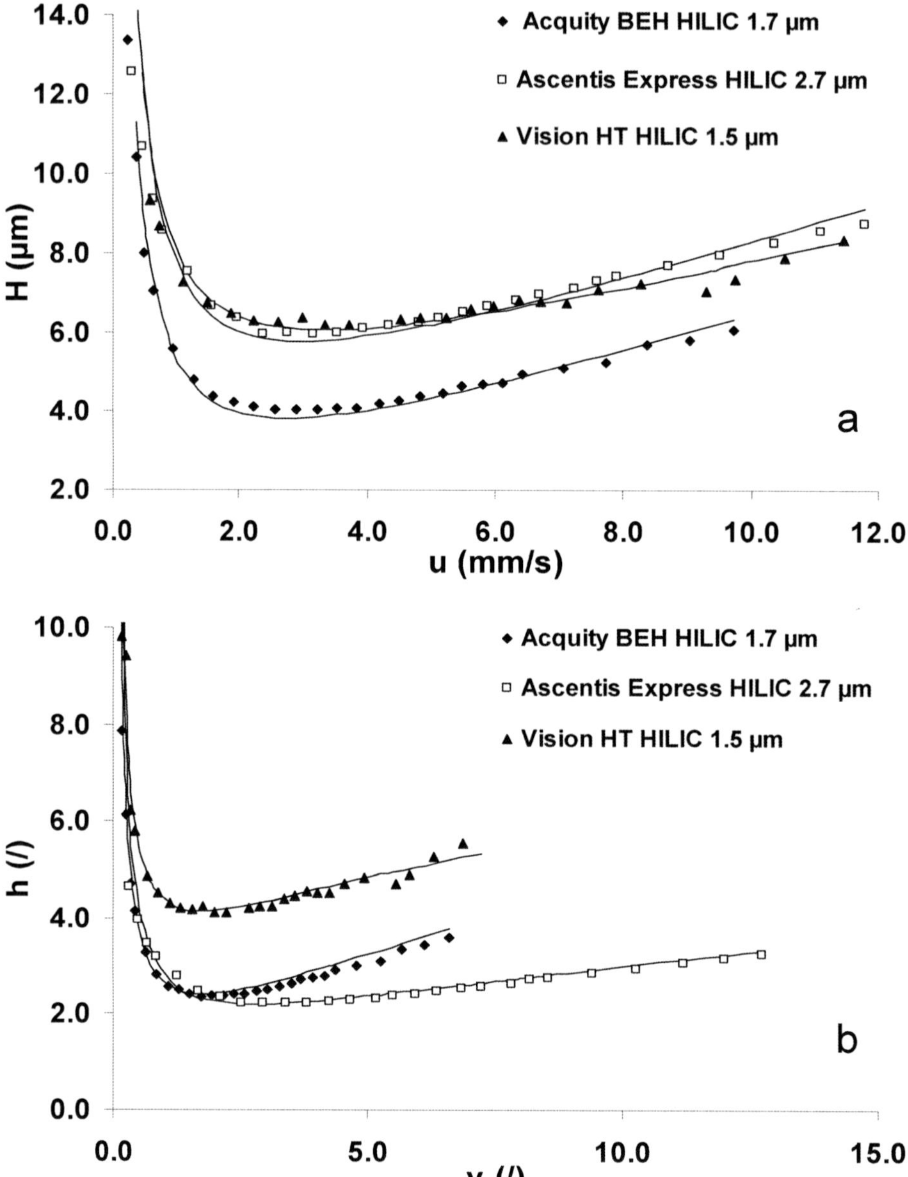

Figure 6.5 van Deemter curves (a) and Knox curves (b) for shell (Ascentis) and sub-2 µm columns for cytosine with $k > 4.5$. Reproduced from reference 26 with permission.

be some packing difficulties in the preparation of narrower bore columns of shell particles,[26] which lead to higher reduced plate height and reduced efficiency compared with 0.46 cm internal diameter (i.d.) columns. Figure 6.6 shows a reduced plot for four basic compounds on a 15 cm Halo silica HILIC

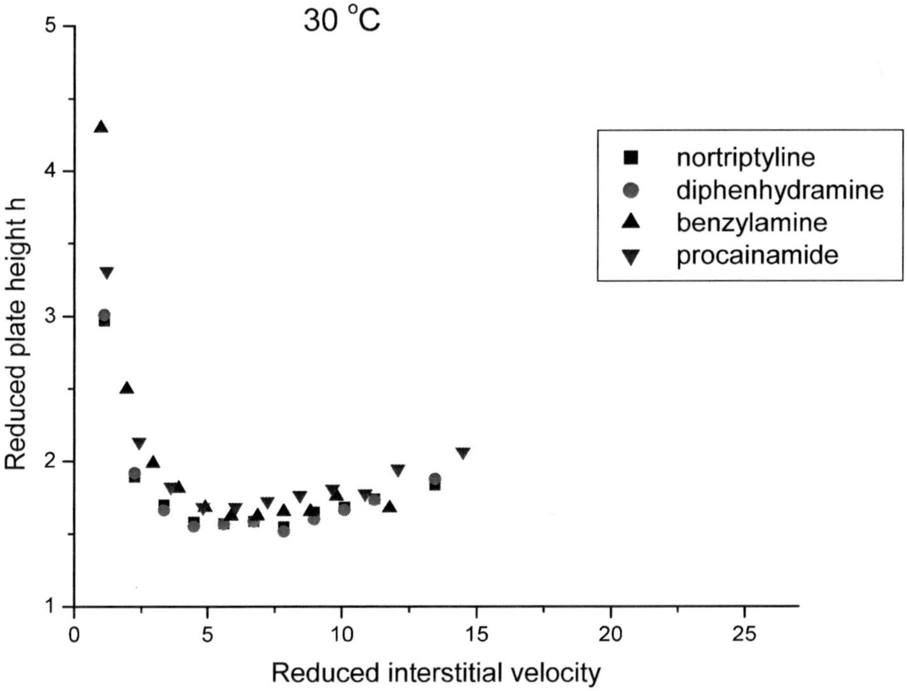

Figure 6.6 Plot of reduced plate height *vs* reduced interstitial velocity (values correspond to flow rates of 0.25–3.0 mL min^{-1}) for four basic compounds on shell particle column (Halo, 15×0.46 cm). Mobile phase, ACN/0.1 M ammonium formate, pH 3.0; temperature, 30 °C. Reproduced from reference 13 with permission.

column using a mobile phase of ACN/0.1M ammonium formate, pH 3.0 (90:10, v/v). The values of h_{min} were as low as 1.5 (considerably lower than the value of ~2.2 found in the other study[26] in which the authors used exclusively 0.21 mm i.d. columns), and may, indeed, be due to the larger diameter format. We believe the manufacturing process of the silica in the Halo and Ascentis range is very similar and thus does not account for the differences in the results of the two studies. The low backpressure generated by the shell column when in combination with the low viscosity of ACN-rich mobile phases allows the possibility of using long coupled columns, even on conventional instruments capable of only 400 bar maximum operating pressure. Figure 6.7 shows the separation of four ionised basic compounds (nortriptyline, diphenhydramine, benzylamine and procainamide), two neutral compounds (phenol and caffeine) and two acid compounds (benzene sulfonic acid and naphthalene-2-sulfonic acid) on a 15 cm porous silica HILIC phase compared with that on one and three linked 15 cm shell particle columns. The pressures noted represent the total system pressure which includes that of the instrument (injector, detector and connecting tubing). The pressure generated by the 15 cm Halo (shell) column is only marginally higher than that of the porous 3 μm column (in line

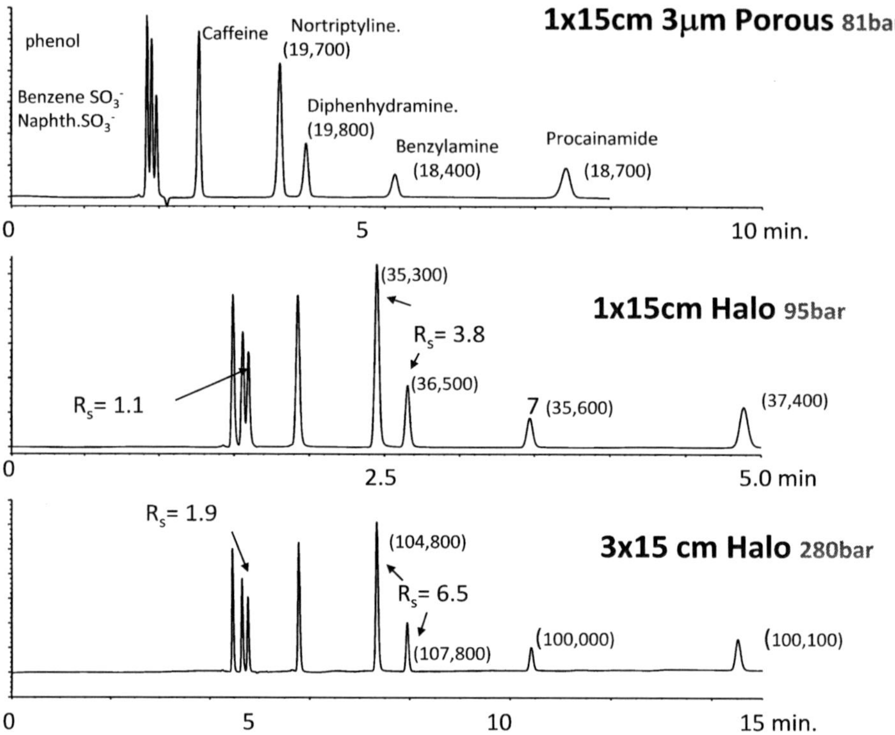

Figure 6.7 Separation of test compounds on 3 μm porous silica and 2.7 μm Halo shell silica columns (15 × 0.46 cm internal diameter). Mobile phase, ACN/0.1 M ammonium formate, pH 3.0 (85:15, v/v); flow rate, 1 mL min^{-1}; temperature, 30 °C. Total plate counts are indicated in brackets.

with its smaller particle size), but gives approximately twice the efficiency. As the total system pressure generated using this single column was <100 bar, three columns could be coupled together, giving >100 000 theoretical plates for the basic compounds with system backpressure still only 280 bar. Thus, additional column length could have been added to increase the efficiency further, while maintaining the backpressure below the maximum of a conventional HPLC system. The resolution of the separation is increased in line with the increase in column efficiency, as shown. It is possible that coupled columns of this type could have applications in difficult separations, such as the determination of closely related impurities in pharmaceuticals.

The reasons for the excellent performance of shell particle columns have been studied extensively, at least for RP versions of these columns. It seems that reduced eddy diffusion (the van Deemter *A* term) is the major factor that gives rise to low plate heights.[27] While the very narrow particle size distributions of these packings could contribute to the small *A* terms, it is not clear whether it is this fact *per se* or the good packing qualities of these materials that is important. However, it appears that the latter may be an extremely important factor.[28] Another factor is the reduced longitudinal diffusion (*B* term) that results from the smaller porous volume of these materials.

6.3 Applications of HILIC in UHPLC

Veuthey and co-workers[26] compared the separation of some drugs using shell or sub-2μm particles on silica columns in the HILIC mode using ACN/ammonium formate buffers, pH 3.0 (see Figure 6.8). Midazolam, bupropion and dextromethorphan, together with their main metabolites (OH-midazolam, OH-bupropion and dextrorphan), could be separated in less than 4 min using either type of phase with backpressures of 375 and 150 bar for the totally porous and shell particles column respectively. The authors[26] pointed out that the order of elution of these solutes was not easily explained according to log D

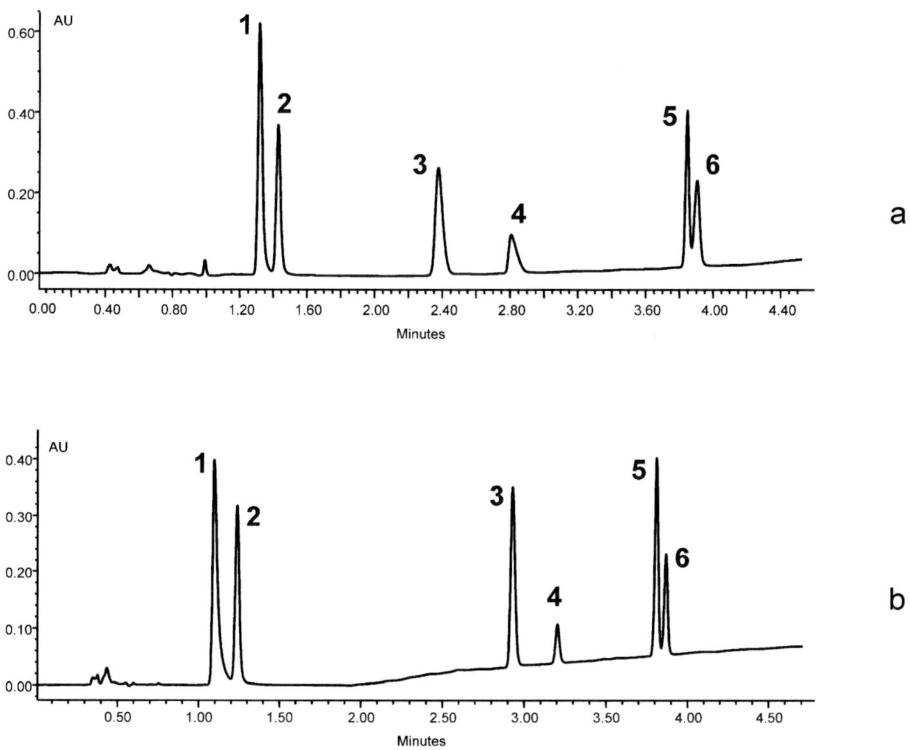

Figure 6.8 Separation of drugs and their metabolites on (a) Acquity, 1.7 μm (10 × 0.21 cm) and (b) Ascentis, 2.7 μm (10 × 0.21 cm) columns. Mobile phase, 10 mM formate buffer, pH 3, and ACN; temperature, 30 °C; flow rate, 0.5 mL min^{-1}. Gradient, 95% ACN for 1.2 min., then 95–80% ACN in 3 min. Peak identities: (1) OH-midazolam, (2) midazolam, (3) bupoprion, (4) OH-bupoprion, (5) dextromethorphan and (6) dextrorphan. Reproduced from reference 26 with permission.

values at pH 3, which predict that dextromethorphan should be eluted between midazolam and bupropion, and dextrorphan between buproprion and OH-bupropion. They proposed that these observations confirmed that partitioning was not the only mechanism operating in HILIC and that ionic interactions, as well as hydrogen bonding, also played an important role.

McCalley[13] showed the fast separation of 10 pyridine and aniline derivatives on a 15 cm shell particle (2.7 μm) silica HILIC column in less than 2 min (see Figure 6.9). The total system backpressure was only 238 bar, achieving 30 200–32 600 plates. Such a separation compares favourably with separations performed with totally porous sub-2 μm RP columns. These basic compounds can still give rise to problems of peak asymmetry, even on some modern RP columns, whereas Figure 6.9 shows very little tailing on the HILIC phase. Nevertheless, separations of some compounds may be problematic on bare silica phases. Catecholamines are hydrophilic biological amines released mainly from the adrenal glands in response to stress. They act as neurotransmitters and hormones, playing an important role in maintaining normal physical activity of the body including heart rate, blood pressure and the reactions of the sympathetic nervous system. High levels of catecholamines in plasma and/or urine can indicate tumours of the adrenal gland, such as pheochromocytoma, or neural tumours, such as neuroblastoma, and are thus routinely monitored in clinical laboratories, typically using RP-LC. However, RP separations require ion pair reagents to increase their retention, which may be detrimental to detection using mass spectrometry. Thus, these solutes would appear to be ideal candidates for analysis by HILIC, in which these reagents are not required.

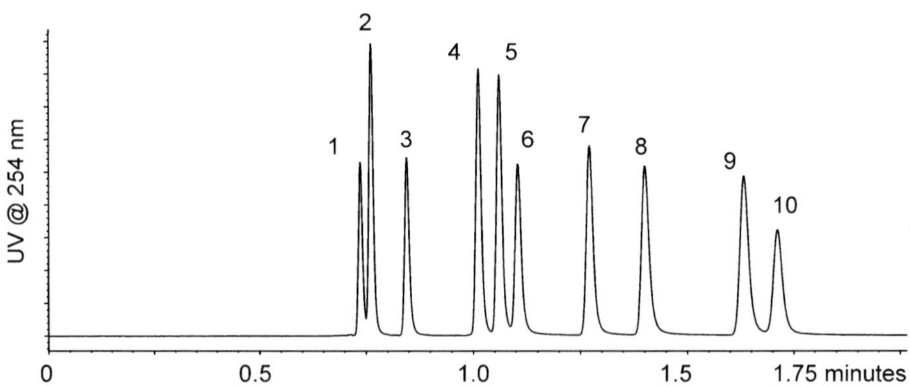

Figure 6.9 Separation of ~0.1 μg of (1) 2-ethylaniline, (2) aniline, (3) 3-butylpyridine, (4) pyridine, (5) 3-methylpyridine, (6) 4-ethylpyridine, (7) 4-methylpyridine, (8) 2,4-dimethylpyridine, (9) 2,4-dimethylpyridine and (10) 2,6-dimethylpyridine on Halo silica column (15 × 0.46 cm). Mobile phase, 80% ACN containing 10 mM ammonium formate, pH 3.0. Flow rate, 2 mL min^{-1}. Reproduced from reference 13 with permission.

Figure 6.10(a) shows the separation of dopamine, adrenaline and noradrenaline using the 2.7 μm Halo shell particle column. Tailing peaks were obtained, together with a disappointingly low efficiency for this type of column. Similar results were obtained on a totally porous silica phase, indicating that the problem was not related to the shell particle structure. Figure 6.10(b) shows the same separation on an amide-bonded phase column, indicating excellent peak shapes. It would appear that, despite the good peak shapes for bare silica reported even for some strongly basic compounds (*e.g.* nortriptyline, see above), there are some solutes which are still best separated on other types of phase. Although this separation was obtained on a 3.5 μm particle size amide phase, it would seem possible to transfer it to a sub-2 μm phase, as one of the same chemistry is available in this small particle size.

The separation of peptides is becoming increasingly important in areas such as proteomics, where proteins are identified by enzymatic cleavage and analysis of the resulting peptides.[29] Peptides are also emerging as an important class of therapeutic agents in the pharmaceutical field.[18] While they are usually analysed by RP-LC in conjunction with mass spectrometric detection, HILIC

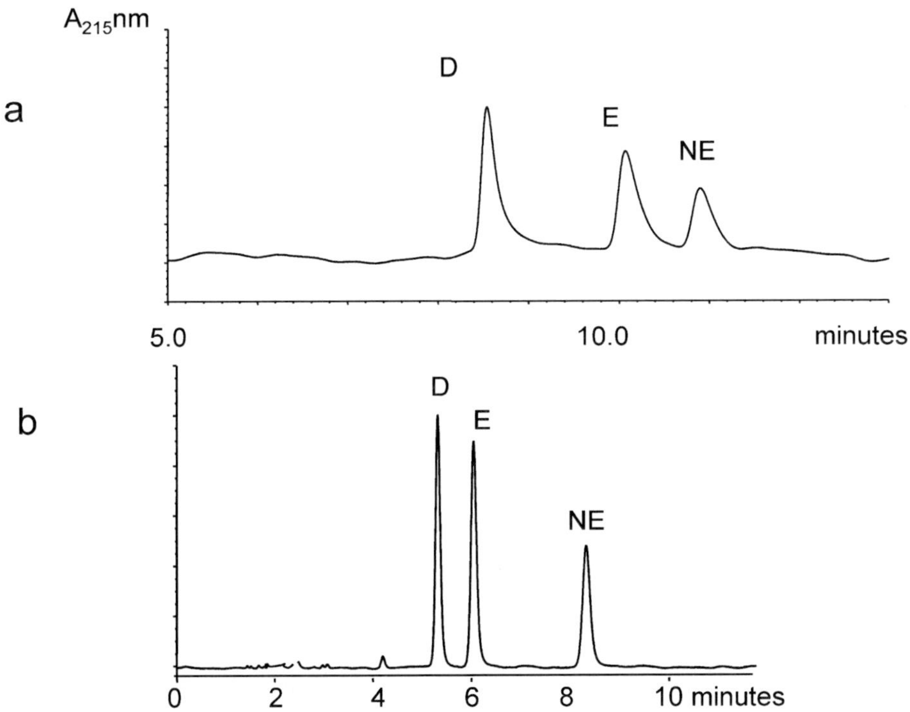

Figure 6.10 (a) Separation of dopamine (D) epinephrine (adrenaline, E) and norepinephrine (noradrenaline, NE) on 2.7 μm Halo silica column (15 × 0.46 cm). Mobile phase, 90% ACN containing 10 mM ammonium formate buffer, pH 3.0, 1 mL min^{-1}. (b) Using a 3.5 μm XBridge amide column (15 × 0.46 cm). Mobile phase, as in (a), except using 85% ACN.

appears to be a useful alternative strategy, as it provides a high degree of orthogonality to RP-LC.[9] Figure 6.11 compares the separation of a mixture of nine peptides using (a) RP-LC and (b) HILIC using 15 cm columns packed with 1.7 µm particles. The correlation coefficient r was –0.35 when plotting HILIC retention times *vs* RP-LC retention times. Note a correlation coefficient closest to zero indicates complete orthogonality of the retentions, and a correlation coefficient of –1 would indicate an inverse correlation, *i.e.* the peaks would be eluted in perfect reverse order in the two separations. Thus the correlation coefficient indicates a low correlation and a high degree of orthogonality; furthermore, it shows that HILIC does not merely elute compounds in the reverse order to RP-LC, which would be undesirable if

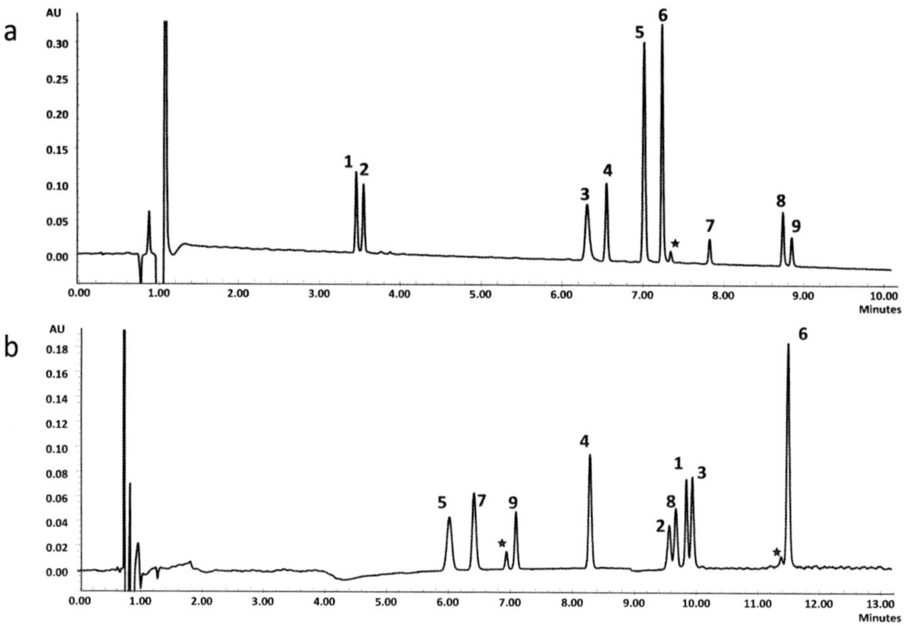

Figure 6.11 Comparison of the chromatographic performance obtained by (a) RPLC *vs.* (b) HILIC for the analysis of a mixture of 9 peptides: (1) lysine vasopressin (20 µg mL^{-1}), (2) arginine vasopressin (12 µg mL^{-1}), (3) peptide D (20 µg mL^{-1}), (4) triptorelin (5 µg mL^{-1}), (5) peptide A (20 µg mL^{-1}), (6) insulin (60 µg mL^{-1}), (7) peptide B (6 µg mL^{-1}), (8) peptide E (25 µg mL^{-1}) and (9) peptide C (6 µg mL^{-1}) dissolved in water for (a) and in isopropyl alcohol (IPA) for (b); *, an impurity present in the synthetic peptides. Conditions: (a) column, Acquity BEH C18 (0.21 cm i.d. × 15 cm, 1.7 µm); flow rate, 400 µL min^{-1}; λ, 214 nm; volume injected, 5 µL; gradient profile, 10 to 90% ACN in 20 min with a temperature of 30°C; (b) column, Acquity HILIC amide (0.21 cm i.d. × 15 cm, 1.7 µm); flow rate, 500 µL min^{-1}; λ, 214 nm; volume injected, 5 µL; gradient profile, 90% ACN for 3 min, then 90 to 62% ACN in 9 min with a temperature of 30°C. Reproduced from reference 18 with permission.

completely different selectivity is sought in the two techniques. Note for example that peptides 8 and 9 are poorly resolved by RP-LC but show considerable separation using HILIC. Conversely, peptides 2 and 8 are not completely resolved by HILIC, but well separated in RP-LC, indicating that the two methods are complementary.

Spagou *et al.*[30] used HILIC–time of flight (TOF)-mass spectrometry with a hybrid silica UHPLC column (10 cm, 1.7 μm particles) to separate highly polar metabolites present in urine after the administration of galactosamine in an investigation of its toxicity. Galactosamine is a well-known hepatotoxin that is widely used to study the mechanism of toxicity. The authors[30] emphasised that HILIC provides complementary information to RP-LC. They pointed out the limitations of RP-LC for metabolomic studies, due to the poor retention of highly polar and ionic analytes, which are very commonly present in urine. The increased mass spectrometric sensitivity because of increased ionisation efficiency resulting from mobile phases containing high organic content, which contribute to the generation of optimum spray conditions, was cited as another important advantage of HILIC in these studies.

Goscinny *et al.*[31] demonstrated a rapid method using the hybrid silica UHPLC column with tandem mass spectrometry for the monitoring of melamine in milk and bakery products. Melamine (2,4,6-triamino-*s*-triazine) was first used as a fertiliser due to its high nitrogen content. More recently it has been used in the production of many commodities including glues, plastics and flame retardants. Unfortunately, contamination of pet foods and later infant food product has been linked with deaths in both animals and humans, necessitating the development of rapid methods for its quantitative analysis to ensure that levels are below the maximum 2.5 mg kg^{-1} regulatory limit. The sample was extracted and purified with acetonitrile/water and the fat removed by partitioning into dichloromethane. Melamine is hydrophilic and highly polar, thus very well suited to analysis using the HILIC technique. Detection was carried out using positive electrospray using the protonated molecular ion at $m/z = 127$. The method was stated to be superior to gas chromatographic procedures, which require derivatisation of melamine.

Glycoprotein characterization and glycosylation profiling are important tasks in the development and production of biopharmaceutical proteins. HILIC has been shown to have good ability to separate positional isomers of these glycans. Gilar and co-workers[32] used a 50×0.21 cm column, 1.7 μm particle size amide phase (Waters glycan column) to separate 2-aminobenzamide-labelled glycans, enabling their detection by fluorescence. Labelled glycans were separated using a gradient of increasing concentration of aqueous ammonium formate buffer in ACN. They showed a gain in peak capacity of the separation by approximately 38% on the 1.7 μm column compared with one of a 3 μm particle size HILIC column of the same material, which was more or less in line with theoretical predictions.

As previously mentioned, applications of HILIC using very small particle columns (sub-2 μm and sub-3 μm shell particles) have been limited by the commercial availability of only bare silica, hybrid silica and amide-bonded hybrid phases. Nevertheless, these phases have been increasingly applied to solve a wide variety of analytical problems in which fast analysis maintaining high efficiency has been required. We expect to see a further growth in the applications of small particle columns using the HILIC technique in the coming years.

References

1. D. V. McCalley, *J. Chromatogr.*, 1983, **260**, 184.
2. A. J. Alpert, *J. Chromatogr.*, 1990, **499**, 177.
3. J. C. Linden and C. L. Lawhead, *J. Chromatogr.*, 1975, **105**, 125.
4. A. J. P. Martin and R. L. M. Synge, *Biochem. J.*, 1941, **35** 1358.
5. D. V. McCalley, *J. Chromatogr. A*, 2007, **1171** 46.
6. D. V. McCalley, *J. Chromatogr. A*, 2010, **1217**, 3408.
7. E. S. Grumbach, D. M. Wagrowski-Diehl, J. R. Mazzeo, B. Alden and P. C. Iraneta, *LC-GC* 2004, **22**, 1010.
8. G. Kesiunaite, E. Naujalis and A. Padarauskas, *J. Chromatogr. A*, 2008, **1209**, 83.
9. H. G. Gika, G. A. Theodoridis and I. D. Wilson, *J. Sep. Sci.*, 2008, **31**, 1598.
10. Z. L. Xiong, Y. Zang, F. Qin, T. Qin, S. Y. Zang and F. M. Li, *J. Chromatogr. B*, 2010, **878,** 23.
11. L. Discenza, C. D'Arienzo, T. Olah and M. Jemal, *J. Chromatogr. B*, 2010, **878**, 1583.
12. J. Bones, S. Mittermayr, N. O'Donoghue, A. Guttman and P. M. Rudd, *Anal. Chem.*, 2010, **82,** 10208.
13. D. V. McCalley, *J. Chromatogr. A*, 2008, **1193**, 85.
14. B. Chauve, D. Guillarme, P. Cléon and J-L. Veuthey, *J. Sep. Sci.*, 2010, **33**, 752.
15. D.V. McCalley, *J. Chromatogr. A*, 2010, **1217,** 4561.
16. T. Yoshida, *J. Biochem. Biophys. Meth.*, 2004, **60,** 265.
17. W. Bicker, J. Y. Wu, M. Lämmerhofer and W.Lindner, *J. Sep. Sci.*, 2008, **31**, 2971.
18. J. Ruta, S. Rudaz, D. V. McCalley, J-L. Veuthey and D. Guillarme, *J. Chromatogr. A*, 2010, **1217**, 8230.
19. P. Hemström and K. Irgum. *J. Sep. Sci.*, 2006, **29,** 1784.
20. X. Liu and C. Pohl, *J. Chromatogr. A*, 2008, **1191**, 83.
21. L. A. Verhaar and B. F. M. Kuster, *J. Chromatogr.*, 1982, **234**, 57.
22. D. V. McCalley and U. D. Neue , *J. Chromatogr. A*, 2008, **1192,** 225.
23. S. M. Melnikov, A. Höltzel, A. Seidel-Morgenstern and U. Tallarek, *Anal. Chem.*, 2011, **83,** 2569.
24. M. Lämmerhofer, M. Richter, J. Wu, R. Nogueira and W. Bicker, *J. Sep. Sci.*, 2008, **31,** 2572.
25. J. Wu, W. Bicker and W. Lindner, *J. Sep. Sci.*, 2008, **31**, 1492.

26. B. Chauve, D. Guillarme, P. Cléon and J-L. Veuthey, *J. Sep. Sci.*, 2010, **33,** 752.
27. F. Gritti and G. Guiochon, *J. Chromatogr. A*, 2010, **1217,** 8167.
28. D. Cabooter, A. Fanigliulo, G. Bellazi, B. Allieri, A. Rottigni and G. Desmet, *J. Chromatogr. A*, 2010, **1217,** 7074.
29. M. Gilar, P. Olivova, A.E. Daly and J.C. Gebler, *Anal. Chem.*, 2005, **77,** 6426.
30. K. Spagou, I. D. Wilson, P. Masson, G. Theodoridis, N. Raikos, M. Coen, E. Holmes, J. C. Lindon, R. S. Plumb, J. K. Nicholson and E. J. Want, *Anal. Chem.*, 2011, **83**, 382.
31. S. Goscinny, V. Hanot, J-F. Halbardier, J-Y. Michelet and J. Van Loco, *Food Control*, 2011, **22,** 226.
32. J. Ahna, J. Bones, Y. Q. Yua, P. M. Rudd and M. Gilar, *J. Chromatogr. B.*, 2010, **878,** 403.

CHAPTER 7

UHPLC/MS Coupling: How to Select a Suitable Configuration?

LUCIE NOVÁKOVÁ[a], MICHAL HOLČAPEK*[b], ROBERT JIRÁSKO[b] AND MIROSLAV LÍSA[b]

[a] Charles University in Prague, Faculty of Pharmacy, Department of Analytical Chemistry, Heyrovského 1203, 500 05 Hradec Králové, Czech Republic; [b] University of Pardubice, Faculty of Chemical Technology, Department of Analytical Chemistry, Studentská 573, 532 10 Pardubice, Czech Republic

*E-mail: Michal.Holcapek@upce.cz

7.1 Introduction

The ultra-high-performance liquid chromatography–mass spectrometry (UHPLC/MS) era started in 2004, when Waters introduced their first Acquity UPLC system on to the market. The researchers and also other manufacturers immediately recognized the importance and potential of new instrument, and nowadays all chromatographic manufacturers provide their own UHPLC systems.[1] The significant reduction in the analysis time often associated with the improvement in the chromatographic resolution impressed the whole chromatographic community and within less than a decade the UHPLC system is considered as a common standard in many analytical laboratories worldwide. The mass spectrometer is one of the most sensitive, universal (various ionization techniques in both polarity modes) and also highly specific (selected reaction monitoring scans for multiple ion reactions) detector, therefore the UHPLC/MS coupling is a method of choice for the most demanding analytical application in chemistry, biology, medicine and

RSC Chromatography Monographs No. 16
UHPLC in Life Sciences
Edited by Davy Guillarme and Jean-Luc Veuthey

Published by the Royal Society of Chemistry, www.rsc.org

related fields. This Chapter is only focused on the mass spectrometric part of this coupling, and potential ionization techniques and mass analyzers are discussed in more detail in this Chapter.

7.2 Selection of Ionization Techniques

The principles of atmospheric pressure ionization (API) techniques are well described in numerous textbooks and monographies,[2,3] therefore we will only focus on the application range of individual techniques. The established standard in (U)HPLC/MS is the use of two complementary ionization techniques: electrospray ionization (ESI) and atmospheric pressure chemical ionization (APCI). Recently, atmospheric pressure photoionization (APPI) has been introduced as the third API technique to cover the whole range of polarities and molecular weights for organic, bioorganic and organometallic compounds. Figure 7.1 shows an example of typical compound classes that are analyzed in metabolomic analysis. On the basis of a Web of Science search, ESI forms approximately 93% of UHPLC/MS applications, whereas APCI and APPI techniques are used in the remaining 5% and 2% papers, respectively. This trend shows that bioanalytical applications (*e.g.*, proteomics, metabolomics) strongly prevail in UHPLC/MS. ESI is usually used for compounds from a medium polarity to ionic species. In contrast to older ionization techniques such as electron ionization, the ionic and non-volatile character of the studied compound is an advantage in ESI, resulting in very good ionization efficiency and excellent sensitivity, for example, in case of polysulfonic acids[4] and ionic metal complexes.[5] ESI can be considered as the most gentle ionization technique, allowing the analysis of large biopolymers

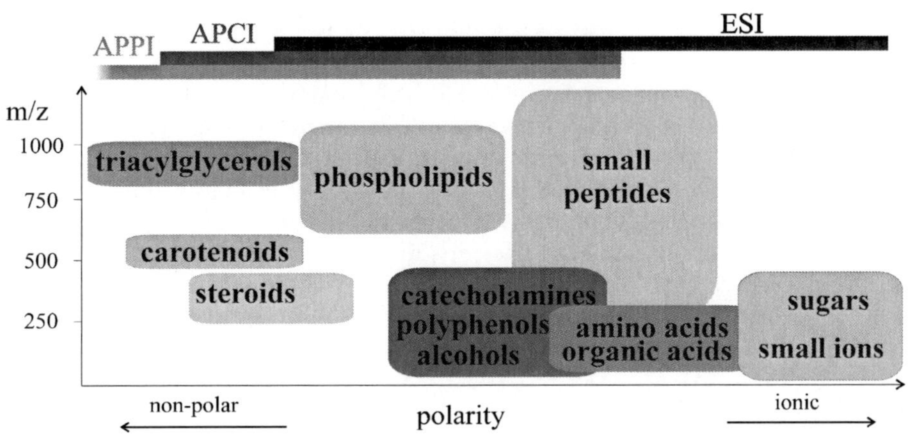

Figure 7.1 Application range of atmospheric pressure ionization (API) techniques for selected compound classes in relation to their molecular weights and polarity.

and their non-covalent complexes, synthetic polymers, organometallic compounds[6] and other labile analytes. The ESI analysis of large proteins[7] due to the presence of multiply charged ions completely changed previous limitations of MS, and J. B. Fenn was awarded by the Nobel Prize for Chemistry in 2002 for this discovery. If a suitable small cation is added into the solution to promote the appropriate adduct formation,[8] *e.g.*, $[M+Na]^+$, $[M+K]^+$, $[M+NH_4]^+$, $[M+Li]^+$ or $[M+Ag]^+$, then ESI can be also applied to non-polar compounds, such as unsaturated hydrocarbons.

APCI and APPI have rather similar application ranges for low-to-medium polar compounds with molecular masses up to approximately 1500–2000 Da. The limit of APPI is slightly wider for highly non-polar analytes (*e.g.*, polyunsaturated aromatic hydrocarbons). APPI may have a softer character compared to APCI, but it depends strongly on the analyte and the type of dopant used, mainly the relation between ionization energies of dopant and analyte.[9] In general, lower relative abundances of ionic adducts and also smaller problems with ion suppression effects are observed in APCI/APPI compared to ESI, because the ionization process takes place solely in a gas phase. Typical application areas of APCI/APPI are drugs and their Phase I metabolites, non-polar lipids, natural phenolic compounds, pesticides and other medium polar organic compounds,[8] but these techniques are not convenient for biopolymers and ionic species. APPI has a stronger tendency to form radical ions due to the different ionization mechanism,[10,11] which depends on the presence or absence of dopant, and also significant differences in the ionization mechanisms have been described for different mobile phase composition.[12,13]

In principle, off-line matrix-assisted laser desorption/ionization (MALDI) can be used in UHPLC/MS, as well as in HPLC/MS, with similar advantages and drawbacks. Narrower chromatographic peaks in UHPLC put even more emphasis on the precise correlation between the *x*,*y*-position of the eluent deposited on the MALDI plate and the retention time, otherwise the chromatographic resolution of UHPLC is sacrificed. This combination does not seem a primary choice for UHPLC/MS coupling, as illustrated by the absence of any reference in the literature. A large number of ambient sampling/ionization techniques have been introduced in recent years.[14] Some of these can be theoretically coupled to UHPLC, but this is evidently not the main stream of development due to rather contradictory concept of such both approaches. On the basis of the current state of knowledge, we do not expect too much research on UHPLC coupling to ambient techniques.

7.3 Overview of Mass Analyzers and Their Main Features

Individual mass analyzers are based on different physical principles using mostly electric and/or magnetic fields. The principal function of mass analyzer is to separate ions based on their mass-to-charge (*m*/*z*) ratios. Each mass analyzer is characterized by several parameters, such as the resolving power (RP), mass accuracy, mass range, linear dynamic range, acquisition speed, sensitivity and tandem mass spectrometry (MS/MS) capability.[15] The RP, as probably the most important parameter for the quality of mass analyzer, is defined as the full width at half maximum (FWHM) of a particular peak divided by its *m/z* value. The mass accuracy is defined as $(m/z_{\text{experimental}} - m/z_{\text{theoretical}}) \times 10^6/(m/z_{\text{theoretical}})$. Nowadays, the six basic types of mass analyzers are quadrupole (Q), ion trap (IT), time-of-flight (TOF), double-focusing magnetic sector, orbitrap and ion cyclotron resonance (ICR). The low-resolution mass analyzers have a RP in the range of thousands (Q and IT), whereas high-resolution analyzers have the RP in the tens or hundreds of thousands (TOF, double-focusing magnetic sector and orbitrap). A RP of higher than a million is referred as the ultra-high resolution (ICR). A useful division of mass analyzers is that of ion beam *vs.* trap-based analyzers.[16] Ion beams are collections of all of the ions moving in well-defined directions by analogy to beams of light (Figure 7.2). Trap-based mass analyzers manipulate ions within a limited analyzer volume, in which ions undergo oscillatory motions under the influence of time-dependent electric fields (Figure 7.3).[17] The *m/z* selective instability is used to expel ions towards the detector (IT) or frequencies of their oscillatory motions are measured (ICR and orbitrap). MS/MS can be performed either in some of the trap-based analyzers (isolation of ions in time) or by the coupling of two analyzers in tandem (isolation of ions in space). The tandem coupling may employ the same type of the analyzer [*e.g.*, triple quadrupole (QqQ)] or two different types of mass analyzers, which is referred as a hybrid instrument (*e.g.*, QqTOF).[18]

Table 7.1 summarizes the basic parameters of individual mass analyzers. At present, the double-focusing magnetic sector mass analyzer almost disappeared from the market except for some special applications (*e.g.*, the analysis of dioxins). Moreover, the acquisition speed of magnetic analyzer is not compatible with fast UHPLC separations, so this analyzer type does not play a role in the UHPLC/MS. The acquisition speed in Table 7.1 is reported in Hz, but the alternative way of scanning speed used by some manufacturers is the number of *m/z* units scanned over 1 s. The aquisition speed strongly depends on the scan type used for the measurement. The reported values are related to the full scan over the common mass range of particular analyzer, which is in fact the slowest scan rate. On the other hand, the selected reaction monitoring on the QqQ analyzer can reach 200–1000 Hz cycle time for one transition.

For Fourier transform (FT) mass analyzers, the increased acquisition speed can be obtained at cost of reduced RP and therefore it is impossible to obtain both performance characteristics (RP *vs.* acquisition speed) in UHPLC/MS experiments. The acquisition speed of both FT mass analyzers (at the best possible RP and mass accuracy) is slower compared to Q, IT and TOF analyzers. The mass range of FT analyzers is mostly limited by the transmission of preceding ion optics similarly as for TOF analyzers in

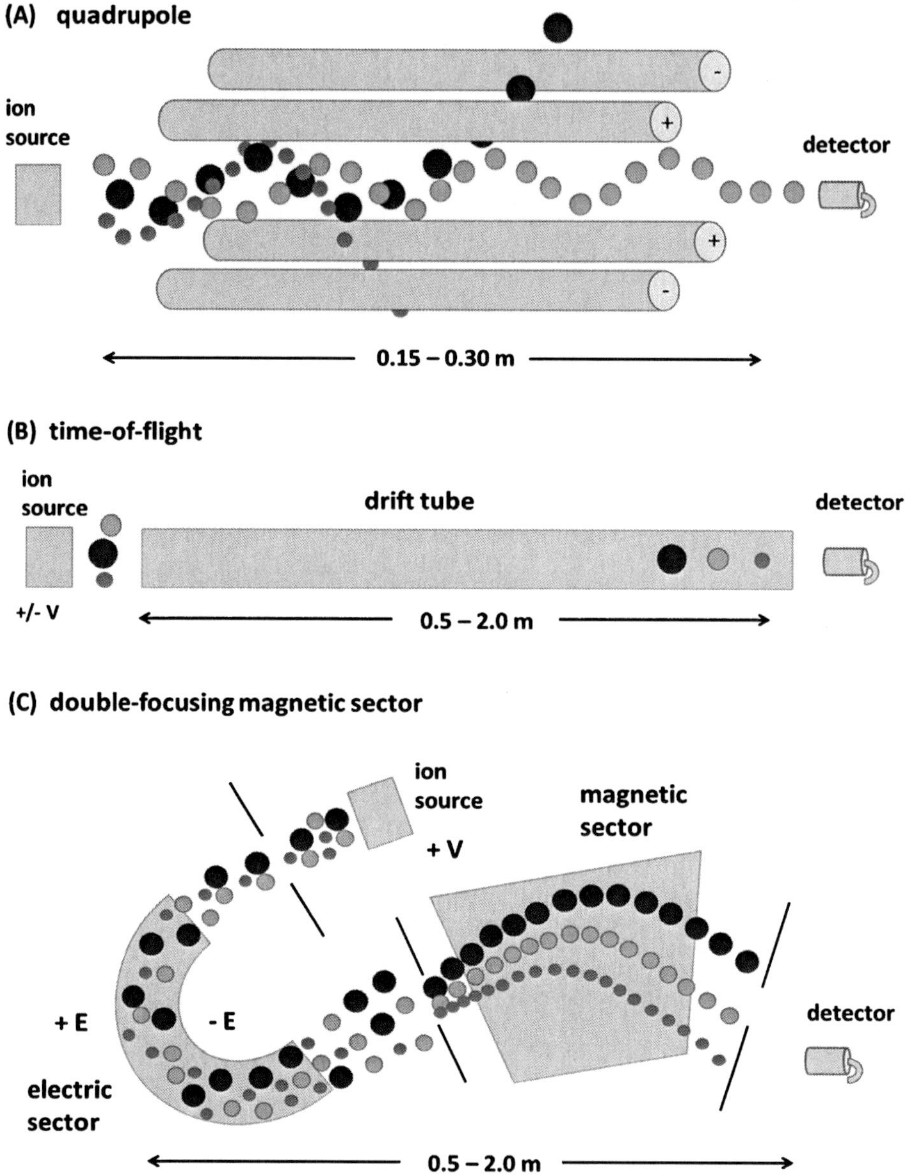

Figure 7.2 Schematic overview of mass analyzers based on the ion beam: (A) quadrupole (Q), (B) time-of-flight (TOF) and (C) double-focusing magnetic sector analyzer. Redrawn and adapted from reference 17 with permission.

QqTOF configuration. In practice, most orbitrap and ICR commercial configurations contain Q or linear IT (LIT) as the first stage. Mass accuracies for low-resolution mass analyzers are in general relatively low (approximately

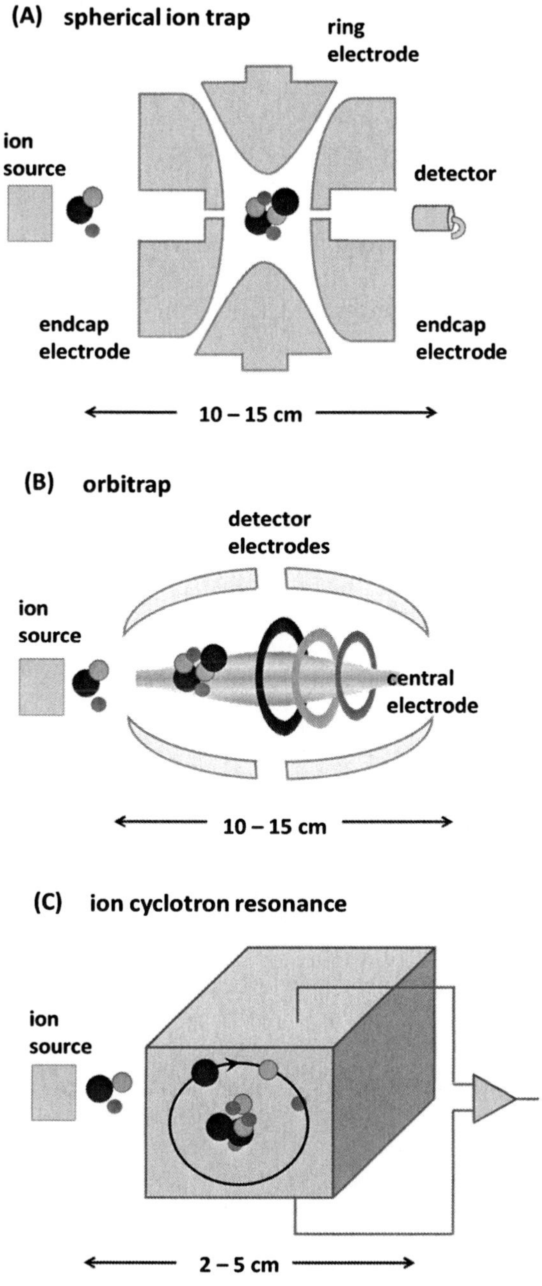

Figure 7.3 Schematic overview of trap-based mass analyzers: (A) spherical ion trap (IT), (B) orbitrap and (C) ion cyclotron resonance (ICR). Redrawn and adapted from reference 17 with permission.

Table 7.1 Overview of basic mass analyzers and their typical characteristics in (U)HPLC/MS. Typical parameters are shown. Maximum values declared in manufacturers specifications are reported in parenthesis, but it may be related to specific conditions or only for one outstanding instrument.

Analyzer	*Mass range* $(\times 10^3)$	*Resolving power*[a] $(\times 10^3)$	*Mass accuracy (ppm)*	*Acquisition speed (Hz)*	*Linear dynamic range*	*Costs*
Quadrupole (Q)	3–4 (20)	3–5 (20)	50–100 (5)	2–10	10^5–10^6	\$–\$\$
Ion trap (IT)	3–4 (6)	3–5 (20)	50–100	2–10 (20)	10^4–10^5	\$–\$\$
Time-of-flight (TOF)	6–20 (40)[b]	15–40 (60)	3 (1)	5–40 (100)	10^4–1–0^5	\$\$–\$\$\$
Magnetic sector	6–15	40–80	5 (1)	0.2–1	10^6–10^7	\$\$\$\$
Orbitrap	4	100–240	2 (1)	1–5 (12)	5×10^3	\$\$\$–\$\$\$\$
Ion cyclotron resonance (ICR)	10	1000 (2500)	1 (0.5)	0.5–2 (10)	10^4	\$\$\$\$–\$\$\$\$\$

[a]The resolving power is calculated according to the full width at half maximum (FWHM) definition, except for the double-focusing magnetic sector analyzer, where the established definition for this particular type of mass analyzer is based on the 10% valley for two neighboring peaks of identical heights. In fact, values stated for magnetic sector should be approximately doubled, if the FWHM definition is used. [b]The mass range of time-of-flight (TOF) analyzer in the MALDI configuration is theoretically unlimited, but this Table shows typical parameters for orthogonally accelerated TOF in the (U)HPLC/MS configuration, where the limitation by preceding ion optics (*e.g.*, the first transmission Q in QqTOF hybrid analyzers) must be taken into account.

50–100 ppm) and insufficient for the elemental composition determination with the exception of Q with hyperbolic rods, where 5 ppm mass accuracy is declared for the external calibration and 3 ppm for the internal calibration. The mass accuracy on TOF-based instruments increases with increasing *m/z* values, whereas the reversed relation is typical for FT instruments.

The cheapest mass analyzers (one dollar symbol in Table 7.1 – \$) on the market are single Q and basic configurations of spherical IT. Prices of QqQ and front-end spherical and linear IT with advanced features (*e.g.*, special types of collision experiments) are slightly higher (\$\$). The next step in the cost requirements is the TOF analyzer (\$\$), especially in QqTOF configuration (\$\$\$). The basic configuration of orbitrap analyzer (\$\$\$) is even slightly cheaper than the best QqTOF, whereas the premier orbitrap configurations are in the range \$\$\$\$. ICR is the superior analyzer type in any respect including the price (\$\$\$\$\$), but the configuration with the 7T magnet is comparable in the price to the best orbitraps (\$\$\$\$). In the case of ICR, higher operating costs should be also kept in mind. Anyway, the cost information should be understood as only rough guide, because such simplified comparison cannot cover all real configurations and special offers.

7.3.1 Basic types of mass analyzers

7.3.1.1 *Quadrupole (Q) analyzers*

Both Q and IT systems use a combination of radio frequency alternating current (AC) and direct current (DC) voltages for the separation of ions.[19,20] These MS systems are probably the most widespread mass spectrometers due to their low price, simple operation and satisfactory performance for less demanding applications.

A Q analyzer employs a combination of DC and AC electric fields as a mass filter. The Q consists of four longitudinally parallel rods. The positive DC voltage is applied on two opposite rods, while the same value of the negative DC voltage is applied on remaining two rods. The radio frequency AC is connected to all four rods, where the amplitude of AC is slightly higher than the value of DC voltage. For any given set of DC and AC potentials, only ions of a specific *m/z* value avoid collisions with poles and successfully pass through the Q filter along the *z*-axis to reach the detector. All other ions have unstable trajectories and they do not reach the detector (Figure 7.2A). The single Q is the simplest mass analyzer, robust and cheap, but it suffers from a low RP, mass accuracy and sensitivity.[21]

The coupling of UHPLC with single Q device was investigated by only few research groups probably due to the limited selectivity of single Q analyzer and the availability of superior tandem instruments. Schappler *et al.*[22] evaluated the performance of single Q in UHPLC/MS analysis of acidic and basic pharmaceuticals under the influence of several variables such as pH value and flow rate of the mobile phase, dwell time and polarity switching. The experiments with polarity switching revealed no loss in sensitivity. Applications of Q in UHPLC/MS were further reported in the qualitative pharmaceutical analysis,[23] the quantitative analysis of simple matrices such as vegetable oils[24] or dietary supplements,[25] as well as in the compound purity determination.[26,27]

7.3.1.2 *Ion Traps (ITs)*

There are basically two kinds of IT: spherical and linear ITs (LITs). Spherical IT uses a three-dimensional Q electric field to store ions of multiple *m/z* values in concentric three-dimensional orbitals. The spherical IT consists of two end-cap electrodes that are electrically isolated from either side of a ring electrode (Figure 7.3A). This three-dimensional device has a hyperbolic cross-sectional surface, which is consistent with those used in the Q technology.[18,19,21] One of the limitations of spherical IT is the space charge effect, which means that only a certain maximum number of ions can be stored in a given volume of spherical IT due to space charge constraints. Once the maximum is reached, newly generated ions of different *m/z* values begin to displace those of other *m/z* values already occupying space in the trap. This results in a loss of resolution with potential *m/z* shift and space charge effects. To prevent this issue, two main techniques of controlling the ion packet injection into spherical IT are employed, *i.e.*, the automatic gain control introduced by Thermo Fisher Scientific and ion-current control used by Bruker Daltonics. The automatic gain control is based on a short pre-scan, where the number of ions is determined and then the appropriate scan time is adjusted to produce an optimal number of ions to reach a maximum sensitivity, but avoid the space charge effects.[21] The principle of ion-current control is very similar, but the number of optimal charges is determined in the previous regular scan instead of short pre-scan, therefore no scanning time is wasted on the pre-scan. Another limitation of spherical IT is that ions generated with an ESI continuous source are not used while spherical IT is processing, which limits the duty cycle substantially.[28] UHPLC-compatible spherical IT was used for metabolite identification,[29] plant analysis[30–32] or peptide characterization.[33] The IT typically works in the MS/MS mode only in the UHPLC/MS coupling due to time constraints.

LIT analyzers are rapidly finding new applications in many areas of MS. The LIT has higher linear dynamic range compared to the spherical IT, which results in a lower risk of space charge effects and better suitability for the quantitation. In the LIT, ions are confined radially by a two-dimensional AC field, and axially by stop potentials applied to end electrodes. In comparison to the spherical IT, LIT has higher injection efficiencies and higher ion storage capacities. The LIT is based on the structure of a transmission Q (an array of four rods with applied voltages). However, instead of filtering ions of all *m/z* values except for those of the desired value, LIT is used for trapping, manipulation of ion trajectories and ion ejection of selected *m/z* values. LIT can be combined with other mass analyzers in hybrid instruments (Section 7.3.2.2) to isolate ions of selected *m/z* values, to perform MS/MS experiments and to study ion-molecular chemistry.[28,34] UHPLC coupled to LIT as a stand-alone instrument has been reported in the scientific literature,[35–37] but the main application of LIT is the coupling with high-resolution mass analyzers in hybrid instruments (see Section 7.3.2.2 for details).

7.3.1.3 Time-of-flight (TOF)

TOF–MS involves the measurement of the flight time of ions in the drift tube (Figure 7.2B). The inherent characteristics of TOF–MS are high sensitivity (all ions are detected), theoretically unlimited mass range and acquisition speed (modern instruments up to 100 Hz). However, early TOF instruments suffered from the low RP and mass accuracy, which has been overcome by three major developments in the TOF technology: the reflectron, delayed extraction and orthogonal acceleration.[38,39] When using a reflectron (sometimes called ion mirror), ions are reflected by a higher electrical potential than the acceleration potential before reaching a detector installed on the source side. The reflectron focuses ions with identical *m/z* values but different kinetic energies, which significantly improves the RP. This focusing results from a deeper penetration in the reflectron for ions with the same *m/z* value but higher velocities. These ions have a longer path of flight in the analyzer than the slower ones.[38–40] The further reduction of kinetic energy spread of individual ions (increase of RP) is achieved by a short delay before the extraction pulse is applied, which is referred to as delayed extraction of ions.

TOF analyzers are directly compatible with pulsed ionization techniques such as laser desorption, the most common being MALDI. The coupling of TOF with continuous ion sources, such as ESI or APCI, causes the transformation of a continuous ion beam into a pulsed one. The orthogonal acceleration is the technique of choice for the coupling of continuous sources with TOF analyzers. At present, high-end TOF instruments have an RP of 40 000–60 000 and mass accuracies of <2 ppm. Routine instruments have a resolution up to 10 000–20 000 and an accuracy of <5 ppm.[39] UHPLC/TOF coupling has been quite widely reported in the scientific literature.[41–45] However, it was currently outstripped by UHPLC/QqTOF hybrid instruments due to MS/MS capabilities (Section 7.3.2.1).

7.3.1.4 Double-focusing Magnetic Sector

In sector instruments, magnetic (B) or electric (E) sectors are used to increase the RP. Magnetic field works as a momentum separator and separates ions according to their *m/z*, while the electric sector focuses ions based on their kinetic energy (Figure 7.2C). In strong contrast to the past time, double-focusing magnet sector instruments are not too widespread and their main disadvantage, namely a low acquisition speed explains why no references have been found for UHPLC/MS coupling with magnetic analyzers. Sector instruments are still widely used in the gas chromatography coupling for specific applications such as the analyses of polychlorinated biphenyls, dioxins and polybrominated compounds used as flame retardants.[39]

7.3.1.5 Orbitrap

The orbitrap mass spectrometer, first described by Alexander Makarov in 2000, is the latest development in trapping devices[46,47] and was commercialized in 2005. The orbitrap analyzer operates by radial trapping of ions around the central spin electrode. An outer barrel-shaped electrode is co-axial with the inner spindle-shaped electrode (Figure 7.3B). A constant electric potential is imposed between these two electrodes. Unlike conventional IT, such as Paul and Penning traps, the orbitrap uses only electrostatic fields (DC) to confine and analyze an injected ion population without AC or magnetic field. *m/z* values are measured from the frequency of harmonic oscillations of ions, along the axis of electric field. This axial frequency is independent of the energy and spatial spread of ions. Ion frequencies are measured non-destructively by acquisition of time-domain image current transients followed by fast FT to obtain mass spectra.[48,49] The orbitrap provides very high RP (100 000–240 000), mass accuracy (typically below 1 ppm) and good dynamic range. In addition, its relatively reasonable cost, simple design and high space charge capacity make it suitable for tackling complex scientific problems. In order to optimize the ion injection and MS/MS capability, the orbitrap is often combined with LIT or a C-trap (Section 7.3.2.2). Recently, the orbitrap has been also coupled with a Q analyzer (Section 7.3.2.1).

7.3.1.6 Ion Cyclotron Resonance (ICR)

ICR is a mass spectrometer with an ultimate performance in terms of ultra-high RP (in the range of 10^6) and mass accuracy (routinely below 1 ppm or even better). Whereas in other mass spectrometers ions are filtered in a magnetic or an electric field or selected by flight time, in ICR they are dispersed by their resonance frequencies.[50] A detailed description of ICR principles has been published in two reviews by Marshall and co-workers.[51,52] Ions undergo a cyclotron motion in a uniform magnetic field. Ions are detected in an ICR cell, which is located inside super-conducting magnet with a fixed field strength (a Penning trap) (Figure 7.3C). Ions arriving at the ICR cell are forced into a circular orbit by the strong magnetic field. Frequencies of cyclotron motion are related to *m/z* values of individual ions. The complication of UHPLC/MS coupling with ICR is the acquisition speed, which typically requires a few seconds per scan to achieve the best performance. The solution is in the compromise between the acquisition speed essential for UHPLC separation and the MS performance.[50] Attempts to couple ICR with UHPLC have been demonstrated only in the tandem arrangement, because this is a common experimental setup for this front-end MS technique (see Section 7.3.2.2).

7.3.2 Tandem Mass Analyzers

The MS/MS employs two stages of mass analysis in order to examine selectively the fragmentation behaviour of isolated ions.[18] The most common approaches for the dissociation experiments include collision-induced dissociation (CID), surface-induced dissociation (SID), infrared multi-photon dissociation (IRMPD), blackbody infrared radiative dissociation (BIRD), electron-capture dissociation (ECD) and electron-transfer dissociation (ETD). CID, wherein an inert gas is introduced into the collision cell and low-energy (1–100 eV) or high-energy (>1 keV) collisions occur between the precursor ion and inert gas atoms (molecules) is the most common ion activation method.[53] The MS/MS is accomplished with trap-based analyzers (IT and ICR) with the isolation in time or by coupling of two analyzers with the isolation in space. The most typical analyzers employed in the first stage of MS/MS are Q and IT.

7.3.2.1 Quadrupole Analyzers in Tandem Mass Spectrometry

7.3.2.1.1Triple Quadrupole (QqQ)

In QqQ, the first quadrupole (Q_1) and the third quadrupole (Q_3) are mass selective devices, while the second quadrupole (q_2) operates only in the AC mode (transmission of all ions) to serve as a collision cell. Q_1 is used for the selection of precursor ions, which are transmitted to the collision cell. The pressure in the collision cell is approximately 10^{-1} Pa due to the presence of collision gas. Q_3 analyzes all ions coming from the collision cell. It is unequivocally the most widely used mass analyzer in quantitative analysis due to its ability to perform highly selective and sensitive scans, such as selective reaction monitoring (SRM).[21] The detection sensitivity decreases when the whole mass range is analyzed in a scanning mode, which is a major drawback of QqQ, so that this analyzer is not a typical solution for the untargeted screening of unknown compounds, where high-resolution mass analyzers are more convenient.

QqQ with high sensitivity and ultra-high acquisition speed in the SRM mode (up to 1000 Hz) are able to perform the quantitative analysis of complex multi-component mixtures and to be coupled with fast UHPLC methods.[54] Furthermore, high-throughput analysis includes the possibility to perform simultaneously the quantitative SRM scans and the qualitative full-scan within one analytical run. QqQ analyzers are clearly the best solution for UHPLC/MS quantitation, as illustrated by numerous references including bioanalytical,[1,54,55] environmental[56] and food analysis[57] applications. The acquisition speed is affected by the total number of SRM transitions and maybe slightly reduced for an excessive number of monitored compounds.

7.3.2.1.2 Quadrupole–Time-of-flight (QqTOF)

QqTOF tandem mass spectrometer can be described in a simplified form as QqQ with Q_3 replaced by a TOF analyzer. In a common QqTOF configuration, an additional quadrupole Q_0 in the transmission regime (AC only) is added to provide collisional damping, thus the instrument consists of three quadrupoles Q_0, Q_1 and Q_2 followed by a TOF mass analyzer with a pulsed acceleration of ions (Figure 7.4). In some commercial instruments, Q are replaced by hexapoles, however, the principle is the same. One of the main advantages of QqTOF instruments over QqQ is the high RP of TOF, typically within 20 000–40 000. As a result, interfering peaks of ions having the same nominal mass can be resolved, the charge state of multiply charged ions can be determined from their isotopic envelope and the signal-to-noise ratio is improved.[41]

The majority of current UHPLC/MS/MS based quantitative assays are performed by means of QqQ in the SRM mode. However, the full-scan approaches offer the advantage of the analysis of virtually unlimited number of analytes simultaneously. Furthermore, the retrospective post-acquisition data evaluation for unexpected analytes by reconstructed ion current chromatograms is a valuable tool in non-targeted screening. Recently, quantitation strategies have shifted towards non-targeted approaches based on high-resolution mass spectrometry (HRMS) coupled to ultra-performance separation techniques instead of conventional targeted SRM methods with QqQ.[58] Several review articles demonstrate the importance of UHPLC/

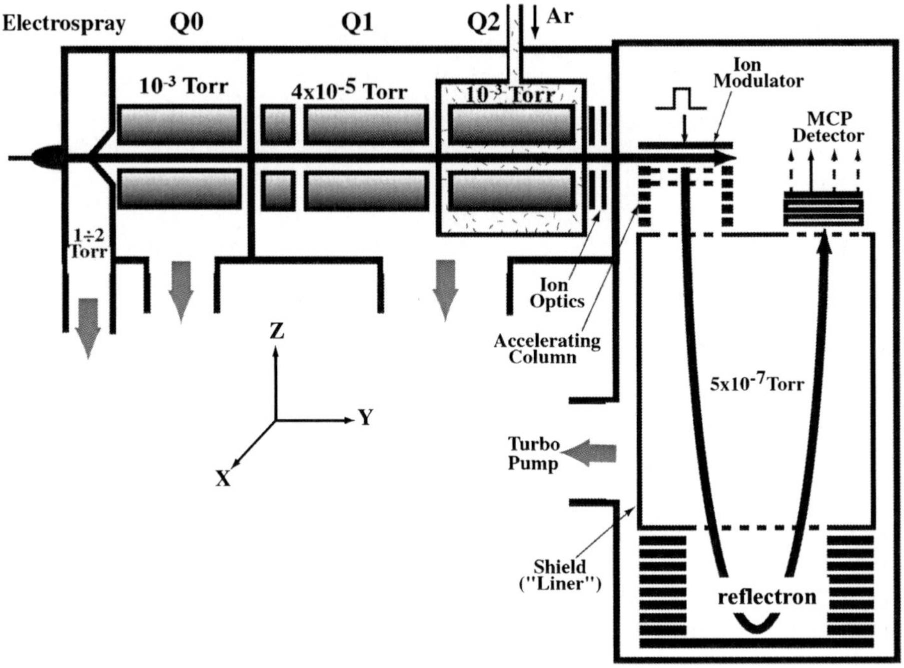

Figure 7.4 Scheme of hybrid QqTOF mass analyzer. Reproduced from reference 41 with permission.

QqTOF in metabolomics,[59] plant analysis,[60] bioanalysis,[39,58] food analysis[57] and other research areas.

7.3.2.1.3 Quadrupole–Linear Ion Trap (QqLIT)

This MS/MS system is based on the QqQ platform, where Q_3 can be operated either in the normal AC/DC mode or as the LIT mode. In the LIT mode, trapped ions are ejected axially in a mass selective fashion using fringe field effects and detected by the standard detector system. This configuration is particularly interesting, because such an instrument retains the classical QqQ functions such as the SRM, product ion, neutral loss and precursor ion scans, but additionally it offers IT experiments. MS/MS experiments are performed in two particular modes: the time-delayed fashion and the selection of multiply charged ions in the trap mode. The combination of QqQ and IT modes realized in the same LC/MS/MS run is unique and offers new possibilities for the quantitative and qualitative analysis.[61,62] The coupling of QqLIT with UHPLC was reported in the quantitative analysis of testosterone hydroxyl metabolites,[63] the pharmacokinetic study of troglitazone[64] and the quantitation of serum apolipoproteins.[65] Further applications of UHPLC or nano-UHPLC with QqLIT profited from information-dependent acquisition. Using these approaches in proteomics, the relationship between protein arginine *N*-methyltransferases 1 and 2 was investigated[66] or the methylation states of histone H3 were revealed.[67] Full-scan MS and MS/MS functions were applied for specific metabolic pattern[68,69] and metabolic profiling.[70]

7.3.2.1.4 Quadrupole–Orbitrap

This configuration is designed as a standalone benchtop instrument suitable mainly for the analysis of small molecules. The ability to separate ions in space and analyze MS and MS/MS with very high resolution in the orbitrap analyzer offers the promise for the efficient multiplexed scan modes not currently used in proteomics research with the trap-based instruments.[71] Due to novelty of this analyzer in orbitrap family, no applications with UHPLC have been published so far.

7.3.2.1.5 Quadrupole–Ion Cyclotron Resonance (QqICR)

QqICR was introduced to remove MS/MS experiments from the Penning trap, because it improves the spectra quality.[50]

7.3.2.2 Ion Traps Combined with Other Analyzers

7.3.2.2.1 Linear Ion Trap–Orbitrap (LIT–Orbitrap)

Ion packets must have a narrow spatial and temporal distribution to ensure the stability and coherency during the image current detection. Three injecting methods can assure these short pulses: (i) electrostatic acceleration lenses, (ii) axial ejection from a LIT and (iii) ejection from a "C"-shaped LIT.[49] The orbitrap operates in a pulsed fashion, hence an external ion accumulation is desirable for interfacing with a continuous API source. To obtain pulses of 100–200 ns, a modified LIT was developed to accumulate and then inject ions from the ESI source into the orbitrap. The need to extract ion packets axially from this LIT with temporal widths shorter than a few hundred nanoseconds necessarily limits the instrument space charge capacity. As a result, the injection of large numbers of ions leads to broad angular, spatial and kinetic energy distributions that limit the performance of the orbitrap. These problems have been minimized using radial (rather than axial) ion ejection from the C-trap. The method proves the fast and uniform injection for large ion populations. The LIT upstream from the C-trap provides increased trapping efficiency, automatic gain control and high-quality accurate mass MS^n data from the mass analysis of fragment ions injected into the orbitrap.[49,72,73] Basic MS/MS scans available on orbitrap-based instruments are shown in Figure 7.5, both for LIT (Figures 7.5B and 7.5C) and Q (Figure 7.5D) coupling.

Numerous applications of UHPLC/orbitrap were published, for example, in mycotoxin analysis,[74] hormone and veterinary drug residue analysis,[75] metabolomics[76,77] and proteomics,[78,79] often using the coupling with nano-UHPLC. The results unequivocally confirmed the advantages of orbitrap in terms of RP, mass accuracy and linear dynamic range. Nano-UHPLC coupled

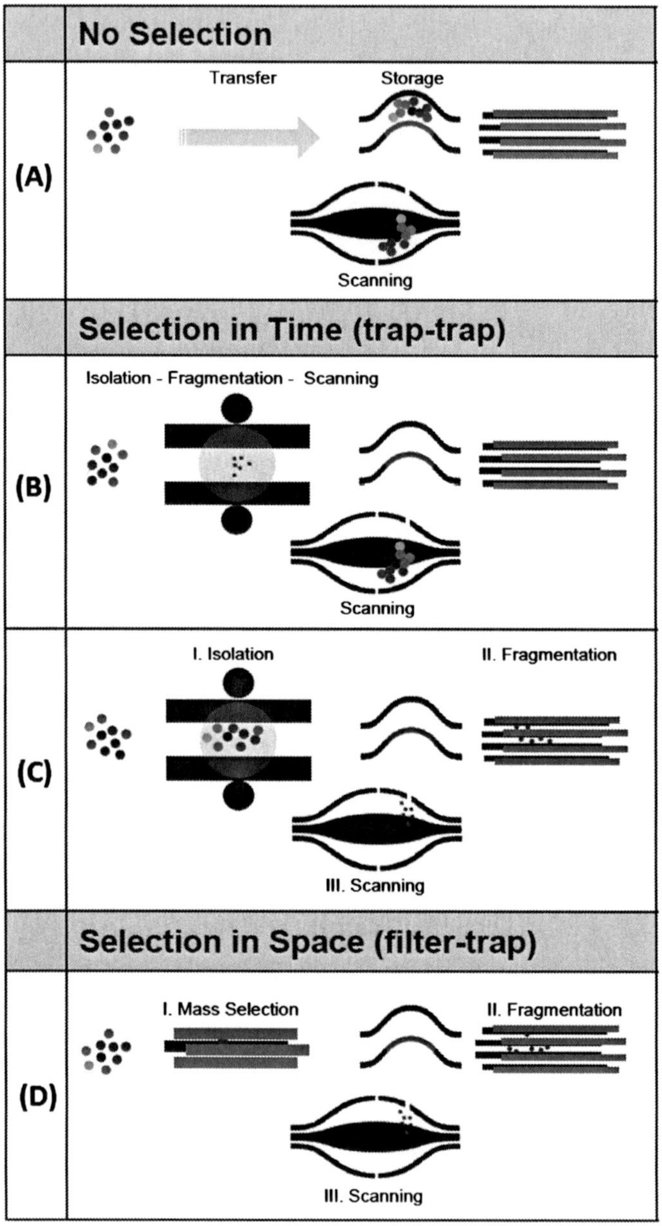

Figure 7.5 Mass spectrometers incorporating the orbitrap analyzer: (A) all ions are collected in the C-trap and injected into the orbitrap analyzer, (B) precursor ions are selected (isolation in time) and fragmented in the LIT, then analyzed in the orbitrap, (C) precursor ions are selected by the LIT, then fragmented in high-energy collision cell and analyzed in the orbitrap, and (D) precursor ions are selected by the Q (isolation in space), then fragmented in high-energy collision cell and analyzed in orbitrap. Reproduced from reference 71 with permission.

to orbitrap was also used for the quantitation,[80] but the peak width was compromised to approximately 7–15 s.

7.3.2.2.2 Linear Ion Trap–Ion Cyclotron Resonance (LIT–ICR)
The intention of LIT coupling with the ICR is similar as for QqICR instrument, *i.e.*, the separation of MS/MS experiments from the detection in the Penning trap to achieve the best performance of ICR. The LIT is used as the mass analyzer, the device for data-dependent ion accumulation and ion cooling, and for collision activation before ion packets are transferred. Mass spectra can be obtained solely with LIT (low RP) or with ICR (ultra-high RP). The LIT increases the duty cycle, if ions are accumulated in the LIT while the ICR is scanning.[50] The coupling of LIT–ICR with UHPLC has already shown great potential[81–84] in metabolomic analysis, in which the previous approaches based on the direct infusion analysis without the chromatographic separation suffered from a number of limitations, such as the discrimination of isobaric species. Other benefits of ICR coupling to UHPLC are the unequivocal association of fragment ions with their parent ions and the accurate relative quantitation.

7.3.2.2.3 Linear Ion Trap–Spherical Ion Trap (LIT–Spherical IT)
Combining the LIT with the spherical IT can help to overcome limitations of spherical IT, such as space charge effects, limited duty cycle and the total number of stored ions. To improve the duty cycle, ions are accumulated in the LIT, whereas the spherical trap performs other functions, such as CID or mass analysis. The duty cycle is improved significantly, when the LIT is used to pre-concentrate analyte ions and remove unwanted ions prior to the MS analysis in spherical trap, which is useful in the analysis of trace ions in the presence of large excess of contaminant ions. The linear dynamic range is extended.[28]

7.3.2.2.4 Linear Ion Trap–Time-Of-Flight (LIT–TOF)
The TOF mass spectrometer has a low duty cycle when coupled with a continuous ion source. The combination of the ion trap with the TOF analyzer can improve the duty cycle and MS/MS capabilities.[28,85]

7.3.3 Scan Types and Their Applications

Conventional MS/MS scanning includes four basic functions (Figure 7.6): product ion scan, precursor ion scan, neutral loss scan and SRM. The typical MS/MS system is QqQ, where all above mentioned scans can be routinely performed. The product ion scan (the old term daughter-ion is discouraged by IUPAC preliminary recommendations) is the most common mode of MS/MS operation. To acquire this scan in the QqQ configuration, the first mass analyzer (Q_1) is set to transmit only the selected precursor ion and the second one (q_2) is used as a collision cell, whereas Q_3 serves for the scanning of a defined *m*/*z* range. The obtained spectrum contains only product ions that are

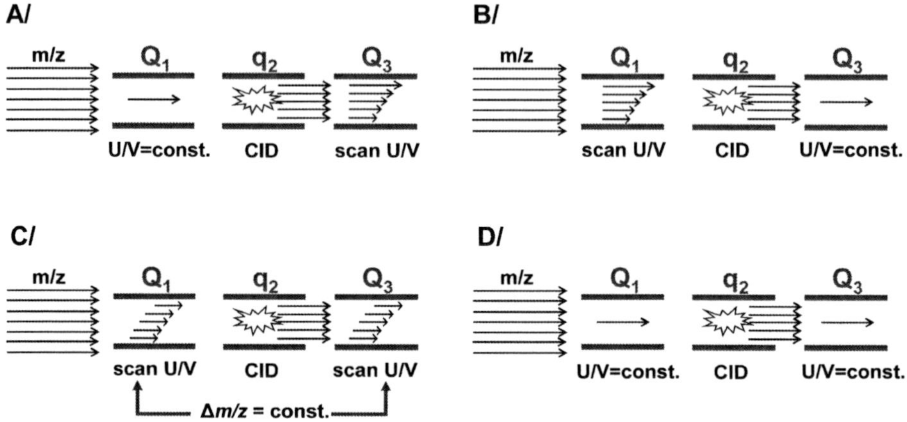

Figure 7.6 Schematic overview of basic scan types in tandem mass spectrometry (MS/MS): (A) product ion scan, (B) precursor ion scan, (C) neutral loss scan and (D) selected reaction monitoring.

formed exclusively from the selected precursor ion, which is useful in the structure elucidation. In the precursor ion scan (the old term parent-ion is discouraged), a spectrum of all precursor ions that might fragment to a common, diagnostic product ion is collected. It is obtained by adjusting the second mass analyzer to transmit a chosen product ion and scanning the first mass analyzer over a certain *m/z* range to transmit only those precursor ions that fragment to yield the chosen product ion. This scan is useful for the identification of a closely related class of compounds in a mixture. In a neutral loss scan, all precursors that undergo the loss of a specified common neutral are monitored. To obtain this information, both mass analyzers are scanned simultaneously, but with a mass offset that correlates with the mass of the specified neutral loss. Similar to the precursor ion scan, this technique is also useful in the selective identification of closely related class of compounds in a mixture. Specific neutral losses might reveal, for example, the presence of carboxylic acid (Δ*m/z* 44), hexose (Δ*m/z* 162), pentose (Δ*m/z* 134) and many others. SRM of one or more ion reactions is used in quantitative measurements of analytes present in complex mixtures. Two mass analyzers are adjusted to monitor one or more chosen precursor–product pairs of the analyte.

7.4 New Developments in Mass Spectrometry Applicable in UHPLC/MS

Recent developments in the field of MS offer several advantages for UHPLC/MS applications. The competition among MS companies and their intensive research efforts result in the continuous improvement of instrumentation. Today, the acquisition rate of QqTOF mass analyzer can be as high as 100 Hz in the full acquisition range, the dynamic range of QqQ is up to 4×10^6, mass accuracies of several types of instruments are below 2 ppm, the mass resolution of orbitrap is over 240 000 at the acquisition speed 1 Hz, *etc*. Ion mobility MS, better parameters associated with the acquisition rate of FT instruments (still in progress), an ESI–APCI multi-mode ionization source, new fragmentation approaches and ultrafast polarity switching can be selected among numerous interesting new features.

Ion mobility spectrometry (IMS) has been developed over the past few decades as a method for the separation and subsequent detection of volatile and semi-volatile organic compounds. In principal, the separation of gas-phase ions at the atmospheric pressure is based on their mobility in the electric field.[86] Concerning the applications of UHPLC/MS, mass spectrometers equipped with the ion mobility (IMS–MS) were commercially introduced in 2007. IMS enables the differentiation of samples by size, shape and charge as well as mass. IMS can be useful for the separation of isobaric compounds, the reduction of high background noise, the separation of endogenous matrix interferences from target analytes and the charge state screening used in proteomics (separation of unwanted singly charged background compounds and cluster ions from doubly charged analyte ions). The reduction of interferences can be further enhanced by the use of suitable chemical modifier.[87] These advantages result in increased dynamic range. In addition to traditional collision-induced dissociation, new fragmentation approaches, *i.e.*, energy-dependent fragmentation (MS^E) or time-aligned parallel (TAP) fragmentation, have been introduced. MS^E can be applied in all tandem mass spectrometers, while the application of TAP is restricted to IMS instruments. Ion mobility separation plays an important role in the TAP approach (Figure 7.7). The precursor ion is fragmented in the trap with subsequent separation of the first generation of product ions by IMS. The second generation of fragment ions are formed in the transfer and they are associated to the first generation based on individual drift-times.[88,89] The combination of IMS–MS with high RP and fragmentation experiments (MS^n, ETD, MS^E or TAP) is a powerful tool for structural analysis.

The improvement of the acquisition speed is observed in all types of mass spectrometers, but the most challenging situation is for FT mass spectrometers with the RP higher than 100 000, *i.e.*, orbitrap and ICR. The new high-field orbitrap platform enables 60 000 RP within the acquisition speed of 4 Hz (four times increase compared with previous models). The maximum RP is higher than 240 000 and the mass accuracy below 0.5 ppm with the internal calibration. Moreover, the new orbitrap instrument enables the combination of multiple fragmentation techniques, *i.e.*, CID and higher-energy collisional dissociation (HCD), that provides superior fragmentation and higher quality MS/MS spectra for many types of samples and optional electron transfer dissociation (ETD) for studying post-translational modifications of proteins. The ultimate RP for the research orbitrap exceeding 600 000 has been reported by Makarov. Concerning ICR, the RP of ICR cell constructed by Nikolajev *et al.*[90] exceeds 20 million (FWHM), which should be commercially available in

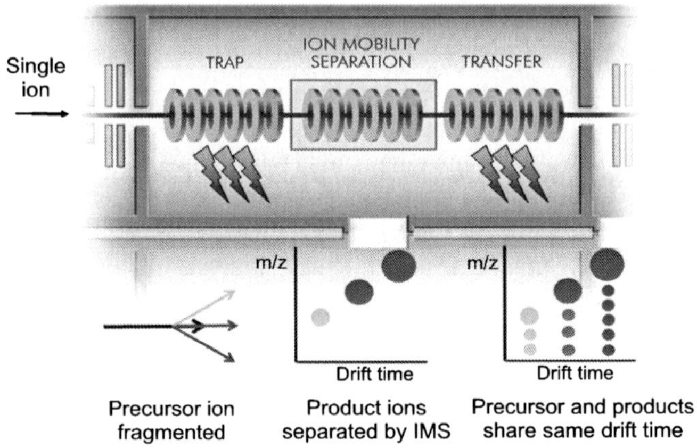

Figure 7.7 Scheme of time-aligned parallel (TAP) fragmentation of a Waters Synapt G2 instrument. Reproduced from reference 89 with permission.

the next year. The low acquisition speed of ICR can be overcome using the chip-based ESI technology TriVersa NanoMate (Advion). UHPLC was connected to the ICR *via* a TriVersa NanoMate using the flow rate of 400 μl min^{-1} and 1:1000 split. The resolution was set up to 50 000 and the maximum loading time for ICR was 250 or 500 ms.[91,92] A nano-UHPLC/MS system was used for the analysis of extremely low amounts of proteins at the flow rate of 400 nl min^{-1} or 250 nl min^{-1}.[93,94]

As already mentioned, the ESI and APCI are the most frequently used ionization sources in UHPLC/MS. However, each ionization technique has own benefits and limitation. ESI/APCI multimode ionization source (ESCi) allows the simultaneous data collection in four ionization modes (ESI$^+$, ESI$^-$, APCI$^+$ and APCI$^-$) during a single UHPLC/MS analysis with the ultra-fast polarity switching corresponding to 20 ms only for the change of ionization technique.[95] The ionization mechanism of this multimode ionization source is a combination of both ionization techniques. When ESCi was used, a significant decrease in the MS response was observed (in average 30–50%, but several compounds showed a 5-fold decrease), which was mainly attributed to the time necessary to perform the ionization as well as polarity switching.[22]

7.5 Conclusions

UHPLC belongs to the group of fast chromatography techniques, therefore the specific demands of UHPLC/MS coupling on parameters of selected mass analyzer are related to the fact that peaks are narrower compared to the conventional HPLC. The UHPLC is based on smaller particle size (sub-2 μm particles), higher pressures (up to 1200 bar) and faster analysis, which subsequently requires the higher acquisition speed of a mass spectrometer to have enough sampling points for the correct peak integration. The peak width in ultra-fast UHPLC analysis can be as low as 1–3 s. The recommended practice in the analytical chemistry is at a minimum of 12 points over the chromatographic peak to perform the right integration and quantitation. Consequently, the minimum acquisition speed of a mass spectrometer in UHPLC detection should be at least 4–12 Hz over the whole mass range without compromising the MS performance. The double-focusing magnet analyzer does not fulfill this requirement and furthermore magnetic analyzers are very rare even in HPLC/MS due to the fact that modern TOF- or orbitrap-based instruments provide a comparable performance at faster acquisition speed, lower space requirements and more favourable price. Two FT mass analyzers (orbitrap and ICR) can reach such acquisition speeds only at a compromised performance, so their use is recommended for peak widths of least 10 s, but not in ultra-fast UHPLC.

Quadrupoles, spherical and linear ion traps have the highest acquisition speeds, all of them at least 10 Hz for latest instruments. The fastest mass analyzer is obviously the TOF with common acquisition speeds 40–50 Hz and recently the fastest QqTOF instrument on the market has reached 100 Hz, which allows numerous parallel SRM scans, even in ultra-fast UHPLC/MS analysis. The acquisition speed and also mass range of the TOF analyzer in QqTOF configuration is restricted by the first Q. Additional advantage of modern TOF-based instruments is rather high RP (40 000–60 000) and mass accuracy (typically, better than 3 ppm with the external calibration and better than 1 ppm with the internal calibration).

Another parameter important for the quantitation is the linear dynamic range, which is at least 5 orders of magnitude for all latest Q, linear IT, TOF and their combinations. It is not surprising that the above stated analyzers are the most common instrumental configurations in UHPLC/MS coupling, but the top-class FT mass analyzers provide additional benefits for the structural elucidation.

Figure 7.8 shows the popularity of individual MS configurations in the UHPLC/MS coupling based on the Web of Science search, and the fastest TOF mass analyzer is clearly the most popular (either in QqTOF hybrid coupling or alone) followed by QqQ as a typical mass analyzer used for the quantitation. The notable number of applications is recorded for IT and orbitrap, but the use of other mass spectrometer is rare at the moment.

Acknowledgements

The financial support of project no. MSM0021627502 sponsored by The Ministry of Education, Youth and Sports of the Czech Republic is acknowledged. We are obliged to numerous colleagues and national company representatives for their consultations on the comparison of individual mass analyzers, mainly to the following individuals (in an alphabetical order): M.

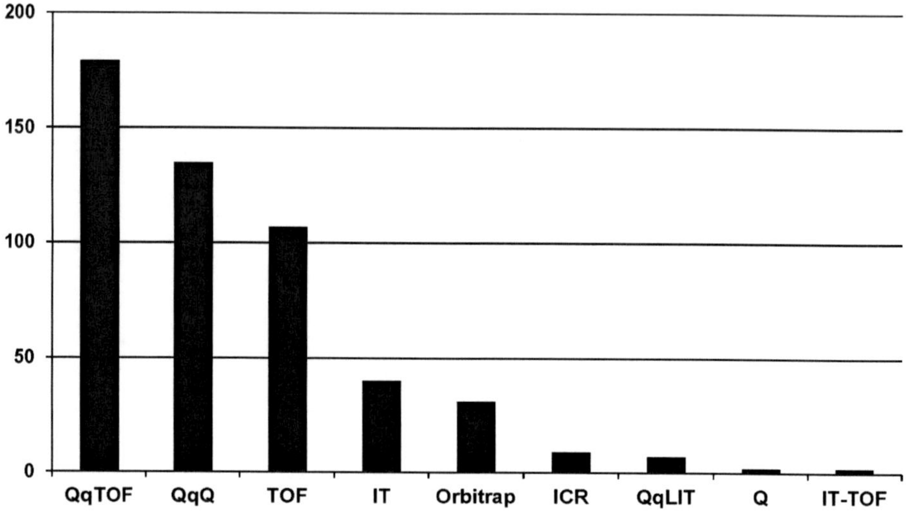

Figure 7.8 Number of UHPLC/MS applications for individual MS configurations based on the Web of Science search.

Boháč, M. Exner, V. Havlíček, T. Korba, K. Lemr, M. Polášek, J. Preisler, M. Procházka, P. Verner and D. Vláčil.

References

1. L. Nováková and H. Vlčková, *Anal. Chim. Acta*, 2009, **656**, 8.
2. R. B. Cole, *Electrospray and MALDI Mass Spectrometry*, John Wiley & Sons Ltd, Hoboken, 2010.
3. E. de Hoffmann and V. Stroobant, *Mass Spectrometry: Principles and Applications*, John Wiley & Sons Ltd, Chichester, 2007.
4. M. Holčapek, P. Jandera and P. Zderadička, *J. Chromatogr. A*, 2001, **926**, 175.
5. R. Jirásko, M. Holčapek, L. Kolářová and T.S. Basu Baul, *J. Mass Spectrom.*, 2007, **42**, 918.
6. R. Jirásko and M. Holčapek, *Mass Spectrom. Rev.*, 2011, **30**, 1013.
7. M. Yamashita and J. B. Fenn, *J. Phys. Chem.*, 1984, **88**, 4451.
8. M. Holčapek, R. Jirásko and M. Lísa, *J. Chromatogr. A*, 2010, **1217**, 3908.
9. D. B. Robb and M. W. Blades, *Anal. Chim. Acta*, 2008, **627**, 34.
10. L. C. Short, K. A. Hanold, S. S. Cai and J. A. Syage, *Rapid Commun. Mass Spectrom.*, 2007, **21**, 1561.
11. J. M. Purcell, R. P. Rodgers, C. L. Hendrickson and A. G. Marshall, *J. Am. Soc. Mass Spectrom.*, 2007, **18**, 1265.
12. J. A. Syage, *J. Am. Soc. Mass Spectrom.*, 2004, **15**, 1521.
13. L. C. Short, S. S. Cai and J. A. Syage, *J. Am. Soc. Mass Spectrom.*, 2007, **18**, 589.

14. G. A. Harris, A. S. Galhena and F. M. Fernández, *Anal. Chem.*, 2011, **83**, 4508.
15. Ch. Dass, *Fundamentals of Contemporary Mass Spectrometry*, Wiley-Interscience, Hoboken, New Jersey, USA, 2007.
16. R. K. Boyd, *Trace Quantitative Analysis by Mass Spectrometry*, John Wiley & Sons Ltd, Chichester, West Sussex, UK, 2008.
17. G. L. Glish and R. W. Vachet, *Nat. Rev. Drug Discov.*, 2003, **2**, 140.
18. E. de Hoffman, *J. Mass Spectrom.*, 1996, **31**, 129.
19. R. E. March, *J. Mass Spectrom.*, 1997, **32**, 351.
20. R. E. March, *Rapid Commun. Mass Spectrom.*, 1998, **12**, 1543.
21. J. T. Watson and O. D. Sparkman, *Introduction to Mass Spectrometry*, 4th edn, John Wiley and Sons Ltd, Chichester, West Sussex, UK, 2007.
22. J. Schappler, R. Nicoli, D. Nguyen, S. Rudaz, J.-L. Veuthey and D. Guillarme, *Talanta*, 2009, **78**, 377.
23. S. A. C. Wren, *J. Pharm. Biomed. Anal.*, 2006, **40**, 571.
24. M. J. Lerma-García, E. F. Simó-Alfonso, A. Méndez, J. L. Llibera and J. M. Herrero-Martínez, *J. Agric. Food. Chem.*, 2010, **58**, 2771.
25. B. Avula, Y.-H. Wang, R. S. Pawar, Y. J. Shukla, T. J. Smillie and I. A. Khan, *J. Pharm. Biomed. Anal.*, 2008, **48**, 722.
26. Z. Zhang, M. B. Wallace, J. Feng, A. Stafford, R. J. Skene, L. Shi, B. Lee, K. Aertgeerts, R. Xu, D. B. Kassel, S. W. Kaldor, M. Navre, D. R. Webb and S. L. Gwaltney, *J. Med. Chem.*, 2011, **54**, 510.
27. W. J. Zuercher, R. G. Buckholz, N. Campobasso, J. L. Collins, C. M. Galardi, R. T. Gampe, S. M. Hyatt, S. L. Merrihew, J. T. Moore, J. A. Oplinger, P. R. Reid, P. K. Spearing, T. B. Stanley, E. L. Stewart and T. M. Willson, *J. Med. Chem.*, 2010, **53**, 3412.
28. D. J. Douglas, A. J. Frank and D. Mao, *Mass Spectrom. Rev.*, 2005, **24**, 1.
29. In the news at Pittcon 2010, *Trends Anal. Chem.*, 2010, **29**, iii.
30. A. K. Kiss and M. Naruszewicz, *Food Chem.*, 2011, doi:10.1016/j.foodchem.2011.09.011.
31. C. Mane, M. Loonis, Ch. Juhel, C. Dufour and C. Malien-Aubert, *J. Agric. Food Chem.*, 2011, **59**, 3330.
32. K. Wang, H. Zhang, L. Shen, Q. Du and J. Li, *J. Pharm. Biomed. Anal.*, 2010, **53**, 1053
33. K. Shimamura, M. Matsuda, Y. Miyamoto, R. Yoshimoto, T. Seo and S. Tokita, *Peptides*, 2009, **30**, 1259.
34. J. C. Schwartz and M. W. Senko, *J. Am. Soc. Mass Spectrom.*, 2002, **13**, 659.
35. K. V. Petrova, R. S. Jalluri, I. D. Kozekov and C. J. Rizzo, *Chem. Res. Toxicol.*, 2007, **20**, 1685.
36. Q. Fang, A. M. Noronha, S. P. Murphy, Ch. J. Wilds, J. L. Tubbs, J. A. Tainer, G.Chowdhury, F. P. Guengerich and A. E. Pegg, *Biochemistry*, 2008, **47**, 10892.
37. Z. Tang, M. V. Martin and F. P. Guengerich, *Anal. Chem.*, 2009, **81**, 3071.
38. A. W. T. Bristow, *Mass. Spectrom. Rev.*, 2006, **25**, 99.

39. J.L. H. Jiwan, P. Wallemacq and M. F. Hérent, *Clin. Biochem.*, 2011, **44**, 136.
40. B. A. Mamzrin, *Int. J. Mass Spectrom.*, 2001, **206**, 251.
41. I. V. Chernuschevich, A. V. Loboda and B. A. Thomson, *J. Mass Spectrom.*, 2001, **36**, 849.
42. E. Zelena, W. B. Dunn, D. Broadhurst, S. Francis-McIntyre, K. M. Carroll, P. Begley, S. OHagan, J. D. Knowles and A. Halsall, *Anal. Chem.*, 2009, **81**, 1357.
43. W. Pongsuwan, T. Bamba, K. Harada, T. Yonetani, A. Kobayashi and E. Fukusaki, *J. Agric. Food Chem.*, 2008, **56**, 10705.
44. S. L. Hughes, J. G. Bundy, E. J. Want, P. Kille and S. R. Stürzenbaum, *J. Proteome Res.*, 2009, **8**, 3519.
45. X. Liang, X. Chen, Q. Liang, H. Zhang, P. Hu, Y. Wang and G. Luo, *J. Proteome Res.*, 2011, **10**, 790.
46. A. Makarov, *Anal. Chem.*, 2000, **72**, 1156.
47. M. Hardman, A. Makarov, *Anal. Chem.*, 2003, **75**, 1699.
48. Q. Hu, R. J. Noll, H. Li, A. Makarov, M. Hardman and R. G. Cooks, *J. Mass Spectrom.*, 2005, **40**, 430.
49. R. H. Perry, G. Cooks and R. J. Noll, *Mass Spectrom. Rev.*, 2008, **27**, 661.
50. W. Schrader and H-W. Klein, *Anal. Bioanal. Chem.*, 2004, **379**, 1013.
51. A. G. Marshall, C. L. Hendrikson and G. S. Jackson, *Mass Spectrom. Rev.*, 1998, **17**, 1.
52. A. G. Marshall and C. L. Hendrikson, *Int. J. Mass Spectrom.*, 2002, **215**, 59.
53. L. Sleno and D. A. Volmer, *J. Mass Spectrom.*, 2004, **39**, 1091.
54. D. Guillarme, J. Schappler, S. Rudaz and J-L. Veuthey, *Trends Anal. Chem.*, 2010, **29**, 15.
55. R. N. Xu, L. Fan, M. J. Rieser and T. A. El-Shourbagy, *J. Pharm. Biomed. Anal.*, 2007, **44**, 342.
56. Z. Liu, J. A. Ogejo, A. Pruden and K. F. Knowlton, *Sci. Total Environ.*, 2011, **24**, 5149.
57. A. K. Malik, Ch. Blasco and Y. Picó, *J. Chromatogr. A*, 2010, **25**, 4018.
58. R. Ramanathan, M. Jemal, S. Ramagiri, Y-Q. Xia, W. G. Humpreys, T. Olah and W. A. Korfmacher, *J. Mass Spectrom.*, 2011, **46**, 595.
59. A. Roux, D. Lison, Ch. Junot and J. F. Heilier, *Clin. Biochem.*, 2011, **44**, 119.
60. J. L. Zhou, L-W. Qi and P. Li, *J. Chromatogr. A*, 2009, **1216**, 7582.
61. J. W. Hager, *Rapid Commun. Mass Spectrom.*, 2002, **16**, 512.
62. G. Hopfgartner, E. Varesio, V. Tschäppät, Ch. Grivet, E. Bourgogne and L. A. Leuthold, *J. Mass Spectrom.*, 2004, **39**, 845.
63. D. Wang and M. Zhang, *J. Chromatogr. B*, 2007, **855**, 290.
64. L.S. New, S. Saha, M. M. K. Ong, U. A. Boelsterli and E. C. Y. Chan, *Rapid Commun. Mass Spectrom.*, 2007, **21**, 982.
65. R. G. Kay, B. Gregory, P. B. Grace and S. Pleasance, *Rapid Commun. Mass Spectrom.*, 2007, **21**, 2585.

66. M. L. Pak, T. M. Lakowski, D. Thomas, M. I. Vhuiyan, K. Hüsecken and A. Frankel, *Biochemistry*, 2011, **50**, 8226.
67. H. Liu, M. Galka, A. Iberg, Z. Wang, L. Li, C. Voss, X. Jiang, G. Lajoie, Z. Huang, M. T. Bedford and S. C. C. Li, *J. Proteome Res.*, 2010, **9**, 5827.
68. L. Tang, R. Singh, Z. Liu and M. Hu, *Mol. Pharmaceutics*, 2010, **3**, 664.
69. L. Tang, R. Singh, Z. Liu and M. Hu, *Mol. Pharmaceutics*, 2009, **6**, 1466.
70. D. P. Q. Goh, A. H. Neo, C. W. Goh, Ch. Ch. Aw, L. S. New, W. S. Chen, Z. Atcha, E. R. Browne and E. C. Y. Chan, *J. Proteome Res.*, 2009, **8**, 5679.
71. A. Michalski, E. Damoc, J-P. Hauschild, O. Lange, A. Wieghaus, A. Makarov, N. Nagaraj, J. Cox, M. Mann and S. Hornig, *Mol. Cell. Proteomics*, 2011, doi:10/1074/mcp.M111.011015.
72. A. Makarov, E. Denisov, A. Kholomeev, W. Baschun, O. Lange, K. Strupat and S. Horning, *Anal. Chem.*, 2006, **78**, 2113.
73. A. Makarov, E. Denisov, O. Lange and S. Hornings, *J. Am. Soc. Mass Spectrom.*, 2006, **17**, 977.
74. M. Zachariášová, O. Lacina, A. Malachová, M. Kostelanská, J. Poustka, M. Godula and J. Hajšlová, *Anal. Chim. Acta*, 2010, **662**, 51.
75. E. van der Heeft, Y. J. C. Bolck, B. Beumer, A. W. J. M. Nijrolder, A. A. M. Stolker and M. W. F. Nielen, *J. Am. Soc. Mass Spectrom.*, 2009, **20**, 451.
76. W. B. Dunn, D. Broadhurst, M. Brown, P. N. Baker, Ch. W. G. Redman, L. C. Kenny and D. B. Kell, *J. Chromatogr. B*, 2008, **891**, 288.
77. R. P. Horgan, D. I. Broadhurst, S. K. Walsh, W. B. Dunn, M. Brown, C. T. Roberts, R. A. North, L. M. McCowan, D. B. Kell, P. N. Baker and L. C. Kenny, *J. Proteome Res.*, 2011, **10**, 3660.
78. S. Houel, R. Abernathy, K. Renganathan, K. Meyer-Arendt, N. G. Ahn and W. M. Old, *J. Proteome Res.*, 2010, **9**, 4152.
79. L. A. Beer, H.-Y. Tang, S. Sriswasdi, K. T. Barnhat and D. W. Speicher, *J. Proteome Res.*, 2011, **10**, 1126.
80. A. Thomas, H. Geyer, M. Kamber, W. Schänzen and M. Thevis, *J. Mass Spectrom.*, 2008, **43**, 908.
81. P. Giavalisco, K. Köhl, J. Hummel, B. Seiwert and L. Willmitzer, *Anal. Chem.*, 2009, **81**, 6546.
82. A. Cuadros-Inostroza, P. Giavalisco, J. Hummel, A. Eckardt, L. Willmitzer and H. Peña-Cortéz, *Anal. Chem.*, 2010, **82**, 3573.
83. F. J. Berendt, K-S. Park and J. S. Trimmer, *J. Proteome Res.*, 2010, **9**, 1976.
84. V. Faca, M. Coram, D. Phanstiel, V. Glukhova, Q. Zhang, M. Fitzgibbon, M. McIntosh and S. Hanash, *J. Proteome Res.*, 2006, **5**, 2009.
85. B. A. Cogllings, J. M. Campbell, D. Mao and D. J. Douglas, *Rapid Commun. Mass Spectrom.*, 2001, **15**, 1777.
86. H. Borsdorf, *Appl. Spectrosc. Rev.*, 2006, **41**, 323.
87. A. B. Anu, P. Dwivedi, M. Tam, L. Matz and H. H. Hill, *J. Mass Spectrom.*, 2008, **43**, 1.

88. J. Sun, A. Baker and P. Chen, *Rapid Commun. Mass Spectrom.*, 2011, **25**, 2591.
89. K. Yu, J. Castro-Perez and J. Shocker, *Waters Application Note*, 2008, 720002542en.
90. E. N. Nikolaev, I. A. Boldin, R. Jertz and G. Baykut, *J. Am. Soc. Mass Spectrom.*, 2011, **22**, 1125.
91. P. Giavalisco, K. Köhl, J. Hummel, B. Seiwert and L. Willmitzer, *Anal. Chem.*, 2009, **81**, 6546.
92. A. Cuadros-Inostroza, P. Giavalisco, J. Hummel, A. Eckardt, L. Willmitzer and H. Peña-Cortéz, *Anal. Chem.*, 2010, **82**, 3573.
93. F. J. Berendt, K-S. Park and J. S. Trimmer, *J. Proteome Res.*, 2010, **9**, 1976.
94. V. Faca, M. Coram, D. Phanstiel, V. Glukhova, Q. Zhang, M. Fitzgibbon, M. McIntosh and S. Hanash, *J. Proteome Res.*, 2006, **5**, 2009.
95. K. Yu, L. Di, E. Kerns, S. Q. Li, P. Alden and R. S. Plumb, *Rapid Commun. Mass Spectrom.*, 2007, **21**, 893.

CHAPTER 8

UHPLC for the Determination of Physicochemical Parameters in Drug Discovery

ALESSANDRA TANIA ZIZZARI, PIERRE-ALAIN CARRUPT AND SOPHIE MARTEL*

School of Pharmaceutical Sciences, University of Geneva, University of Lausanne, 30 quai Ernest Ansermet, 1211 Geneva 4, Switzerland
*E-mail: Sophie.martel@unige.ch

8.1 Introduction

Drug research is a complex and time-consuming process. Indeed, a period of 12 to 20 years is required from the discovery of new chemical entities (NCEs) to the marketing approval of a new effective drug. Compound attrition during the whole process of drug research is extremely high. On average, only approximately 10 out of more than 10 000 synthesized chemical entities reach clinical testing in healthy volunteers, and only one will reach market.[1] Undesirable pharmacokinetic (PK) properties have been recognized as one of the major factors leading to the attrition of NCEs from drug development.[2] Thus, there have been great efforts to perform PK studies as early as possible in the discovery process, preferably before a drug candidate enters into the lead optimization phase, so that only compounds with high potency and suitable PK properties are selected for development.[3] Physicochemical properties, such as solubility, ionization, lipophilicity and permeability, can be used as predictors of absorption, distribution, metabolism and excretion (ADME) properties.[4–6] Compared with *in vivo* ADME procedures, these properties can

RSC Chromatography Monographs No. 16
UHPLC in Life Sciences
Edited by Davy Guillarme and Jean-Luc Veuthey

Published by the Royal Society of Chemistry, www.rsc.org

be screened *in vitro* for a significant number of compounds with reduced expenses, time and animal use.

Solubility is one of the reasons of attrition in drug discovery and development processes. Good solubility is the most desired property for a drug candidate; thus, this property is checked throughout the development process. Indeed, insufficient solubility compromises other assays (*e.g.*, false positive or false negative results) and can lead to artificially low activity and unreliable results, which may be used to generate erroneous structure–activity relationships. Moreover, poor solubility often implies low bioavailability and may ultimately affect the development of a compound.[6,7]

Lipophilicity has been shown to be a fundamental property for predicting both PK (membrane permeability, tissue distribution, protein binding and metabolism) and pharmacodynamic (binding affinity) drug properties. Lipophilicity is an indispensable tool in predicting the transport and activity of drugs.[8] Increasing the lipophilicity of a compound generally increases permeability, protein binding and tissue distribution; however, solubility is usually decreased.[9] Highly lipophilic compounds are the preferred targets for metabolism and often have high clearance rates and elevated plasma protein binding, which result in low bioavailability.

Ionization is also very important because it influences many physicochemical and PK properties. Indeed, the ionization constant(s) (pK_a) of a compound influence properties such as pH profile solubility, lipophilicity, permeability, protein binding and stability. For ionizable compounds, the ionized form is more soluble, less lipophilic and less permeable through biological membranes than the neutral form.[6] Therefore, ionization influences *in vivo* ADME. Moreover, the knowledge of pK_a in industry is very useful for drug design, (quantitative) structure–activity relationship [(Q)SAR] model development and formulation strategies.

Because drugs encounter several different barriers in living systems (*e.g.*, the gastro-intestinal track and the blood–brain barrier), their activity depends on their ability to cross biological membranes and reach the active site at a suitable concentration. Different mechanisms (para- or trans-cellular) are involved in the permeability of compounds through biological membranes, including passive diffusion, facilitated transport, active uptake, endocytosis and efflux; however, approximately 80% of drugs are known to be absorbed by passive diffusion processes. *In vitro* permeability investigations can also help with understanding mechanism in cell-based bioassays and support the interpretations of *in vivo* PK results.

There is an increasing need to develop strategies for physicochemical profiling.[10] Because of the increasing number of proposed compounds and the reduction of the available amount of material at the early discovery stage, methods need to be fast, use minimal amounts of compound, be highly automated and have a low cost. After several years of high-throughput physicochemical profiling, many pharmaceutical companies realized that there was a need for higher quality data to assist in making more relevant decisions in compound selection.[7] The challenge in drug discovery consists of the development of rapid, generic and automated methods with low limits of detection that provide reliable results for compound selection. Many methods to assess solubility, ionization, lipophilicity and passive permeability values based on solute quantification have been miniaturized and parallelized using multi-well technology, and these techniques have benefitted from the advantages of liquid chromatography coupled with mass spectrometry (LC/MS), because LC/MS is insensitive to impurities and valid for compounds without a chromophore. Furthermore, because LC/MS has an increased sensitivity compared with traditional UV detection, the range of accessible physicochemical values is increased. In addition, LC is also used to indirectly determine physicochemical and PK parameters.

Ultra-high-performance LC (UHPLC) is a technique that is not currently widely used for the determination of physiochemical properties; however, many HPLC approaches have been developed in which the throughput could be increased with UHPLC. The use of small particles within the columns allows for optimal separation at high linear velocities (because of the low mass transfer resistance of these supports); thus, UHPLC can attain high resolution with a short analysis time. This makes UHPLC an interesting technique for physicochemical profiling in drug discovery. In the present review, strategies used to measure solubility, ionization, lipophilicity and permeability in the drug discovery stage are briefly described, and emphasis is given to the applications of LC (i) in compound quantification-based methods and (ii) for the indirect measurements of physicochemical properties (Figure 8.1). HPLC methods are reviewed and the advantages derived from UHPLC transfer are discussed.

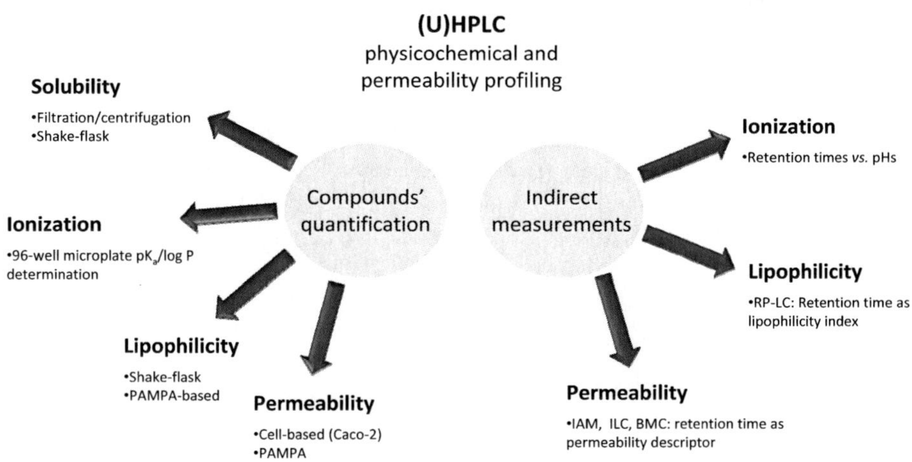

Figure 8.1 Use of (U)HPLC in physicochemical and permeability profiling in drug discovery stage. BMC, bio-partitioning micellar chromatography; IAM, immobilized artificial membrane; ILC, immobilized liposome chromatography; PAMPA, parallel artificial membrane permeability assay.

8.2 Solubility

8.2.1 Some Definitions

Solubility can be loosely defined as the amount of compound that can be dissolved in a defined quantity of a given solvent at a defined temperature; however, several definitions have to be considered. For example, the unbuffered solubility considers the quantity of a compound in a saturated solution (often in water, known as "water solubility") at the resulting (but not controlled) pH of the solution, which could be very different from the pH of water because of self-buffering of ionizable compounds. Buffered solubility is the solubility of a compound in a solution with a controlled pH, whereas intrinsic solubility corresponds to the solubility of the neutral form of an ionizable compound.[7]

Two kinds of solubility values can be obtained depending on the experimental assay. Kinetic solubility is the solubility obtained by the re-precipitation of a compound, which implies that the compound was first pre-dissolved in a co-solvent or pre-dissolved in its charged form (the most soluble form) by pH adjustment. The solubility value corresponds to the concentration of a compound that remained in solution after precipitation. Thermodynamic solubility is the concentration of a compound in equilibrium with an excess of solid compound at the end of the dissolution process. This latter solubility is also called "true solubility".

In general, thermodynamic solubility is more relevant because it corresponds to the solubility of a compound when equilibrium is reached. Kinetic solubility, however, seems more interesting in the drug discovery stage because it corresponds to the solubility in specific conditions of bioassays and is obtained in a faster manner. In other words, when a compound is initially pre-dissolved in co-solvent, the final aqueous solution is supersaturated, which leads to a higher solubility than the "true (thermodynamic) solubility". Furthermore, the crystalline aspects and the dissolution rates of compounds are not taken into consideration in kinetic solubility measurements (*i.e.*, the energy needed to break the crystal lattice is not considered due to the pre-dissolution of a compound in a strong organic solvent). Therefore, kinetic solubility measurements, influenced by parameters such as incubation time, mixing conditions, co-solvent type and concentration, polymorphism and purity, lead to solubility values that are markedly different from the thermodynamic solubility (*i.e.*, experimental parameters are important because the system does not reach equilibrium). If this difference could be considered as a drawback *a priori*, determining the kinetic solubility could be an advantage when researchers want to know the solubility of compound under particular conditions, such as those that would be used in future biological assays.[7,11] In the discovery stages, the main goals are to prioritize/flag compounds, build (Q)SAR models and optimize further biological assays. Later, decisions must be made about the potential development of a reduced number of compounds. During the development process, the most important

Table 8.1 Solubility in drug discovery and drug development processes

	Discovery	*Development*
Compounds tested		
Number	100–1000	10
Quantity available	Few mg	>1 g
Purity	Limited	Improved
Solid state	Amorphous or partially crystalline (not characterized)	Stable, crystalline material (characterized)
Distribution	Generally DMSO stock solutions	Generally solid form
Methods		
Type of solubility measured	Kinetic (fully dependent on experimental conditions) thermodynamic-like	Thermodynamic ("true") solubility
Throughput	High	Low
Automation	Fully	Only partially automated
Format	96-well or 384-well microplates	Small scale (single tube)
Incubation time	Minutes	Hours to days
Media	Aqueous	>20 (*e.g.*, aqueous, organic, biorelevant media, formulations and excipients)
Data generated and intended purpose	Solubility in screening bioassay media to avoid misinterpretation/optimize conditions	Solubility and dissolution in biorelevant media Evaluation of formulations
	Rank-order hits	Characterization and optimization of solid state
	Flag compounds	Selection of promising compounds
	Data for structure activity relationships	Development of adequate strategies to overcome solubility problems

questions focus on formulation strategies and solid form properties. Therefore, measuring solubility is a challenge, and the solubility information at each step has to be carefully defined according to phase properties and requirements (Table 8.1).[11,12]

8.2.2 Traditional Methods for Solubility Measurements

The gold standard technique to measure thermodynamic solubility is the shake-flask method, in which an excess of powdered compound is added to the solvent of choice and shaken for several hours or days to reach equilibrium. Then, the concentration of the compound in the solution is determined by UV or LC (MS) detection after filtration or centrifugation. Although the technique seems simple, several key points have to be carefully controlled because many error-prone manual steps are required. Even if the original shake-flask method has been adapted to be faster during the discovery process, this technique is generally applied in advanced stages of development because of numerous drawbacks that are incompatible with high-throughput necessities.[7] At the early stages of drug discovery, most of the employed methods provide kinetic solubility because they are developed to be useful during the whole discovery process, during which compounds are often in dimethyl sulfoxide (DMSO) solution. These latter methods can be divided into three classes: (i) filtration- or centrifugation-based methods,[13–16] (ii) nephelometric or turbidimetric methods[13,17–20] and (iii) potentiometric approaches. In filtration- and centrifugation-based methods, an aliquot of DMSO stock solution is added to a volume of solvent, and the precipitate that appears after a defined incubation time is removed by filtration or centrifugation. The amount of compound remaining in solution is then quantified. In nephelometric or turbidimetric methods, aliquots of DMSO stock solution are added until precipitation occurs. The precipitation is used as an indicator and is detected by optical measurement (UV or direct light-scattering detector). Approximated solubilities correspond to compound concentrations just before the first precipitation. Potentiometric approaches have recently been improved in terms of throughput and are mainly useful for pH-solubility profiles. A compound is added into solution and dissolved under its ionized (more soluble) form by pH adjustment. Thus, solubility is determined based on the difference in values of the pK_a of a compound measured in the absence and the presence of precipitate.[21–24]

Some high-throughput methods also allow for the determination of thermodynamic (or thermodynamic-like) solubility values, and these methods can also be classified into three groups: (i) methods adapted from kinetic protocols, (ii) methods using solid compounds from dried stock solutions to avoid the formation of metastable solutions and (iii) automated and miniaturized shake-flask methods.[7,25–28] Methods are adapted from kinetic protocols because the main source of high kinetic solubility values is the formation of metastable solutions; thus, researchers proposed increasing the incubation times, shaking intensity or temperature or decreasing the DMSO percentages.[7,16,19,29,30] The method of using solid compounds from dried stock solutions to avoid the formation of metastable solutions can be used even if the crystalline form generated does not necessarily correspond to the interesting form. Small volumes of DMSO,[7] acetonitrile (MeCN)[31] or methanol (MeOH)/dimethyl ether[32] stock solutions are evaporated under pressure, and the resulting solid material is hydrated with an appropriate volume of buffer before being filtrated and analyzed.

8.2.3 (U)HPLC as a Quantification Method for Solubility Measurements

For all of the methods described above, except nephelometry/turbidimetry, the quantification of a compound in solution is a crucial parameter in terms of sensitivity and throughput. Enhancing the reliability of a screening method includes increasing the throughput, sensitivity and accuracy of the analytical quantification. An increasing number of compounds with higher molecular weights have been investigated as potential drugs, and this has resulted in decreased solubility;[33] thus, chromatographic techniques have been the tools of choice for microscale measurements.

UV detection is often used to evaluate the amount of compound in solution; however, this technique fails when solvent or co-solvent overlay the absorption spectrum of a compound, when impurities or degradation products are present, and when a calibration curve has to be performed for all of the tested solvents prior to the solubility measurements.[7,34]

Many authors have reported the use of HPLC-UV as a quantification technique in solubility measurements to obtain higher quality results. For example, a new miniaturized device was proposed for microscale solubility measurements based on the circulation of a saturated solution of the compound to be tested and an "online" filtration through a syringe filter. In this experiment, 20 µl of filtrate was collected at two different times (6 h and 24 h) and directly injected into the HPLC-UV system. Because the collected volume for injection after 6 h was very low, it was not necessary to replace it, and the small volume loss had no significant influence on the solubility measurement after 24 h. Furthermore, due to the separative technique used for quantification, solubilities could be determined in different co-solvent systems, such as cremophor/, polysorbate 80/ or PEG 400/water mixtures.[35] Indeed, HPLC-UV is clearly more sensitive for the quantification of compounds compared with UV-based or nephelometric techniques.[36] For low solubility values, LC techniques are mandatory and, as already mentioned, are useful for impure compounds (where UV completely failed) and for solubility in various solvents with absorbance spectra that can overlay the spectrum of the tested compound. Compared with UV detection or chemiluminescent nitrogen detection (CLND), as used for thermodynamic solubility measurements by Bhattachar and colleagues,[11] the use of HPLC-UV decreases the throughput. Indeed, for better results, an adaptation of the chromatographic set-up is required for each compound. Therefore, a fast-gradient HPLC-UV method (runtime of 1.2 min) was reported to meet the throughput requirements of the drug discovery stage.[27] With this method, 24 compounds in three different aqueous media or 48 compounds in one medium can be determined in one run, together with single point calibration of each compound on the same 96-well plate.[27] The next labour-intensive step is the chromatogram analysis, but the analysis can be optimized with the use of customized visual basic for applications (VBA) reporting templates. The solubility range of the fast-gradient HPLC-UV method was 1 to 200 µl ml^{-1}, which is appropriate for lead selection in the lead optimization drug discovery stage.

Quantification by UHPLC should considerably diminish the major drawback of classic HPLC (*i.e.*, the analysis time). To date, several columns with high chemical stability have been used in various quantification studies of pharmaceutical compounds, and these columns could be advantageously applied to HTS solubility protocols.[37–39] Furthermore, the direct transfer of existing HPLC methods to UHPLC is possible without requiring more time.[40,41]

8.3 Ionization

8.3.1 Traditional Methods for pK_a Measurements

Several strategies exist to measure the ionization constants of NCEs in drug screening processes,[42,43] and all of the methods are based on the variation of the behaviour of a compound with regards to its ionization state. The gold standard is the potentiometric method, which measures the variation of the electrical potential with a pH change. Even if a co-solvent can be used to perform titrations,[44] the potentiometric method remains applicable for relatively soluble compounds (and if the compound remains in solution during the entire titration process), requires a relatively high amount of compound (around several mg) and can be categorized as low-to-medium throughput. UV methods have also been used, with the limitation that ionized and non-ionized forms should possess different UV spectra (at least one of the forms should have an absorbing group) and a high purity. Capillary electrophoresis (CE) has been used to perform accurate pK_a determination by comparing the effective mobility of a compound with a neutral marker for a series of pH buffers. In addition to its usefulness for impure compounds, CE is compatible with the small amounts of compounds that are normally available in the drug discovery stage and offers a relatively high throughput, particularly when coupled with MS detection. The level of throughput, however, is decreased for poorly soluble compounds because organic modifiers have to be added, and the number of measurements has to be increased for each pK_a determination.

8.3.2 Indirect pK_a Measurements Based on Liquid Chromatography

In the mid-1990s, HPLC was proposed as an alternative to the potentiometric approach.[45,46] Based on variations in the retention times between the ionized and neutral forms of a compound and plots of the retention times *versus* mobile phase pH, the pK_a can be determined as the inflection point of the curve. The HPLC technique has been successfully applied for the determination of ionization constants of parabens,[47] polyphenolic acids,[48] sulfonamides,[49,50] sulfonylureas,[50] the anti-hypertensives enalapril, lercanidipine and ramipril,[51] tetracycline antibiotics[52] or auxins.[53] Numerous buffers had to be prepared for each pK_a determination. In addition, many stationary phases were used for pK_a determination by HPLC; however, some stationary phases were limited in terms of the pH range that was exploitable for the mobile phase. Therefore, only strong acids (mainly carboxylic acids) and, to a lesser extent, weak bases were studied (pK_a between 3 and 7) because mobile phase pH values ranged from 2 to 9, but most of the buffers did not exceed a pH value of 7. In addition, working with a purely aqueous mobile phase was difficult because reversed-phase LC (RPLC) was employed. In cases of RPLC, apparent pK_a values were obtained depending on the percentage of co-solvent in the mobile phase (Figure 8.2). Originally, the idea was to obtain the pK_a of compounds in MeCN/water or MeOH/water systems; however, studies have shown that variations in apparent pK_a values measured with mobile phases containing different percentages of organic modifier (*e.g.*, MeCN and MeOH) were linearly correlated with the co-solvent fraction.[54] Therefore, it was possible to obtain pK_a in pure water by extrapolation using HPLC. Furthermore, the presence of co-solvent is important for poorly soluble compounds. Because of potential interactions between particular compounds and the stationary phase, derived aqueous pK_a values were not as accurate as values obtained by other techniques.[42] If this LC-based approach required time-consuming steps (buffer preparation, large number of chromatographic methods and injections), this approach would become useful when coupled with MS detection because the sample-pooling approach drastically increases the throughput.[42]

A variant of the "pure" chromatographic strategy is the HPLC-UV/DAD (diode array detection) method, which combines chromatographic and UV methods to achieve a greater precision. The HPLC-UV/DAD method consists of recording the UV absorbance at the maximum chromatographic peak

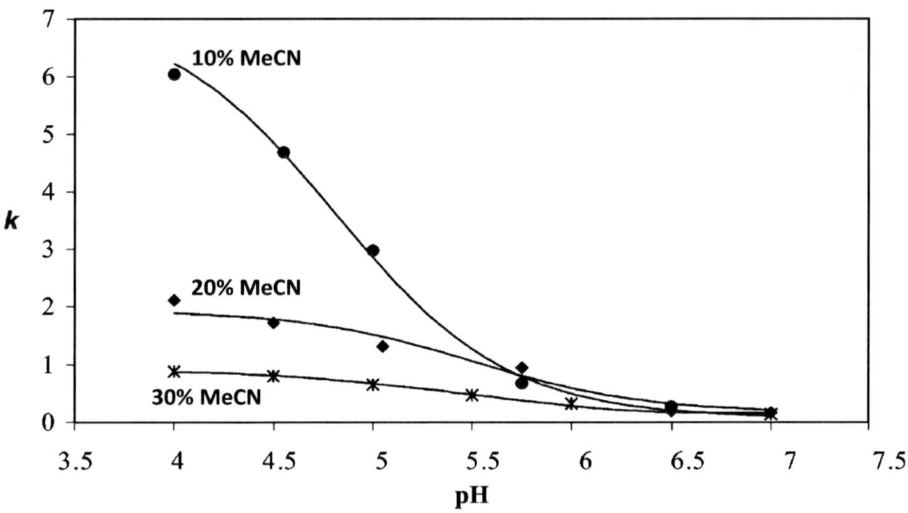

Figure 8.2 Retention factor k of caffeic acid *vs.* pH of mobile phases for different percentages of acetonitrile: (●) 10%, (♦) 20% and (*) 30%. Apparent pK_a values obtained were 4.33, 5.17 and 5.33 respectively. Reproduced from reference 48 with permission of Elsevier.

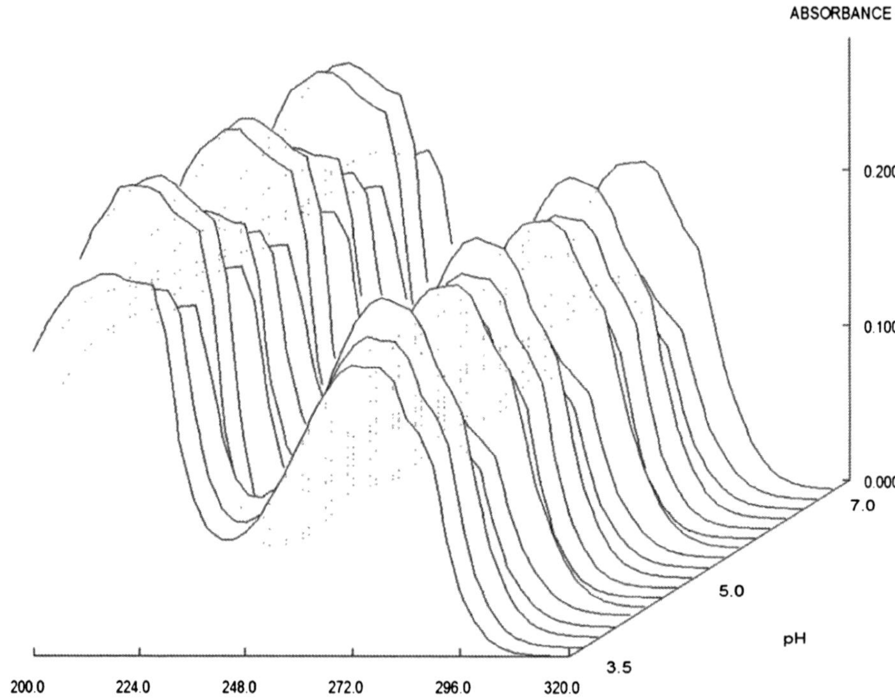

Figure 8.3 Absorption spectra of indole-3-lactic acid at the maximum of LC peaks with 30% methanol in mobile phase. Reproduced from reference 53 with the permission of the Brazilian Chemical Society.

(Figure 8.3).[48,53] Therefore, it is possible to obtain pK_a values from both techniques (chromatographic and UV) in a single experiment. Combining chromatographic and UV techniques allows scientists to utilize the advantages of both techniques. For example, the UV approach is more accurate than chromatographic methods, but UV detection has drawbacks for impure samples. Impurities, however, can be separated by chromatography, and the UV spectra can be monitored at different pH values to precisely evaluate the pK_a values of compound(s). Many studies have compared experimental pK_a values obtained by HPLC-UV and HPLC-UV(DAD), and results were often more precise with HPLC-UV(DAD) than with simple HPLC-UV methods.[48,49,52]

An alternative chromatographic approach was developed to obtain pK_a values by pH-gradient RPLC. In this approach, the pH of the eluent containing a constant proportion of organic solvent was linearly decreased (for basic compounds) and increased (for acidic compounds) by mixing two buffers of low and high pH, respectively. Linearly altering the pH led to a functional increase in analyte dissociation and a decrease in retention time.[55] The determination of pK_a was obtained after a complex mathematical treatment (theory and equations can be found elsewhere[55]). In other words, two initial organic solvent/buffer gradients at different gradient times were performed at a fixed pH value (*i.e.*, the pH at which the investigated analyte was in its non-ionized form) to evaluate an appropriate organic modifier concentration such that an acceptable retention time range can be provided for each analyte. Then, a pH gradient was performed, starting at a pH where the compound was under its neutral form and ending at a pH where the compound was fully ionized. An additional injection was also performed at the end of the pH gradient to determine the retention time of the ionized form. The pK_a values determined with this method were highly correlated to the pK_a values measured by titration, but different and specific correlations were obtained for acidic compared with basic compounds. The effect of MeOH in the mobile phase was studied to evaluate if the Yasuda–Shedlovsky linear relationship was applicable to reach the aqueous pK_a. Although pK_a values determined by pH-gradient RPLC were close to the values obtained by traditional methods, the results showed that they were still different. This was explained by the specific interactions with silanol groups inherent to chromatography stationary phases. The direct injection of a mixture of compounds also allowed an increase in the throughput.[55]

To the best of our knowledge, neither the HPLC nor the pH-gradient RPLC approaches was transferred to a UHPLC system. The isocratic HPLC method is not expected to save time because the limiting steps are the preparation of a large number of buffers (around 8–10 buffers to cover the largest pH range) and the number of chromatographic methods. Interestingly, the pH-gradient strategy has the advantage of a more rapid chromatographic analysis. Indeed, gradient times in the original (HPLC) method were 20 and 60 min for the first two organic modifier gradients compared with 12 to 30 min for the pH gradient, depending on the tested stationary phase and the type of analyte (acid or base). The use of UHPLC stationary phases could markedly reduce the analysis time (*i.e.*, by a factor 9, in theory) for pK_a determination because three gradients are applied for one compound, and this number increases for the determination of aqueous pK_a values by Yasuda–Shedlovsky extrapolation. Thus, UHPLC could significantly reduce the time for pK_a determination. Furthermore, UHPLC could also be used to extend the obtainable pK_a range because numerous UHPLC phases have extended pH range compatibility.

HPLC has also been used to simultaneously determine the lipophilicity profile (log D_{oct} *vs.* pH) and the pK_a,[56,57] which enhances the throughput during the drug discovery stage. This approach allows the determination of pK_a and log P_{oct} in half of a day.[56] If correlations between retention time obtained by HPLC and log P_{oct} for a series of acids and bases were not sufficient, then the use of UHPLC should largely improve the technique because the number of stationary phases in UHPLC have been shown to be particularly efficient for the log P determination in gradient mode (please see the lipophilicity part).

8.3.3 (U)HPLC as a Quantification Method for pK_a Measurements

pK_a and lipophilicity were also simultaneously measured by HPLC coupled to a 96-well microplate injector.[58] In this case, the lipophilicity profile was performed by mixing 0.5 ml of *n*-octanol and 0.5 ml of buffer containing the compound to be tested (at different pH values) in a 96-well plate. Twelve different buffers were prepared (one in each column of the plate), and eight compounds at twelve different pH values were analyzed in one microplate. Both phases (*n*-octanol and aqueous phases) were analyzed by HPLC in gradient mode, and the area under the curve at the maximum wavelength of each well was used to determine the log D_{oct}/log P_{oct} of the analytes. This profile allowed for pK_a determination. Because two injections per well are needed (one injection for the aqueous phase and another for the organic phase), a total of 192 injections are needed to determine the pK_a and lipophilicity profile of eight compounds. In this context, the use of UHPLC could be a way to enhance the throughput, even if the limiting step remains the "shake-flask", where twelve buffers have to be prepared, and *n*-octanol and buffer have to be mutually saturated and the plate shaken for several hours.

8.4 Lipophilicity

Lipophilicity is defined by the tendency of a compound to partition into a non-polar lipid matrix *vs.* an aqueous matrix. It represents the molecular parameter of choice in numerous (Q)SARs of different classes of compounds.[59]

Lipophilicity is quantified by the partition coefficient (log *P*), which is the logarithm of the partition coefficient of a compound between an organic phase (*e.g.*, *n*-octanol) and an aqueous phase (*e.g.*, buffer) at a pH value at which the compound is under its neutral form:

$$\log P = \log \frac{[compound_{organic}]}{[compound_{aqueous}]} \tag{8.1}$$

For ionizable compounds, different electric forms can co-exist, depending on the pH of the environment; thus, the lipophilicity is expressed by the distribution coefficient log *D*:

$$\log D = \log \frac{[X_{(organic)} + XH_{(organic)} + XH_{2(organic)} + \ldots]}{[X_{(aqueous)} + XH_{(aqueous)} + XH_{2(aqueous)} + \ldots]} \tag{8.2}$$

8.4.1 Traditional Methods

The gold standard for assessing lipophilicity is the shake-flask approach, which measures compound partition between immiscible non-polar and polar liquid phases. Traditionally, *n*-octanol has been used as the non-polar phase and aqueous buffer as the polar phase. Although this procedure may remain the reference for lipophilicity determination,[60] there are several drawbacks: the procedure is time consuming, sensitive to impurities and requires a large amount of compound.

When dealing with ionizable compounds, the dual-phase potentiometric titration is also considered as a reference method. The potentiometric titration method provides an interesting way to obtain the partition coefficient by comparing the aqueous pK_a measured with and without organic solvent. The dual-phase titrations also provide the distribution profile (log D *vs.* pH), but the same limitations of the shake-flask method also apply to dual-phase potentiometric titration.[34]

CE is a separative technique that is also used for log P_{oct} determination. Different CE modes have been considered, including micellar electrokinetic chromatography (MEKC),[61] micro-emulsion electrokinetic chromatography (MEEKC)[62,63] and vesicle/liposome electrokinetic chromatography (VEKC/LEKC).[64] These electrokinetic techniques use buffers containing micelles, nanometer-sized oil droplets, liposomes or vesicles as the pseudo-stationary phase. The principle of the separation mechanism is based on the partitioning of the analytes between the aqueous and the pseudostationary phases. The major drawbacks are low reproducibility and a limited range of accessible log P.[65]

8.4.2 Indirect Lipophilicity Measurements Based on Liquid Chromatography

LC approaches are considered to be the best alternatives to the other lipophilicity assessment strategies for several reasons: they only require a small amount of compound, they can be automated to speed up the assessment, they are insensitive to the impurities and they have a good level of throughput. Many systems have been proposed in this field using HPLC, and this has been covered in several recent reviews.[60,66,67] A large amount of literature on lipophilicity determination based on reversed-phase high-pressure LC (RP-HPLC) has been published. In this part of the Chapter, we will focus on strategies based on UHPLC technology for the log P_{oct} determination.

The determination of log P_{oct} by RPLC is based on the partitioning of a solute between a polar mobile and an noin-polar stationary phase. Depending on the system considered, the retention factor (log k) was correlated with the partition coefficient parameter (in general with log P_{oct}):

$$\log P_{oct} = a \log k + b \tag{8.3}$$

where a and b are the linear regression constants specific to the considered mobile and stationary phases.

8.4.2.1 Measurement of Retention Factors

The isocratic and the gradient modes were evaluated for the measurement of log k in the determination of log P. The gradient-based methods were developed to speed up the throughput, but the mathematical treatment remains complex and less widespread.

In the isocratic mode, the retention factor (log k) for the investigated compound is determined by the following formula:

$$\log k = \log [(t_r - t_0)/t_0] \tag{8.4}$$

where t_r and t_0 are the retention times of the solute and an unretained compound.

When working with short columns at high mobile phase linear velocities, it is necessary to apply corrections to the experimental t_r and t_0 values to discard the influence of the chromatographic system and obtain accurate log k values.

The extra-column volume (V_{ext}) and the injection delay (V_{delay}) can be experimentally determined and used in the following equation:

$$\log k = \left(\frac{t_r - t_{delay} - V_{ext}/F}{t_0 - t_{delay} - V_{ext}/F} - 1 \right) \tag{8.5}$$

where t_r and t_0 are the retention times of the solute and an unretained compound, t_{delay} is the injection delay, V_{ext} is the extra-column volume and F is the mobile phase flow rate.

Linear regression relationships between log P_{oct} and isocratic log k values have been reported for some classes of analytes,[68–70] and studies have shown that extrapolated log k_w led to better correlations with log P_{oct}.[65,71,72] In isocratic conditions, log k_w is obtained by extrapolation to 100% water by plotting isocratic log k values as a function of the proportion of organic modifier. The relationship between isocratic log k values and organic modifier concentration depends on the experimental conditions. With MeOH, isocratic log k values are generally linearly correlated with the organic modifier percentage in the mobile phase:

$$\log k = \log k_w - S\varphi \tag{8.6}$$

where k is the retention factor, S is a constant for a given solute and organic modifier and log k_w is the retention factor extrapolated to 100% water as the mobile phase. When MeCN is used, the relation between isocratic log k and the percentage of organic modifier in the mobile phase is quadratic:

$$\log k = \log k_w + B\varphi + A\varphi^2 \tag{8.7}$$

In gradient mode, the retention time (t_r) can be expressed with a linear gradient separation[73] as follows:

$$t_r = \frac{t_0}{b} \cdot \log(2.3 \cdot k_0 \cdot b + 1) + t_0 + t_D \tag{8.8}$$

where t_0 is the retention time of an unretained compound, k_0 is the k value at the beginning of the gradient (for $\varphi = \varphi_0$), t_D is the system dwell time for gradient elution (min), which can be experimentally determined,[74] and b is the gradient steepness parameter, which is described by the following relationship:

$$b = \frac{t_0 \cdot \Delta\varphi \cdot S}{t_G} \tag{8.9}$$

where t_G is the gradient time from the beginning to the end of the gradient (min), and $\Delta\varphi$ is the change in φ during the gradient [equal to $(\%B_{final} - \%B_{initial})/100$], which ranges from 0 to 1. Ideally, the parameter b should remain constant throughout the gradient run and have the same value for all the compounds eluted during the chromatographic process.

As described by eqns (8.6), (8.8) and (8.9), the lipophilicity measurement in the gradient mode is based on the determination of log k_0 and S (*i.e.*, both terms are necessary to obtain log k_w). To obtain an accurate value of log k_w, two gradient runs with different t_G values are needed.[63,75–78] Because of the complex mathematical treatment and the quadratic relationship between log k and the percentage of organic modifier observed for MeCN, only gradients with MeOH were performed.

8.4.2.2 Log P *Determination*

Many HPLC stationary phases were developed to determine the lipophilicity, and some were more successful than others. The stationary phases were mostly based on silica, which has the well-known drawbacks of secondary interactions with residual silanol, which interferes with the partition mechanism,[79] and low stability at high pH. These drawbacks were overcome by developing a new generation of columns that are deactivated silica-based, embedded and end-capped stationary phases with reduced secondary interactions or polymer-based stationary phases with increased stability over the whole pH range.[66,72,80] To speed up analysis, short columns and high flow rates were also proposed for lipophilicity determination, and these parameters allowed for an extended log P range.[81–83] Nevertheless, these methods generated a significant diminution of the chromatographic performance and accuracy of the results with relative problems of overloading and short column lifespans (Figure 8.4A).[84] Monolithic supports showed promising results that were almost equivalent to the results obtained with silica-based stationary phases, and the analysis times were reduced 10-fold;[85] however, the applicability of

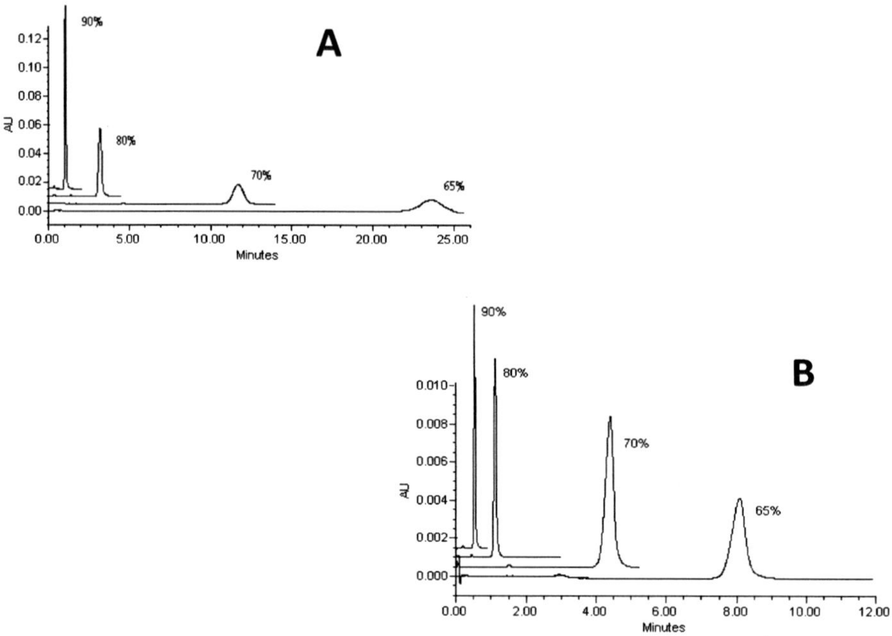

Figure 8.4 Comparison between HPLC-UV (A) and UHPLC-UV (B) methods for the log *P* determination of the highly lipophilic compound tetrachlorodibenzodioxin (log $P = 7.15$). Conditions for LC-UV analysis: column, Discovery® RP amide C16 20×4 mm, 5 μm; isocratic mode: 90 to 65 % MeOH. Conditions for UHPLC-UV analysis: column, Hypersil Gold Javelin HTS 10×2.1 mm, 1.9 μm; isocratic mode: 90 to 65% MeOH.

monolithic columns is limited because of their poor resistance at high pH and the lack of available chemistries and dimensions.

A large number of new silica or hybrid stationary phases that use 1.5–2.0 μm particles have been developed in recent years, and their diversity in terms of chemistry is excellent (C8, C18, phenyl and others, which allows for chemical stability and the possibility of using a wide pH range).[86] The introduction of hybrid supports provides high chemical stability and allows their use throughout almost the entire pH range, typically 1–12. In addition, these new stationary phases have permitted the successful analysis of basic compounds at pH values far above their pK_a values.[87] Moreover, fascinating perspectives in analytical chemistry and in the determination of log P_{oct} of very lipophilic compounds have been provided. A study by Henchoz *et al.*[87] used UHPLC coupled with UV detection for the log P_{oct} determination of a series of structurally diverse compounds. They noted that 25 min per compound was required to obtain log P_{oct} in the isocratic mode, whereas approximately 20 min, including time for column equilibration, was needed for the gradient mode. This approach was also applied to basic drugs (β-blockers and local anesthetics) because of the stability of the UHPLC column in basic pH conditions.[87] Henchoz *et al.*[87] also observed a good relationship between isocratic log k at 50% MeOH (log k_{50}) *vs.* log P_{oct}, which was suitable for the estimation of log $P_{oct,}$ in approximately 5 min. Another study determined lipophilicity by UHPLC using two conventional columns (Acquity BEH Shield RP18 from Waters and Hypersil™ Gold Javelin HTS from Thermo Scientific) in the isocratic mode with MeOH or the gradient mode with MeCN and MeOH.[88] For a series of 52 diverse compounds covering a large log P range between 0 and 8, we observed good correlations between log P_{oct} and log k_w values with both columns, and we obtained high resolution peaks compared with HPLC strategies (Figure 8.4B). This shows that these UHPLC columns are more suitable for highly lipophilic compounds than traditional HPLC methods (Figure 8.5). The results also show that accurate log P_{oct} values were achieved in the isocratic mode with both Hypersil™ Gold Javelin HTS with MeCN and Acquity BEH Shield RP18 with MeOH as the organic modifier. Interestingly, the Hypersil™ column with MeOH resulted in faster determination of log P in the isocratic mode (6 to 8 min depending on the lipophilicity of the tested compound), whereas the Acquity column resulted in an accurate and faster determination of log P in the gradient mode (20 min for log P determination of one compound). Although the treatment remains more complex than the isocratic mode, the advantage of the gradient mode is that the analysis time remains constant regardless of the lipophilicities of the tested compounds. In the isocratic mode, run times have to be adapted to the expected lipophilicities of the tested compounds. These promising UHPLC columns open the possibility of obtaining higher log P values with moderate analysis time, which is critical in the pharmaceutical as well as in the environmental field because highly lipophilic compounds are known to be more prone to bioaccumulation and can significantly contribute to environmental pollution.

The coupling of UHPLC with MS devices provides an additional gain in terms of sensitivity, selectivity and resolution compared with UV detection

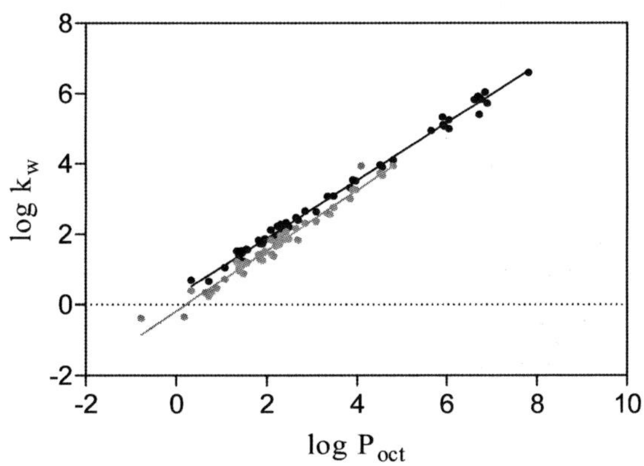

Figure 8.5 Correlations between log P_{oct} and extrapolated log k_w obtained by HPLC with a discovery® RP amide C16 (●)[80] and obtained by UHPLC with an Acquity BEH Shield RP18 (●).[88]

(*i.e.*, *m*/*z* information).[70] The coupling of UHPLC and MS has opened new perspectives in lipophilicity determination of mixtures of compounds with different masses and lipophilicities.[89–91]

8.4.3 (U)HPLC as a Quantification Method in Lipophilicity Measurements

The shake-flask method is a direct measurement of *n*-octanol-buffer partitioning and remains the "gold standard" technique, particularly during method validation.[92]

Different strategies have been applied to speed up and automate the shake-flask procedure, such as the use of LC/MS, robotic systems and 96-well plates.[93,94] Alelyunas *et al.* proposed a high-throughput *n*-octanol–water lipophilicity measurement based on a 96-well shake-flask and LC/UV/APPI (atmospheric pressure photo-ionization)/MS approach[95] that eliminates DMSO and generates solid-like material that can be used for log *P*/log *D* measurements (the removal of DMSO minimizes the co-solvent effect on the measured values). Both the *n*-octanol and buffer phases were quantified using UHPLC coupled to an APPI mass spectrometer. The throughput of the method was 2 days for a batch of 96 compounds, and it has been validated on 72 compounds with diverse ionization and log *D* values ranging from −2 to +6.

The parallel artificial membrane permeability assay (PAMPA) is an *in vitro* tool that was originally developed to evaluate the permeability of NCEs through biological membranes (*i.e.*, intestinal[96,97], blood–brain[98] or cutaneous barriers;[99] see Section 8.5). In 2005, Faller and colleagues[100] reported a high-throughput assay that was designed to measure the *n*-octanol/water partition coefficients. The assay was carried out in 96-well microfilter plates, and the diffusion of compounds between two aqueous compartments separated by a thin *n*-octanol liquid layer was measured. The measurements were shown to correlate with log P_{oct}. To derive the partition coefficient, the amount of compound that remained in the donor compartment and appeared in the acceptor compartment after a defined incubation time was measured. The assay allows for the determination of 96 partition coefficients within the range of −2 to +5 (and up to 8 when coupled with MS detection) in one single experiment. The PAMPA method can also be extended to other water/partition solvent systems.

In 2008 Ottaviani *et al.*[101] focused on the application of artificial membranes for the evaluation of partition coefficients. In this study, *ortho*-nitrophenyl octyl ether (*o*-NPOE) was immobilized on polyvinylidene fluoride (PVDF) supporting filters. After incubation, the microplate was centrifuged to ensure the removal of *o*-NPOE, which interfered with the compounds UV absorption. Interestingly, the results showed that *o*-NPOE could form an artificial membrane, which created the possibility of deriving partition coefficients.

To the best of our knowledge, UHPLC has not been used for the determination of lipophilicity by PAMPA. Based on experience, however, existing methods can be modified to enhance current techniques (see Section 8.5). For example, UHPLC measurements could provide advantages by rapidly separating compounds on a column and checking the residual membrane. Modifying and improving existing techniques can improve our ability to measure the lipophilicities of compound mixtures.

8.5 Permeability

8.5.1 Indirect Permeability Measurements Based on Liquid Chromatography

Immobilized artificial membranes (IAM, silica-based stationary phase with phospholipids covalently bound),[102,103] immobilized liposome chromatography (ILC, liposomes immobilized into gel beads)[104] and biopartitioning micellar chromatography [BMC, based on a RP C18 stationary phase and a mobile phase containing polyoxyethylene (23) lauryl ether (Brij35)][105–107] are systems that have been developed to reproduce biomembrane and predict passive permeability using HPLC. In some cases, these kinds of systems have shown reasonable results in terms of permeability prediction, even if the retention time on the column does not reflect transport across the membrane;[108,109] however, there are some limitations in terms of retention times for lipophilic compounds and stationary phase stability. Recently, short IAM columns have appeared on the market. These columns (1–3 cm *versus* 10–12 cm) allow greater throughput, but they do not really offer any enhancements in terms of reliability. To date, IAM, ILC and BMC have not been transferred to UHPLC.

8.5.2 (U)HPLC as a Quantification Method in Permeability Measurements

Several cell-based *in vitro* permeability measurement methods have been developed (*e.g.*, Caco-2, MDCK and BBCEC) to predict permeability through different biological barriers. These methods have not only provided information about passive diffusion but also about active transport, depending on the transporters expressed in the considered cells. Studies had already shown that LC/MS detection could be a useful tool for cell-based permeability models. For example, one study showed that the Caco-2 cell model could measure samples in duplicate (with a throughput of 20 samples in 24 h) to predict gastrointestinal absorption.[110] Although cellular models have been useful, they are time consuming because cells have to grow for several days, and their manipulation requires extensive knowledge and practice. Cellular assays are not only labour intensive but also expensive. Therefore, from a throughput standpoint, the use of HPLC/MS detection (and particularly UHPLC/MS detection) is not sufficient to analyze a greater number of compounds.

PAMPA was proposed by Kansy *et al.*[96] in 1998, and a recent review reported the different models that were derived from this pioneering study.[111] PAMPA is a high-throughput *in vitro* assay system that is used to evaluate passive diffusion through biological membranes in pharmaceutical research. Several studies have indicated that PAMPA permeability is correlated with both Caco-2 cell permeability and human intestinal absorption.[112,113] Depending on the artificial membrane that is used, permeability predictions through other membranes, such as the blood–brain and percutaneous barriers,[99,114] can be achieved. In contrast to cell-based assays, the PAMPA methods, which are based on artificial membranes supported on filters, are cheaper and allow automation; thus, this methodology is faster than other systems. In a PAMPA, a "sandwich" is formed from two plates: a donor microfilter plate supporting the artificial membrane and a receiver plate. Samples dissolved in DMSO stock solutions are diluted with buffer and dispensed into donor wells. Usually, buffer solution is added to the acceptor compartment system. After an appropriate incubation time, the plates are separated, and aliquots are removed from both the donor and the receiver wells to determine the solute concentration on both sides of the membrane. In a PAMPA, the effective (log P_e) or apparent (log P_a) permeability coefficients are determined from a single time-point sampling, which allows high-throughput measurements. Avdeef[115] demonstrated that equations used to calculate permeability coefficients can be deduced in several ways depending on the experimental conditions and the design of the *in vitro* assay. Although quantification is generally carried out with a UV plate reader, HPLC and HPLC/MS have also been used for additional sensitivity. In regards to PAMPA, the most delicate and time-consuming step is the sample analysis.

LC coupled to MS detection has proved its efficiency to increase the sensitivity of PAMPA experiments, particularly for poorly soluble compounds or compounds that have no or low UV chromophore.[116] Nevertheless, HPLC/MS detection considerably decreases the throughput of the assay and should only be used for problematic compounds. With the commercialization of UHPLC, analysis with LC/MS or LC/MS/MS can be significantly improved.[117,118]

In 2007, Mensch *et al.*[119] employed UHPLC/MS/MS for permeability assessment. Column injections were performed in gradient conditions. Using this methodology, they achieved a 4-fold increase in throughput compared with the generic HPLC/MS strategy. Another study[120] showed that UHPLC/MS/MS enabled the determination of compound concentrations in the reference, donor and acceptor compartments in triplicate in approximately 40 min at a pH of 2 (*i.e.*, 4-fold faster than HPLC/MS procedures).

Carrara *et al.*[121] attempted to set up a high-throughput method to assess passive blood–brain barrier penetration using UHPLC for quantification. Attention was focused on the effects of the solvent and the influence of the phospholipids that composed the artificial membrane on blood–brain barrier PAMPA results. The high-throughput nature of the assay was also maximized by using a cassette mode.[119–121] Standards were tested in four sets of six compounds, and the permeability results were not significantly different from the permeability coefficients obtained when the compounds were incubated separately. In addition, the determination of permeability coefficients decreased to 4.5 min for six compounds when mixing both multiple-incubations and UHPLC/MS detection strategies. A PAMPA performed with cassette incubation has also been reported by Balimane *et al.*,[122] and UHPLC/MS was also used in brain penetration studies of complex mixtures,[119] even when particular attention should be paid to compounds with the same elution times, similar ion fragmentation patterns or with low permeability.[65,121,123]

8.6 Conclusions

Chromatographic approaches are widely recognized as useful methods for physicochemical profiling and permeability studies in the drug discovery phase. In particular, LC allows for accurate and rapid indirect determination of ionization and lipophilicity parameters. To the best of our knowledge, HPLC approaches for indirect pK_a measurements have not been transferred to a UHPLC system; however, this could save time, especially when using the gradient pH mode. Interestingly, the introduction of columns packed with sub-2-μm porous particles used under ultra-high-pressure conditions has reduced analysis time in lipophilicity measurements and increased performance, which has enhanced the detectable log P range.

More sensitive quantifications were obtained when LC was applied to traditional strategies for solubility, ionization, lipophilicity and permeability measurements. Indeed, LC permits researchers to detect and use lower concentrations and achieve better accuracy with new synthesized compounds that often present low solubility. In drug discovery, HPLC has generally only been used for problematic compounds because of the decreases in throughput that are associated with HPLC quantification. UHPLC combined with MS detection, however, results in shorter analysis times and allows for the analysis of compound mixtures.

Therefore, UHPLC is the most promising strategy to achieve high-throughput and/or high-resolution physicochemical property determinations and permeability studies in a drug discovery environment. Indeed, UHPLC will probably replace conventional LC in the different assays.

References

1. J. I. Drayer and J. P. Burns, in *Burger's Medicinal Chemistry and Drug Discovery*, ed. M. E. Wolff, John Wiley & Sons, New York, 1995, p. 251.
2. T. Kennedy, *Drug Discov. Today*, 1997, **2**, 436.
3. B. Testa, H. van de Waterbeemd, G. Folkers and R. H. Guy, *Pharmacokinetic Optimization in Drug Research: Biological,*

Physicochemical and Computational Strategies, Wiley-VHCA, Zurich, 2001.
4. D. A. Smith, B. C. Jones and D. K. Walker, *Med. Res. Rev.*, 1996, **16**, 243.
5. E. H. Kerns, *J. Pharm. Sci.*, 2001, **90**, 1838.
6. E. H. Kerns and L. Di, *Drug Discov. Today*, 2003, **8**, 316.
7. J. Alsenz and M. Kansy, *Adv. Drug Deliv. Rev.*, 2007, **59**, 546.
8. S. DeVito, in *Handbook of Property Estimation Methods for Chemicals*, ed. R. S. Boethling and D. Mackay, Lewis Publishers, Boca Raton, 2010, p. 261.
9. H. van de Waterbeemd, D. A. Smith, K. Beaumont and D.K. Walker, *J. Med. Chem.*, 2001, **44**, 1313.
10. E. H. Kerns and L. Di, *Drug Discov. Today Technol.*, 2004, **4**, 343.
11. S. N. Bhattachar, L. A. Deschenes and J. A. Wesley, *Drug Discov. Today*, 2006, **11**, 1012.
12. P. S. Burton and J. T. Goodwin, *Comb. Chem. High Throughput Screening*, 2010, **13**, 101.
13. L. Pan, Q. Ho, K. Tsutsui and L. Takahashi, *J. Pharm. Sci.*, 2001, **90**, 521.
14. T. M. Chen, H. Shen and C. Zhu, *Comb. Chem. High Throughput Screening*, 2002, **5**, 575.
15. A. A. Agbodjan, H. Bui and M. G. Khaledi, *Langmuir*, 2001, **17**, 2893.
16. B. Bard, S. Martel and P. A. Carrupt, *Eur. J. Pharm. Sci.*, 2008, **33**, 230.
17. C. A. Lipinski, F. Lombardo, B. W. Dominy and P. J. Feeney, *Adv. Drug Deliv. Rev.*, 1997, **23**, 3.
18. C. D. Bevan, *Anal. Chem.*, 2000, **72**, 1781.
19. K. A. Dehring, H. L. Workman, K. D. Miller, A. Mandagere and S. K. Poole, *J. Pharm. Biomed. Anal.*, 2004, **36**, 447.
20. T. A. Fligge and A. Schuler, *J. Pharm. Biomed. Anal.*, 2006, **42**, 449.
21. A. Avdeef, *Pharm. Pharmacol. Commun.*, 1998, **4**, 165.
22. A. Avdeef, C. M. Berger and C. Brownell, *Pharm. Res.*, 2000, **17**, 85.
23. A. Avdeef and C. M. Berger, *Eur. J. Pharm. Sci.*, 2001, **14**, 281.
24. M. Stuart and K. Box, *Anal. Chem.*, 2005, **77**, 983.
25. A. Glomme, J. März and J. B. Dressman, *J. Pharm. Sci.*, 2005, **94**, 1.
26. J. P. Jeanie Cherng, J. Gonzalez-Zugasti, N. R. Kane, M. Cima and A. V. Lemmo, *JALA*, 2004, **9**, 228.
27. P. Mansky, W-G. Dai, S. Li, C. Pollock-Dove, K. Daehne, L. Dong and G. Eichenbaum, *J. Pharm. Sci.*, 2007, **96**, 1548.
28. M. C. Wenlock, R. P. Austin, T. Potter and P. Barton, *J. Lab. Autom.*, 2011, **16**, 276.
29. K. Sugano, T. Kato, K. Suzuki, K. Keiko, T. Sujaku and T. Mano, *J. Pharm. Sci.*, 2006, **95**, 2115.
30. K. Obata, K. Sugano, M. Machida and Y. Aso, *Drug Dev. Ind. Pharm.*, 2004, **30**, 181.
31. H. Tan, D. Semin, M. Wacker and J. Cheetham, *JALA*, 2005, **10**, 364.

32. D. Roy, F. Ducher, A. Laumain and J. Y. Legendre, *Drug Dev. Ind. Pharm.*, 2001, **27**, 107.
33. C. A. Lipinski, *J. Pharmacol. Toxicol. Meth.*, 2000, **44**, 235.
34. A. Avdeef and B. Testa, *Cell. Mol. Life Sci.*, 2002, **59**, 1681.
35. X. Q. Chen and S. Venkatesh, *Pharm. Res.*, 2004, **21**, 1758.
36. B. Hoelke, S. Gieringer, M. Arlt and C. Saal, *Anal. Chem.*, 2009, **81**, 3165.
37. U. Chiuminatto, F. Gosetti, P. Dossetto, E. Mazzucco, D. Zampieri, E. Robotti, M. C. Gennaro and E. Marengo, *Anal. Chem.*, 2010, **82**, 5636.
38. E. Deconinck, P.Y. Sacré, S. Baudewyns, P. Courselle and J. De Beer, *J. Pharm. Biomed. Anal.*, 2011, **56**, 200.
39. E. Gracia-Lor, J. V. Sancho and F. Hernández, *J. Chromatogr. A*, 2011, **1218**, 2264.
40. D. Guillarme, J. Ruta, S. Rudaz and J. L. Veuthey, *Anal. Bioanal. Chem.*, 2010, **397**, 1069.
41. J. L. Veuthey, D. Guillarme and S. Rudaz, *Bioanalysis*, 2010, **3**, 5.
42. H. Wan and J. Ulander, *Expert Opin. Drug Metab. Toxicol.*, 2006, **2**, 139.
43. H. Wan and A. G. Holmen, *Comb. Chem. High Throughput Screening*, 2009, **12**, 315.
44. A. Avdeef, K. J. Box, J. E. A. Comer, M. Gilges, M. Hadley, C. Hibbert, W. Patterson and K. Y. Tam, *J. Pharm. Biomed. Anal.*, 1999, **20**, 631.
45. E. Bosch, P. Bou, H. Allemann and M. Rosés, *Anal. Chem.*, 1996, **68**, 3651.
46. H. Y. Ando and T. Heimbach, *J. Pharm. Biomed. Anal.*, 1997, **16**, 31.
47. T. Angelov, A. Vlasenko and W. Tashkov, *J. Liq. Chromatogr. Relat. Technol.*, 2008, **31**, 188.
48. J. L. Beltran, N. Sanli, G. Fonrodona, D. Barron, G. Özkan and J. Barbosa, *Anal. Chim. Acta*, 2003, **484**, 253.
49. N. Sanli, S. SanlI, G. Ozkan and A. Denizli, *J. Braz. Chem. Soc.*, 2010, **21**, 1952.
50. M. Manderscheid and T. Eichinger, *J. Chromato. Sci.*, 2003, **41**, 323.
51. M. Gumustas, S. SanlI, N. Sanli and S. A. Ozkan, *Talanta*, 2010, **82**, 1528.
52. S. SanlI, N. Sanli and G. Alsancak, *J. Braz. Chem. Soc.*, 2009, **20**, 939.
53. M. Kelen and N. Sanli, *J. Braz. Chem. Soc.*, 2009, **20**, 133.
54. J. Barbosa, J.L. Beltran and V. Sanz-Nebot, *Anal. Chim. Acta*, 1994, **288**, 271.
55. R. Kaliszan, P. Wiczling and M. J. Markuszewski, *J. Chromatogr. A*, 2004, **1060**, 165.
56. P. Wiczling, P. Kawczak, A. Nasal and R. Kaliszan, *Anal. Chem.*, 2006, **78**, 239.
57. P. Wiczling, M. Waszczuk-Jankowska, M. J. Markuszewski and R. Kaliszan, *J. Chromatogr. A*, 2008, **1214**, 109.
58. P. C. Chiang and Y. Hu, *Comb. Chem. High Throughput Screening*, 2009, **12**, 250.

59. M. Bajda, S. Boryczka and B. Malawska, *Biomed. Chromatogr.*, 2007, **21**, 123.
60. S. Gocan, G. Cimpan and J. Comer, in *Advances in Chromatography*, ed. E. Grushka and N.Grinberg, Taylor & Francis Group, Boca Raton, 2006, p. 79.
61. S. Terabe, K. Otsuka and T. Ando, *Anal. Chem.*, 1985, **57**, 834.
62. Y. Ishihama, Y. Oda and N. Asakawa, *Anal. Chem.*, 1996, **68**, 1032.
63. R. Kaliszan, A. Nasal and M. J. Markuszewski, *Anal. Bioanal. Chem.*, 2003, **377**, 803.
64. C. W. Huie, *Electrophoresis*, 2006, **27**, 60.
65. Y. Henchoz, S. Romand, J. Schappler, S. Rudaz, J. L. Veuthey and P. A. Carrupt, *Electrophoresis*, 2010, **31**, 952.
66. A. Nasal and R. Kaliszan, *Curr. Computer-Aided Drug Des.*, 2006, **2**, 327.
67. R. Nicoli, S. Martel, S. Rudaz, J. L. Wolfender, J. L. Veuthey, P. A. Carrupt and D. Guillarme, *Expert Opin. Drug Del.*, 2010, **5**, 475.
68. A. Pagliara, E. Khamis, A. Trinh, P. A. Carrupt, R. S. Tsai and B. Testa, *J. Liq. Chromatogr.*, 1995, **18**, 1721.
69. C. M. Du, K. Valko, C. Bevan, D. Reynolds and M. H. Abraham, *Anal. Chem.*, 1998, 70, 4228.
70. Y. Henchoz, D. Guillarme, S. Martel, S. Rudaz, J. L. Veuthey and P. A. Carrupt, *Anal. Bioanal. Chem.*, 2009, **394**, 1919.
71. K. Valko and P. Slegel, *J. Chromatogr.*, 1993, **631**, 49.
72. S. Martel, D. Guillarme, Y. Henchoz, A. Galland, J. L. Veuthey, S. Rudaz and P. A. Carrupt, in *Drug Properties: Measurement and Computation*, ed. R. Mannhold, Wiley-VCH, Weinheim, 2008, p. 331.
73. L. R. Snyder, J. W. Dolan and J. R. Gant, *J. Chromatogr. A*, 1979, **165**, 3.
74. J. W. Dolan, *LC-GC N. Am.*, 2006, **24**, 458.
75. S. K. Poole and C. F. Poole, *J. Chromatogr. B*, 2003, **797**, 3.
76. A. Nasal, D. Siluk and A. Kaliszan, *Curr. Med. Chem.*, 2003, **10**, 381.
77. N. C. Dias, M. I. Nawas and C. F. Poole, *Analyst*, 2003, **128**, 427.
78. R. Kaliszan, P. Haber, T. Baczek and D. Siluk, *Pure Appl. Chem.*, 2001, **73**, 1465.
79. J. Nawrocki, *J. Chromatogr. A*, 1997, **779**, 29.
80. X. Liu, H. Tanaka, A. Yamauchi, B. Testa and H. Chuman, *Helv. Chim. Acta*, 2004, **87**, 2866.
81. K. Valko, C. M. Du, C. Bevan, D. P. Reynolds and M. H. Abraham, *Curr. Med. Chem.*, 2001, **8**, 1137.
82. N. Gulyaeva, A. Zaslavsky, P. Lechner, M. Chlenov, A. Chait and B. Zaslavsky, *Eur. J. Pharm. Sci.*, 2002, **17**, 81.
83. L. Ayouni, G. Cazorla, D. Chaillou, B. Herbreteau, S. Rudaz, P. Lanteri and P. A. Carrupt, *Chromatographia*, 2005, **62**, 251.
84. S. F. Donovan and M. C. Pescatore, *J. Chromatogr. A*, 2002, **952**, 47.
85. T. Welerowicz and B. Buszewski, *Biomed. Chromatogr.*, 2005, **19**, 725.
86. L. Novakova and H. Vlckova, *Anal. Chim. Acta*, 2009, **656**, 8.

87. Y. Henchoz, D. Guillarme, S. Rudaz, J. L. Veuthey and P. A. Carrupt, *J. Med. Chem.*, 2008, **51**, 396.
88. A. Guillot, Y. Henchoz, C. Moccand, D. Guillarme, J. L. Veuthey, P. A. Carrupt and S. Martel, *Chem. Biodivers.*, 2009, **6**, 1828.
89. G. Camurri and A. Zaramella, *Anal. Chem.*, 2001, **73**, 3716.
90. Y. Zhao, J. Jona, D. T. Chow, H. Rong, D. Semin, X. Xia, R. Zanon, C. Spancake and E. Maliski, *Rapid Commun. Mass Spectrom.*, 2002, **16**, 1548.
91. E. H. Kerns, L. Di, S. Petusky, T. Kleintop, D. Huryn, O. McConnell and G. Carter, *J. Chromatogr. B*, 2003, **791**, 381.
92. S. Martel, V. Gasparik and P. A. Carrupt, in *Hit and Lead Profiling*, ed. B.Faller and L.Urban, Wiley-VCH, Weinheim, 2009.
93. D. M. Wilson, X. Wang, E. Walsh and R. A. Rourick, *Comb. Chem. High Throughput Screening*, 2001, **4**, 511.
94. N. Gulyaeva, A. Zaslavsky, P. Lechner, A. Chait and B. Zaslavsky, *J. Peptide Res*, 2003, **61**, 71.
95. Y. W. Alelyunas, L. Pelosi-Kilby, P. Turcotte, M. B. Kary and R. C. Spreen, *J. Chromatogr. A*, 2010, **1217**, 1950.
96. M. Kansy, F. Senner and K. Gubernator, *J. Med. Chem.*, 1998, **41**, 1007.
97. F. Wohnsland and B. Faller, *J. Med. Chem.*, 2001, **44**, 923.
98. L. Di, E. H. Kerns, K. Fan, O. J. McConnell and G. T. Carter, *Eur. J. Med. Chem.*, 2003, **38**, 223.
99. G. Ottaviani, S. Martel and P. A. Carrupt, *J. Med. Chem.*, 2006, **49**, 3948.
100. B. Faller, H. P. Grimm, F. Loeuillet-Ritzler, S. Arnold and X. Briand, *J. Med. Chem.*, 2005, **48**, 2571.
101. G. Ottaviani, S. Martel, C. Escalara, E. Nicolle and P. A. Carrupt, *Eur. J. Pharm. Sci.*, 2008, **35**, 68.
102. A. Taillardat-Bertschinger, P. A. Carrupt, F. Barbato and B. Testa, *J. Med. Chem.*, 2003, **46**, 655.
103. C. Y. Yang, S. J. Cai, H. Liu and C. Pidgeon, *Adv. Drug Deliv. Rev.*, 1996, **23**, 229.
104. L. Escuder-Gilabert, J. J. Martinez-Pla, S. Sagrado, R. M. Villanueva-Camanas and M. J. Medina-Hernandez, *J. Chromatogr. B*, 2003, **797**, 21.
105. M. Molero-Monfort, L. Escuder-Gilabert, R.M. Villaneuva-Camanas, S. Sagrado and M. J. Medina-Hernandez, *J. Chromatogr. B*, 2001, **753**, 225.
106. L. Escuder-Gilabert, M. Molero-Monfort, R. M. Villanueva-Camanas, S. Sagrado and M. J. Medina-Hernandez, *J. Chromatogr. B*, 2004, **807**, 193.
107. Y. Martin-Biosca, M. Molero-Monfort, S. Sagrado, R. M. Villanueva-Camanas and M. J. Medina-Hernandez, *Eur. J. Pharm Sci.*, 2003, **20**, 209.
108. P. Stenberg, U. Norinder, K. Luthman and P. Artursson, *J. Med. Chem.*, 2001, **44**, 1927.
109. T. Osterberg, M. Svensson and P. Lundahl, *Eur. J. Pharm. Sci.*, 2001, **12**, 427.

110. K. Bellman, C. J. Decker, Li Jiang, Y. Li, Z. Liu, V. Sanders, A. Shokri, M. Steel, and J. Worley, *AAPS*, 2007 (November).
111. K. Sugano, M. Kansy, P. Artursson, A. Avdeef, S. Bendels, L. Di, G. F. Ecker, B. Faller, H. Fischer, G. Gerebtzoff, H. Lennernaes and F. Senner, *Nat. Rev. Drug Discovery*, 2010, **9**, 597.
112. C. Li, L. Nair, T. Liu, F. Li, J. Pichardo, S. Agrawal, R. Chase, X. Tong, A. S. Uss, S. Bogen, F. G. Njoroge, R. A. Morrison and K. C. Cheng, *Biochem. Pharmacol.*, 2008, **75**, 1186.
113. M. Fujikawa, R. Ano, K. Nakao, R. Shimizu and M. Akamatsu, *Bioorg. Med. Chem.*, 2005, **13**, 4721.
114. G. Ottaviani, S. Martel and P. A. Carrupt, *J. Med. Chem.*, 2007, **50**, 742.
115. A. Avdeef, *Absorption and Drug Development*, John Wiley & Sons, Hoboken, 2003.
116. H. Liu, C. Sabus, G. T. Carter, C. Du, A. Avdeef and M. Titeler, *Pharm. Res.*, 2003, **20**, 1820.
117. L. Novakova, L. Matysova and P. Solich, *Talanta*, 2006, **68**, 908.
118. M. I. Churchwell, N. C. Twaddle, L. R. Meeker and D. R. Doerge, *J. Chromatogr. B*, 2005, **825**, 134.
119. J. Mensch, M. Noppe, J. Adriaensen, A. Melis, C. Mackie, P. Augustijns and M. E. Brewster, *J. Chromatogr. B*, 2007, **847**, 182.
120. K. S. Hakala, L. Laitinen, A. M. Kaukonen, J. Hirvonen, R. Kostiainen and T. Kotiaho, *Anal. Chem.*, 2003, **75**, 5969.
121. S. Carrara, V. Reali, P. Misiano, G. Dondio and C. Bigogno, *Int. J. Pharm.*, 2007, 345, 125.
122. P. V. Balimane, E. Pace, S. Chong, M. S. Zhu, M. Jemal and C. K. Van Pelt, *J. Pharm. Biomed. Anal.*, 2005, **39**, 8.
123. P. V. Balimane, E. Pace, S. Chong, M. Zhu, M. Jemal and C. K. V. Pelt, *J. Pharm. Biomed. Anal.*, 2005, **39**, 8.
124. B. Bard, Development of New Experimental Tools for the Fast Determination of Solubility and Lipophilicity, PhD Thesis, University of Geneva, 2008.

CHAPTER 9

UHPLC in Modern Bioanalysis

LUCIE NOVÁKOVÁ

Charles University in Prague, Faculty of Pharmacy, Department of Analytical Chemistry, Heyrovského 1203, 500 05 Hradec Králové, Czech Republic
E-mail: nol@email.cz

9.1 Introduction

The measurement of drug concentrations in biological matrices is an important aspect of the drug development process for those products containing new active substances, as well as for line extensions and generic products. Such data may be required to support new applications as well as variations to authorized drug products. The results of toxicokinetics, pharmacokinetics and bioequivalence studies are used to make critical decisions supporting the safety and efficacy of a medicinal drug substance or product. It is therefore of key importance that the applied bioanalytical methods are well characterized and fully validated in order to yield reliable results. The details concerning this topic, appropriate legislation and acceptance criteria have been discussed recently.[1] Requirements for such methods are often accompanied by an increasing number of biological samples requiring fast quantitative analysis, together with a decrease in the desired quantitation levels, as the bioavailability of many drugs is at a low level and thus target concentrations are very low. Consequently, appropriately designed, reliable, fast, effective and sensitive bioanalytical methods are needed.

Modern approaches in bioanalysis used for the quantitative determination of drugs and their metabolites in biological materials have changed substantially over the years. In the beginning high-performance liquid chromatography (HPLC) with ultraviolet (UV) detection was the gold

RSC Chromatography Monographs No. 16
UHPLC in Life Sciences
Edited by Davy Guillarme and Jean-Luc Veuthey

Published by the Royal Society of Chemistry, www.rsc.org

standard, and later in 1990s it was replaced by HPLC with tandem mass spectrometry (MS/MS) of the triple quadrupole type. HPLC-MS/MS based on selected reaction monitoring (SRM) continues to be the best tool to date; however, there has been an important change happening recently, which is the shift of HPLC separation to ultra-high performance liquid chromatography (UHPLC) separation. In addition, high-resolution mass spectrometry (HRMS) is finding its place in a field of bioanalysis.[2]

Challenges in bioanalytical laboratories are represented by several issues:[3]

- the development of the fast LC-MS method, enabling the separation of closely related compounds (*e.g.* analytes and metabolites) from endogenous components and to quantify them reliably with high sensitivity using appropriate internal standards (ISs)
- the choice of suitable sample preparation steps to isolate the analytes and remove interfering compounds
- full method validation according to rigorous criteria set by appropriate authorities, such as the FDA (Food and Drug Administration), the ICH (International Conference on Harmonization of Technical Requirements for Registration of Pharmaceuticals for Human Use) or the EMEA (European Medicines Agency), including parameters of method accuracy, precision, linearity, method selectivity, matrix effects, sample carry-over, analyte stability and sensitivity expressed by LOD (limit of detection) and LOQ (limit of quantitation).[1,4]

UHPLC with MS/MS detection appears to be a suitable technique in terms of sensitivity, selectivity and peak assignment for rapid determination of analytes at low concentrations in complex matrices.[3] The features of UHPLC in modern bioanalysis will be discussed in following Sections with the emphasis to UHPLC-MS and sample preparation.

9.2 UHPLC in Bioanalysis

In the bioanalytical field, rapid method development and high-throughput bioanalytical assays are critical not only for reducing cost but also for shortening the drug development time cycle. Although the feasibility of high-throughput bioanalysis was demonstrated earlier with liquid chromatography (LC) coupled to MS/MS, it has only become more practical with recent advances in LC, such as sub-2 μm particle columns and UHPLC systems that can operate at ultra-high pressure (up to 1300 bar nowadays).[5] Full implementation of such highly efficient chromatographic separation (peak width of 1 s or less) into analytical methods is further challenged by the requirements for detectors. The narrow peaks produced by fast UHPLC separation require minimization of the system extra-column volumes generally, with small tubing internal diameter (I.D.) and detection cell volume. Furthermore, a fast acquisition rate to ensure reliable peak integration is another very important feature.

In capillary UHPLC on-column detection is used, which evidently leads to low sensitivity, as the path-lengths for detection are equivalent to the column internal diameters of capillaries (30–150 μm).[6] Commercial UHPLC instruments are often equipped with a modified UV detector, where the flow cell volume (typically 0.5–2.0 μl) is much lower than that for conventional HPLC. The conventional flow cell of the SS (stainless steel) cylinder type cannot provide the performance needed for UHPLC detection, as such a smaller flow cell would reduce the path-length upon which the signal depends (as predicted by Lambert–Beer's Law). Therefore, in order to maintain the path-length, dedicated light-guided flow cells must be used. These are teflon (Waters) or fused silica (Agilent) capillaries, which prolong the path-length of light transmission based on the difference of refractive indexes of fluid and cladding.[7]

The detection software must be capable of achieving both a fast detector time constant (<0.1 s) and a high data acquisition rate to ensure that there are enough data points for a narrow peak (typically >20 Hz, with 200 Hz being currently maximum acquisition rate). The same applies for MS detectors, which are nowadays the most currently used in connection to UHPLC. Low dwell times, inter-channel and inter-scan delays are required in order to obtain a sufficient number of data points per peak (>15 points). Other detectors, including fluorescence detector or ELSD (evaporative light scattering detector), are currently available for coupling with UHPLC systems.[3,7]

9.2.1 UHPLC with Ultraviolet and Fluorescence Detection

Although UHPLC-MS/MS is currently dominating the field of bioanalytical methods, several approaches still employ more simple and less expensive detection such as UV, PDA (photodiode array) or fluorescence detection[8–16] as demonstrated in Table 9.1. Due to the lower selectivity of such detection in complex matrices, more thorough sample preparation steps providing cleaner extracts and pre-concentration are often required. Simple and fast protein precipitation (PP) is typically insufficient, while solid phase extraction (SPE), liquid–liquid extraction (LLE) or a combination of LLE and PP have been successfully applied (see Table 9.1).

UHPLC provided short analysis times which were decreased to 5 min in all cases, while multi-analyte separation of 22 compounds including β-blockers, isoflavones and their metabolites took 8 min using PDA detection (Figure 9.1).[8] Twelve *N*-acyl homoserines and their lactones were separated within 1 min at 60 °C on a 100 mm BEH C18 column. UHPLC-PDA was found to be a suitable tool for the optimization of the sample preparation from complex matrix and for further analysis of samples of bacterial supernatants.[15] PDA detection was further applied for the analysis of fat-soluble vitamins and coenzyme Q10, with the analysis time still less than 2 min.[11] Fluorescence detection was applied for the analysis of doxorubicin in cell cultures.[9] Both fluorescence and PDA detection were applied in UHPLC determination of

Table 9.1 An overview of UHPLC bioanalytical methods using FD or UV detection.

Determined substances	*Matrix*	*Sample preparation pre-concentration factor*	*Stationary phase*	*Mobile phase*	*Detection*	*Analysis time (min)*	*Method sensitivity*	*Year and reference*
β-Blockers, isoflavones and their metabolites (22 analytes)	Urine	SPE 3 ×	Hypersil Gold (50 × 2.1 mm, 1.9 μm)	0.05% TFA/ ACN Gradient elution	PDA	8	LOD = 10.7–66.9 ng mL^{-1}	2011[8]
Doxorubicin and 3 metabolites	Cell cultures	PP Dilution of sample	Capcell Pak C18 (50 × 2.0 mm, 2.0 μm)	50 mM sodium phosphate, buffer pH 2.0/ACN (65:27)	FD λ_{ex} = 470 nm λ_{ex} = 590 nm	3	LOD = 3.5–7.4 pg $injection^{-1}$	2010[9]
Daptomycin, rifampicin	Plasma	PP Dilution of sample	Acquity UPLC BEH C18 (100 × 2.1 mm, 1.7 μm)	0.05% TFA/ MeOH Gradient elution	UV, 220 nm, 342 nm	4.5	LOQ = 2 μg ml^{-1}	2010[10]
4 fat-soluble vitamins, coenzyme Q10	Serum	PP, LLE 3.3 ×	Acquity UPLC BEH Shield RP 18 (50 × 2.1 mm, 1.7 μm)	ACN, water/ MeOH, 2-propanol Gradient elution	PDA	2	LOD = 4–78 ng mL^{-1}	2009[11]
Valsartan and its metabolite, chlortalidone, fluvastatin	Plasma	SPE NA	Acquity UPLC BEH C18 (50 × 2.1 mm, 1.7 μm)	10 mM ammonium formate, pH 4.1, 0.01% formic acid/ACN Gradient elution	UV, 220 nm (80 Hz)	2.5	LOQ = 20–110 ng mL^{-1}	2009[12]

Table 9.1 (*Continued*)

Determined substances	*Matrix*	*Sample preparation pre-concentration factor*	*Stationary phase*	*Mobile phase*	*Detection*	*Analysis time (min)*	*Method sensitivity*	*Year and reference*
Ascorbic acid, dehydroascorbic acid, uric acid	Plasma	Dilution	Hypersil Gold (20 × 2.1 mm, 1.9 μm) (100 × 2.1 mm, 1.9 μm)	MeOH/150 mM chloroacetic acid, 2 mM disodium EDTA, pH 3.0, with NaOH Gradient elution	PDA, 265 nm ECD	2.5	LOD = 0.5 μg ml^{-1}	2009[13]
4 tocopherols	Human colostrum, milk	LLE 18 ×	Acquity UPLC BEH C18 (50 × 2.1 mm, 1.7 μm) Zorbax Exlipse Plus C18 RRHT (50 × 2.1mm, 1.8μm)	ACN/MeOH (60:40)	PDA (20 Hz) FD (10 Hz) λ_{ex} = 297 nm λ_{ex} = 328 nm	1.6	LOD_{PDA} = 21–34 ng mL^{-1} LOD_{FD} = 10–22 pg mL^{-1}	2009[14]
N-Acyl homoserine lactones, *N*-acyl homoserines (12 analytes)	Bacterial supernatant	SPE 10 ×	Acquity UPLC BEH C18 (100 × 2.1 mm, 1.7 μm)	10 mM ammonium formate, pH 4.25/ACN Gradient elution	PDA (20 Hz)	1	LOD = 0.11–1.64 μg mL^{-1}	2007[15]
8 *N*-Acyl homoserine lactones	Bacterial supernatant	SPE NA	Acquity UPLC BEH C18 (100 × 2.1 mm, 1.7 μm)	Water/ACN Gradient elution	PDA (20 Hz)	1.5	LOD = 0.4–1.0 μmol L^{-1}	2006[16]

ACN, acetonitrile; ECD, electrochemical detection; FD, fluorescence detection; LLE, liquid liquid extraction; LOD, limit of detection; LOQ, limit of quantitation; MeOH, methanol; PDA, photodiode array; PP, protein precipitation; SPE, solid phase extraction; TFA, trifluoroacetic acid; UV, ultraviolet detection.

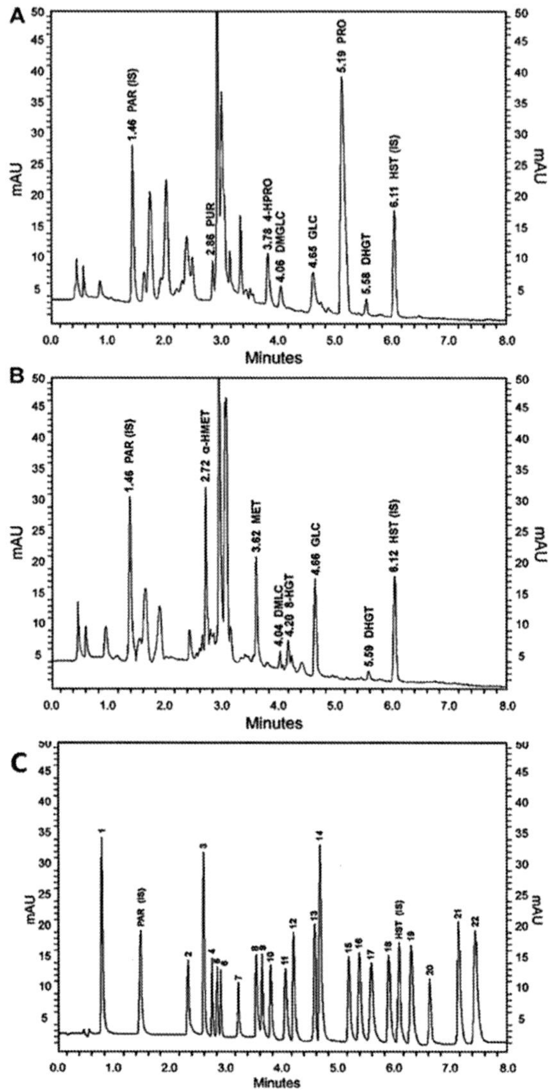

Figure 9.1 A typical chromatogram of UHPLC-UV bioanalytical method. The separation of isoflavones, their metabolites and β-blockers in human urine of patients on a diet rich in isoflavones and treated with (A) propranolol, (B) methoprolol and (C) standard mixture is shown. Peaks are designated as follows: (1) milrinone (MIL), (2) sotalol (SOT), (3) daidzein 7,4′diglucoside (GLU-DA), (4) α-hydroxymetoprolol (α-HMET), (5) o-desmethylmetoprolol (O-DMMET), (6) puerarin (PUR), (7) glycitin (GLY), (8) 8-hydroxydaidzein (8-HAD), (9) metoprolol (MET), (10) 4-hydroxypropranolol (4-HPRO), (11) desmethylglycitein (DMGLC), (12) 8-hydroxygenistein (8-HGT), (13) daidzein (DA), (14) glycitein (GLC), (15) propranolol (PRO), (16) 5′-hydroxyphenylcarvedilol (5′HCAR), (17) dihydrogenistein (DHGT), (18) genistein (GT), (19) o-desmethylcarvedilol (O-DMCAR), (20) carvedilol (CAR), (21) dihydrobiochanin (DHBIO), (22) biochanin (BIO); internal standards (IS) included paracetamol (PAR) and hesperetin (HST). Figure reprinted from reference 8 with permission.

tocopherols in human colostrum and milk. Chromatographic separation was achieved in less than 1.6 min, with a sensitivity of 21–34 ng ml^{-1} for PDA detection and 10–22 pg ml^{-1} for fluorescence detection.[14] In all of the studies presented, fast-scanning UHPLC detectors were used, enabling acquisition rates of 10 or 20 Hz and high-sensitivity detection in combination with analyte pre-concentration during the sample preparation steps. Only one bioanalytical study employed electrochemical detector for the determination of ascorbic, dehydroascorbic and uric acids in urine.[13] Aerosol detectors, such as ELSD or CAD (charged aerosol detector) coupled to UHPLC have not yet been employed in quantitative bioanalysis. The aerosol detectors are certainly too universal and thus not sufficiently selective towards complex matrix components.

9.2.2 UHPLC with Mass Spectrometry

9.2.2.1 Reliable Quantitation using LC-MS

The aim of quantitative analysis is to provide accurate and reliable determination of the amount of a target analyte(s) in complex sample. Quantitative analysis is performed to provide absolute or relative quantification, where the amount or concentration of the analyte is determined relative to another analyte or another sample. A reliable quantitative LC-MS method consist of three main steps, where each one must be carefully addressed during method development: (1) sample preparation (discussed in detail in Section 9.3), (2) efficient and fast chromatographic separation and (3) sensitive and selective MS/MS detection.[17,18] The choice of appropriate data processing is also important, and finally method validation is necessary before it can be applied to biological samples.[1,4]

In some cases, maximized LC separation is not necessary, because MS/MS selectivity is very high; however, separation from impurities and other compounds in complex mixture might improve method sensitivity and reduce matrix effects. On the other hand, the separation is crucial when analysing isobaric compounds or chiral substances. UHPLC was employed for the analysis of warfarin and its metabolites in human plasma, using two-dimensional separation by reversed phase (RP)-UHPLC and chiral HPLC separation mode. This two-dimensional method enabled separation of all enantiomeric pairs within 17 min due to very narrow peaks eluted in the first RP-UHPLC dimension, which was impossible in any previous experiment using either chiral or RP mode in HPLC (Figure 9.2).[19]

In order to obtain reliable quantitative information without separation, the molecular weight of the analytes should differ at least by three units. The differences between two analytes should not be exactly the weight of an obvious metabolite (*e.g.* +16, −14 *etc.*) [20] or the weight of the expected adduct (such as +17 for $[M+NH_4]^+$).[21] The same applies for the analytes that may undergo interconversion reactions such as lactone > hydroxy-acid, amine >

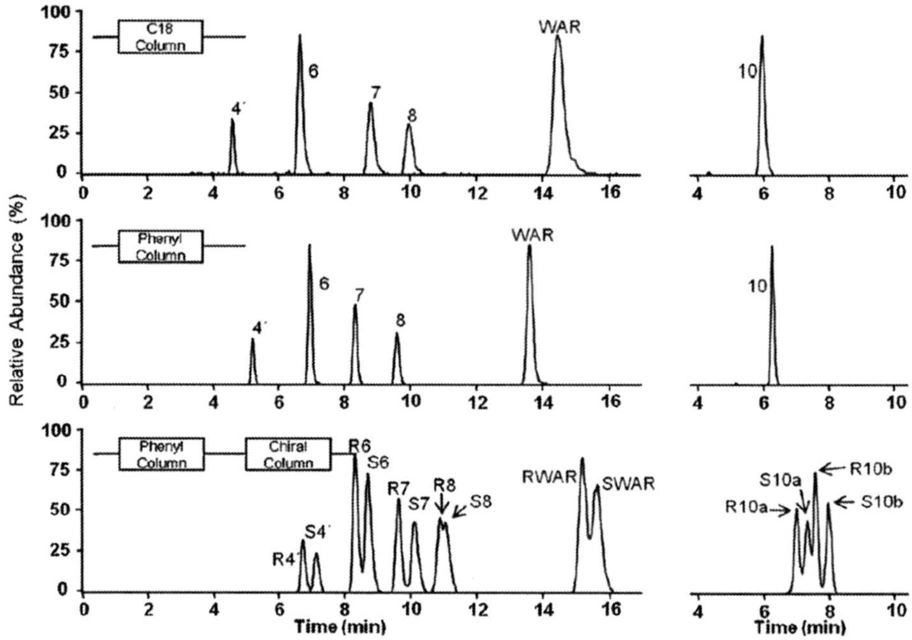

Figure 9.2 Ultra-high performance separation of warfarin (WAR), its hydroxyl-metabolites and isomers. The separation is shown on a BEH C18 column (upper panel), on a BEH Phenyl column (middle panel) and on a BEH Phenyl and a Chirobiotic column in series (lower panel). Numbers indicate sites of hydroxylation, and R or S indicates stereochemistry. Figure reprinted from reference 19 with permission.

N-oxide, carboxylic acid > acylglucuronide *etc.*[21] Separation efficiency might be enhanced by using some of the available fast LC approaches, including UHPLC, monolith columns, core-shell columns or high-temperature LC, as reviewed by several research groups.[7,22,23]

High sensitivity and selectivity of MS detection is enabled by the selected ion monitoring (SIM) or SRM modes of data acquisition, therefore they are widely used in quantitative LC-MS analysis. In the SIM mode, the ion current from only one or a few selected ions is/are repetitively recorded and accumulated. The data system spends more time in recording selected *m/z* values, which significantly enhances detection sensitivity *versus* a full scan. On the other hand, the specificity may be questionable. In SIM experiments the most abundant ion, often the molecular ion, is selected for monitoring. Such an ion must originate exclusively from the analyte. In SRM experiments a specific precursor–product pair (transition) is monitored using MS/MS. If several SRM transitions are monitored, the experiment is designated as a multiple reaction monitoring (MRM). The MRM cycle is defined by the summation of three parameters: dwell times of all SRM channels, inter-channel delay times between successive SRM channels and inter-scan delay times. The precursor–product transition in SRM must be unique to the analyte and should provide a strong ion current signal, and a stable isotopically labelled IS (SIL-IS) of the same compound should be available. Using SRM for MS/MS quantitation provides increased detection specificity due to the structural link and virtual elimination of chemical noise. In some cases, primary SRM transition is used for quantitation and a secondary one for confirmation purposes.[24] A triple-quadrupole instrument is ideally suited for SRM experiments, although magnetic sector and some hybrid analysers employing quadrupole may also be used (*e.g.* Q-trap, Q-TOF…). Sometimes narrow-mass-range and full-scan mode are applied for quantitative analysis to reliably confirm the analyte structure simultaneously with quantitative results.

Several calibration approaches might be used based on requirements for the levels of method accuracy and precision.[25] Limited accuracy and precision is obtained when using the *external standard method* because the standards and the samples are analysed at two different times and, within this time window, the MS response might drift. Moreover, the matrix of calibration standards is difficult to match with the sample matrix. For very complex matrices, the *standard addition method* is more convenient, using whole calibration curves or a single point calibration, which is much less precise. Unfortunately, this method is time-consuming and offers an appreciable compensation only with diluted samples in a uniform matrix. Furthermore, the availability of appropriate blanks might not always be accomplished. The *IS method* is the most widely used approach in MS quantitation, as this method yields a high level of accuracy and precision. The IS added to the calibration solutions, as well as to unknown samples, compensates for any fluctuation in the MS response and the sample losses that might occur in sample preparation and chromatographic steps, as well as for matrix effects. The IS should have the same physical, chemical and chromatographic properties as the analyte (ideally eluted at the same retention time), as well as MS behaviour (ionization, fragmentation), the molecular weight should be distinct from that of the analyte and finally it should not be a constituent of the sample. Two kinds of IS are commonly used in quantification by LC-MS/MS, namely structural analogues or SIL-ISs. SIL-ISs are believed to yield better method precision and accuracy because they show almost identical behaviour to the analyte in sample preparation and chromatography, as well as in MS. SIL-ISs should compensate for matrix effects, however, this issue needs to be verified during method validation. Important drawbacks are following: their use is rather expensive and for many compounds they are not commercially available.[25] Finally, the developed quantitative LC-MS/MS method must be properly validated in order to assure its reliability and suitability for intended purpose.[1,4]

9.2.2.2 Matrix Effects and Some Strategies to Reduce Them

The main obstacle of quantitative LC-MS/MS analysis is represented by matrix effects. The matrix co-extracted with the analytes can alter the signal response causing either ion suppression or enhancement, which results in poor accuracy, linearity and inter- and intra-day precision of the method. The exact mechanism of matrix effects is unknown, but it probably originates from the competition between an analyte and the co-eluting undetected (not shown in the SRM channel) matrix components. Besides the matrix type, matrix effects are also component-dependent.[26,27] There are three main strategies to assess matrix effects: (1) post-column infusion, (2) post-extraction addition and (3) a comparison of the slopes of calibration curves.[26–28] In the first case (1) a pump is used to deliver a constant flow of the analyte into the LC eluent at a point after the chromatographic column and before the mass spectrometer (Figure 9.3A). A sample extract free from the analyte is then injected under the required chromatographic conditions and the response of the infused analyte is recorded. Any endogenous matrix component that elutes from the column and induces matrix effects can be seen as a suppression or enhancement of the infused signal.[26,29]

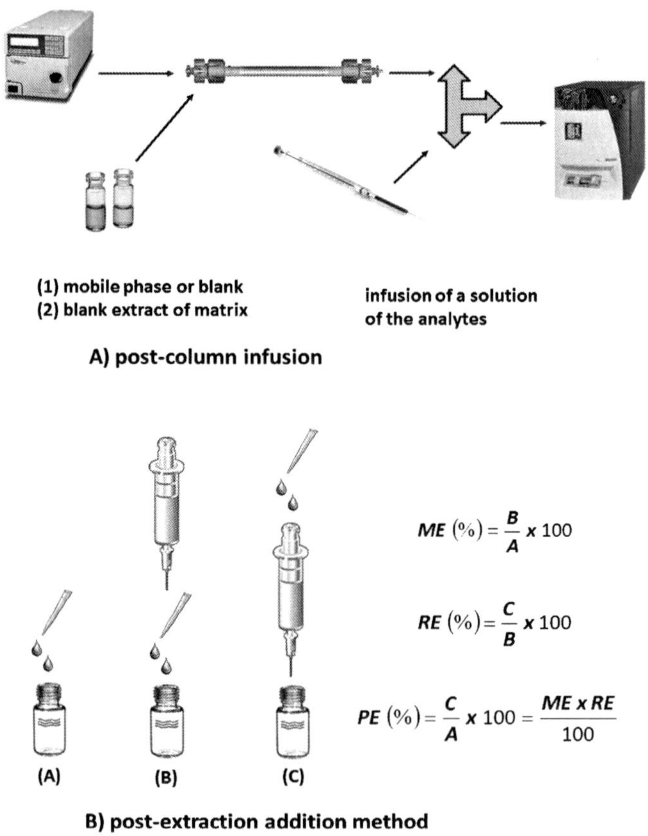

Figure 9.3 The approaches to determine matrix effects: (A) post-column infusion method and (B) post-extraction addition method. ME, matrix effects; PE, process efficiency; RE, recovery.

In the post-extraction addition (2), a sample extract with the analyte of interest added after the extraction (matrix-matched standard) is compared with a pure solution containing the same amount of target analyte (Figure 9.3B). A difference in response (%) between the matrix-matched standard and the pure solution indicates ion suppression or enhancement. The entity of the matrix effect (ME) is defined as 100 – ME (%). If ME (%) = 100, no matrix effect is present, if ME (%) > 100 there is a signal enhancement and if ME (%) < 100 a signal suppression is observed. Furthermore, parameters of recovery efficiency (RE) and process efficiency (PE) might be calculated using this approach.[26,28]

In order to obtain a robust and reliable LC-MS/MS method, there is a need to remove or minimize the presence of the matrix effects. The source of an interfering matrix component may come from the current sample being injected, a previously injected sample (as late-eluting interference) or build-up and overload of an analytical column. Several strategies were described to reduce matrix effects:[25–31]

- improvement of sample preparation step (enhancement of selectivity)
- sample dilution, if possible
- improvement in chromatographic separation [gradient elution, change in separation mode, *e.g.* hydrophilic interaction LC (HILIC), two-dimensional chromatographic separation or nano-LC]
- quantitation using SRM experiment instead of SIM
- atmospheric pressure chemical ionization (APCI), atmospheric pressure photoionization (APPI) or direct electron ionization MS (EI-MS) ionization modes, which are less susceptible to matrix effects
- negative ion mode is less prone to matrix effects
- echo-peak technique
- application of an appropriate calibration approach (matrix-matched calibration, standard addition calibration, IS calibration with SIL-ISs, *etc.*).

According to recently published guidelines for the validation of a bioanalytical method, an assessment of the matrix effect became a part of full method validation.[1] Because UHPLC enhances the overall chromatographic resolution, co-elution either with matrix or with other analytes is reduced, which leads to a decrease of matrix effects as demonstrated in several studies.[25,30,32]

SIL-ISs are believed to be the most appropriate ISs in the quantitative bioanalytical LC-MS/MS methods. They should compensate for variability in chemical derivatization, sample extraction and LC-MS/MS analysis due to their nearly identical chemical and physical properties to the unlabelled analyte. Hence, the analyte-to-IS area ratio should be constant, despite any variations in sample processing or analysis. However, several authors demonstrated an unexpected behaviour, in which the analyte-to-IS peak area ratio changed with another batch of plasma or with a slight change of analytical conditions.[33–36]

In practical examples, it was demonstrated that, under certain conditions, high level of matrix suppression affected the ionization of analyte and its isotopically labelled analogue differently due to a slight difference in retention time between the analyte and SIL-IS. This phenomenon was induced by the deuterium isotope effect due to a small change in lipophilicity when exchanging hydrogen atom for deuterium. Wang *et al.* [33] observed an unexpected change in the analyte-to-IS peak area ratio with different commercially supplied human plasma samples in the analysis of carvedilol enantiomers. Jemal *et al.*[35] described similar phenomenon for various samples of urine and sample volumes used for extraction when analyzing mevalonic acid. Lindegardh *et al.*[36] showed that matrix effects originating from sample clean-up using SPE were responsible for ion suppression in the analysis of piperaquine in plasma. The suppression in this study[36] was induced by phosphate salt residues or residues of triethylamine remaining from SPE if the eluates were not evaporated to complete dryness. When the amount of triethylamine in the sample was low, the region of ion suppression co-eluted with the tails of piperaquine and piperaquine-D6. When the amount of triethylamine in the sample was higher, the region of ion suppression co-eluted at or around peak maxima. The deuterated IS is less lipophilic than the parent compound, thus eluting slightly earlier in the chromatogram, which resulted in ion suppression of approximately 50%. In all reported cases, the matrix effect seriously affected method accuracy and precision. Due to the high-resolving power of UHPLC, there is more chance to separate the analyte from its deuterated IS and thus there could potentially be more problems with the use of a deuterium SIL-IS compared to HPLC.

For this reason, ^{13}C-, ^{15}N- or ^{17}O-labelled compounds are considered to be more appropriate than deuterium-labelled ISs.[34,37] SIL-ISs must always co-elute with the analyte in order to assure the stable analyte IS peak area ratio and thus sufficient method accuracy and precision. The assessment of matrix effects in different batches of biological material has become an important part of validation of bioanalytical method.[1]

9.2.2.3 UHPLC with Tandem Mass Spectrometry with a Triple Quadrupole Analyzer

Currently, LC-MS/MS with a triple quadrupole analyzer has become the best tool for quantitation in the field of bioanalysis. The introduction of soft-ionization techniques, such as electrospray (ESI) and APCI, has simplified the coupling of MS with LC.[38] The separation nowadays is more and more widely performed by UHPLC instead of HPLC due to a substantial decrease of analysis time, and an increase in the separation efficiency and sensitivity.[2,7,24] A triple quadrupole analyser is the best suited for SRM experiments, allowing high sensitivity and wide linear range. An overview of recent UHPLC-MS/MS bioanalytical applications[39–56] using triple quadrupoles for quantitation is displayed in Table 9.2. In most bioanalytical methods only one quantitative SRM transition was used, while secondary SRM transition was used less often.[43,47,50,54,56] Secondary or tertiary SRM transition was crucial in methods for both screening and quantitative purposes to reliably confirm the structure.[43,47]

Newly developed UHPLC-MS/MS methods follow the trend of fast LC separation. Most UHPLC separations although multi-analyte mixtures are now performed within 1–5 min using fast gradient elution on 50–100 mm sub-2 μm column (see Table 9.2). An example of the analysis of tamoxifen and its structurally similar metabolites within 8 min, which a challenging separation, is shown in Figure 9.4.[55]

Fast UHPLC separation resulting in very narrow peaks might induce poor repeatability in peak area and low sensitivity due to insufficient number of data points across the peaks. At least 15 data points are usually required to define an LC peak for accurate and reliable quantitation. To accommodate the narrow UHPLC peak widths, it is straightforward to reduce dwell time when many SRM transitions have to be monitored simultaneously. On one hand, this can enhance the number of acquired data points and positively affect measurement precision, whereas, on the other hand, this can reduce sensitivity.[3,57] Typical dwell times employed in quantitative UHPLC-MS/MS analysis were within 20–100 ms (see Table 9.2). Fast scanning triple quadrupole devices are therefore convenient for coupling with UHPLC. When more SRM transitions have to be simultaneously monitored, the dwell times, inter-scan and inter-channel delays need to be decreased or scheduled MRM experiments might be employed. Scheduled MRM divides analysis time in time segments and thus less SRM transitions are simultaneously monitored, which allows keeping higher dwell-times and thus elevated sensitivity (Figure 9.5).[58,59] The sensitivity of developed UHPLC-MS/MS bioanalytical methods reaches typically ng mL^{-1} levels or lower. Low dwell times, inter-scan and inter-channel delays are especially important when using polarity switching mode.[60]

9.2.2.4 UHPLC with High-resolution Mass Spectrometry (HRMS)

HPLC-HRMS was originally used exclusively for structural elucidation and/or metabolite identification in bioanalysis. Recently, it has been applied also for quantitative assay of small molecules in a complex matrix coupled to UHPLC separation. The main reason for using UHPLC-HRMS for quantitative analysis is to maximize information from an analysed sample and to simplify the method development.[2] While in triple quadrupole based methods the optimization of parameters for Q1 and Q3 *m/z* values, collision energy and interface voltages is necessary, the ability to perform full-scan acquisition for quantitative analysis would eliminate the need for this optimization. The information can be even mined from the data post-acquisition leading to so-called quantitative–qualitative (Quan–Qual) workflows. Moreover, triple quadrupole-based assays are only able to detect specified compounds and the information about other components, such as metabolites, might be lost.

Table 9.2 An overview of typical recent UHPLC-MS/MS bioanalytical methods using QqQ for quantitation.

Determined substances	*Matrix/ sample preparation*	*Stationary phase*	*Mobile phase*	*Detection*	*Dwell time*	*Analysis time (min)*	*Method sensitivity*	*Reference and year*
Maackiain, maackiain-7-glucuronide maackiain-7-sulfate	Plasma/PP	Acquity UPLC BEH C18 (50 × 2.1 mm, 1.7 μm)	ACN/water Gradient elution	ESI-MS/MS QqQ	100 ms 1 SRM	3	LOD = 4 nmol L^{-1}	2011[39]
Quinine, (3*S*)-3-hydroxyquinine	Urine/dilution	Acquity UPLC BEH C18 (50 × 2.1 mm, 1.7 μm)	10 mM AmBi, pH 10.0/MeOH Gradient elution	ESI-MS/MS QqQ	5 ms 1 SRM	1	LOD = 1 ng mL^{-1}	2011[40]
Nifedipine	Plasma/LLE	Acquity UPLC BEH C18 (50 × 2.1 mm, 1.7 μm)	10 mM AmAc/ACN (25:75)	ESI-MS/MS QqQ	100 ms 1 SRM	1.5	LOD = 0.1 ng mL^{-1}	2011[41]
Cyclosporin A, tarcrolimus	Blood/DBS	Acquity UPLC BEH C18 (50 × 2.1 mm, 1.7 μm)	2 mM AmAc, 0.1% formic acid, MeOH Gradient elution	ESI-MS/MS QqQ	100 ms 1 SRM	1.8	LOD = 0.5–8.5 μg L^{-1}	2011[42]
34 antidepressants Multi-analyte screening and quantitation	Plasma/LLE	Hypersil Gold Phenyl (100 × 2.1 mm, 1.9 μm)	10 mM AmF, pH 3.4/ACN, 0.1% formic acid Gradient elution	APCI-MS/MS QqQ	2 ms 2–3 SRMs	15	LOD = 3.8–100 μg L^{-1}	2011[43]
Octreotide	Plasma/SPE	Acquity UPLC BEH C18 (100 × 2.1 mm, 1.7 μm)	0.1% formic acid, ACN, MeOH Gradient elution	ESI-MS/MS QqQ	200 ms 1 SRM	4.5	LOQ = 25 pg mL^{-1}	2011[44]
Tamoxifen and 3 metabolites	Plasma/LLE	Acquity UPLC BEH C18 (100 × 2.1 mm, 1.7 μm)	0.2 mM AmF, ACN, 0.1% formic acid Gradient elution	ESI-MS/MS QqQ	50 ms 1 SRM	5	LOQ = 0.19–1.86 ng mL^{-1}	2011[45]
13 antidepressants and antipsychotics	Serum/PP	Zorbax SB C8 (50 × 2.0 mm, 1.8 μm)	0.1% formic acid, MeOH Gradient elution	ESI-MS/MS QqQ	25 ms 1 SRM	2.2	LOD = 1.0–5.5 nmol L^{-1}	2011[46]
13 benzodiazepines	Urine/PP	Acquity UPLC BEH C18 (50 × 2.1 mm, 1.7 μm)	0.1% formic acid, MeOH Gradient elution	ESI-MS/MS QqQ	30–250 ms 2 SRMs	3	LOD = 0.5–2.0 ng mL^{-1}	2011[47]

Table 9.2 (*Continued*)

Determined substances	*Matrix/ sample preparation*	*Stationary phase*	*Mobile phase*	*Detection*	*Dwell time*	*Analysis time (min)*	*Method sensitivity*	*Reference and year*
Ethosuximide	Plasma/SPE	Hypersil Gold C18 (100 × 2.1 mm, 1.9 μm)	10 mM AmAc/ACN (30:70)	ESI-MS/MS QqQ	100 ms 1 SRM	1.8	LOD = 0.25 μg mL^{-1}	2010[48]
Acetylcholine, choline, betaine, dimethylglycine	Urine, plasma, serum/PP	Acquity UPLC BEH HILIC (100 × 2.1 mm, 1.7 μm)	15 mM AmF, pH 3.5/ACN Gradient elution	ESI-MS/MS QqQ	50 ms 1 SRM	6	LOD = 0.35–340 nmol L^{-1}	2010[49]
Fentanyl, norfentanyl	Urine, blood/SPE	Acquity UPLC BEH C18 (50 × 2.1 mm, 1.7 μm)	10 mM ammonium bicarbonate, pH 9.0/MeOH Gradient elution	APCI-MS/MS QqQ	25 ms 2 SRMs	3.5	LOQ = 0.25–5 pg mL^{-1}	2010[50]
Mycophenolic acid and its glucuronide	Serum, plasma/ dilution	Acquity UPLC BEH C8 (50 × 2.1 mm, 1.7 μm)	MeOH, 0.1% formic acid, 2 mM AmAc Gradient elution	ESI-MS/MS QqQ	100 ms 1 SRM	1.5	LOD = 0.014–1.85 mg L^{-1}	2010[51]
Genistein and 4 metabolites (phase II)	Blood/PP	Acquity UPLC BEH C18 (50 × 2.1 mm, 1.7 μm)	2.5 mM AmAc, pH 7.4/ACN Gradient elution	ESI-MS/MS QqQ-IT	100 ms 1 SRM	3	LOD = 0.78–6.25 nmol L^{-1}	2010[52]
Levetiracetam	Plasma/PP	Acquity UPLC BEH C18 (100 × 2.1 mm, 1.7 μm)	10 mM AmF, pH 3.5/MeOH Gradient elution	ESI-MS/MS QqQ	100 ms 1 SRM	2.5	LOD = 0.06 μg mL^{-1}	2010[53]
Doxazosin, verapamil	Serum/LLE	Zorbax Eclipse XDB C18 (50 × 4.6 mm, 1.8 μm)	5 mM AmF, 0.02% formic acid, ACN Gradient elution	ESI-MS/MS QqQ-LIT	75 ms 2 SRMs	1.5	LOQ = 1 ng mL^{-1}	2010[54]
Tamoxifen and 7 metabolites	Plasma/PP	Acquity UPLC BEH C18 (30 × 2.1 mm, 1.7 μm)	20 mM AmF, ACN, 0.1% formic acid Gradient elution	ESI-MS/MS QqQ	20 ms 1 SRM	8	LOQ = 0.1–0.75 ng mL^{-1}	2010[55]
Adefovir	Plasma/LLE	Acquity UPLC BEH HILIC (50 × 2.1 mm, 1.7 μm)	MeOH/water/ formic acid (85:15:0.2)	ESI-MS/MS QqQ	200 ms 3 SRMs	3	LOQ = 1.02 ng mL^{-1}	2010[56]

ACN, acetonitrile; AmAc, ammonium acetate; AmF, amonium formate; APCI, atmospheric pressure chemical ionization; CSF, cerebrospinal fluid; ESI, electrospray ionization; LIT, linear ion trap; LLE, liquid liquid extraction; LOD, limit of detection; LOQ, limit of quantitation; MeOH, methanol; MS/MS, tandem mass spectrometry; PP, protein precipitation; QqQ, triple quadrupole; SPE, solid phase extraction.

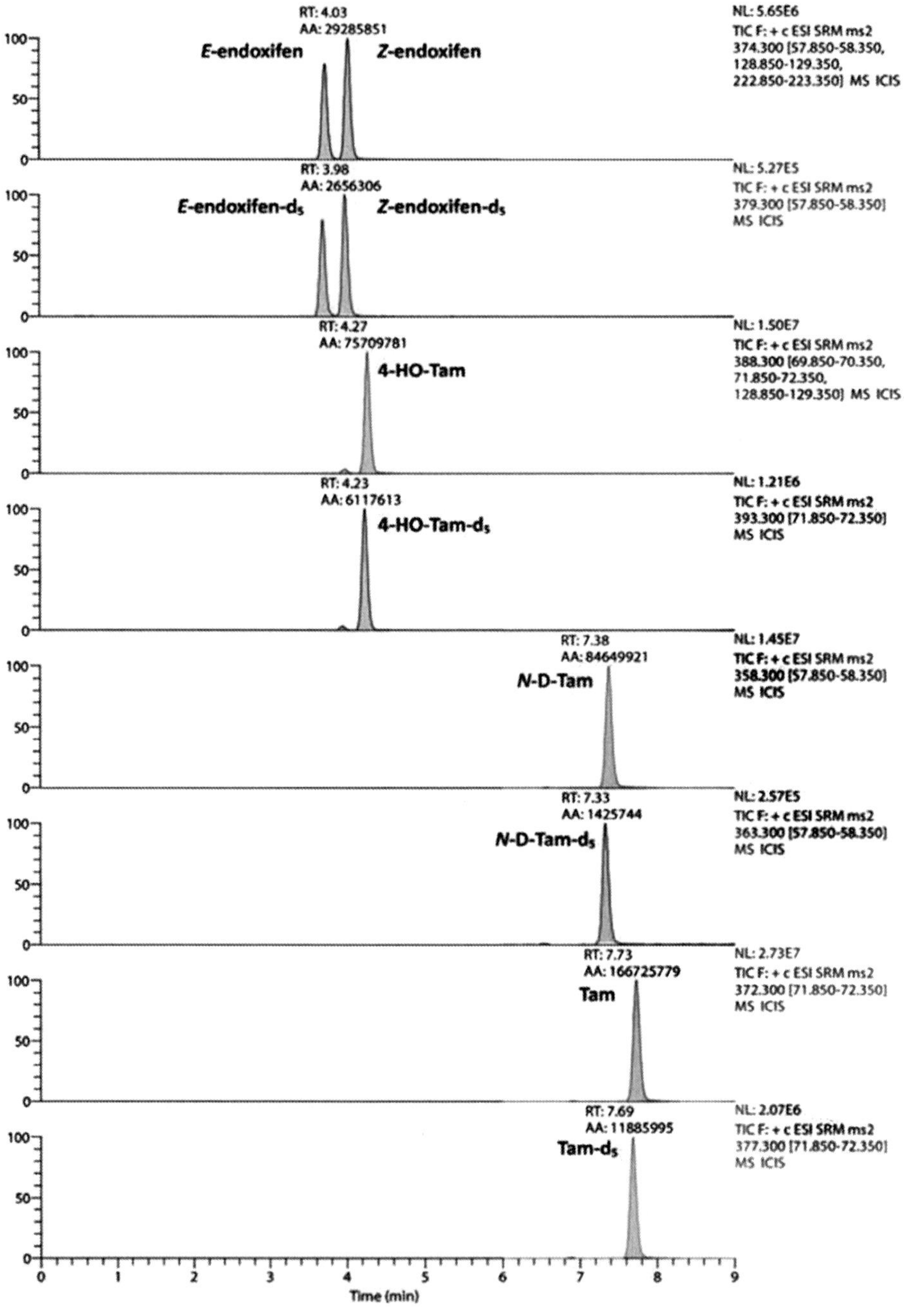

Figure 9.4 A typical chromatogram of UHPLC-MS/MS bioanalytical method using QqQ for quantitation. Tamoxifen, its three major metabolites and corresponding deuterium-labelled internal standards were separated within 8 min. Tam, tamoxifen; *N*-D-Tam, *N*-desmethyltamoxifen; 4-OH-Tam, Z-4-hydroxytamoxifen. Figure reprinted from reference 55 with permission.

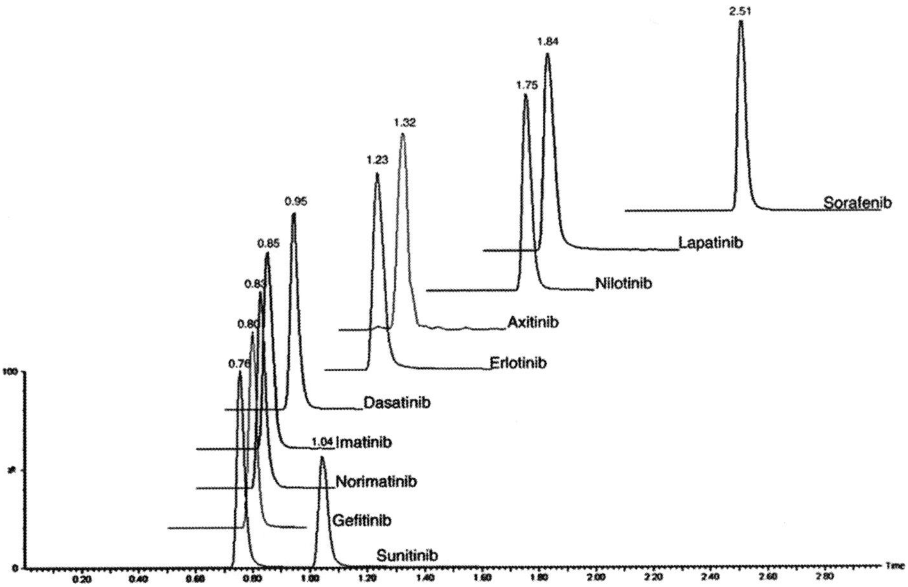

Figure 9.5 A typical chromatogram of UHPLC-MS/MS bioanalytical method using QqQ with the demonstration of scheduled MRM experiment in analysis of tyrosine kinase inhibitors. Figure reprinted from reference 59 with permission.

Quantitative approaches in bioanalysis require high sensitivity, selectivity, wide dynamic range and high scan speed.[61]

The desirable requirements for HRMS instrument according to Ramanathan *et al.*[2] are: mass resolution ≥ 30 000, mass accuracy of at least 5 ppm, scan speed ≥ 5 Hz, linear dynamic range of at least three orders of magnitude and MS/MS capability is also convenient. The MS/MS system that can currently best meet these requirements is the hybrid quadrupole-time-of-flight (Q-TOF) system.[2] The resolution of a TOF analyzer is limited by the initial velocity distribution of the ions. This distribution might be compensated by refectron or by the delayed-extraction of ions. MS/MS using TOF analysers is feasible either with post-source decay or, more easily, with instruments having a quadrupole and a collision cell, thus in a Q-TOF arrangement.[62] A Q-TOF hybrid analyser permits both MS and MS/MS experiments and provides an accurate mass for both precursor and product ions, which offers high resolution, sensitivity and identification reliability.[63] However, most of conventional Q-TOF analysers offer a resolution of 10 000 or 20 000, therefore further improvement in the instrumentation are needed. Such example is represented by newly introduced Q-TOF Maxis Impact *(*Bruker*)*, which is able to provide high resolution of 30 000 at 50 Hz acquisition rate in full-scan mode, at *m*/*z* 300. Maximum resolution is obtained at *m*/*z* 1200 (40 000, 50 Hz), but it decreases less significantly with the change of *m*/*z* compared to orbitrap. Currently, the fastest Q-TOF instrument enabling scanning speed up to 100 Hz in MS/MS mode with a high resolution (>35 000 for an *m/z* 900 and 25 000 for an *m/z* 195) is a TripleTOF 5600 (AB-Sciex). A similar acquisition rate of 100 Hz might be obtained by a Citius LC-HTR (LECO), with a maximum resolution of 100 000. The orbitrap analyser is another option for such HRMS quantitative bioanalysis.[61] Orbitraps have the advantage of higher resolution but suffer from a slow data acquisition, whereas TOF analysers display opposite characteristics. A resolution of 30 000 can be attained with an orbitrap, however, typically only with an acquisition rate lower than 5 Hz. With orbitrap, the resolution is also strongly dependent on the *m/z* ratio. For the orbitrap Q-Exactive, which is the best suited for coupling with UHPLC, the resolution corresponds to 25 000 at 7 Hz for an *m/z* 400, while it is much higher for an *m/z* 200 corresponding to 35 000. At 12 Hz the obtained resolution for an *m/z* 400 is 12 500, while it is 17 500 for an *m/z* 200. MS/MS experiments using an orbitrap are possible due to the presence of quadrupole, linear ion trap in front of the orbitrap or the higher-energy collisional dissociation (HCD) collision cell.[62]

An overview of UHPLC-HRMS methods[64–70] applied in bioanalysis is displayed in Table 9.3. Several approaches were used for quantitation. Most of studies employed full-scan mode at a high resolution (approximately 30 000) with post-acquisition data processing using an extracted ion chromatogram (XIC) of the mono-isotopic mass of target compound.[2,65,67,70] When a large mass extraction window is used, the selectivity is poor and accurate quantitation is not possible. If the extraction window is narrowed down to 10 mDa, the selectivity is improved, allowing accurate quantitation. An example of comparison of various extraction windows is shown for the orbitrap analyser (Figure 9.6).[61] Other approaches used a base peak chromatogram[68] or SRM transition for quantitation.[64,66] The latter approach, however, lacks the above discussed advantages and was reported to be less sensitive compared to triple quadrupole.[63] Several studies[2,69,71] compared UHPLC-HRMS quantitation with triple quadrupole quantitation. Ramanathan *et al.*[2] reported similar LOQs and a linear range for 25 tested compounds. O'Connor *et al.*[71] demonstrated a good correlation between HPLC-MS/MS and UHPLC/TOF-MS in metabolic stability assay of 14 test compounds. In another study,[69] it was found, that UHPLC/Q-TOF analysis was five times less sensitive and two times faster compared to HPLC-triple quadrupole for the quantitation of a drug candidate in rat plasma.

Coupling of UHPLC with HRMS brings several advantages, such as short analysis time, in most cases down to 5 min (Table 9.3) and higher sensitivity plus structural information.[70] Wang *et al.*[68] reported an approximate four times faster separation and an approximate 2-fold higher sensitivity, when using UHPLC/Q-TOF compared to HPLC/Q-TOF. At present time, triple quadrupole instruments still represent the gold standard for LC-MS analysis in bioanalytical laboratories and probably will represent it also in the future due to fast instruments capable of hundreds SRM transition in a single LC-MS/MS run. However, this targeted analysis with the major disadvantage

Table 9.3 An overview of UHPLC-MS/MS bioanalytical methods using high-resolution MS for quantitation.

Determined substances	*Matrix/sample preparation*	*Stationary phase*	*Mobile phase*	*Detection*	*Linearity*	*Analysis time (min)*	*Method sensitivity*	*Reference and year*
Daptomycin	Plasma/PP	Acquity UPLC BEH C18 (100 × 2.1 mm, 1.7 μm)	0.1% formic acid/ACN Gradient elution	ESI-MS/MS Q-TOF SRM[a]	0.01–10 μg mL^{-1}	3.5	LOQ = 0.01 μg mL^{-1}	2011[64]
25 pharmaceutical compounds	Plasma/PP	Acquity UPLC HSS C18 (50 × 2.1 mm, 1.8 μm)	0.005% formic acid/ACN Gradient elution	ESI-MS QqQ Q-TOF MS; MS/MS scans XIC (10 mDa)	3–4 orders	–	LOD = 0.2–2 ng mL^{-1}	2011[2]
3 tetracycline antibiotics	Urine/SPE	ACQUITY UPLC BEH C18 (50 × 2.1 mm, 1.7 μm)	0.1% formic acid/ACN Gradient elution	ESI-MS Q-TOF XIC (20 mDa), EDC	0.5–10 ng mL^{-1}	5	LOD = 0.089–0.138 ng mL^{-1}	2010[65]
Etoposide, piperine analogue	Mice plasma/ SPE	Acquity UPLC BEH C18 (50 × 2.1 mm, 1.7 μm)	MeOH/water (72:28)	ESI-MS/MS Q-TOF SRM	2–1000 ng mL^{-1}	3.5	LOQ = 1.0–2.0 ng mL^{-1}	2010[66]
Synthetic mixture of drugs Verapamil	Microsomal incubation samples Rat plasma	Hypersil Gold (100 × 2.1 mm, 1.9 μm)	0.1% formic acid/ACN Gradient elution	ESI-MS/MS orbitrap MS; MS/MS scans XIC (10 mDa)	NA	5	Absolute sensitivity 10 × less compared to QqQ	2009[60]

Table 9.3 (*Continued*)

Determined substances	*Matrix/sample preparation*	*Stationary phase*	*Mobile phase*	*Detection*	*Linearity*	*Analysis time (min)*	*Method sensitivity*	*Reference and year*
Celecoxib, etoricoxib	Serum, synovial fluid/LLE	Acquity UPLC BEH C18 (150×2.1 mm, 1.7 μm)	0.1% formic acid/MeOH Gradient elution	ICP-MS ESI-MS/MS Q-TOF MS; MS/MS scans	0.48–10 μg mL^{-1}	7	LOD = 0.45 ng mL^{-1}	2009[67]
Scoparone and 4 metabolites	Rat bile, urine, plasma, faeces/LLE	Acquity UPLC BEH C18 (50×2.1 mm, 1.7 μm)	0.1% formic acid/ACN Gradient elution	ESI-MS Q-TOF BPI	NA	4	LOD = 0.22–0.32 ng mL^{-1}	2007[68]
Drug candidate and 8 metabolites	Rat plasma/PP	Acquity UPLC BEH C18 (50×2.1 mm, 1.7 μm)	0.25 mM ammonium formate, pH 3.0/ACN Gradient elution	ESI-MS QqQ Q-TOF XIC (25 mDa)	6–2000 ng mL^{-1}	2.5	LOD = 5 ng mL^{-1}	2006[69]
Histamine and its 4 metabolites	Mice hair/ Extraction by acidified MeOH	Acquity UPLC BEH C18 (100×2.1 mm, 1.7 μm)	20 mM AmF/ ACN Gradient elution	ESI-MS Q-TOF XIC	1.0–25 pmol	8	LOD < 1 pmol	2006[70]

[a]SRM experiment employs a sum of more fragment ions in MS/MS. ACN, acetonitrile; AmAc, amonium acetate; AmF, amonium formate; APCI, atmospheric pressure chemical ionization; BPI, base peak intensity; EDC, enhanced duty cycle; ESI, electrospray ionization; ICP, inductively coupled plasma; LLE, liquid liquid extraction; LOD, limit of detection; LOQ, limit of quantitation; MeOH, methanol; MS/MS, tandem mass spectrometry; PP, protein precipitation; SPE, solid phase extraction; TOF, time of flight; XIC, extracted ion chromatogram.

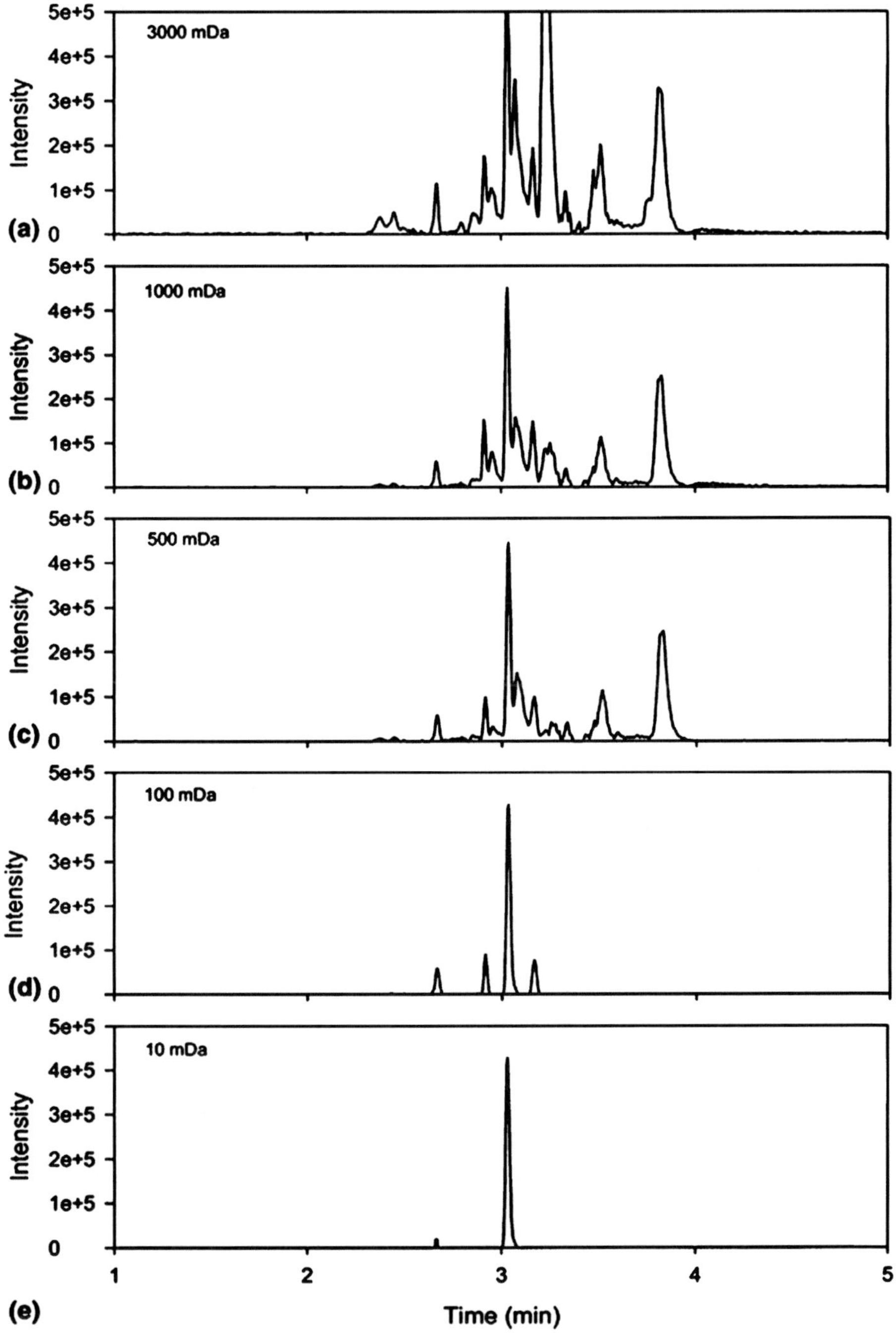

Figure 9.6 A demonstration of the importance of extraction window in HRMS quantification. Extracted ion chromatograms with the same data file using decreasing extraction windows: (a) 3000 mDa, (b) 1000 mDa, (c) 500 mDa, (d) 100 mDa and (e) 10 mDa. Improved selectivity is obtained without loss of sensitivity using high-resolution full-scan data acquisition. Figure reprinted from reference 61 with permission.

of blindness for non-targeted analytes is already being replaced by HRMS instruments in the case of screening/quantitative (Quan–Qual) methods, for example, as shown in Figure 9.7 for the analysis of histamine and its metabolite.[70] Quan–Qual bioanalytical methods are performed in a full-scan using high-resolution accurate mass measurement with the possibility of post-acquisition data processing for both targeted, as well as non-targeted compounds. [62]

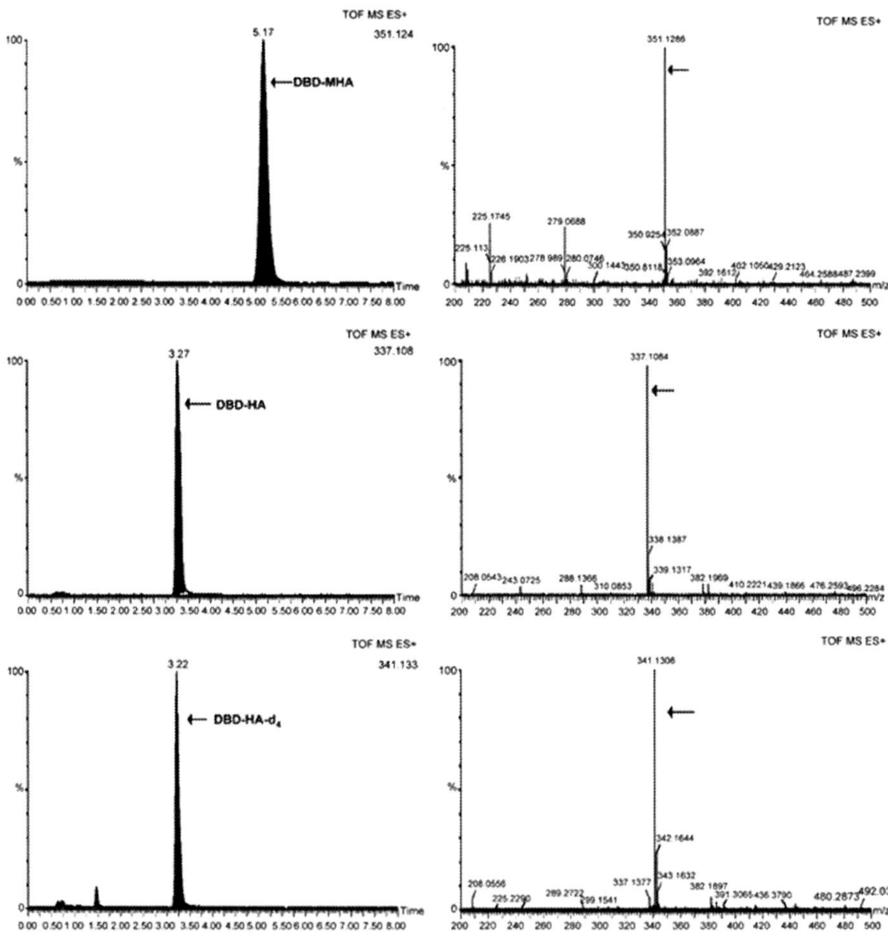

Figure 9.7 A demonstration of application of UHPLC-HRMS in bioanalysis. Typical mass chromatograms and mass spectra obtained from histamine (HA), methylhistamine (MHA) and corresponding deuterium-labelled internal standard HA-*d*4 (IS) in the hair shaft of mice. Both qualitative and quantitative (Quan–Qual) data can be obtained within one analytical run. Figure reprinted from reference 70 with permission.

9.2.2.5 *UHPLC-MS/MS Compared to HPLC-MS/MS in Bioanalysis*

UHPLC should provide substantial advantages in terms of increased separation efficiency, speed of analysis and sensitivity. Several papers compared developed methods under both HPLC and UHPLC conditions.[24,60,72] Churchwell *et al.*[24] compared UHPLC-ESI-MS/MS and HPLC-ESI-MS/MS for the analysis of β-agonists, tamoxifen metabolites, soy isoflavones and *Ephedra* alkaloids. The increase of sensitivity for UHPLC was found to be analyte-dependent and could be as high as 10-fold, whereas the analysis time was shortened up to 5-fold. In HPLC-MS/MS method, dwell times of 100 ms and inter-scan/inter-channel delay times of 10 ms were applied, whereas in UHPLC-MS/MS dwell times of 20 ms and inter-scan times of 5 ms were necessary. A typical demonstration of increased analysis speed when using UHPLC compared to HPLC is demonstrated in Figure 9.8.[24]

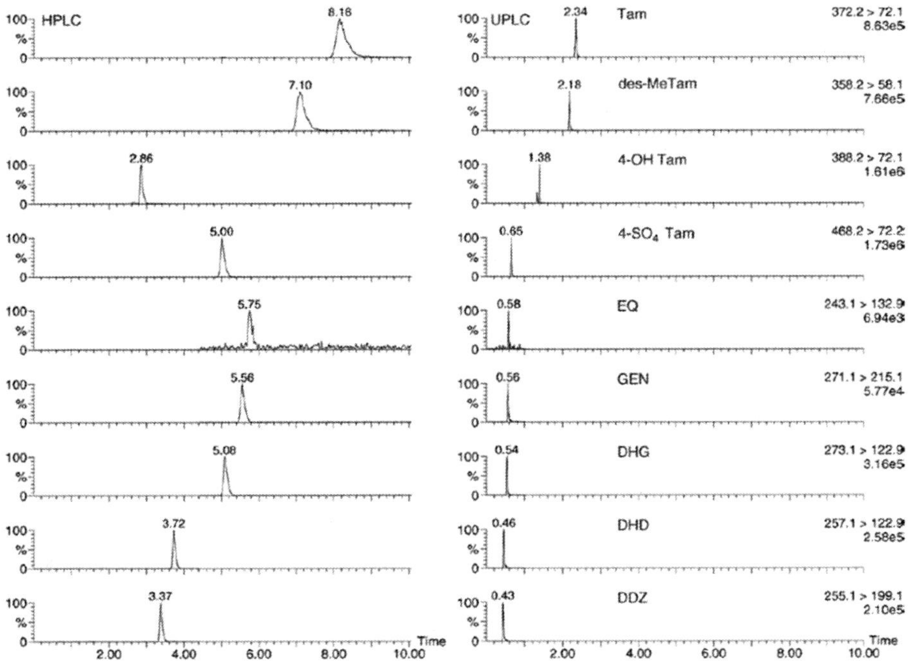

Figure 9.8 A comparison of HPLC-MS/MS and UHPLC-MS/MS analysis of soy isoflavones and its metabolites, and tamoxifen and its metabolites. The MS acquisition for both analyses consisted of 19 MRM transitions for equol (EQ), daidzein (DDZ), dihydrodaidzein (DHD), genistein (GEN), dihydrogenistein (DHG), 4-sulfoxy-tamoxifen (4-SO_4-Tam), 4-hydroxy-tamoxifen (4-OH Tam), N-des methyl-tamoxifen (des-MeTam), tamoxifen (Tam) and their corresponding internal standards for UHPLC (right panel) and HPLC (left panel). HPLC: 100 ms dwell, 10 ms inter-channel delay and 10 ms inter-scan times. UHPLC: 20 ms dwell, 5 ms inter-channel delay and 5 ms inter-scan times. Figure reprinted from reference 24 with permission.

Yu *et al.*[60] compared a quantitative method for the analysis of a mixture of five compounds in rat plasma by ESI-HPLC-MS/MS and UHPLC-MS/MS methods using SRM in both positive and negative ion modes during a single injection. UHPLC-MS/MS method offered a 3-fold decrease in retention time, up to a 10-fold increase in detected peak height and a 2-fold decrease in peak width. Similarly, in the UHPLC-MS/MS method a shorter dwell and delay times were necessary (a dwell time of 5 ms for ESI^+ and 20 ms for ESI^-, inter-scan and inter-channel delays of 5 ms), whereas longer times were suitable in HPLC-MS/MS (a dwell time of 20 ms for both ESI^+ and ESI^-, inter-scan and inter-channel delays of 10 ms). The method sensitivity, expressed as a LOQ, was again found to be compound-dependent with a 5-fold increase for UHPLC on average and up to 10-fold for a few compounds.[60]

Finally, the HPLC-MS method for the determination of edelfosine in both biological samples (plasma and tissue samples) and in lipid microparticulate system was compared with UHPLC-MS/MS method. The UHPLC-MS/MS method enabled a 3-fold decrease in analysis time and a 2-fold decrease in peak asymmetry, with a substantial increase in sensitivity from 0.2 μg mL^{-1} up to 0.075 μg mL^{-1}.[72]

9.3 Sample Preparation for UHPLC in Bioanalysis

Samples of biological materials are not directly compatible with LC analyses due to their complexity and protein content. Indeed, the proteins might irreversibly adsorb to the stationary phase, resulting in a substantial loss of column efficiency and an increase in back pressure. The sample preparation step impacts on all of the later assay steps and is hence critical for unequivocal identification, confirmation and quantification of analytes. It includes both the isolation and/or pre-concentration of compounds of interest from various matrices and interfering coumpounds, as well as making the analytes more suitable for separation and detection. Sample preparation techniques must be chosen and optimized with regards to the method. As stated in the ICH guidelines, the validated method must be appropriate for the intended purposes.[4] An appropriate technique should be chosen according to the extraction time, selectivity, number of steps, solvent consumption and the possibility of using an on-line method arrangement. Highly selective sample preparation is an important step for minimizing the matrix effects and ionization alterations that could occur with ESI-MS detection. The above-stated reasons make the sample preparation step the most important integral part of bioanalytical methods.[3,7,73]

The sample preparation step typically takes approximately 80% of the total analysis time. It is often done off-line, therefore becoming a limiting step of fast bioanalysis. Convenient sample preparation techniques should employ small amounts of the sample as well as simple methods which are "just enough" prior to analysis, as more steps could introduce more errors.

Currently, a huge number of sample preparation techniques are available. Based on the history and frequency of their use, they can be divided in two main groups. (1) Conventional sample preparation methods are widely spread across analytical laboratories, well-established and easily automated. They are quite well-optimized and reproducible, therefore method development and validation is usually not problematic. (2) The second group includes more recently developed techniques belonging to modern approaches in the field of sample preparation. Such methods should facilitate the sample preparation step with the main requirements for the decrease of sample and solvent consumption, shortening sample preparation time, reducing number of steps, reducing analysis cost and automatization.[7,73]

Although fast chromatographic methods are currently preferred in bioanalysis, the sample preparation step still regularly employs conventional LLE, PP and SPE. In contrast with ultra-fast chromatographic analyses, conventional sample preparation approaches are highly labour-intensive and time-consuming, consisting of many steps. For this reason, many new sample preparation techniques have been developed over the past decade to deal with the demand for faster, low-cost methods with improved selectivity and sensitivity.[7,73] An overview and comparison of properties of sample preparation approaches is shown in Table 9.4.

9.3.1 Conventional Approaches of Sample Preparation in Bioanalysis

Conventional approaches of sample preparation in bioanalysis are simultaneously the most widely used techniques in routine practice and include:

(1) Protein precipitation (PP)
(2) Liquid-liquid extraction (LLE)
(3) Solid phase extraction (SPE).

The manual operations associated with these processes are highly labour-intensive and time-consuming, consisting of many steps. Using these methods, sample preparation time is far greater than with ultra-fast chromatographic techniques. While ultra-fast chromatographic analysis takes about several minutes, such a sample preparation step lasts typically 15–30 min depending on sample type. Modern trends introduced in conventional sample preparation techniques include parallel sample processing in 96-well plates and direct injection of the biological sample using an on-line SPE, LLE or PP arrangement.[7,73]

9.3.1.1 Protein Precipitation (PP)

PP is a traditional sample preparation technique for the treatment of plasma. It is considered to be the fastest and the simplest extraction approach applicable for both hydrophilic and hydrophobic compounds with minimal development requirements and costs, therefore it is highly convenient when

Table 9.4 A comparison of conventional and modern sample preparation approaches.

Sample preparation technique		*Extraction time (min)*		*Selectivity*	*Multi-step process*	*Automation*	*Solvent consumption*
		Off line	*On-line*				
Conventional techniques	LLE	15–25	<5	Medium	Yes	+	Very high
	SPE	15–35	<5	Medium	Yes	+	Relatively high
	PP	<10	<5	Low	No (centrifugation)	+	High
Microextraction techniques	SD-LPME	5–60	–	Medium	Adsorption/desorption	–	Very low
	HF-LPME	10–120	–	Medium	Adsorption/desorption	–	Very low
	DLLME	<10	–	Medium	No	–	Very low
	SPME	10–60	10–30	Medium	Adsorption/desorption	+	No
	SBSE	30–240	–	Medium	Adsorption/desorption	–	No
	MEPS	7–10	1–5	Medium	Yes	+	Very low
	DPX	15– 40	<5	Medium	Yes	+	Very low
On-line extraction approaches	RAM	–	2–10	Medium	No (centrifugation)	+	No
	TFC	–	<1	Medium	No	+	No
	In-tube SPME	–	20–30	Medium	Yes (repeated cycles)	+	No
Selective extraction approaches	MIP	15–35	<5	High	Yes	+	Relatively high
	IA-SPE	15–35	<5	High	Yes	+	Relatively high

fast method development is desired. Method development includes only the choice of precipitation agent and optimization of its volume with the regard to the sample clean-up and LOQ. Miscible organic solvents (acetonitrile, methanol) or, less commonly, other precipitation agents (acids, salts, metal ions) have been used for PP. The most efficient protein precipitants were found to be acetonitrile, trichloroacetic acid and zinc sulphate with precipitant-to-plasma volume ratios of 2:1 and greater.[74] However, after PP, the supernatant is considered to remain relatively unclean, since it might still contain a significant amount of unprecipitated plasma components. Therefore, selectivity is low and might induce analyte co-precipitation or MS signal suppression. Additional centrifugation in order to separate the resultant protein precipitates from the analyte provides sufficient cleaning for most HPLC-MS analyses.[20,39]

Although PP is not excellent in terms of selectivity, it has been used in recent applications employing UHPLC-MS/MS[20,39,46,47,49,52,53] due to inherent advantages including speed, ease of use, no requirements for method development and low sample consumption (10–100 μl). No considerable or very low matrix effects were observed in published studies in spite of a less selective sample preparation step, which was in agreement with the requirements of appropriate guidelines.[1,4] To speed-up and automate the procedure, the 96-well plate arrangement of PP is highly convenient. Such an arrangement is fully automatic and reduces the time due to the omission of the evaporation/reconstitution step.[20]

9.3.1.2 Liquid–liquid Extraction (LLE)

LLE was one of the first sample preparation techniques and continues to be widely used for biological sample analysis, especially in routine laboratories. An easy principle, being the transfer of analyte from the aqueous sample to a water-immiscible solvent based on the octanol/water partition coefficient, makes LLE quite easy and straightforward to optimize. Nevertheless, some shortcomings, such as emulsion formation, poor phase separation, the use of large sample volumes and toxic organic solvents, unsuitability for hydrophilic compounds and above all, the production of a large volume of environmental pollutants, make LLE expensive, time-consuming and environmentally harmful. Despite this, LLE is still widely used for the sample preparation of biological fluids, which is demonstrated by several newly developed UHPLC-MS/MS methods which employ this methodology in sample preparation.[41,43,45,54,56] A fully automated 96-well plate arrangement meets the criteria for modern, fast and efficient sample preparation step better due to much shorter extraction time, lower solvent and sample consumption.[75]

In order to eliminate some of the above-mentioned drawbacks of LLE, Extrelut columns (manufactured by Merck) were developed in 1980s as a solid phase support for LLE. Extrelut columns contain a chemically inert matrix, which acts as a holder after the sample is applied to the column. The extraction is performed by 2–3 times the sample volume of organic solvent and the residue is either evaporated to dryness, reconstituted and injected into the chromatographic system or injected directly into an on-line version. The advantages include the prevention of the formation of emulsions and the possibility of automation. However, in principle, the procedure is still an LLE method with the main drawbacks, such as the use of toxic organic solvents (often more than 10 ml for the elution of one sample) and, above all, the production of a large volume of environmental pollutants, which makes LLE expensive and environmentally harmful. The technique is also quite time-consuming, as 5–20 min are needed after sample loading until an extraction can be completed, therefore this extraction approach has not been yet implemented in UHPLC-MS/MS analytical method.[7]

Another way to perform LLE is salting-out liquid–liquid extraction (SALLE)[76] or LLE using sub-zero temperatures, including the addition of acetonitrile and cooling down to −20 °C.[77] The salt-induced phase separation between the water miscible solvent (acetonitrile, isopropanol) and the aqueous solution may be induced by the addition of a variety of inorganic and organic electrolytes. A constant and potentially large amount of salt must be added to obtain constant recovered volumes between the two phases.[76] Advantages of SALLE with acetonitrile have been demonstrated, including the ease of biological sample clean-up, as well as analyte enrichment. However, the use of magnesium sulphate, ammonium sulphate, sodium chloride, calcium chloride, potassium carbonate and calcium sulphate can cause problems for subsequent LC-MS analysis. Recently, Wu *et al.*[78] reported the utilization of SALLE using MS-compatible salts. According to the authors,[78] a small portion of the salt may dissolve into the acetonitrile phase and the impact of the salts on subsequent LC-MS analysis should be insignificant. It is questionable if a 1 M concentration of ammonium acetate does not induce an excessive concentration of this salt in the acetonitrile phase, therefore further investigations of this factor should be undertaken. Miniaturisation of SALLE was recently first reported in the scientific literature to provide some advantages related to microextraction approaches.[79]

9.3.1.3 Solid Phase Extraction (SPE)

SPE has established itself as an important sample preparation technique in the field of bioanalysis due to the following advantages: high recovery, effective pre-concentration, need for less organic solvent (compared to LLE), no foaming, no formation of emulsions, ease of operation and important possibility of automation.[80]

In SPE, the analytes to be extracted are partitioned between a solid phase and a liquid phase (sample). The analytes must have a greater affinity for the solid phase than for the sample matrix. Partition may involve non-polar, polar or ionic interactions. A wide range of currently available SPE sorbents ensure various selectivities. The choice of sorbent is the key factor in SPE, because this can control parameters such as selectivity, affinity and capacity. The choice depends strongly on the analytes and their physico-chemical properties, the type of sample matrix and interactions with both the sorbent and the analyte. SPE sorbents range from chemically bonded silica with the C8 and C18 organic group, graphitized carbon and ion-exchange materials, polymeric materials [polystyrene–divinylbenzene (PS–DVB), cross-linked styrene-divinylbenzene, polymethacrylate, cross-linked methacrylate–DVB (MA–DVB) and many others], mixed-mode sorbents [containing both non-polar and strong cation (SCX) or anion exchanger (SAX), immunosorbents, molecularly imprinted polymers (MIP), as well as restricted access materials (RAMs)] and also recently developed monolith sorbents.[80,81] Silica-based sorbents have several disadvantages as compared with polymeric sorbents. Silica sorbents are unstable in a broader pH range and contain silanols, which can cause the irreversible binding of some groups of compounds, *e.g.* tetracyclines. Conventional SPE cartridges are easy to handle by using vacuum- or positive-pressure manifold. However, it is not easy to control the flow-rate and, in addition, care should be taken to prevent the column from drying out prior to sample application. This is still a drawback of many commercial cartridges except for polymeric, *e.g.* Oasis (Waters) or Abselut (Varian).[80]

SPE might be performed in various formats including column cartridges, discs and well-plates. The on-line configuration of SPE utilizes a 96-well plate format for SPE, automated with a robotic liquid-handling system, facilitating high-throughput analyses of biological samples.[80] New trends in SPE, including various miniaturizations, are discussed in more detail in Section 9.3.2.1. Conventional SPE is currently widely used in bioanalysis. Recently developed UHPLC-MS/MS methods often prefer PP due to its speed and simplicity, nevertheless SPE is an important sample preparation technique to be coupled with UHPLC as well.[44,48,50] Manual SPE is, however, time-consuming and relatively expensive, as SPE cartridges are manufactured for single use only. They sometimes have poor batch-to-batch reproducibility and still require a relatively large amount of organic solvent. These disadvantages do not prevent SPE from being one of the most widely used sample preparation technique in routine bio-analytical laboratories, although with UHPLC separation, faster sample preparation (see Table 9.2) or the on-line 96-well plate SPE[59] arrangement are preferred.

9.3.2 Modern Approaches for Sample Preparation in Bioanalysis

Modern approaches for sample preparation allow facilitate this step, which is the most time-consuming step of the whole procedure of complex sample analysis. Recent investigations have focused on the development of new modern sample preparation techniques which meet the following requirements as much as possible:[7,82]

- decrease in time of sample preparation procedure
- decrease in solvent consumption/or solvent-less techniques

- decrease in the amount of sample needed for analysis (miniaturization of techniques)
- decrease in the cost of the sample preparation
- enhancement of the selectivity of sample preparation
- decrease in the number of steps
- possibility of on-line coupling and automation.

Based on the above stated requirements, modern approaches for sample preparation before chromatographic analysis might be divided into three groups:

(1) Microextraction approaches
(2) On-line techniques enabling direct injection of biological material
(3) Sample preparation techniques with high selectivity.

A complete review of the current status and recent advances in sample preparation techniques was conducted by Chen *et al.*,[83] in which the detailed classification of all sample preparation techniques is outlined and selected techniques are discussed in detail. Similarly, the aspect of the sample preparation step was also discussed in a review article of our group[7] and in several other papers.[73,82] The characteristics of conventional and modern sample preparation approaches are compared in Table 9.4.

9.3.2.1 *Microextraction Techniques*

The main advantage of microextraction approaches is a minimal or practically zero solvent consumption. Microextractions may be based both on liquid–liquid phase extraction, including various liquid–liquid microextractions (LLME), or on SPE, including solid phase microextraction (SPME), stir-bar sorptive extraction (SBSE), microextraction by packed sorbent (MEPS), monolith spin extraction or disposable pipette tips extraction (sometimes designated as DPX).[7,82,83] A detailed overview of microextraction techniques is shown in Figure 9.9.

9.3.2.1.1 Liquid–liquid microextraction (LLME)

Various LLME approaches were developed to replace conventional LLE because of several disadvantages, such as the formation of emulsions and the use of large volumes of toxic organic solvents. LLME techniques should use negligible volumes of extracting solvent and take a minimum number of steps. The three main groups of miniaturized technologies include single-drop microextraction (SDME), hollow-fibre liquid phase microextraction (HF-LPME) and dispersive liquid–liquid microextraction (DLLME).[82–84] SDME techniques are based on the use of a micro-drop (1–10 μl) of extracting solvent, which is immiscible with water. It may be accomplished in a direct immersion (DI-SDME), head space (HS-SDME), continuous-flow microextraction (CFME) or liquid–liquid–liquid microextraction (LLLME) arrangement

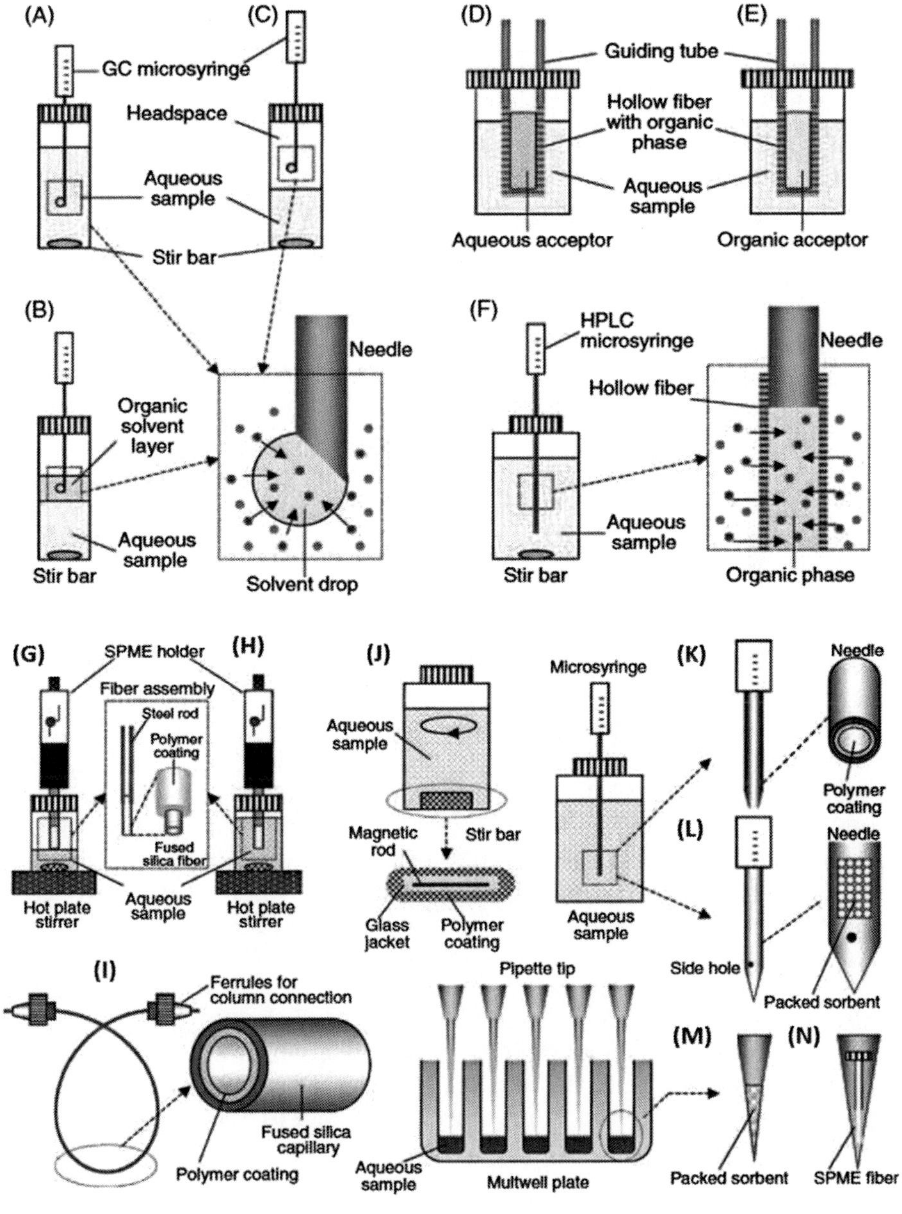

Figure 9.9 An overview of microextraction sample preparation techniques. (1) LLE based approaches, including direct-immersion single-drop microextraction (A and B), head-space single-drop microextraction (C) and variants of hollow-fibre liquid-phase microextraction (D–F). (2) SPE based approaches, including head-space solid phase microextraction (G), direct immersion solid phase microextraction (H), in-tube solid phase microextraction (I), stir bar sorptive extraction (J), solid phase dynamic extraction (K), microextraction by packed sorbent (L), sorbent-packed in-tip solid phase microextraction (M), fibre-packed in tip solid phase microextraction (N). Figure reprinted from reference 82 with permission.

(some arrangements are shown in Figure 9.9). In HF-LPME, a hydrophobic porous hollow fibre is used to protect and expose a certain volume of extracting solvent to the sample. Here, an organic solvent is immobilized in the wall pores of the hollow fibre, providing a supported liquid membrane (SLM), and an aqueous acceptor solution is held within its lumen.[82,84,85] When the acceptor solution is an organic solvent, HF-LPME is two-phase extraction, which is directly compatible with gas chromatography (GC). Three-phase sampling mode (HF-LLLME), where the acceptor solution is aqueous, is compatible with HPLC or CE (Figure 9.9). The sample-to-acceptor ratio is very high, therefore analyte enrichments are also elevated, without the need for evaporation and reconstitution.[83,86] DLLME is based on a ternary component solvent system in which an appropriate mixture of extraction solvent and disperser solvent is quickly injected into the aqueous sample with a syringe. The mixture is then gently shaken and a cloudy solution is formed. After centrifugation, the fine particles of extraction solvent are sedimented in the bottom of a conical test tube. The resultant sedimented phase is taken out with a microsyringe and injected into chromatographic system for analysis.[84,87]

LLME methods have many advantages such as simplicity, cost effectiveness and negligible solvent consumption. There are also some drawbacks of individual techniques, such as prolonged extraction time and the need for careful and elaborate manual operation for SDME and HF-LPME.[83–86] For DLLME the main advantages are its simplicity of operation, rapidity (equilibrium is reached very quickly, as opposed to LLME; see also Table 9.4), low cost, high recovery, a high enrichment factor and environmental friendliness. Nevertheless, this technique is limited to the use with only a small number of extracting solvents.[87] LLME techniques are quite commonly used in the modern sample preparation of biological fluids. The most widely used technique is HF-LPME. Its principles and applications have been reviewed by Pedersen-Bjergaard and Rasmussen,[86] in which 34% of summarized applications were bioanalytical methods for the determination of drugs in various biological materials, in addition to environmental and food applications. LLLME has been used less commonly in the analysis of biological fluids and other SDME techniques (DI-SDME, CFME and HS-SDME) were mostly reserved for the determinations of environmental samples and trace elements. Similarly, for DLLME only a few bioanalytical methods were found as the technique was reserved for environmental applications.[7] LLME techniques coupled to UHPLC have been so far reported only in environmental[88] or food analysis.[89]

9.3.2.1.2 Fibre Solid Phase Microextraction (Fibre SPME)

SPME was developed by Pawliszyn and co-workers[90,91] as a simple and effective adsorption/absorption and desorption technique which eliminates the need for solvents. Extraction may be performed in two arrangements: fibre SPME and in-tube SPME.[82,83,92] Fibre SPME is based on a modified syringe which contains stainless steel microtubing within its needle. Inside there is a fused-silica fibre tip which is coated with an organic polymer, typically polydimethylsiloxane (PDMS) or many other materials, such as polyacrylate, carboxen-PDMS, carbowax-divinylbenzene, *etc*. This coated fibre can be moved inside and outside by a plunger. Using such simple equipment, all of the steps—extraction, pre-concentration, derivatization and transfer to the chromatographic system—are integrated into one device. Initially, either DI-SPME or HS-SPME (see Figure 9.9) has been applied in bioanalytical methods with or without an *in situ* derivatization step.[93] To date, there has been a great technological development in SPME to enhance the utility of this technique in the field of bioanalysis. New biocompatible coating phases, *in vivo* SPME sampling devices and fully automated 96-well format multi-fibre SPME have been introduced.[92,94]

Advantages and disadvantages of fibre SPME in bioanalysis have been widely discussed by Ulrich.[93] Some important advantages include ease of use and automation, no requirement for solvents, minimal equipment requirements, good linearity and relatively high sensitivity. Some drawbacks are represented by the longer time needed for extraction, generally lower recoveries than those of LLE and SPE, and several principal disadvantages such as limited capacity of SPME fibre, long time needed for desorption (much longer than in the injection of LLE or SPE extracts), the occurrence of carryover effects and, finally, the fragility of SPME fibre.[93] In spite of these limitations, SPME has become a useful tool in bioanalytical laboratories, especially due to novel possibilities using *in vivo* SPME sampling (Figure 9.10)[95] and high-throughput devices.[96] Probably due to the slow extraction process, no effort has been put into coupling

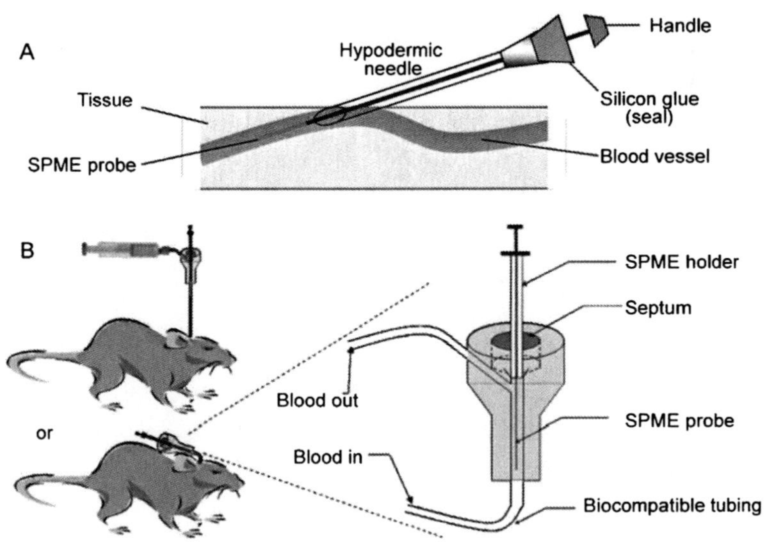

Figure 9.10 A demonstration of SPME *in vivo* sampling device and interface connection to the carotid artery. Figure reprinted from reference 95 with permission.

fibre SPME to an UHPLC system. The lack of commercially available coatings with different selectivity and relatively slow process are limiting factors that slow down the process of the use of this technique routinely in laboratories. However, fibre SPME has become a complementary technique, which offers an interesting alternative to more conventional systems.[92]

9.3.2.1.3 Stir-bar Sorptive Extraction (SBSE)

SBSE is a sorptive and solvent-less extraction technique based on the same principles as SPME (Figure 9.9). Instead of a polymer-coated fibre, a large amount of extracting phase, typically PDMS, is coated on a stir-bar. The extraction of an analyte from the aqueous phase into an extraction medium is controlled by the partitioning coefficient of the analyte between the silicone phase and the aqueous phase. The amount of extraction phase in SBSE is 50–250 times greater than in fibre SPME (typically 0.5 μl of extraction phase for 100 μl volume PDMS fibre). SBSE of a liquid sample is performed by placing a suitable amount of sample in a head space vial or a container. A coated stir-bar is added and the sample is stirred for 30–240 min. The extraction time is determined by the sample volume, the stirring speed and the stir-bar dimensions, and must be optimized for given applications. After the extraction, the stir-bar is removed and wiped very gently with the tissue to remove water droplets. Thermal desorption or liquid desorption (back-extraction with a small volume of liquid solvent) are used for subsequent connection with GC or LC respectively. This process provides high sensitivity, since the complete extract can be analysed. In contrast to SPME, the desorption process is slower because the extraction phase is extended, thus desorption needs to be combined with cold trapping and re-concentration.[84,97] The typical extraction optimization steps include the pH of sample/aqueous phase, phase ratio, stirring speed, sample temperature, extraction time and effect of salting out.[73]

Since only the non-polar PDMS coating is available as an extraction phase, SBSE was prevalently used for low polarity analytes. The problems of polar compounds analysis in biological materials (*e.g.* metabolites) may be solved by *in situ* derivatization. It is clear that further developments in stir-bar coatings and designs could extend the applicability of the method. Neither SBSE has been ever coupled to UHPLC as a sample preparation technique in bioanalysis. The main drawback of SBSE is the length of time needed for extraction, typically 30–150 min. For this reason SBSE may be impractical for high-throughput bioanalytical laboratories.[98]

9.3.2.1.4 Microextraction by Packed Sorbent (MEPS)

MEPS is a miniaturized version of SPE, wherein the extraction is performed in the solid packing material inserted into the barrel of a syringe as a plug or between the needle and the barrel as a cartridge (Figure 9.9). The bed dimensions are scaled from a conventional SPE bed and, in this way, MEPS can be adapted to SPE methods by scaling the reagents and sample volumes.

Sample preparation in MEPS takes place on a 1–4 mg packed bed. Any sorbent material (similar to conventional SPE) can be used either as the packing bed or as a coating including typically silica based (C2, C8, C18, SCX), PS-DVB, RAMs or MIPs sorbents. The extraction protocol is identical with SPE, including conditioning, sample load, washing step/s and elution achieved by the movement of the syringe plunger. The sample can be drawn and ejected several times from the same vial if pre-concentration of the analyte is required.[82,99]

The extraction is performed either manually, with the help of a hand-held automated analytical syringe (eVol, provided by SGE) or by an autosampler. The analytes are eluted with an organic solvent such as methanol or a LC mobile phase (20–50 μl) directly into the instrument injector, which enables full automation coupled to GC, LC or CZE. The packed MEPS syringe (100–250 μl) may be used several times handling small sample volumes (10 μl of plasma, urine or water), as well as large volumes (1000 μl). A single MEPS cartridge was reported to be used more than 100 times with plasma or urine samples and more than 400 times for water samples.[100] This is great advantage compared to SPE, where the reported re-use of SPE cartridges was only a few times (3–4 times),[101,102] ignoring the recommendations of manufacturer that SPE cartridges are designed for a single use.

Compared with manual LLE and SPE, MEPS reduces sample preparation time, organic solvent consumption and analysis cost. Compared to SPME, the MEPS technique is more robust, as in SPME the sampling fibre is quite sensitive to the nature of the sample matrix.[99] Some drawbacks of MEPS technique include the possibility of bubble formation and some other difficulties connected to off-line arrangement, as on-line coupling is not possible with every LC system. In off-line MEPS, the speed of plunger movement is crucial for the recovery of analytes and repeatability. Too high speed of movement does not enable the adsorption of the analyte to the MEPS support and leads to misleading recovery results and unrepeatability.[103]

MEPS was one of the first microextraction techniques employed with UHPLC applications.[103] Its applications so far are not wide, but this technique has a great potential to become a routine sample preparation technique in bioanalytical laboratories, similar to conventional SPE.

9.3.2.1.5 Monolith Spin Extraction

Another quite recent approach for the sample preparation in bioanalysis based on SPE was introduced by Namera *et al.*[104] in 2008. Monolithic silica disks manufactured in the laboratory and chemically modified by C18 were packed into spin columns. Typical sample preparation steps of SPE, including conditioning, sample loading, washing and elution of target compounds, were all achieved by the centrifugation of the spin column. The extraction column can prepare more samples simultaneously using centrifugation, without the need for evaporation and reconstitution step. In a monolith spin column, the flow rate is controlled by monitoring the rotation speed and time. The amount of the drug that is adsorbed from the samples depends on the duration for which the sample is in contact with the sorbent, as well as on sample pH, similarly as in SPE. The main advantages of monolithic spin column extraction are the possibility of its use with low sample volumes as well as the minimal requirements of elution solvent. The elution solvent can be injected directly into a chromatographic system without the need for evaporation. The operation is very simple.[104,105] This approach, however, is quite new, thus its wider applicability must be further verified.

9.3.2.1.6 Disposable Pipette Tips Extraction (DPX)

DPX is another variant of microextraction based on SPE (Figure 9.9). In the original form, standard pipette tips are loaded with a free-flowing sorbent powder, which is free to disperse, enabling the easy movement of the solvent in or out. In its modified form, the standard tip contains a dispersible sorbent loosely placed between two frits. In DPX, the sample is mixed with material and every sorbent particle has contact with the analyte several times. This leads to fast and efficient extraction, in which only a little material is needed to retain the analytes compared to conventional SPE cartridges (a greater amount of sorbent is needed, as the analyte becomes in contact with the sorbent particles only once). The sample preparation steps are the same as in SPE, including conditioning, sample loading and mixing sample with the sorbent, washing and elution of the target compounds. All of the steps are achieved by aspiration/withdrawing through the tip either manually or by automated devices. Commercially available DPX tips include RP sorbents (C18, polystyrene–divinylbenzene co-polymer), ion-exchange sorbents (SAX or WAX) and graphitized carbon, titanium oxide or zirconium oxide.[73,82] Monolithic stationary phases were also used to perform pipette tips extraction using a 96-well plate arrangement.[106]

For the moment, the scientific literature on DPX in bioanalysis is quite limited; however, its simplicity in operation, variety of sorbents and ability to automate will probably enable widespread use of this technique in bioanalytical laboratories in the future.[73,82] The compatibility of this sample preparation technique with UHPLC is evident, although they have not been coupled so far. DPX demonstrates low solvent and sample consumption, therefore low cost, ease of automation and straightforward method development due to similarity to conventional SPE.

9.3.2.2 On-line Sample Preparation Techniques – Direct Injection of Biological Samples

One of the most important requirements on sample preparation technique is the ease of automation with the aim to decrease the necessity of intervention by the operator and to reduce substantially time requirements. Thus, most of conventionally used sample preparation techniques, including SPE, LLE and PP, are easily automated at present. Concerning microextraction techniques, the possibility to fully automate the process addresses only selected techniques such as SPME and MEPS or DPX, which are widely used in connection with LC. While in an off-line arrangement, only a part of extract is injected, whereas, in an on-line arrangement, the whole extracted volume is injected into chromatographic system. This leads to an enhanced sensitivity of on-line techniques. A miniaturization of the system is often required to prevent an overload. Typical techniques that represent on-line coupling of sample preparation into the chromatographic system use a switching valve or another way, enable direct injection of biological sample, but cannot be easily practiced independently in an off-line arrangement. They include extraction using RAM cartridges, in-tube SPME and turbulent flow chromatography (TFC). Besides ease of automation, the on-line techniques also reduce analysis time and cost due to solvent-free operation and provide better precision and sensitivity than manual off-line techniques.[107]

9.3.2.2.1 On-line Extraction Using Restricted Access Materials (RAMs)

RAMs, introduced in 1991 by Desilets *et al.*,[108] are biocompatible sample preparation supports which enable the direct injection of biological fluids into a chromatographic system. RAM sorbents are able to fractionate a biological sample into a protein matrix and analyte fraction, based on molecular weight cut-off. The outer surface of RAM particles, which is in contact with biological matrix components such as proteins and nucleic acids, possesses a special chemistry to prevent adsorption of these molecules. Macromolecules can be excluded by a physical barrier by means of the pore diameter or by a chemical diffusion barrier created by a protein (or polymer) network at the outer surface of the particle. Thus, the basis of RAMs is the simultaneous size exclusion of macromolecules and extraction/enrichment of low-molecular-weight compounds into the interior phase *via* partition. RAMs can be classified according to the protein-exclusion mechanism into the following two groups: RAM with a physical barrier (RP, alkyl-diol-silica material, porous silica with combined ligand) or RAM with a chemical barrier (semi-permeable surface, protein-coated silica, mixed functional materials or shielded hydrophobic phase).[109] The application of RAM to direct and repeated analyses of drugs in biological fluids using a column switching technique has become well established, especially with the commercialization of RAM supports.[107] It is highly recommended to remove cryoproteins and specific impurities to prevent damage of the cartridge and clogging.[110]

Direct on-line hyphenation of UHPLC with any sample pre-treatment is highly challenging and therefore not many coupled systems have been described in the scientific literature so far. There are two main reasons: (1) UHPLC has extremely high pressure (up to 1300 bar) and low flow-rate requirements, which are not directly compatible with typical on-line extraction methods, and (2) the large amounts of strong solvent (typically methanol) required for elution cannot be directly introduced into UHPLC system while maintaining retention and peak shape. UHPLC-MS/MS coupled to RAM was reported in analysis of 13 bile acids in human serum. The problem of compatibility between a large volume of strong solvent used for RAM elution and the requirement for a weak solvent was solved by using an auxiliary pre-column cross-flow of 0.1% aqueous formic acid. The whole procedure took 30 min using 100 μl of serum for the injection.[110]

9.3.2.2.2 Turbulent Flow Chromatography (TFC)

The principle of TFC in sample preparation is the separation of small analyte molecules from the macromolecular matrix due to the low diffusion coefficients of proteins. Such enhanced separation efficiency is the result of the onset of turbulence, which is responsible for an increase of the mass transfer rate through formation of "eddies" in the mobile phase. The generation of turbulent flow requires short, narrow-bore columns, packed with large-sized particles (typically 50 mm × 1.0 mm, 20–60 μm) and flow rates in the range of 4–5 ml min^{-1}. Proteins are eluted as a sharp peak within 0.5 min using a pure water or buffer mobile phase. The retained analytes are subsequently eluted from the extraction column using organic mobile phase on to an analytical column for the chromatographic separation. The coupling is most commonly performed with the help of multiport valves and one or more pumps.[107] On-line coupling of fast LC to TFC increases separation efficiency and analysis throughput.

UHPLC-MS/MS coupled to TFC was reported in ultra-fast analysis of diclofenac in urine in pharmacokinetic studies. Analytes were extracted within 30 s using 10 μl of the whole blood sample after dilution and centrifugation, while the chromatographic separation took 1.5 min, which allowed high sample throughput.[111] A longer analysis time of 7 min was needed for UHPLC analysis and TFC isolation of verticine and its derivatives from rat plasma.[112]

9.3.2.2.3 In-tube Solid Phase Microextraction (In-tube SPME)

In-tube SPME, firstly introduced by Eisert and Pawliszyn[113] in 1997, uses the inner surface of capillary column (Figure 9.9) instead of a fibre to facilitate on-line hyphenation with chromatography and to overcome some problems related to fibre SPME, such as fragility, low sorption capacity *etc.* The extraction might be performed using polymer-coated open tubular capillary, fibre-packed, sorbent-packed or monolithic capillary.[92] A capillary column is used as the injection loop in a standard autosampler. During in-tube SPME, the sample is repeatedly drawn and ejected through an internally coated capillary for extraction, followed by desorption of the extracted analytes from the capillary by a mobile phase for the transport to chromatographic column. This approach requires more complex instrumentation than regular SPME, but higher sensitivity can be obtained by using a longer tube, in which more sorbent is exposed.[78,113] The main disadvantage of in-tube SPME is a requirement for very clean samples, because the capillary can be easily clogged. Therefore, a filtration step is required before the analysis.[92] In-tube SPME has not yet been coupled to UHPLC as a sample preparation technique in bioanalysis, probably due to reasons generally concerning the challenge of coupling on-line sample preparation techniques with UHPLC as stated above and furthermore due to instrumental and pressure requirements (the connection of the capillary does not withstand common UHPLC system pressure).

9.3.2.3 Sample Preparation with High Selectivity

9.3.2.3.1 Molecularly Imprinted Polymers (MIPs)

MIPs are stable synthetic polymers with molecular recognition abilities, provided by the presence of a template during their synthesis. Therefore, they are excellent in terms of selectivity for sample preparation. These materials are obtained by the polymerization of functional and cross-linking monomers around a template molecule, leading to a highly cross-linked three-dimensional network polymer. After polymerization, the template is extracted and binding sites with shape, size and functionalities complementary to the target analyte are established. The resulting polymers are stable, robust and resistant to a wide range of pH values, solvents and temperature. MIPs have been widely employed in on-line, as well as off-line, SPE and recently there is a growing interest in the use of MIPs with other sample preparation techniques, such as microextractions (*e.g.* SPME, SBSE and others).[114]

Due to the substantial advantages of MIPs, including high selectivity, stability, reusability, ease of use and low cost preparation, they have been used effectively as sorbents in the clean-up and selective enrichment of analytes from different samples. However, some features still need an improvement. Currently, the MIPs used for SPE are typically prepared by a non-covalent imprinting technique, which gives quite low yields of specific binding sites. This results in a low sample load capacity and high levels of non-specific binding. Therefore, advances are expected in the development of new MIP synthesis methods to improve their capacity and selectivity.[115] SPE using MIPs, also designated as MISPE, coupled to UHPLC has been so far applied only in the field of environmental analysis.[116]

9.3.2.3.2 Immunoaffinity SPE (IA-SPE)

Immuno-based sample preparation techniques use the principle of molecular recognition. Due to high affinity and high selectivity of antigen–antibody interaction, they have a great potential for sample preparation. The key reagent for immuno-extraction is the antibody. As many compounds of interest (antigens) are small molecules, it is necessary to couple them to a carrier protein in order to generate an immune response. The target analyte that is coupled to the carrier protein is known as a hapten.[117]

The immunosorbents are obtained by linking the antibodies to a solid support, which is then packed into the SPE cartridge or pre-column. Both polyclonal and monoclonal antibodies have been used for immobilization. The binding of analytes to antibodies is the result of a good spatial complementarity, which is a function of the sum of the intermolecular interactions. Therefore, cross-reactivity of antibodies with structural analogues is often observed. While it is considered as a negative feature in immunoassay, it is exploited in IA extraction, as it is usually followed by separation technique. In the biological field, it is often interesting to determine a group of related compounds and their metabolites.[118] The coupling of IA-SPE to LC-MS/MS was reported to bring several advantages such as analyte enrichment and an increase in sensitivity and selectivity.[119] According to recently published review concerning IA-SPE, this sample preparation technique was coupled with UHPLC only in analysis of cytokinins in plant material.[120] The drawbacks including difficult, expensive and time-consuming synthesis of antibodies with no certainty of success probably prevents wider use of IA-SPE in routine bioanalytical laboratories.[114]

9.3.2.3.3 Aptamers

Aptamers are artificial single-stranded DNA or RNA sequences (more recently, peptides) that fold into secondary and tertiary structures allowing them to bind to certain targets with extremely high specificity. Aptamers are generated by an *in vitro* selection process called SELEX (systematic evolution of ligands by exponential enrichment), permitting the identification of unique RNA/DNA molecules from very large populations of random sequence oligomers (DNA or RNA libraries). These molecules bind to the target molecules (small molecules, proteins and even entire cells) with very high affinity and specificity. Aptamers show a very high affinity for their targets, with dissociation constants typically ranging from the micromolar to low picomolar, comparable to those of some monoclonal antibodies, sometimes even better. Due to extremely high affinity of aptamers to their specific targets, their use in affinity chromatography is expected for separation and purification of small molecules, isoforms and proteins.[121] Aptamers have been reported for use in the selective isolation of cocaine from biological fluids by means of high-selectivity binding.[122] Their potential for wider application in sample preparation must be further explored.

9.4 Conclusions

UHPLC-MS/MS has a great potential in the field of bioanalysis due to the high separation efficiency and high selectivity and sensitivity. The high separation efficiency is very important in the analysis of complex matrices, such as biological materials, that contain a low concentration of target analyte(s). UHPLC is very helpful in the separation of complex mixtures of structurally similar compounds, as well as in separation of target analytes from matrix components to prevent matrix effects. The determination of matrix effect was newly established as an inherent part of full method validation. At present time, triple quadrupole instruments still represent the gold standard in bioanalytical laboratories and probably will represent it also in the future due to fast instruments capable of hundreds SRM transition in a single run. Recently, the targeted analysis is already being replaced by HRMS instruments in the case of screening/quantitative methods. Quan–Qual bioanalytical methods are usually performed in a full-scan using high-resolution accurate mass measurement with the possibility of post-acquisition data processing for both targeted, as well as non-targeted, compounds. Due to the enhanced selectivity of both the above-stated approaches, UHPLC coupled to conventional detectors such as UV or fluorescence detection (FD) has only been used in several limited bioanalytical applications, while universal aerosol detectors have never been applied.

The sample preparation steps prior to UHPLC analysis have still regularly employed conventional techniques such as PP, LLE or SPE. PP was the most frequently used technique due to its inherent advantages including speed, ease of use, no requirements for method development and low sample consumption, which is in a good agreement with the requirements for a fast and environmentally friendly UHPLC method. On the other hand, the manual arrangement of SPE and LLE is a time-consuming and labour-intensive multi-steps procedure, which is in great contrast with fast UHPLC separation. New trends in sample preparation techniques, including various microextractions and on-line approaches, have not been widely used as a sample preparation step prior to UHPLC analysis so far. However, their application in ultra-fast and environmentally friendly UHPLC bioanalytical assays is highly expected in the near future.

References

1. Committee for Medicinal Products for Human Use, *Guideline on Validation of Bioanalytical Methods*, EMEA, London, 2010.
2. R. Ramanathan, M. Jemal, S. Ramagiri, Y-Q. Xia, W. G. Humpreys, T. Olah and W. A. Korfmacher, *J. Mass Spectrom.*, 2011, **46**, 595.
3. D. Guillarme, J. Schappler, S. Rudaz and J-L. Veuthey, *Trends Anal. Chem.*, 2010, **29**, 15.
4. International Conference on Harmonization of Technical Requirements for Registration of Pharmaceuticals for Human Use (ICH). *Q2 (R1): Text on Validation of Analytical Procedures*, US FDA Federal Register, 2005.
5. G. Liu, H. M. Snapp, Q. C. Ji and M. E. Arnold, *Anal. Chem.*, 2009, **81**, 9225.
6. N. Wu and A. M. Clausen, *J. Sep. Sci*, 2007, **30**, 1167.
7. L. Nováková and H. Vlčková, *Anal. Chim. Acta*, 2009, **656**, 8.
8. I. Baranowska, S. Magiera and J. Baranowski, *J. Chromatogr. B*, 2011, **879**, 615.
9. K. Sakai-Kato, E. Saito, K. Ishikura and T. Kawanishi, *J. Chromatogr. B*, 2010, **878**, 1466.

10. E. Gikas, F. N. Bazoti, P. Fanourgiakis, E. Perivolioti, A. Roussidis, A. Skoutelis and A. Tsarbopoulos, *J. Pharm. Biomed. Anal.*, 2010, **51**, 901.
11. E. M. Paliakov, B. S. Crow, M. J. Bishop, D. Norton, J. George and J. A. Bralley, *J. Chromatogr. B*, 2009, **877**, 89.
12. G. Irate, O. Gonzales, N. Ferreirós, M. I. Maguregui, R. M. Alonso and R. M. Jiménez, *J. Chromatogr. B*, 2009, **877**, 3045.
13. X. Li and A. A. Franke, *J. Chromatogr. B*, 2009, **877**, 853.
14. C. Moltó-Puigmartí, A. I. Castellote and M. C. Lopéz-Sabater, *J. Chromatogr. A*, 2009, **1216**, 4388.
15. M. Englmann, A. Fekete, Ch. Kuttler, M. Frommberger, X. Li, I. Gebefügi, J. Fekete and P. Schmitt-Koplin, *J. Chromatogr. A*, 2007, **1160**, 184.
16. X. Li, A. Fekete, M. Englmann, Ch. Götz, M. Rothballer, M. Frommberger, K. Buddrus, J. Fekete, Ch. Cai, P. Schröder, A. Hartmann, G. Chen and P. Schmitt-Kopplin, *J. Chromatogr. A*, 2006, **1134**, 186.
17. F. M. Lagerwerf, W. D. van Dongen, R. J. J. M. Steenvoorden, M. Honing and J. H. G. Jonkman, *Trends Anal. Chem.*, 2000, **19**, 418.
18. R. N. Xu, L. Fan, M. J. Rieser and T. A. El-Shourbagy, *J. Pharm. Biomed. Anal.*, 2007, **44**, 342.
19. D. R. Jones, G. Boysen and G. P. Miller, *J. Chromatogr. B*, 2011, **879**, 1056.
20. T. Bueters, J. Dahlström, K. Kvalvågnes, I. Betnér, S. Briem, *J. Pharm. Biomed. Anal.*, 2011, **55**, 1120.
21. M. Jemal and Z. Ouyang, *Rapid Commun. Mass. Spectrom.*, 2000, **14**, 1757.
22. D. Guillarme, J. Ruta, S. Rudaz and J-L. Veuthey, *Anal. Bioanal. Chem.*, 2010, **397**, 1069.
23. T. J. Causon, K. Broeckhoven, E. F. Hilder, R. A. Shellie, G. Desmet and S. Eelting, *J. Sep. Sci.*, 2011, **34**, 877.
24. M. I. Churchwell, N. C. Twaddle, L. R. Meeker and D. R. Doerge, *J. Chromatogr. B*, 2005, **825**, 134.
25. H. Trufelli, P. Palma, G. Famiglini and A. Cappiello, *Mass Spectrom. Rev.*, 2011, **30**, 491.
26. A. Cappielo, G. Famiglini, P. Palma, E. Pierini, V. Termopoli and H. Trufelli, *Anal. Chem.*, 2008, **80**, 9343.
27. P. J. Taylor, *Clin. Biochem.*, 2005, **38**, 328.
28. F. Gosetti, E. Mazzucco, D. Zampieri and M. C. Gennaro, *J. Chromatogr. A*, 2010, **1217**, 3929.
29. R. Dams, M. A. Huestis, W. E. Lambert and C. M. Murphy, *J. Am. Soc. Mass Spectrom.*, 2003, **14**, 1290.
30. E. Chambers, D. M. Wagrowski-Diehl, Z. Lu and J. R. Mazeo, *J. Chromatogr. B*, 2007, **852**, 22.
31. S. Souverain, S. Rudaz and J-L. Veuthey, *J. Chromatogr. A*, 2004, **1058**, 61.

32. J. C. Van De Steene and W. E. Lambert, *J. Am. Soc. Mass Spectrom.*, 2008, **19**, 713.
33. S. Wang, M. Cyronak and E. Yang, *J. Pharm. Biomed. Anal.*, 2007, **43**, 701.
34. E. Stokvis, H. Rosing and J. H. Beijnen, *Rapid Commun. Mass Spectrom.*, 2005, **19**, 401.
35. M. Jemal, A. Schuster and D. B. Whigan, *Rapid Commun. Mass Spectrom.*, 2003, **17**, 1723.
36. N. Lindegardh, A. Annerberg, N. J. White and N. P. J. Day, *J. Chromatogr. B*, 2008, **862**, 227.
37. E.I. Fukusaki, K. Harada, T. Bamba and A. Kobayashi, *J. Biosci. Bioeng.*, 2005, **99**, 75.
38. R. Bakhtiar and T. K. Majumar, *J. Pharmacol. Toxicol. Methods*, 2007, **55**, 227.
39. S. Gao, Z. Yang, T. Yin, M. You and M. Hu, *J. Pharm. Biomed. Anal.*, 2011, **55**, 288.
40. J. Heaton, N. Rahmioglu, K. R. Amadi, C. Legido-Quigley and N. W. Smith, *J. Pharm. Biomed. Anal.*, 2011, **55**, 494.
41. D. Wang, K. Jiang, S. Yang, F. Qin, X. Lu and F. Li, *J. Chromatogr. B*, 2011, **879**, 1827.
42. E. Hinchliffe, J. E. Adaway and B. G. Keevil, *J. Chromatogr. B*, 2012, **883–884**, 102.
43. D. Remane, M. R. Meyer, D. K. Wissenbach and H. H. Maurer, *Anal. Bioanal. Chem.*, 2011, **400**, 2093.
44. O. A. Ismaiel, T. Zhang, R. Jenkins and H. T. Karnes, *J. Chromatogr. B*, 2011, **879**, 2081.
45. L. Binkhorst, R. H. J. Mathijssen, I. M. G. Moghaddam-Helmantel, P. de Brujin, T. van Gelder, E. A. C. Wiemer and W. J. Loos, *J. Pharm. Biomed. Anal.*, 2011, **56**, 1016.
46. J. Hasselstrøm, *J. Chromatogr. B*, 2011, **879**, 123.
47. D. S. Ming and J. Heathcote, *J. Chromatogr. B*, 2011, **879**, 421.
48. M. Bhatt, S. Shah and Shivprakash, *J. Chromatogr. B*, 2010, **878**, 1605.
49. S. H. Kirsch, W. Herrmann, Y. Rabagany and R. Obeid, *J. Chromatogr. B*, 2010, **878**, 3338.
50. R. Verplaetse and J. Tytgat, *J. Chromatogr. B*, 2010, **878**, 1987.
51. J. Kuhn, Ch. Götting and K. Kleesiek, *Talanta*, 2010, **80**, 1894.
52. Z. Yang, W. Zu, S. Gao, H. Xu, B. Wu, K. Kulkarni, R. Singh, L. Tang and M. Hu, *J. Pharm. Biomed. Anal.*, 2010, **53**, 81.
53. M. I. Blonk, B. C. van der Nagel, L. S. Smit and R. A. A. Mathot, *J. Chromatogr. B*, 2010, **878**, 675.
54. L. Chytil, B. Štrauch, J. Cvačka, V. Marešová, J. Widimský Jr, R. Holaj and O. Slanař, *J. Chromatogr. B*, 2010, **878**, 3176.
55. E. Dahmane, T. Mercier, B. Zanolari, S. Cruchon, N. Guignard, T. Buclin, S. Leyvraz, K. Zaman, C. Csajka and L. A. Decosterd, *J. Chromatogr. B*, 2010, **878**, 3402.

56. Z. Xiong, Y. Zhang, F. Quin, S. Yang and F. Li, *J. Chromatogr. B*, 2010, **878**, 2111.
57. J. Schappler, R. Nicoli, D. Nguyen, S. Rudaz, J. L. Veuthey, D. Guillarme, *Talanta,* 2009, **78**, 377.
58. J. J. V. Bussche, L. Vanhaecke, Y. Deceuninck, K. Verheyden, K. Wille, K. Bekaert, B. Le Bizec and H. F. De Brabander, *J. Chromatogr. A*, 2010, **1217**, 4285.
59. S. Bouchet, E. Chauzit, D. Ducint, N. Castaing, M. Canal-Raffin, N. Moore, K. Titier and M. Molimard, *Clin. Chim. Acta*, 2011, **412**, 1060.
60. K. Yu, D. Little, R. Plumb and B. Smith, *Rapid Commun. Mass Spectrom.*, 2006, **20**, 544.
61. K. P. Bateman, M. Kellmann, H. Muenster, R. Papp and L. Taylor, *J. Am. Soc. Mass Spectrom.*, 2009, **20**, 1441.
62. J-L. H. Jiwan, P. Wallemacq and M. F. Hérent, *Clin. Biochem.*, 2011, **44**, 136.
63. I. V. Chernushevich, A. V. Loboda and B. A. Thomson, *J. Mass Spectrom.*, 2001, **36**, 849.
64. F. N. Bazoti, E. Gikas, A. Skoutelis and A. Tsarbopoulos, *J. Pharm. Biomed. Anal.,* 2011, **56**, 78.
65. H. Jin, A. P. Kumar, D- H. Paik, K-Ch. Ha, Y-J. Yoo and Y-I. Lee, *Microchem. J.*, 2010, **94**, 139.
66. B. S. Sachin, I. A. Najar, S. C. Sharma, M. K. Verma, M. V. Reddy, R. Anand, R. K. Khajurria, S. Koul and R. K. Johri, *J. Chromatogr. B*, 2010, **878**, 823.
67. H. G. Gika, A. Theodoridou, F. Michopoulos, G. Theodoridis, E. Diza, L. Settas, P. Nikolaidis, Ch. Smith and I. D. Wilson, *J. Pharm. Biomed. Anal.*, 2009, **49**, 579.
68. X. Wang, H. Lv, H. Sun, L. Liu, W. Sun and H. Cao, *Rapid Commun. Mass Spectrom.*, 2007, **21**, 3883.
69. D. O'Connor and R. Mortishire-Smith, *Anal. Bioanal. Chem.*, 2006, **385**, 114.
70. H. Kawanishi, T. Toyo'oka, K. Ito, M. Maeda, T. Hamada, T. Fukushima, M. Kato and S. Inagaki, *J. Chromatogr. A*, 2006, **1132**, 148.
71. D. O'Connor, R. Morsishire-Smith, D. Morisson, A. Davies and M. Dominguez, *Rapid Commun. Mass Spectrom.*, 2006, **20**, 851.
72. A. E.-H. de Mendoza, M. A. Campanero, F. Mollinedo and M. J. Blanco-Príeto, *J. Chromatogr. B*, 2009, **877**, 4035.
73. P. L. Kole, G. Venkatesh, J. Kotecha and R. Sheshala, *Biomed. Chromatogr.*, 2011, **25**,199.
74. C. Polson, P. Sarkar, B. Incledon, V. Raguvaran and R. Grant, *J. Chromatogr. B*, 2003, **785**, 263.
75. M. Yadav, P. Contractor, V. Upadhyay, A. Gupta, S. Guttikar, P. Singhal, S. Goswami and P. S. Shrivastav, *J. Chromatogr. B*, 2008, **872**, 167.
76. Y. Nagaosa and K. Sakata, *Talanta*, 1998, **46**, 647.

77. M. Yoshida, A. Akane, M. Nishikawa, T. Watabiki and H. Tsuchihashi, *Anal. Chem.*, 2004, **76**, 4672.
78. H. Wu, J. Zhang, K. Norem and T. A. El-Shourbagy, *J. Pharm. Biomed. Anal.*, 2008, **48**, 1243.
79. J. Liu, M. Jiang, G. Li, L. Xu and M. Xie, *Anal. Chim. Acta*, 2010, **679**, 74.
80. M. C. Henion, *J. Chromatogr. A*, 1999, **856**, 3.
81. T. Kumazawa, Ch. Hasegawa and X-P. Lee, *Forensic Toxicol.*, 2010, **28**, 61.
82. H. Kataoka, *Anal. Bioanal. Chem.*, 2010, **396**, 339.
83. Y. Chen, Z. Guo, X. Wang and Ch. Qiu, *J. Chromatogr. A*, 2008, **1184**, 191.
84. F. Pena-Pereira, I. Lavilla and C. Bendicho, *Spectrochim. Acta B*, 2009, **64**, 1.
85. A. A. Nuhu, Ch. Basheer and B. Saad, *J. Chromatogr. B*, 2011, **879**, 1180.
86. S. Pedersen-Bjengaard and K. E. Rasmussen, *J. Chromatogr. A*, 2008, **1184**, 132.
87. A. V. Herrera-Herrera, M. Asensio-Ramos, J. Hernández-Borges and M. A. Rodríguez-Delgado, *Trends Anal. Chem.*, 2010, **7**, 728.
88. J-H. Guo, X-H. Li, X-L. Cao, Y. Li, X-Z. Wang and X-B. Xu, *J. Chromatogr. A*, 2009, **1216**, 3038.
89. P. P. Bolaños, R. R.Gonzáles, A. G. Frenich and J. L. Martínez Vidal, *J. Chromatogr. A*, 2008, **1208**, 16.
90. R. P. Beraldi and J. Pawliszyn, *Water Pollut. Res. J. Can.*, 1989, **24**, 179.
91. C. L. Artur and J. Pawliszyn, *Anal. Chem.*, 1990, **62**, 2145.
92. H. Kataoka, K. Saito, *J. Pharm. Biomed. Anal.*, 2011, **54**, 926.
93. S. Ulrich, *J. Chromatogr. A*, 2000, **902**, 167.
94. D. Vukovic, X. Zhang, E. Cudjoe and J. Pawliszyn, *J. Chromatogr. A*, 2010, **1217**, 4041.
95. F. M. Musteata, I. de Lannoy, B. Gien and J. Pawliszyn, *J. Pharm. Biomed. Anal.*, 2008, **47**, 907.
96. R. Vatinno, D. Vukovic, C. G. Zambonin and J. Pawliszyn, *J. Chromatogr. A*, 2008, **1201**, 215.
97. M. Cruz-Vera, R. Lucena, S. Cárdenas and M. Valcárcel, *Trends Anal. Chem.*, 2009, **28**, 1164.
98. F. David and P. Sandra, *J. Chromatogr. A*, 2007, **1152**, 54.
99. M. Abdel-Rehim, *Anal. Chim. Acta*, 2011, **711**, 119
100. Z. Altun, M. Abdel-Rehim and L. G. Blomberg, *J. Chromatogr. B*, 2004, **813**, 129.
101. Y. Che, J. M. Potter and P. J. Ravenscroft, *Ther. Drug. Monit.*, 1992, **14**, 267.
102. X. H. Chen, J. P. Franke, J. Wijsbeek and R. A. Zeeuw, *J. Chromatogr.*, 1993, **619**, 137.
103. H. Vlčková, D. Solichová, M. Bláha, P. Solich and L. Nováková, *J. Pharm. Biomed. Anal.*, 2011, **55**, 301.

104. A. Namera, A. Nakamoto, M. Nishida, T. Saito, I. Kishiyama, S. Miyazaki, M. Zajata, M. Yashiki and M. Nagao, *J. Chromatogr. A*, 2008, **1208**, 71.
105. M. Kanamori-Kataoka, H. Kato, H. Uzawa, S. Otha, Y. Takei, M. Furuno and Y. Seto, *J. Mass. Spectrom.*, 2011, **46**, 821.
106. L. G. Blomberg, *Anal. Bioanal. Chem.*, 2009, **393**, 797.
107. W. M. Mullet, *J. Biochem. Biophys. Methods*, 2007, **70**, 263.
108. C. P. Desilets, M. A. Rounds and F. E. Regnier, *J. Chromatogr.*, 1991, **544**, 25.
109. S. Souverain, R. Rudaz and J. L. Veuthey, *J. Chromatogr. B*, 2004, **801**, 141.
110. K. Bentayeb, R. Batle, C. Sánchez, C. Nerín and C. Domeño, *J. Chromatogr. B*, 2008, **869**, 1.
111. N. H. Schebb, B. Inceoglu, T. Rose, K. Wagner and B. D. Hammock, *Anal. Methods* 2011, **3**, 420.
112. G-Z. Xin, J-L. Zhou, L-W. Qi, Ch-Y. Li, P. Liu and H-J. Li, *J. Chromatogr. B*, 2010, **878**, 435.
113. R. Eisert and J. Pawliszyn, *Anal. Chem.*, 1997, **69**, 3140.
114. E. Turiel and A. Martín-Esteban, *Anal. Chim. Acta*, 2010, **668**, 87.
115. Ch. He, Y. Long, J. Pan, K. Li and F. Liu, *J. Biochem. Methods*, 2007, **70**, 133.
116. P. Lucci, O. Núñez and M. T. Galceran, *J. Chromatogr. A*, 2011, **1218**, 4828.
117. D. Stevenson, *J. Chromatogr. B*, 2000, **745**, 39.
118. M-C. Henion and V. Pichon, *J. Chromatogr. A*, 2003, **1000**, 29.
119. D. R. Dufield and M. R. Radabaugh, *Methods*, 2011, doi:10.1016/j.ymeth.2011.08.012.
120. D. Tsikas, *J. Chromatogr. B*, 2010, **878**, 133.
121. A. B. Iliuk, L. Hu and W. A. Tao, *Anal. Chem.*, 2011, 83, 4440.
122. M. N. Stojanovic, P. de Prada and D. W. Landry, *J. Am. Chem. Soc.*, 2001, **123**, 4828.

CHAPTER 10

Ultra-high Pressure Liquid Chromatography coupled to Mass Spectrometry in Doping Control Analysis

F. BADOUD[a,b,c], M. SAUGY[a,b,c] AND J.-L. VEUTHEY*[a,c]

[a] School of Pharmaceutical Sciences, University of Geneva, University of Lausanne, 20 Bd d'Yvoy, 1211 Geneva, Switzerland; [b] Swiss Laboratory for Doping Analyses, University Center of Legal Medicine, Geneva and Lausanne, Centre Hospitalier Universitaire Vaudois and University of Lausanne, Ch. des Croisettes 22, 1066 Epalinges, Switzerland; [c] Swiss Centre for Applied Human Toxicology, University of Geneva, CMU, Rue Michel-Servet 1, 1211 Geneva, Switzerland
*E-mail: jean-luc.veuthey@unige.ch

10.1 Introduction

The fight against doping relies heavily on the analytical strategies used to highlight doping agent misuse. The analyses are strictly regulated to avoid any false-positives and to detect all true-positive samples. Based on the results, an athlete can be incriminated and banned from competition. Particular attention is thus paid to obtaining excellent data quality. The most appropriate strategies and techniques must be applied for the unequivocal determination of a doping case.

Various analytical approaches have been employed to detect prohibited substances in athletes' urine. Historically, gas chromatography coupled to

RSC Chromatography Monographs No. 16
UHPLC in Life Sciences
Edited by Davy Guillarme and Jean-Luc Veuthey

Published by the Royal Society of Chemistry, www.rsc.org

mass spectrometry (GC-MS) has been the method of choice for screening and confirming all classes of doping agents. Liquid chromatography coupled to mass spectrometry (LC-MS) or tandem MS (MS/MS) has been used as a very selective and sensitive complementary approach for bioanalytical samples. Recently, the development of fast liquid chromatography techniques, such as ultra-high pressure liquid chromatography (UHPLC), with the possibilities of either high-throughput or high-resolution methods, has been of great interest. Analyses for doping cases are still challenging, as the targeted substances are found at very low levels in the biological matrix, which requires even more sensitive and selective techniques that can operate within a reasonable time-delivery response. It is interesting to evaluate the breakthrough of UHPLC-MS(/MS) in the analysis of samples in doping cases.

The current state of the art of UHPLC applications in the field is presented in this Chapter, with a focus on (i) high-throughput development, (ii) enhanced chromatographic resolution for confirmatory or targeted and untargeted metabolite profiling, and (iii) the use for peptides and proteins analysis. For each part, a description of the analytical strategy involving sample preparation, separation conditions and MS hyphenation will be presented because of the contribution of each to the overall analytical process.

10.1.1 Doping Control Guidelines

Doping analyses are regulated by the World Anti-Doping Code[1] in conjunction with five International Standards, namely the Prohibited List, Testing, Laboratories, Therapeutic Use Exemptions (TUEs), and Protection of Privacy and Personal Information. The Prohibited List[2] and the International Standard for Laboratories (ISL)[3] are directly related to the analyses of forbidden substances, while the other standards give procedures for sample traceability and the use authorisation delivery, for example. The Prohibited List contains all of the substances and methods banned in sport either in-competition or in- and out-of-competition. The doping agents are classified into ten classes (S0–S9), three methods (M1–M3) and two classes of compounds prohibited only in particular sports (P1 and P2), and include more than 200 entities. The ISL procedure requires uniform and harmonised results and reporting from all of the accredited laboratories. This document recommends a two-step analytical process involving an initial testing procedure (*i.e.* screening analysis) and a confirmatory analysis.

The purpose of the screening step is to identify any biological samples that may contain a prohibited substance or indicate the use of a forbidden method. The prohibited substance can be a listed doping agent itself, a metabolite or a marker of the use of a forbidden compound.

When a sample is presumed suspect at the screening step, a confirmatory analysis must be performed to identify unequivocally or quantify the presence of a particular analyte and/or its metabolite(s), or indirect marker of doping, as presented in Figure 10.1.

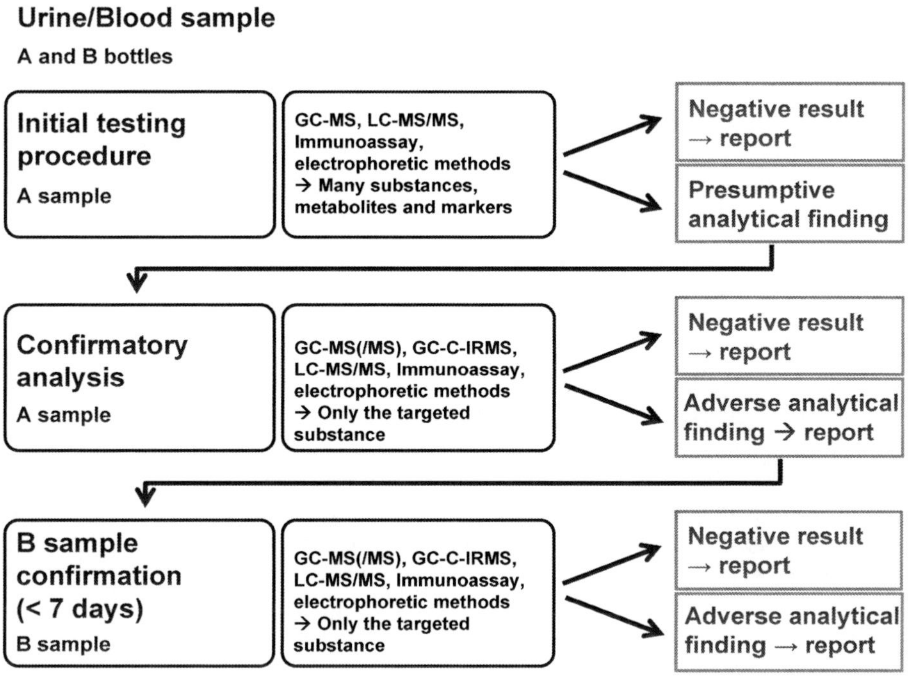

Figure 10.1 Workflow of analytical strategies for doping control analyses.

Strict standards are established to confirm the presence of a doping agent in a biological matrix, including the requirement of the highest number of identification criteria, such as matching in chromatographic retention time, relative mass spectrometric intensity of a minimum of three ions per analyte and comparison with reference material.[4,5] For consistency of measurement, minimal detection levels, expressed as concentrations, were established for all analytical methods. Therefore, all laboratories accredited by the World Anti-Doping Agency (WADA) must reach Minimum Required Performance Levels (MRPLs) for each class of doping agents to homogenise the sensitivity of inter-laboratory methods.[6] This is particularly true concerning the biological passport, as the longitudinal follow-up is not performed by the same laboratory. The WADA has established the proficiency test programme to estimate the performance of all accredited laboratories and improve the uniformity of the results.

10.1.2 Analytical Strategies

For the majority of the doping agents, the analyses are focused on the direct detection of the prohibited substances.[7] The mere presence of the compound or a specific metabolite in the biological matrix constitutes a violation of anti-doping rules. Quantitative measurement is performed for some substances (*i.e.*

threshold substances), which constitute doping agents only when they exceed a defined threshold (*e.g.* carboxy-tetrahydrocannabinol, ephedrine and its derivatives).

Nevertheless, common analytical strategies, such as GC-MS or LC-MS(/MS) cannot distinguish endogenous compounds, such as testosterone, from their synthetic analogues. The detection of anabolic androgenic steroid misuse in urine is instead determined by analysing the steroid profile, which comprises the testosterone precursor and metabolites (*e.g.* testosterone, epitestosterone, dihydrotestosterone, androsterone, etiocholanolone, 5-dehydroepiandrosterone (DHEA) and other specific metabolites). The administration of endogenous anabolic steroids, such as testosterone or DHEA, alters one or more of the parameters of the urinary steroid profile. Elevated levels or increased ratios of specific pairs of steroids or urinary metabolites that are part of the steroid profile may indicate an intake of steroids and would require analytical confirmation. The differentiation between endogenous steroids and their synthetic analogues is performed by GC coupled to isotope-ratio mass spectrometry (GC-C-IRMS), which measures the $^{13}C/^{12}C$ ratio. This parameter allows the differentiation between the exogenous origin of steroid hormone as compared with the endogenous compounds due to less ^{13}C for synthetic compounds.

Proteins are screened for their direct recombinant form in urine or blood by immunoassays or electrophoretic methods. However, as it is difficult to differentiate between some endogenous and exogenous substances, such as natural and recombinant human growth hormone (hGH)[7] or erythropoietin (EPO),[8] indirect methods are often performed. Indeed, the athlete's biological passport is aimed at the detection of blood doping (including EPO administration) by monitoring individual variations of specific, sensitive and robust biomarkers. Parameters such as haematocrit, reticulocytes or haemoglobin are indirect markers of blood doping because their physiological values are altered under EPO exposure. In the athlete's biological passport, a statistical approach based on large-scale population empirical tests is used in determining normal and suspect biomarker variability.[9] This multiparametric approach involves twelve different indirect markers [abnormal blood profile score (ABPS)] modified by recombinant EPO misuse or from blood transfusion. The results of indirect approaches are given as a probability or confidence interval.[7]

In addition, untargeted approaches, such as metabolomics or proteomics, are also developed for anti-doping purposes because they explore the global metabolite or protein information after exposure to a doping agent. Untargeted methods have been employed more widely in animals for revealing doping, in particular for anabolic practices in equine sports[10] or for agricultural meat production.[11,12] The discrimination between non-exposed and exposed animals was made by studying their urinary signatures. Such approaches were applied to human urine for the detection of anabolic agent misuse (steroidomics),[13] hGH misuse detection[7] or gene doping.[14]

10.2 GC-MS Analyses

GC-MS was historically applied to all of the prohibited substances and has remained the gold standard technique. Indeed, even if time-consuming derivatisation and hydrolysis steps are required for most of the doping agents excreted in urine, GC-MS provides excellent chromatographic resolution and the possibility to build robust databases due to the use of electron impact (EI) sources. GC-MS, working in full-scan mode, is an efficient approach for comprehensive screening, allowing high separation power and universality.[15] However, this method possesses drawbacks, including the incompatibility of some of the prohibited substances with this analytical approach due to thermal instability of the derivate (*e.g.* mesocarb metabolites)[16] and time-consuming sample preparation, including complex derivatisation for higher volatility and sensitivity. Additional care must be taken during the derivatisation step to ensure complete hydrolysis and exclude possible unspecific enzyme reactions. This phenomenon was observed for the deconjugation reaction with *Helix pomatia* glucuronidase, but not with *Escherichia coli* β-glucuronidase mixtures for the anabolic steroids.[17] In addition, the need to hydrolyse and derivate urine or blood analytes into volatile species is not compatible with screening of unknown substances because the rate of derivatisation of the new entity cannot be estimated accurately for each untargeted substance. The hydrolysis of urine metabolite conjugates leads to a loss of information from phase II metabolism. Nevertheless, GC-MS is still the technique of choice for the analysis of anabolic steroids and, particularly when coupled to MS/MS, for the improvement in selectivity.[18,19]

10.3 LC-MS(/MS) Analyses

In the last decade, LC-MS(/MS) has become an essential tool for doping control analyses in terms of the selectivity, sensitivity and rapidity of an analytical run.[20] It was quickly considered to be a complementary approach to GC-MS, immunoassays or electrophoretic techniques for routine analyses. In addition, the evolution of LC-MS(/MS) techniques combined to even more sensitive MS instruments fulfils doping control analysis requirements.

LC-MS(/MS) is an ideal complementary method, as it offers excellent selectivity for analytes and is directly compatible with urine or blood matrices, without the need for analyte derivatisation and/or hydrolysis.

This technique has found increased application and has become an interesting tool for doping control analyses because of the development of atmospheric pressure ionisation (API) sources, including electrospray (ESI), atmospheric pressure chemical ionisation (APCI) and atmospheric-pressure photo-ionisation (APPI), allowing easy hyphenation to MS. API sources are soft ionisation techniques compared with the strong ionisation obtained through EI. With API sources, fragmentation is low, and MS/MS is required to obtain enough fragments for selectivity and also to improve sensitivity.

LC-MS/MS methods were initially developed for doping agents such as diuretics and corticosteroids, first and foremost. Subsequently, applications were generalised to all of the classes of substances on the Prohibited List.[21]

Various column chemistries are commercialised to analyse a wide range of analytes with different physico-chemical properties on a single system. LC can be performed in reversed-phase mode (RPLC), normal-phase mode (NPLC) or, recently, hydrophilic interaction mode (HILIC). RPLC is the most commonly used separation mode for biological sample analysis and uses silica-based particles with various bonding such as C_{18}, C_8, C_4, phenyl or pentafluorophenyl (PFP) functions. NPLC, which involves a polar stationary phase with a non-polar mobile phase, is used most often with polar compounds, but presents several disadvantages for routine use. The mobile phase consists of an organic solvent (*e.g.* hexane or heptane mixed with a slightly less apolar solvent such as isopropanol or ethyl acetate). The technique is, however, not applicable for routine drug testing, because the analytes must be readily soluble in organic solvents, which is not always possible for all pharmaceuticals, such as phase I and II metabolites or acids and amines, and also as it uses large quantities of apolar solvents as the mobile phase.

Recently, the HILIC technique was developed for the analysis of polar compounds. It uses polar stationary phase chemistries in combination with a hydro-organic mobile phase mixture, comparable with those used in RPLC. This technique is very efficient for the analysis of urine metabolites because matrix effects can be reduced and sensitivity improved. Furthermore, the large amount of acetonitrile in the mobile phase facilitates the ionisation process.[22] However, this strategy is still in its infancy in the field of anti-doping analysis because of the requirement for longer equilibration times and the complexity of the retention mechanism. Up to now, only a few applications have been developed for stimulants and nicotine.[23–25]

While LC-MS/MS is a sensitive and selective technique for a wide range of analytes, it suffers from several limitations, including reduced chromatographic efficiency in comparison with GC-MS, susceptibility to matrix effects and inter-instrument variability in the fragmentation pattern. Despite these disadvantages, LC-MS/MS is considered an indispensable tool for the detection of low analyte amounts in complex biological matrices. The sensitivity and selectivity of this technique is useful for many bioanalytical applications. Nevertheless, improvements in the peak capacity (the number of peaks that can be separated during the gradient time) and the sensitivity, as well as a reduction in the analysis time, still constitute challenges in the detection of doping agents.

10.3.1 UHPLC-MS(/MS) Analyses

Various approaches have been proposed during the two last decades to reduce analysis time and improve chromatographic resolution.[26–28] The strategies involve the development of monolithic columns, the use of high-temperature

LC (HTLC), the recent development of shell particles and the use of columns packed with porous sub-2-μm-sized particles (UHPLC).[29]

Monolithic supports allow excellent permeability and efficient separation because of their bimodal mesoporous and macroporous structure. Therefore, the generated backpressure is very low and high flow rates can be applied, leading to very fast separation. This technique has, nevertheless, not been applied in the anti-doping field because of the limited number of providers, as well as the available columns chemistries and dimensions.

HTLC is a very attractive approach for fast LC because the high mobile phase temperature (> 90 °C) reduces the viscosity of the solvents and decreases water polarity (dielectric constant). However, this strategy is not easy to implement in the doping area for routine sample analysis because of the instability of columns and analytes at high temperatures.

Shell-particle technology and UHPLC are more appropriate to the field. Both strategies present valuable performance in terms of efficiency and throughput. Firstly, the development of sub-3 μm shell particles (also known as core-shell, fused-core or superficially porous particles) has led to the generation of new columns with reduced backpressure that are compatible with conventional HPLC systems. This development is very recent, and it has therefore not expanded into doping analysis.[30] Fused-core technology seems very promising to enhance chromatographic performance and reduce analysis time without the need for expensive high-pressure pumps, as shown for the separation of a mixture of steroids in Figure 10.2.

Secondly, UHPLC can lead to shorter analysis time while keeping the performance constant, as depicted in Figure 10.3. UHPLC also achieves enhanced chromatographic resolution while maintaining conventional analysis times. This technique is of particular interest for anti-doping analyses, but requires special instrumentation, with devices and columns that can withstand high pressure (>1000 bar) and particularly low extra-column and dwell

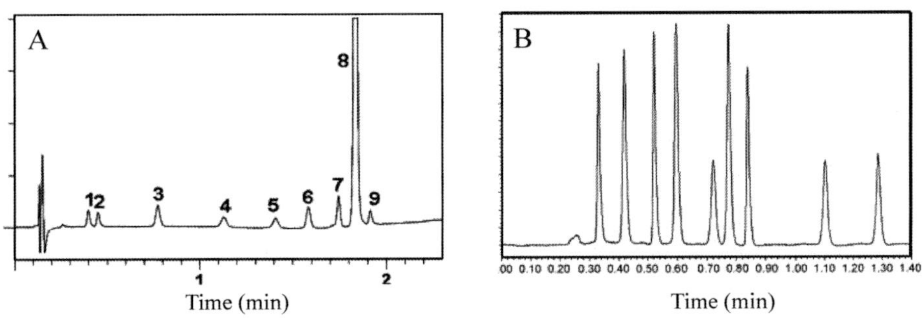

Figure 10.2 Separation of steroids on a fused-core column (Kinetex C18 2.6 μm) of (A) 50 mm × 2.1 mm at 0.95 mL min^{-1} and 40 °C providing a backpressure of 335 bar, and (B) 150 mm × 3.0 mm at 1.8 mL min^{-1} and 60 °C providing a backpressure of 516 bar. Reprinted from reference 31 with permission.

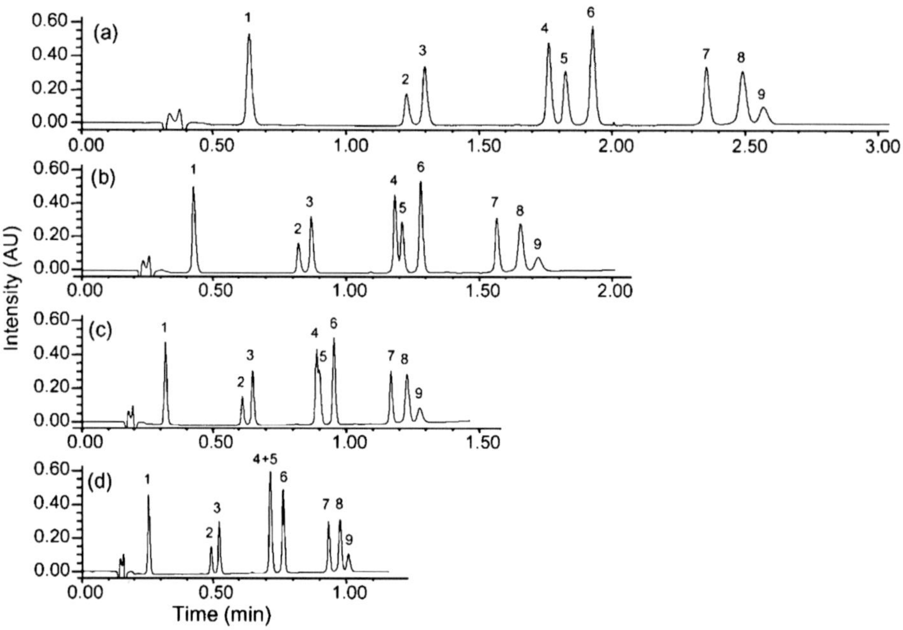

Figure 10.3 Separations of doping agents at 30 °C and flow rates of (a) 400, (b) 600, (c) 800 and (d) 1000 μL min^{-1} on a BEH Shield RPC18 (50 mm × 2.1 mm, 1.7 μm). Reprinted from reference 32 with permission.

volumes. A large panel of stationary phase chemistries has been developed, and the robustness of these columns has been proven. Therefore, UHPLC is today the most appropriate technique for routine drug testing in the anti-doping field. Such an approach can drastically improve the performance of high-throughput screenings or enhance the analytical resolution for more complex separations within a conventional analysis time.

10.4 Application of UHPLC-MS(/MS) for Drug Testing in Sports

UHPLC-MS(/MS) emerged in the anti-doping field in 2006 with the development of fast initial testing procedures to increase sample throughput. Several multi-residue screening methods were developed by using columns packed with porous sub-2 μm particles. The analyte detection and the separation from closely related compounds or endogenous compounds from the matrix remain challenging for several classes of doping agents. Because the developed method must fulfil the WADA requirements, the entire analytical process, including sample preparation, chromatographic analysis, MS detection and data treatment, must be optimised to ensure accurate and robust results.

10.4.1 UHPLC-MS(/MS) in Initial Testing Procedure

In the anti-doping field, there is a particular need for fast chromatographic approaches while obtaining excellent separation efficiency, because results are generally required within 24 to 48 h after major sporting events and also for cost-effectiveness.

The screening test aims to detect the highest number of prohibited substances in a single run. Urine is the commonly used biological matrix because it can be collected in a non-invasive manner, with generally longer detection windows than blood because compounds are subjected to phase I and II metabolism before urine excretion. Previously, sample preparation was generally dedicated to a class of analytes that presented similar physico-chemical properties.[33] Currently, the tendency is to develop the most generic procedure at the screening step, including the sample clean-up, the separation method and the MS detection.

UHPLC-MS(/MS) has thus greatly evolved in this area. Hence, several anti-doping laboratories have developed high-throughput screening methods by combining UHPLC with MS or MS/MS, as reported in Table 10.1.

10.4.1.1 Sample Preparation

Sample preparation is often the bottleneck for high-throughput screening. Analytes and metabolites found in the targeted matrix have various physico-chemical properties, and their simultaneous extraction may be difficult. Several approaches for screening have been developed, but only a few of them led to generic and fast sample preparation. Because sufficient sensitivity is required to confidently identify substances, it is often preferred to extract and pre-concentrate the analytes found at low concentration in the matrix, even at the initial testing step. For example, a rapid alkaline liquid–liquid extraction (LLE) at pH 9.5 was proposed for stimulants, synthetic anabolic steroids or diuretics after hydrolysis (1 h at 50 °C).[34–36] Even if LLE allows for clean extracts, a large amount of organic solvents is used, and this technique is less compatible with multi-analyte detection. Other authors have preferred to use solid phase extraction (SPE) as a sample clean-up method, even if extracts are often dirty.[37] For that purpose, SPE with hydrophilic/lipophilic properties were selected (*e.g.* Oasis by Waters or Absolut Nexus by Varian). Polymeric sorbents are used for multi-component analyses as various stationary phases are commercialised with the possibilities of mixed-mode cation or anion exchange. Those materials provide the possibility of simultaneously extracting acidic, neutral and basic compounds. Moreover, they offer greater pH stability and enhanced retention than C_{18}-bonded silica sorbents.[38] Selective sample preparations have thus been successfully developed using these polymers, allowing excellent sensitivity. Extraction protocols for this material with[39,40] or without[41] hydrolysis of conjugates have been proposed. Despite generic procedures, SPE requires time, particularly when a large number of samples have to be screened simultaneously and when preceded by a hydrolysis step.

Table 10.1 Description of UHPLC-MS(/MS) methods for screenings analyses.

Compounds	*Matrix*	*Sample preparation*	*Hydrolysis*	*LC system*	*Columns*	*Temperature*
Stimulants, anti-oestrogens synthetic, anabolic steroids	urine	LLE alkaline (pH 9.5)		Agilent 1100 (Agilent Technologies, Waldbronn, Germany)	Zorbax C18 (50 x 2.1 mm, 1.8 μm) (Agilent Technologies, Waldbronn, Germany)	
Corticosteroids, synthetic anabolic steroids, β2-agonists	urine	LLE alkaline (pH 9.5)	β-glucuronidase	Acquity UPLC (Waters, Milford, USA)	BEH C18 (50 x 2.1 mm, 1.7 μm) (Waters, Milford, USA)	35°C
Diuretics, β-blockers, some stimulants, steroids	urine	Hydrophilic Lipophilic polymeric SPE		Acquity UPLC (Waters, Milford, USA)	Shield RP 18 (100 x 2.1 mm, 1.7 μm) (Waters, Milford, USA)	45°C
Diuretics, masking agents, central nervous system stimulants and opiates	urine	dilute and shoot (1:1)		Acquity UPLC (Waters, Milford, USA)	Shield RP 18 (50 x 2.1 mm, 1.7 μm) (Waters, Milford, USA)	60°C
β2-agonists, glucocorticosteroids, thiazide, narcotics and stimulants	urine	Mixed-mode cation exchange polymeric SPE	β-glucuronidase/ arylsulfatase	Acquity UPLC (Waters, Milford, USA)	BEH C18 (100 x 2.1 mm, 1.7 μm) (Waters, Milford, USA)	
Anabolic agents, agents with anti-oestrogenic activity, β2-agonists, β-blockers, cannabinoids, diuretics, glucocorticosteroids, enhancement of oxygen transfer, narcotics, stimulants	urine	Mixed-mode cation and anion exchange SPE	β-glucuronidase	Agilent 1200 series rapid resolution (Agilent Technologies, Waldbronn, Germany)	Zorbax Eclipse Plus rapid resolution HT C18 (50 x 2.1 mm, 1.8 μm) (Agilent Technologies, Waldbronn, Germany)	40°C
agents with anti-oestrogenic activity, β2-agonists, β-blockers, diuretics, enhancement of oxygen transfer, narcotics, stimulants, α-reductase inhibitor	urine	dilute and shoot (1:1)		Acquity UPLC (Waters, Milford, USA)	BEH C18 (50 x 2.1 mm, 1.7 μm) (Waters, Milford, USA)	30°C
Diuretics, anti-estrogen, β-blockers, corticosteroids, anabolic agents, β2-agonists, stimulants, releasing peptide, gene doping agent, narcotics, SARMs*	plasma	Protein precipitation, centrifugation		Accela UPLC (Thermo Finnigan, San Jose, USA)	Hypersil Gold C18 column (50 x 2.1 mm, 1.7 μm) (Thermo Finnigan, San Jose, USA)	
Diuretis, probenecid and stimulants	urine	LLE alkaline (pH 9.5)		Acquity UPLC (Waters, Milford, USA)	BEH C18 (100 x 2.1 mm, 1.7 μm) (Waters, Milford, USA)	

*Selective androgen receptor modulator

Table 10.1 (*Continued*)

Mobile phase	*Gradient*	*Flow rate*	*Injection volume*	*Ionization*	*Mass Spectrometry*	*LOD*	*References*
A: 0.1% acetic acid B: ACN with 0.1% acetic acid	15% to 60% B in 5 min and 100% B in 7 min. Total 7 min	300 μL/min	5 μL	ESI +	API 4000 in MRM mode (AB SCIEX, Foster City, USA)	100–200 ng/mL (stimulants), 1–30 ng/mL (glucocorticoids and steroids)	34
A: 0.1% formic acid B: ACN with 0.1% formic acid	5% to 30% B in 2 min, constant for 1.3 min, increased to 90% in 2.9 min, finally set at 100% in 0.7 min. Total 7.5 min	350 μL/min	20 μL	ESI +	LCT Premier (Waters, Milford, USA)	30 ng/mL	35
A: 0.1% formic acid B: MeOH with 0.1 % formic acid	5% to 98% B in 3 min, hold 1 min, re-equilibrate for 2 min at 5% B. Total 6 min	500 μL/min		ESI + and -	API 3200 in MRM mode (AB SCIEX, Foster City, USA) and Micromass Quattro Premier (Waters, Milford, USA)	7 ng/mL (carboxy-finasteride) – 158 ng/mL (canrenoic acid) (MRPL)	41
A: 10mM ammonium acetate B: MeOH	5% B to 40% in 4 min, increased to 95% at 5 min and held 1 min. Return to initial conditions for 2.5 min. Total 7.5 min	400 μL/min	5 μL	ESI with fast polarity switching	MRM Micromass Quattro Premier (Waters, Milford, USA)	1 ng/mL (benzoylecgonine) – 50 ng/mL (methylclothiazide)	42
A: 0.01% formic acid B: ACN with 0.01 % formic acid	0% B to 40% in 4 min and from 4 to 10 min to 100% with a final hold for 2 min. Total 13 min	400 μL/min	20 μL	ESI + ESI- (thiazide)	MicroTOF (Bruker Daltonics, Bremen, Germany)	0.5 ng/mL (clenbuterol) – 50 ng/mL (carbuterol)	39
A: 0.1% formic acid and 2.5 mM ammonium formate B: 90% ACN with 0.1 % formic acid and 2.5 mM ammonium formate	10% B for 1 min and increased to 40% in 2 min, to 70% in 1 min, to 90% in 2 min and held at 90 % for 0.5 min and then back to 10% in 0.5 min. Re-equilibrate for 2 min. Total 8 min	400 μL/min	3 μL	ESI + and ESI- in 2 runs	Bruker Daltonics MicroTOF (Bruker Daltonics, Bremen, Germany)	MRPL	40
A: 0.1% formic acid B: ACN with 0.1 % formic acid	5% to 95% B in 6 min with 3 min equilibration time. Total 9 min	400 μL/min	10 μL	ESI + and ESI- in 2 runs	QTOF Premier (Waters, Milford, USA)	10 ng/mL (xipamide) -250 ng/mL (adrafinil) (MRPL)	43
A: 0.2% formic acid B: ACN	1% B for for 1 min, raised to 100% B in 11 min, held for 0.5 min and re-equilibrated for 2.5 min. Total 15 min	200 μL/min	5 μL	ESI + ESI-	Orbitrap Exactive (Thermo Finnigan, San Jose, USA)	10 ng/mL	33
A: 0.01% formic acid B: ACN with 0.01% formic acid	5% B for for 0.6 min, raised to 90% B in 3.8 min, held for 0.2 min and re-equilibrated for 1 min. Total 5 min	600 μL/min	5 μL	ESI + ESI-	MRM Micromass Quattro Premier (Waters, Milford, USA)	50 – 200 ng/mL	36

Therefore, to make sample preparation compatible with fast UHPLC-MS(/MS), a rapid and easy-to-use sample preparation is often chosen, with no selective discrimination of analytes. The dilute-and-shoot technique developed by several groups is more suitable when sensitivity and matrix effects are not a major issue.[42,43]

Matrix effects remain the most important drawback of LC-ESI-MS applications and refer to the alteration of ionisation (ion enhancement or suppression) or extraction recoveries (lower recovery) due to the presence of interfering compounds from the matrix. Two strategies are commonly assessed for the qualitative or quantitative evaluation of matrix effects. The qualitative approach consists of a post-infusion method described by Bonfiglio *et al.*,[44] while the quantitative post-extraction method was proposed by Matuszewski *et al.*[45] Evaluation of this phenomenon is required by the Food and Drug Administration (FDA) guidelines for bioanalytical method validation[46] and is a mandatory step in the validation process. In this context, an exhaustive classification system of possible alteration of extraction or ionisation was proposed.[47] The effect on global process efficiency can thus be assessed and quantified.

Following dilute-and-shoot, the matrix effect can only be attributed to the ionisation step, and can be important in determining the appropriate dilution factor of a biological sample. Thörngren *et al.*[42] studied the modification of signals by post-infusion of four substances and determined that the signal alteration ranged from 50% to 130%. Other authors quantified matrix effects for all of the targeted analytes (103) and determined that 74% of analytes were altered during ionisation, either by signal suppression or enhancement.[43] However, for most of the substances on the Prohibited List, the MRPL was easily reached, and excellent limits of detection (LODs) were obtained by 2-fold dilution of urine samples. Because of this simple sample treatment, a batch of 50 samples can be prepared in less than 1 h. This strategy can be easily automated, which can further reduce the time for sample preparation.

A similar strategy was developed to screen multiple classes of doping agents in plasma samples, including diuretics, anti-oestrogens, β-blockers, corticosteroids, anabolic agents or stimulants. A simple protein precipitation (PP) method was employed, which consisted of the dilution of a plasma sample with an excess of acetonitrile to precipitate proteins. This approach only required a small sample volume, which is in agreement with the disposable volume for blood analysis, and allows potential re-analysis for confounding new substances. This strategy provides LOD values as low as 10 ng mL^{-1} for the targeted analytes with excellent selectivity towards the plasma matrix.[33] Possible signal suppression from interfering compounds in the matrix was studied by the post-column infusion of two model compounds, one in the positive ESI mode (stanozolol) and the other in the negative ESI mode (furosemide). A prohibited time window was considered at the beginning of the chromatogram, at which strong ion suppression was observed. As no analyte was detected in this initial volume, no signal alteration was considered for the targeted compounds.

10.4.1.2 Separation Conditions

For high-throughput screening, fast analysis combined with sufficient resolution is required to minimise analyte or matrix component co-elution. The chromatographic runs are usually performed within a reasonable analysis time and in a non-specific gradient mode, allowing the elution of analytes of various physico-chemical properties. The analyses are generally performed using acidic elution conditions, such as 0.1% formic acid or 0.1% acetic acid in 6 to 13 min on 50- to 100-mm C18 or Shield C18 columns (embedded polar group, carbamate). Gradients of less than 7 min are generally performed on columns with a length of 50 mm, an inner diameter (I.D.) of 2.1 mm and a particle size of 1.7 μm,[34–36] while longer columns are used for extended gradient times.[48] However, for the development of a generic plasma screening, a gradient of 15 min at a flow rate of 200 μL min^{-1} through a C18 column (50 × 2.1 mm, 1.7 μm), including the re-equilibration time, was proposed. In this particular set of conditions, it may be feasible to shorten the gradient time and increase the flow rate to obtain higher peak capacity, as the latter is inversely proportional to column dead time.[49]

Generally, a gain in time by a factor of 2 to 5 can be achieved by using UHPLC as compared with conventional HPLC or GC methods. As an example, Murray *et al.*[41] compared HPLC to UHPLC methods for the analysis of some diuretics, β-blockers, stimulants and steroids. The UHPLC-MS method was found to be two times faster than conventional HPLC and could run 240 samples in 24 h.

In addition, the chromatographic resolution can be maintained at a constant or even improved due to sub-2 μm particles. As presented in Figure 10.4, even with fast gradient conditions (7.5 min), the critical pair of corticosteroid isomers, dexamethasone and betamethasone, could be sufficiently resolved through a column with a length of 50 mm.[35]

10.4.1.3 MS Hyphenation

MS-based strategies are essential because MS allows the confirmation of analytes based on spectral information. Therefore, it is necessary to hyphenate UHPLC to MS for a throughput increase and/or resolution improvement. However, due to the sharpness of the chromatographic peaks (commonly 2–4 s for bioanalytical methods), the instrument scan-rate must be high enough to obtain a minimum of 10 to 15 points per peak. As shown in Table 10.2, the mass analysers presenting the highest acquisition rates are quadrupole, triple quadrupole and time-of-flight (TOF) instruments.

Screenings for doping control analysis are mainly developed with triple quadrupole instruments, which are sensitive, selective and cost-effective mass

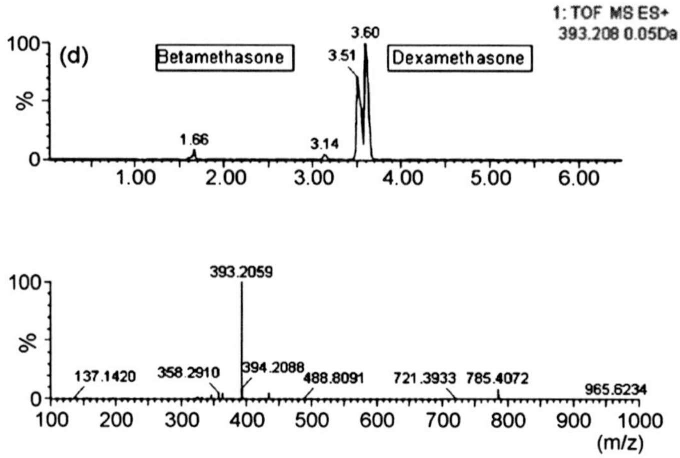

Figure 10.4 UHPLC-TOF-MS extracted ion chromatogram (mass window of $\pm$0.05 Da) of a 30 μg L^{-1} fortified female urine sample for betamethasone and dexamethasone (*m/z* 393.2059) obtained in ESI positive mode. Adapted from reference 35 with permission.

analysers. For multi-residue screening, triple quadrupole MS instruments are generally operated in the multiple reaction mode (MRM). Triple quadrupole instruments have fast scanning capabilities but are limited for high-throughput screening by the number of acquisitions per segment time. As presented in Table 10.2, triple quadrupole instruments can generally assume 150 MRM per s with dwell times as low as 5 or even 1 ms, while new generation instruments can theoretically handle 500 MRM per s.

The coverage of a huge panel of analytes can be achieved by using fast polarity switching, which allows ESI+ and ESI− acquisition in a single analytical run.[41,42] Polarity switching could be performed in 700 ms on older quadrupole ion trap (QTRAP) instruments,[34] while polarity switching of 15 to 20 ms can be achieved with modern triple quadrupole instruments.

Current instruments are also promising in terms of sensitivity, maintenance and stability, but require more attention to solvent purity and rigor against instrument contamination.[50]

For screening purposes, various MS/MS scan modes have been successfully applied to the detection of doping agents. As an example, the precursor ion scan mode was applied for the screening of known and unknown corticosteroids. This scan mode allows screening for fragments belonging to structural relatives of glucocorticosteroids.[51] A similar approach is performed by neutral loss scanning, which involves following a loss of similar neutral fragments for a class of analytes (*e.g.* fluoxymesterone metabolites with a neutral loss of HF).[52]

In addition, half of the fast multi-residue testing methods were proposed using TOF-MS or QTOF-MS. The obvious advantages of coupling UHPLC with TOF or QTOF for this kind of application have been recognised, despite its high cost and less accurate quantitation abilities.[53] TOF-MS or QTOF-MS have fast acquisition capabilities. With recent developments, an acquisition of 100 Hz (10 ms per scan) was possible while maintaining a resolution of 40 000 full-width-at-half-maximum (FWHM). It allows sensitive full mass spectra acquisition with high selectivity because of its high mass accuracy and resolution. Moreover, acquired data can be processed retroactively to identify unknown compounds without sample reinjection that were not *a priori* suspected. Several initial testing methods were developed using UHPLC coupled to TOF-MS.[35,39,40] Even so, the information provided by the TOF-MS spectra is not sufficient to obtain enough diagnostic ions as only in-source fragmentation can occur. Fragmentation of ions directly in the source can lead to irreproducibility between spectral patterns. Therefore, the QTOF-MS instrument provides an additional benefit as compared with TOF-MS because it is possible to simultaneously obtain full mass range spectra in a function and MS/MS experiments in a second function. Such a method was developed for the screening of hundreds of prohibited compounds, including a pre-confirmatory step based on the simultaneous acquisition of MS/MS spectra with a minimum of three diagnostic ions per analyte.[42] Moreover, UHPLC-QTOF-MS combines the possibility of using a targeted approach for known compounds, whose standards are available, and accurate mass profiling for unknown or unidentified metabolites.[54] By combining the dilute-and-shoot sample treatment and a generic gradient with the information provided by QTOF data, retrospective analyses were achievable because no analyte was excluded.

UHPLC hyphenated to Orbitrap technology was also employed for fast comprehensive screening of plasma samples. This mass spectrometer possesses longer scan times ($\leq$2 s per scan) to obtain a resolution of 50 000–100 000 FWHM for a mass range between *m*/*z* 100 and 1000, allowing a mass accuracy of 1 ppm.[33] The width of a UHPLC peak is conventionally 2–4 s, and a scan speed of at least five spectra is required to obtain a minimum of 10 points per peak. Therefore, the rapidity and high sensitivity (due to thin peaks) of UHPLC is partially lost when coupled to this type of devices.[39] Because the mass resolution is correlated with the acquisition rate, lower mass resolution is expected with faster cycle times. As an example, a resolution of 7 500 FWHM was observed at a scan time of 0.1 s, providing 20 data points for a peak width of 2 s. Interfacing UHPLC with Orbitrap is therefore currently used by working with longer acquisition times to obtain high-resolution mass spectra (30 000 FWHM) for identification purposes.[55]

Nevertheless, Orbitrap can work in two simultaneous modes, namely MS and higher-energy collisional dissociation (HCD) experiments. This acquisition mode is very generic, as no selection of a product ion is needed. This allows collection of rich information for identifying unknown compounds using exact mass measurements of (de)protonated molecular ions and fragments. This strategy was applied for the plasma screening of 32 targeted analytes of various anti-doping classes, but can also give information on unknown peaks because of accurate mass and fragment measurements and the

Table 10.2 Performance comparison of mass spectrometers.

Mass Spectrometer	*Provider / Model*	*Mass Range (Da)*	*Resolving power (FWHM)*	*Mass Accuracy (ppm)*
Single quadrupole	Agilent[b] 6100	3'000	-	-
	Thermo[e] MSQ Plus	3'000	-	-
	Waters[c] SQD	2-2'048	-	500-2000
	Perkin Elmer[i] Flexor SQ 300	20-3'000	-	-
Triple quadrupole	Agilent[b] 6400	5-3'000	(0.7 Da for m/z 500) ~700	200
	Thermo[e] Quantum Access Max	10-3'000	7'500	-
	Thermo[e] Quatum Vantage	30-1'500	7'500	-
	Waters[c] Xevo TQS	2-2'048	-	2'000
	AB Sciex[f] API 4000	3-3'000	-	-
	Shimadzu[h] 8030	10-2'000	~1'000	-
Time-of-flight	Agilent[b] 6200	20-20'000	> 20'000	< 1
	Waters[c] LCT Premier	20-30'000	> 12'000	3
	Bruker[d] Daltonics	-	> 16'500	< 2
	Leco[g] Citius LC-HRT	-	100'000	
Quadrupole time-of-flight	Agilent[b] 6500	20-20'000	> 40'000	< 1 (MS); < 2(MS/MS)
	Waters[c] Xevo G2	20-100'000	20'000	1
	Waters[c] Synapt G2	20-100'000	> 40'000	< 1
	AB Sciex[f] Triple TOF		> 40'000	< 2
	Bruker[d] Daltonics micro TOFQII		> 17'500	1-2
	Bruker[d] MaXis	50-20'000	60'000	0.6-0.8
Ion Trap and Quadrupole ion trap	AB Sciex[f] Qtrap 5500	5-2800	2'000	
	Thermo[e] LTQ	50-2'000	-	-
	Bruker[d] Daltonics amaZon	50-30'000	20'000	< 200
Orbitrap	Thermo[e] Exactive	50-4'000	> 100'000	< 2 or < 5 (internal or external calibration)
	Thermo[e] LTQ Orbitrap	50-6'000	> 100'000	1-5
Fourier Transform	Thermo[e] LTQ FT Ultra	50-4'000	100'000 (m/z 400)	< 1.2 ppm (external calibration)
			> 750'000 (m/z 400)	< 1 ppm (internal calibration)

[a]The information presented was gathered from advertising and from providers' Websites in June 2011; [b]Agilent Technologies, Waldbronn, Germany; [c]Waters, Milford, USA; [d]Bruker Daltonics, Bremen, Germany; [e]Thermo Finnigan, San Jose, USA; [f]AB SCIEX, Foster City, USA; [g]LECO Corporation, St-Joseph, USA; [h]Shimadzu Corporation, Kyoto, Japan; [i]Perkin Elmer, West Chester, USA.

Table 10.2 (*Continued*)

Acquisition Speed (amu/s) or (spectra/s)	*Dwell Time (ms)*	*Dynamic Range*	*Polarity Switching (ms)*
10'000	-	-	20
12'000	-	-	? (Fast)
10'000	5	10^6	20
10'000	-	10^5	-
5'200 (150 MRM/s)	1	10^6	30
5'000	2	10^6	< 25
5'000	2	10^6	< 25
> 10'000	5	10^6	20
2'400	20	$4 \cdot 10^6$	700
15'000 (500 MRM/s)	1	10^6	15
40 spectra/s	-	10^4	-
20 spectra/s	-	10^4	300
-	-	-	-
100 spectra/s	-	-	-
20 or 10 spectra/s (MS or MS/MS)	-	-	-
30 spectra/s	-	10^4	-
20 or 10 spectra/s (without or with pDRE)	-	10^4	-
50-100spectra/s	-	10^4	100
	-	-	-
10-30 spectra/s	-	10^5	-
20'000 Da/s or 4'000 amu/s (LIT) or 2'400 (QqQ)	-	10^6	-
-	-	-	$< 1'000$
52'000 amu/s	-	-	-
1 spectra/s (10 spectra/s for 10'000 FWHM)	-	10^3	-
1 spectra/s	-	10^3	-
1 scan/s	-	-	-
> 1scan/s	-	-	-

possibility of determining the elemental composition. Moreover, storing such data is promising for retrospective analysis without sample reinjection.

10.4.2 UHPLC-MS(/MS) for Confirmatory, Metabolite Profiling and Quantitative Approaches

UHPLC applications have also been developed to take advantage of high peak capacities. Indeed, it is commonly assessed that a 3-fold gain in resolution could be reached when transferring a conventional HPLC method to UHPLC by keeping the column length constant.[29] Higher plate counts can be attributed to columns packed with sub-2 μm particles.[56] This feature can be useful for obtaining higher selectivity towards matrix constituents or for resolving positional isomers. This is particularly valuable for the analysis of complex matrices, where a large diversity of phase I and II metabolites are present.

Moreover, a gain in sensitivity can be observed because the peak shapes are thinner and, due to higher peak capacity, less signal suppression is expected. The sample preparation is generally dedicated to a single or a class of analytes to be confirmed, determined or quantified, and identification criteria must be reached by MS. UHPLC-MS(/MS) was thus applied in the anti-doping area, as reported in Table 10.3, for confirmatory analysis, metabolite studies and quantitative measurement for threshold substances, where high chromatographic resolution and sensitivity are mostly required.

10.4.2.1 Sample Preparation

For confirmatory analysis of a presumptive analytical finding, the sample preparation is aimed at being very selective towards a single doping agent, a

Table 10.3 Description of UHPLC-MS(/MS) methods for confirmatory, metabolites and quantitative analyses.

Compounds	*Matrix*	*Sample preparation*	*Hydrolysis*	*LC system*	*Columns*	*Temperature*
Salbutamol	urine	dilute and shoot (1:1)		Acquity UPLC (Waters, Milford, USA)	BEH C18 (100 × 2.1 mm, 1.7 μm) (Waters, Milford, USA)	45°C
Stanozolol and metabolites	urine	LLE or SPE	β-glucuronidase	HPLC (Thermo Finnigan, San Jose, USA) Acquity UPLC (Waters, Milford, USA)	Omnispher C18 (100 x 2 mm, 3 μm) (Varian, Sint-Katelijne-Waver, Belgium) BEH C18 (50 x 2.1 mm, 1.7 μm) (Waters, Milford, USA)	
Agents with anti-oestrogenic activity, β2-agonists, β-blockers, diuretics, enhancement of oxygen transfer, narcotics, stimulants, α-reductase inhibitor	urine	Mixed mode anion and cation exchange SPE	-	Acquity UPLC (Waters, Milford, USA)	BEH C18 (100 x 2.1 mm, 1.7 μm) (Waters, Milford, USA)	30°C
Phase II steroids (androgenic and oestrogenic)	Animal urine	LLE and anion exchange SPE	-	Acquity UPLC (Waters, Milford, USA)	BEH C18 (100 × 2.1 mm, 1.7 μm) (Waters, Milford, USA)	50°C
Nicotine and metabolites	urine	Dilute and shoot (1:40)	-	Acquity UPLC (Waters, Milford, USA)	BEH HILIC (50 × 2.1 mm, 1.7 μm) (Waters, Milford, USA)	30°C
Phase II steroids (androgenic)	urine	Hydrophilic Lipophilic polymeric SPE	-	Acquity UPLC (Waters, Milford, USA)	BEH C18 (150 × 2.1 mm, 1.7 μm) (Waters, Milford, USA)	25°C
Testosterone and metabolites	urine	Alkalinisation and LLE	-	Acquity UPLC (Waters, Milford, USA)	Eclipse Plus C18 (50 x 2.1 mm, 1.8 μm) (Agilent Technologies, Waldbronn, Germany)	

Table 10.3 (*Continued*)

Mobile phase	*Gradient*	*Flow rate*	*Injection volume*	*Ionization*	*MS*	*LOD*	*References*
A: formic acid 0.01% B: ACN with 0.01 % formic acid	2% for 1 min to 80% in 1 min, return to initial conditions in 1.2 min. Total 3.2 min	600 μL/min		ESI +	Quattro Premier in MRM mode (Waters, Milford, USA)	200 (LOQ)	60
A: ammonium acetate 1mM and 0.001% acetic acid B: MeOH with ammonium acetate 1mM and 0.001% acetic acid	30% B for 8 min, inceased to 55% in 7 min, and to 95% in 14.5 min, maintained 1 min and return to initial conditions in 5 min. Total 34 min	250 μL/min 300 μL/min	20 μL	ESI +	TSQ Quantum (Thermo Finnigan, San Jose, USA) QTOF Premier (Waters, Milford, USA)		57
A: formic acid 0.1% B: ACN with 0.1% formic acid	5% to 95% B in 3 min, 1.5 equilibration time. Total 4.5 min	400 μL/min	5 μL	ESI + ESI -	QTOF Premier (Waters, Milford, USA)	MRPL	58
A: formic acid 0.1% B: ACN with 0.1% formic acid	5% B to 0.3 min, increased to 43% at 9.6 min, and to 100% from 10.5 to 12.5 min; equilibrate to 5% B between 13 and 16.5 min. Total 16.5 min	600 μL/min	3 μL	ESI -	Xevo TQ-MS (Waters, Milford, USA)		10
A: ACN B: 10 mM ammonium formate (pH 3.0)	1% B for 0.7 min to 25% B in 1.9 min, to 60% over 0.35 min, held for 0.25 min and returned to initial conditions for 1.8 min. Total 5 min	800 μL/min	2 μL	ESI +	Xevo TQ-S (Waters, Milford, USA)	1	62
A: 0.1% formic acid B: ACN with 0.1% formic acid	5% B to 37% over 25 min, increased to 95% B for 3 min; equilibrate to 5% B for 8 min. Total 36 min	300 μL/min	20 μL	ESI -	QTOF Premier (Waters, Milford, USA)		59
A: 0.01% formic acid and ammonium formate 1 M B: MeOH with 0.01% formic acid and ammonium formate 1 M	45% B to 65% over 3.5 min, increased to 95% B to 4.5 min, maintained for 0.5 min; equilibrate to 45% B for 3 min. Total 8 min	300 μL/min	10 μL	ESI +	Quattro Premier in MRM mode (Waters, Milford, USA)	0.1-20 ng/mL	67

major metabolite or an indirect marker. It requires a strict clean-up to remove matrix interfering compounds, and analyte pre-concentration is often essential to gain more sensitivity. Therefore, selective extraction techniques are generally performed. SPE is either developed for a specific analyte and its major metabolites or as a generic and selective protocol for compounds with various physico-chemical properties. For selective extraction, a LLE protocol with hydrolysis of the conjugated part of phase II metabolites was performed for primary detection of metabolites, while an SPE protocol was performed for confirmatory procedure by LC-ESI-MS/MS analysis in the positive mode, as it is considered more efficient than LLE to obtain clean extracts.[57] To perform a selective extraction of a wide range of analytes, reversed-phase mixed-mode extraction supports were selected for sample preparation.[58] Thus, reversed-phase cation-exchange cartridges are used for neutral and basic compounds, while acidic substances are extracted by using reversed-phase anion-exchange cartridges. This strategy presents obvious advantages of analyte pre-concentration, low sample volume and high throughput using the selected 96-well plate format. Generic and selective sample preparation by SPE is also employed for metabolite profiling of a targeted class of compounds, such as anabolic androgenic agents in animal or human urine.[10,59]

Nevertheless, for the quantitative measurement of threshold substances, sample preparation can be drastically simplified in function of their pharmacokinetics with late-generation MS. Indeed, simple sample dilution before injection might be sufficient to obtain excellent selectivity and sufficient sensitivity. As an example, a method was developed for the β_2-agonist salbutamol by diluting a urine sample with water in a 1:1 proportion.[60] Because this substance is only prohibited in sport above a concentration of 1 $\mu g\ mL^{-1}$,[61] sensitivity was not the bottleneck, as high doses are necessary to attain such concentrations in urine. Another approach based on simple 40-fold dilution was recently developed for the detection and quantification of nicotine and metabolites in urine. The dilution factor was optimised to obtain sufficient sensitivity while avoiding matrix effects. Interestingly, a limit of quantification (LOQ) of 10 ng mL^{-1} can be reached for all analytes due to the sensitivity of UHPLC-MS/MS.[62]

10.4.2.2 Separation Conditions

For confirmatory analysis, separation conditions are generally developed for a single analyte and/or its major metabolites. Therefore, chromatographic methods developed with sub-2 μm particles provide fast gradients on short columns with respect to sensitivity and resolution.

A fast UHPLC confirmatory method was developed with a runtime of 4.5 min through a BEH C18 50 mm column, which was sufficient for the confirmation of 103 prohibited compounds independently.[58] Indeed, as the targeted analyte is presumptively known at the screening test and the sample preparation dedicated to the doping agent, less selectivity can be afforded during the chromatographic run.

For metabolite studies or metabolite profiling approaches, the chromatographic method must be as selective as possible to avoid co-elution of analytes together or with interfering compounds from the matrix. This is even more difficult when studying endogenous compounds such as testosterone. Hence, high chromatographic resolution is mandatory to obtain very thin peaks and, thus, elevated peak capacity.

Methods for metabolomics require high resolution, increased sensitivity and generally long runs to collect richer information.[63,64] UHPLC is particularly useful for such an approach because it offers better retention time repeatability and a higher signal-to-noise ratio (S/N).[65,66]

For quantitative confirmatory approaches, selectivity is generally improved to obtain clean analyte mass spectrum without co-elution, while the analysis time is not found detrimental.

Ventura *et al.*[60] confirmed the presence of salbutamol using a 3.2 min gradient on a 100 mm BEH C18 column length (100 × 2.1 mm, 1.7 μm). In that case, the separation time could be further reduced by using a 50 mm column length because only one compound was monitored. In addition, when working on a shorter column, the pressure is reduced and the column lifetime extended. Fabregat *et al.*[67] proposed a protocol with a total run time of 8 min through a 50-mm column length for the detection and quantification of testosterone and three metabolites using UHPLC-MS/MS. Because of the sensitivity of the method, the analytes could be quantified at low levels in urine with excellent selectivity. A fast quantitative strategy was also developed for the analysis of eight compounds, including nicotine and its major metabolites in urine, using a 5 min gradient through a 50 mm column length at 800 μL min^{-1}.[62] A BEH HILIC column was selected for the purpose, allowing high chromatographic selectivity and improved sensitivity, thanks to the huge proportion of organic solvent used in the HILIC mode.

The high resolution of UHPLC was used for the separation of isomers pairs, such as cathine/phenylpropanolamine and ephedrine/pseudoephedrine. Because these analytes are substances that are prohibited only above a defined threshold, they must be isolated to quantify each of them accurately. Therefore, after extraction by SPE on mixed-mode cation-exchange cartridges, an isocratic method with 5% acetonitrile was proposed to separate the isomers at baseline in only a 5 min analytical run.

However, long gradient times are sometimes unavoidable to separate the targeted analytes from each other and for the accurate quantification of matrix constituents. Consequently, a 36-min gradient separation method was developed through a 150-mm column length for analysing 11 endogenous testosterone precursors and metabolites in athletes' urine.[59]

10.4.2.3 MS Hyphenation

High selectivity is required for confirmatory analyses to reach ideal identification criteria by MS/MS. Indeed, the spectrum of a pure standard or a standard spiked in the matrix must match the spectrum of the suspected analyte in the number and intensity ratio of diagnostic ions (a minimum of three), while no interfering ion should be present in the spectra. MS/MS is generally selected for confirmatory purposes to obtain the ion transitions with maximum sensitivity and selectivity. A 10-fold increase in sensitivity can usually be observed as compared with precursor ion scanning. Product ion scanning is commonly dedicated to the elucidation of the structure of a particular analyte and is mostly used for confirmatory purposes. The spectrum obtained corresponds to a scan of fragments formed from a selected precursor ion, as depicted in Figure 10.5, and offers assistance for structural information in the case of unknown identification.[57]

However, for possible structure identification, a QTOF-MS analysis was performed in the product ion mode for accurate mass determination and for structural confirmation purposes. An approach was developed based on the acquisition of two simultaneous functions, one at low collision energy and the second one at an analyte-dedicated collision energy for obtaining rich and stable fragmentation, as presented in Figure 10.6.[58] Because two functions are acquired simultaneously with QTOF-MS, the precursor ion can be conserved in the first function, while it is fragmented by working in MRM mode with a triple quadrupole. Therefore, no restricted number of diagnostic ions can be

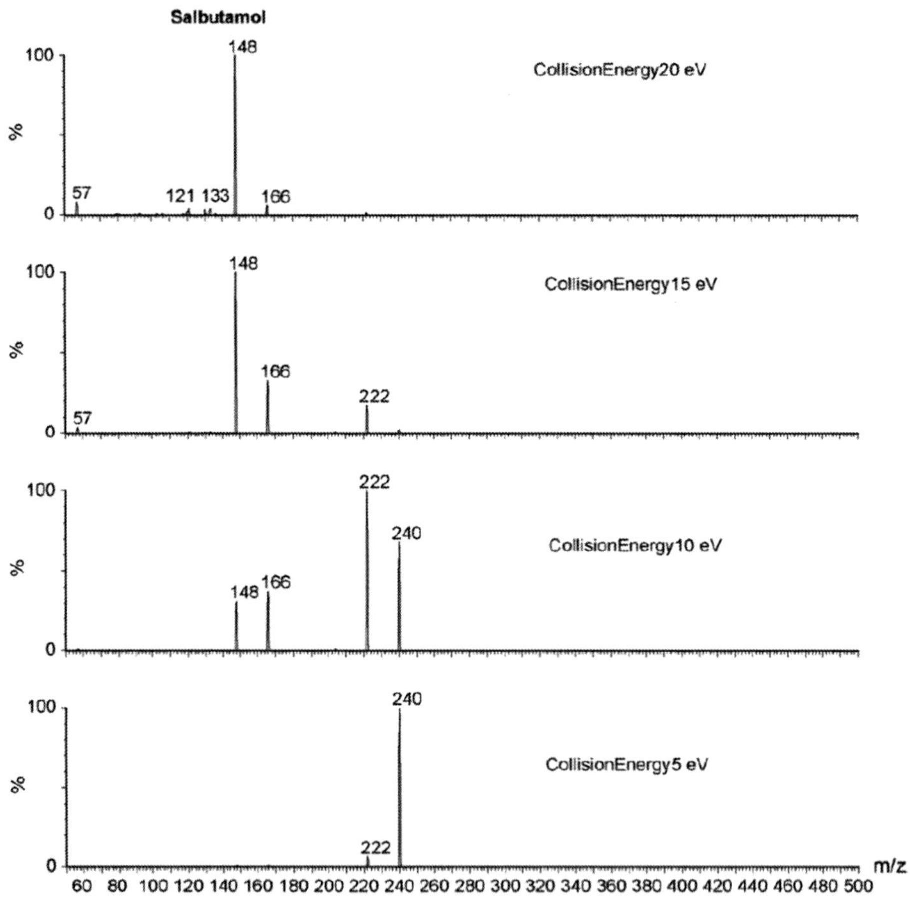

Figure 10.5 Product ion mass spectra of the pseudo-molecular ion $[M+H]^+$ of salbutamol (*m/z* 240) at different collision energies. Reprinted from reference 60 with permission.

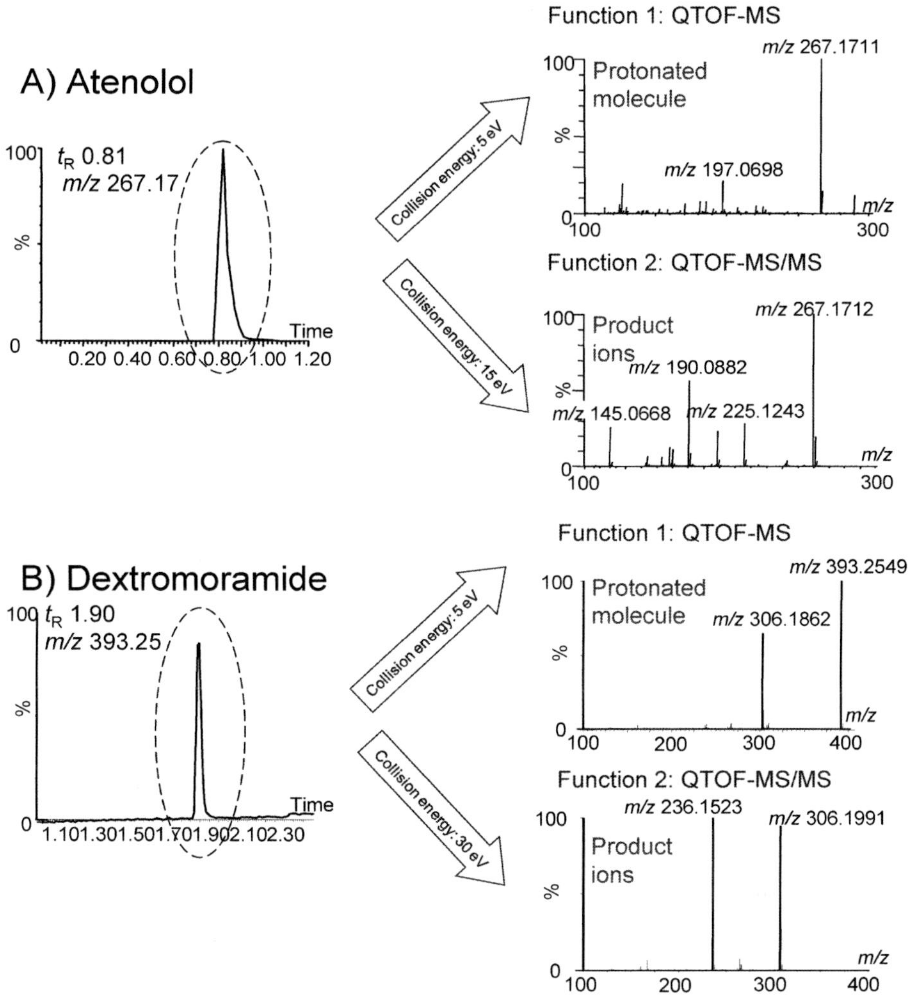

Figure 10.6 QTOF-MS and MS/MS spectra for atenolol (A) and dextromoramide (B). (A) Chromatogram obtained at low collision energy of 5 eV to obtain the protonated molecule of atenolol (*m/z* 267.17). Higher collision energy of 15 eV was applied to obtain product ions from the precursor ion (*m/z* 225.12, *m/z* 190.08 and *m/z* 145.06). (B) A low collision energy of 5 eV allowed to keep the protonated molecule of dextromoramide (*m/z* 393.25) in the function 1, whereas, in the function 2, product ions (*m/z* 306.19 and *m/z* 239.15) were obtained with a higher collision energy of 30 eV from the precursor ion. Reprinted from reference 58 with permission.

obtained for confirmation with a QTOF instrument, and accurate mass measurement can further assess the elemental composition of an ion.

This acquisition mode can also work without selecting a precursor ion, namely MS^E mode, with the acquisition of two parallel functions at low and high energy. Untargeted profiling can thus be performed, and because all analytes, even those of closely related structures, have different fragmentation patterns, it is also possible to use, instead of a fixed collision energy, ramped collision energy (*e.g.* from 5 to 70 eV). The use of ramped energy allows conservation of the (de)protonated molecular ion together with fragmented ions in the same spectrum. Moreover, no loss of structural information will occur because each molecule reaching the collision cell will be fragmented according to its structure. This strategy allows exact mass measurement and fragmentation in a single run, thus providing more information from the MS signal.

High-resolution instruments, such as the LTQ-Orbitrap or Exactive, have recently been introduced for doping control and present valuable opportunities for unknown structural elucidation for confirmatory analysis because they can reach a resolution as high as 100 000 FWHM with a mass accuracy of 1–2 ppm. Some applications are dedicated to small molecules, but because of the lower scan speed, as reported in Table 10.2, the Orbitrap is not the instrument of choice for UHPLC coupling. Indeed, to take advantage of a resolution of 100 000 FWHM, the cycle time will be 1 scan per s. This is not sufficient to define a UHPLC peak, which is normally 2–4 s in width. Therefore, QTOF is generally preferred for low-molecular-mass compounds to obtain high resolution within a reasonable analysis time.

For quantitative purposes, the MRM mode is the preferred method. Triple quadrupole MS is known to be very accurate, stable and robust for quantitation. Additionally, the S/N ratio can be significantly increased by selecting a transition between the chosen precursor and the product ions. In this context, Pozo *et al.*[57] selected two transitions per analyte (stanozolol and 18 metabolites) to increase the sensitivity of the analysis. For the quantification of nicotine and its metabolites in urine, Marclay *et al.*[62] took advantage of the high sensitivity and selectivity of a late-generation triple quadrupole in MRM mode, allowing for a simple 40-fold dilute-and-shoot sample pre-treatment.

10.4.3 UHPLC-MS(/MS) for Peptide and Protein Analysis

The introduction of peptidic and glycoprotein hormones in the Prohibited List provides new analytical challenges for their detection in the biological matrix. These biomolecules must be separated from their endogenous analogues, detected and quantified at low physiological levels. This class of compounds remains difficult to analyse, and only a few applications have been developed by LC-MS, while electrophoretic techniques and immunoassays are still currently preferred.[68–71] However, for some molecules, immunoassays suffered from cross-reactivity, leading to uncertain results. MS-based methods for macromolecules are now expanding, as this is an invaluable tool for peptides or proteins identification. Nano-UHPLC is an adequate strategy to combine with MS for the detection of peptides and proteins and its use is growing for doping control, as reported in Table 10.4. Strict sample preparations are required, allowing extract purification, analyte pre-concentration and compatibility with injection into the separation system.

10.4.3.1 Sample Preparation

Applications for doping control analyses involve mainly peptidic hormones, such as gonadotropin-releasing hormone (GnRH), insulin-like growth factors 1 and 2 (IGF-1 and IGF-2), synthetic insulins and corticotrophin (ACTH). These peptides have molecular masses less than 10 kDa. The extraction procedure for the small hexapeptide growth hormone-releasing peptide 2 (GHRP-2) (molecular mass of 817 Da) was a single SPE method, which was sufficient to obtain a clean extract with excellent recoveries.[72] Because of its low molecular mass, this peptide was also introduced to the fast plasma screening method described previously, in which the sample preparation consisted of simple PP and

Table 10.4 Description of UHPLC-MS(/MS) methods for peptides and proteins analyses.

Compounds	*Matrix*	*Sample preparation*	*LC system*	*Columns*	*Temperature*
Hemopexin, albumin, orosomucoid 1, transferrin, carbonic anhydrase 1	urine	Ultrafiltration and 2-D PAGE, Trypsin digestion	nano-UHPLC (Waters, Milford, USA)	BEH130C18 peptide column (100 mm x 100 μm; 1.7 μm)	
Prostaglandin H2 isomerase albumin, orosomucoid 1, transferrin, immunoglobulin	urine	Ultrafiltration and 2-D PAGE, Trypsin digestion	nano-UHPLC (Waters, Milford, USA)	BEH130C18 peptide column (100 mm x 100 μm; 1.7 μm)	
Synacthen (ACTH)	urine	Hydrophilic lipophilic polymeric SPE, antibody coated magnetic beads	nano-UHPLC (Waters, Milford, USA)	BEH130C18 peptide column (100 mm x 75 μm; 1.7 μm)	
GnRH	urine	Hydrophilic lipophilic polymeric SPE, antibody coated magnetic beads	nano-UHPLC (Waters, Milford, USA)	BEH130C18 peptide column (150 mm x 75 μm; 1.7 μm)	
GHRP-2	urine	Hydrophilic lipophilic polymeric SPE	nano-UHPLC (Waters, Milford, USA)	BEH C18 (50 x 2.1 mm, 1.7 μm)	25°C
IGF-1, IGF-2	urine	Hydrophilic lipophilic polymeric SPE, antibody coated magnetic beads	nano-UHPLC (Waters, Milford, USA)	Trapping column: Symmetry C18, (20 mm x 180 μm, 5 μm) Column: BEH130C18 peptide column (150 mm x 75 μm, 1.7 μm)	
Human, synthetic and animal insulin	urine	Hydrophilic lipophilic polymeric SPE, antibody coated magnetic beads, in-tube protein precipitation	nano-UHPLC (Waters, Milford, USA)	Trapping column: Symmetry C18, 5 μm, 180 μm x 20 mm) Column: BEH130C18 peptide column (100 mm x 75 μm, 1.7 μm)	
Animal insulins, GnRH, IGF-1, synachten	urine	Hydrophilic lipophilic polymeric SPE, antibody coated magnetic beads	nano-UHPLC (Waters, Milford, USA)	Trapping column: Symmetry C18, 5 μm, 180 μm x 20 mm) Column: BEH130C18 peptide column (100 mm x 75 μm, 1.7 μm)	

Table 10.4 (*Continued*)

Mobile phase	*Gradient*	*Flow rate*	*Injection volume*	*Ionization*	*MS*	*LOD*	*References*
A: 0.1% formic acid B: ACN with 0.1% formic acid	3% B for 3 min, to 40% in 45 min, elevated to 80% in 3 min , then 97% in 2 min and re-equilibrated to 3% for 15 min. Total 65 min	750 nL/min		nano ESI+	LTQ Orbitrap (Thermo Finnigan, San Jose, USA)		78
A: 0.1% formic acid B: ACN with 0.1% formic acid	3% B for 3 min, to 40% in 45 min, elevated to 80% in 3 min , then 97% in 2 min and re-equilibrated to 3% for 15 min. Total 65 min	750 nL/min		nano ESI+	LTQ Orbitrap (Thermo Finnigan, San Jose, USA)		79
A: 0.1% formic acid B: ACN with 0.1% formic acid	1 min at 3% B, increased to 100% in 20 min, re-equilibration for 14 min. Total 35 min	750 nL/min	1 μL	nano ESI+	LTQ Orbitrap (Thermo Finnigan, San Jose, USA)	3 pg/mL	75
A: 0.1% acetic acid with 0.01% TFA B: ACN with 20% A	10% B to 40% in 10 min, to 80% in 10 min, hold 2 min and reequilibration at 10% for 38 min. Total 60 min	750 nL/min		nano ESI+	LTQ Orbitrap (Thermo Finnigan, San Jose, USA)	5 pg/mL	73
A: 0.1% TFA B: ACN	10% B for 1 min, linear to 35% in 7 min, linear to 80% in 8 min, decreased to 10% in 0.1 min. Equilibrate for 2 min. Total 10 min	500 μL/min	10μL	ESI+	TQD mass spectrometer (Waters, Milford, USA)	50 pg/mL	72
A: 0.1% formic acid B: ACN	Trapping column load: 3% A for 3 min at 5 μL/min Gradient: 3% B to 70% in 22 min. Equilibrate at 3% B for 13 min. Total 35 min	750 nL/min			LTQ Orbitrap (Thermo Finnigan, San Jose, USA)	50 pg/mL	74
A: 0.1% formic acid B: ACN with 0.1% formic acid	Trapping column load: 3% A for 3 min at 5 μL/min Gradient: 3% B to 70% in 22 min. Equilibrate at 3% B for 13 min. Total 35 min	750 nL/min	1 μL	nano ESI+	LTQ Orbitrap (Thermo Finnigan, San Jose, USA)	0.5 fmol/mL	76
A: 0.1% formic acid B: ACN with 0.1% formic acid	Trapping column load: 3% A for 3 min at 5 μL/min Gradient: 3% B to 90% in 21 min. Equilibrate at 3% B for 13 min. Total 35 min	750 nL/min	1 μL	nano ESI+	LTQ Orbitrap (Thermo Finnigan, San Jose, USA)	1-5 pg/mL	77

centrifugation before injection.[33] For higher-molecular-mass peptides (>1000 Da), such as GnRH or insulins, the sample preparation strategy was more elaborate. The main challenge for macromolecule analysis is sensitivity. This necessitates high selectivity from interfering compounds for qualitative and quantitative analysis. Therefore, several sample treatments are performed successively to attain adequate extract purification. GnRH,[73] IGF-1 and -2[74] and ACTH[75] were extracted by SPE followed by affinity extraction using antibody-coated magnetic beads, while in-tube PP was subsequently performed for the detection of synthetic insulins.[76] However, despite a drastic loss of analytes during the three-step procedure (recovery of 33%), very clean extracts were obtained. Immunoaffinity purification offered highly selective extraction and led to an increase in sensitivity by a factor of at least 20 compared with other developed approaches. Recently, Thomas *et al.*[77] proposed a comprehensive peptide screening involving animal insulins, GnRH, IGF-1 and corticotrophin. The sample preparation was based on SPE with HLB cartridges followed by antibody-coated magnetic beads purification, allowing clean extract for all peptides.

A very efficient approach for doping control was presented by Kohler and co-workers[78,79] for studying the effect of endurance exercise on the urinary proteome that is constituted of proteins with molecular masses greater than 10 kDa. Here, the sample preparation consisted of ultrafiltration through a 10 kDa membrane to concentrate the mixture of proteins contained in a urine sample. This first step was followed by a two-dimensional polyacrylamide gel electrophoresis (PAGE) separation, which consisted of the separation of protein mixtures by isoelectric focusing in a first dimension and by size in the second dimension. Each isolated spot of protein was excised and digested with the protease trypsin. This bottom-up approach allowed the direct analysis of the generated peptides in the nano-UHPLC-MS system.

10.4.3.2 Separation Conditions

A separation technique is required for the quantification and/or the unequivocal identification of macromolecules in complex matrices and is often dedicated to a few analytes. A column packed with sub-2 μm particles was selected for the detection of peptides of small size, such as the hexapeptide GHRP-2. It was recently shown that the UHPLC strategy with 2.1 mm I.D. columns was fully compatible with protein sizes reaching 40 kDa by adding an adequate amount of trifluoroacetic acid (TFA) and by working at elevated temperatures (≤70 °C).[80] Above 40 kDa, stationary phases with large pores were necessary (300 Å) to avoid peak tailing and broadening.

Nevertheless, nanoscale UHPLC is often preferred for protein separation because of its additional benefits. Indeed, an improvement in sensitivity can be obtained due to the ratio between the injected volume and the column diameter. For example, a factor of 8000 can theoretically be obtained by reducing the column diameter from the conventional LC (4.6 mm) to a nano-LC (50 μm). However, this is only possible if the injected volume remains constant, which is difficult to obtain. Moreover, peptides and proteins are expensive standards, and the minimum amount should be injected. Eventually, a gain in sensitivity of only approximately 100 can be expected. Nano-UHPLC separations were thus applied for several anti-doping approaches regarding peptide and protein detection in urine. Generally, 0.1% formic acid was added to the mobile phase, except for the separation of GnRH, in which a mixture of 0.1% acetic acid and 0.01% TFA was selected. Indeed, as presented elsewhere,[80] the adding a small amount of TFA as an ion-pairing agent, the chromatographic performance for peptides could be improved. The separations were performed in 35–60 min through 100- to 150-mm column lengths with 75–100 μm I.D. and sub-2 μm particle sizes at a flow rate of 750 nL min^{-1}. Such low flow rates are difficult to maintain stably during the chromatographic run, making it necessary to use appropriate instrumentation. In addition, low tubing volume is a prerequisite to avoid, as much as possible, extra-column band broadening. Nano-UHPLC separations are now expanding for peptidic hormones, and sensitivities as low as the physiological concentration can be reached. As an example, LODs of 50 pg mL^{-1} and 3 pg mL^{-1} were obtained for insulin[76] and ACTH[77] in urine, respectively. The sensitivity of both methods was thus found to be adequate, as it corresponds to normal levels of insulin after injection and physiological levels of ACTH, ranging from 2.5 to 150 ng mL^{-1}.[81]

10.4.3.3 MS Hyphenation

Nano-UHPLC methods for peptides and proteins were developed predominantly because of the possibility of coupling to MS. Indeed, selective and sensitive MS spectra are required for the unambiguous identification of targeted macromolecules. High-resolution instruments, such as TOF, QTOF or LTQ-Orbitrap, are generally preferred for routine analyses of peptides and proteins. The ionisation of intact macromolecules is commonly performed by ESI-MS, leading to multi-charged ions due to an excess of charge density in droplets. This makes them compatible with a mass spectrometer having a low mass range. The molecular ion mass is then obtained by a mathematical deconvolution procedure. High-resolution MS is preferentially selected for identification of macromolecules because it allows exact mass measurement and presents a broad mass range. In the case of the bottom-up approach, a minimum of two fragmented peptides including a representative one and a covering sequence of at least 10% is recommended.[53] Peptides are introduced into the mass spectrometer directly after sample preparation and UHPLC separation. Nano-ESI sources are used by the majority because multi-charged ions are formed, and those ions could be further fragmented in the collision cell for confident identification. Such a top-down approach was selected for major applications in the anti-doping area regarding corticotrophin, GnRH, GHRP-2, insulins and IGF-1 and 2 by using nano-ESI in the positive mode. For the comprehensive peptide screening involving animal insulins, GnRH, IGF-1 and corticotrophin, developed by Thomas *et al.*,[77] the LTQ-Orbitrap was operated in two modes: the full scan mode, which provides high resolution (30 000 FWHM) with high mass accuracy acquisition, and the MS/MS mode, which provides ion transitions with high selectivity. Complete and confident information was reached by combining the selectivity of the MRM transition with accurate mass measurements.

10.5 Conclusions

UHPLC is a powerful technique providing many benefits for implementation in the anti-doping field. This separation method is even more potent when coupled with MS for excellent selectivity and sensitivity in the biological matrix. First, applications were conducted for the development of fast screening procedures that enable a reduction in analysis time and improved cost-effectiveness with simplified methods. This allows to generic and simple sample preparation with sensitive and selective MS acquisition towards the urine or blood matrices. However, last-generation triple quadrupole or (Q)TOF mass analysers are required for compatibility in cycle time because they possess fast data acquisition rates. In contrast, various anti-doping applications require higher chromatographic resolution and increased sensitivity with fewer constraints in analysis time. UHPLC-MS was also a great strategy for confirmatory analysis, metabolite study or even metabolomics, in which high peak capacity can be obtained with longer column lengths. Complex separation, including the analysis of positional isomers, can thus be resolved with excellent sensitivity. Nano-UHPLC was also an attractive approach for the detection of peptides and proteins because it allows sensitivity and hyphenation with MS to obtain specific and confident identification criteria.

UHPLC-MS has thus found its application in the anti-doping field for the detection of compounds from each class of doping agents found in the Prohibited List. It is a promising tool because the fight against doping necessitates ever more sensitive and selective techniques to detect low amounts of doping agents in complex matrices, while avoiding any false-positive case.

References

1. World Anti-doping Agency (WADA), The World Anti-Doping Code. Identification criteria for qualitative assays, Technical Document TD2010IDCR, Montreal, 2010, www.wada-ama.org (accessed June 2011).
2. World Anti-Doping Agency (WADA), The World Anti-Doping Code. The 2011 Prohibited List, Montreal, 2011, www.wada-ama.org (accessed June 2011).
3. World Anti-Doping Agency (WADA), International Standard for Laboratories V6.0, Montreal, 2009, www.wada-ama.org (accessed June 2011).
4. World Anti-doping Agency (WADA), The World Anti-Doping Code. Identification criteria for qualitative assays incorporating column chromatography and mass spectrometry, Technical Document TD2010IDCR, Montreal, 2010, www.wada-ama.org (accessed June 2011).
5. L. Rivier, *Anal. Chim. Acta*, 2003, **492**, 69.
6. World Anti-doping Agency (WADA), The World Anti-Doping Code. Minimal Required Performance Limits, Technical Document

TD2010MRPL, Montreal, 2010, www.wada-ama.org (accessed June 2011).
7. S. T. Bennett, *Clin. Chem.*, 2008, **54**, 1265.
8. E. Diamanti-Kandarakis, P. A. Konstantinopoulos, J. Papailiou, S. A. Kandarakis, A. Andreopoulos and G. P. Sykiotis, *Sports Med.*, 2005, **35**, 831.
9. P-E. Sottas, M. Saugy and C. Saudan, *Endocrinol. Metab. Clin. N. Am.*, **39**, 59.
10. S. Anizan, D. Di Nardo, E. Bichon, F. Monteau, N. Cesbron, J-P. Antignac and B. Le Bizec, *Anal. Chim. Acta*, 2011, **700**, 105.
11. I. Riedmaier, C. Becker, M. W. Pfaffl and H. H. D. Meyer, *J. Chromatogr. A*, 2009, **1216**, 8192.
12. J. C. W. Rijk, A. Lommen, M. L. Essers, M. J. Groot, J. M. Van Hende, T. G. Doeswijk and M. W. F. Nielen, *Anal. Chem.*, 2009, **81**, 6879.
13. J. Boccard, F. Badoud, E. Grata, S. Ouertani, M. Hanafid, G. Mazerolles, P. Lantéri, J-L. Veuthey, M. Saugy and S. Rudaz, *Forensic Sci. Int.*, 2011, **213**, 85.
14. A. Baoutina, I. E. Alexander, J. E. J. Rasko and K. R. Emslie, *J. Gene Med.*, 2008, **10**, 3.
15. H. H. Maurer, *Clin. Chem. Lab. Med.*, 2004, **42**, 1310.
16. K. R. Mueller, J. Grosse, R. Lang and D. Thieme, *J. Chromatogr. B: Biomed. Sci. Appl.*, 1995, **674**, 1.
17. C. Ayotte, in *Doping in Sports: Biochemical Principles, Effects and Analysis*, ed. D. Thieme and P. Hemmersbach, Springer-Verlag, Heidelberg, 2010, vol. 195, p. 77.
18. M. K. Parr, G. Fußhöller, N. Schlörer, G. Opfermann, H. Geyer, G. Rodchenkov and W. Schänzer, *Toxicol. Lett.*, 2011, **201**, 101.
19. P. Van Eenoo, W. Van Gansbeke, N. De Brabanter, K. Deventer and F. T. Delbeke, *J. Chromatogr. A*, 2011, **1218**, 3306.
20. M. Thevis, A. Thomas and W. Schänzer, *Anal. Bioanal. Chem.*, 2011, **401**, 1.
21. R. Kazlauskas, in *Doping in Sports: Biochemical Principles, Effects and Analysis*, ed. D. Thieme and P. Hemmersbach, Springer-Verlag, Heidelberg, 2010, vol. 195, p. 154.
22. W. Jian, R. W. Edom, Y. Xu and N. Weng, *J. Sep. Sci.*, 2010, **33**, 681.
23. A. Vonaparti, E. Lyris, I. Panderi, M. Koupparis and C. Georgakopoulos, *Drug Test. Anal.*, 2009, **1**, 365.
24. M. Kolmonen, A. Leinonen, T. Kuuranne, A. Pelander and I. Ojanperä, *J. Chromatogr. B*, 2010, **878**, 2959.
25. F. Marclay and M. Saugy, *J. Chromatogr. A*, 2010, **1217**, 7528.
26. L. Kortz, C. Helmschrodt and U. Ceglarek, *Anal. Bioanal. Chem.*, 2011, **399**, 2635.
27. P. W. Carr, X. Wang and D. R. Stoll, *Anal. Chem.*, 2009, **81**, 5342.
28. D. Guillarme, J. Ruta, S. Rudaz and J-L. Veuthey, *Anal. Bioanal. Chem.*, 2010, **397**, 1069.

29. D. T. T. Nguyen, D. Guillarme, S. Rudaz and J-L. Veuthey, *J. Sep. Sci.*, 2006, **29**, 1836.
30. M. Mazzarino, X. de la Torre, F. Botrè, N. Gray and D. Cowan, *Drug Test. Anal.*, 2010, **2**, 311.
31. E. Oláh, S. Fekete, J. Fekete and K. Ganzler, *Journal of Chromatography A*, 2010, **1217**, 3642.
32. D. T. T. Nguyen, D. Guillarme, S. Heinisch, M.-P. Barrioulet, J.-L. Rocca, S. Rudaz and J.-L. Veuthey, *J. Chromatogr. A*, 2007, **1167**, 76.
33. A. Thomas, S. Guddat, M. Kohler, O. Krug, W. Schänzer, M. Petrou and M. Thevis, *Rapid Commun. Mass Spectrom.*, 2010, **24**, 1124.
34. M. Mazzarino and F. Botrè, *Rapid Commun. Mass Spectrom.*, 2006, **20**, 3465.
35. M. E. Touber, M. C. van Engelen, C. Georgakopoulus, J. A. van Rhijn and M. W. F. Nielen, *Anal. Chim. Acta*, 2007, **586**, 137.
36. R. R. Ventura, M. Roig, N. Montfort, P. Sáez, R. Bergés and J. Segura, *Eur. J. Mass Spectrom.*, 2008, **14**, 191.
37. J. Henion, *Anal. Chem.*, 1998, **70**, 650A.
38. V. F. Samanidou, I. P. Imamidou and I. N. Papadoyannis, *J. Liq. Chrom. Rel. Technol.*, 2002, **25**, 185.
39. R. J. B. Peters, J. E. Oosterink, A. A. M. Stolker, C. Georgakopoulos and M. W. F. Nielen, *Anal. Bioanal. Chem.*, 2010, **396**, 2583.
40. M. Kolmonen, A. Leinonen, T. Kuuranne, A. Pelander and I. Ojanperä, *Drug Test. Anal.*, 2009, **1**, 250.
41. G. J. Murray and J. P. Danaceau, *J. Chromatogr. B*, 2009, **877**, 3857.
42. J-O. Thörngren, F. Östervall and M. Garle, *J. Mass Spectrom.*, 2008, **43**, 980.
43. F. Badoud, E. Grata, L. Perrenoud, L. Avois, M. Saugy, S. Rudaz and J. L. Veuthey, *J. Chromatogr. A*, 2009, **1216**, 4423.
44. R. Bonfiglio, R. C. King, T. V. Olah and K. Merkle, *Rapid Commun. Mass Spectrom.*, 1999, **13**, 1175.
45. B. K. Matuszewski, M. L. Constanzer and C. M. Chavez-Eng, *Anal. Chem.*, 2003, **75**, 3019.
46. Food and Drug Administration, Guidance for Industry: Bioanalytical method validation, US Department of Health and Human Services, FDA, Centre for Drug Evaluation and Research, Rockville, 2001.
47. I. Marchi, V. Viette, F. Badoud, M. Fathi, M. Saugy, S. Rudaz and J-L. Veuthey, *J. Chromatogr. A*, 2010, **1217**, 4071.
48. D. Guillarme, J. Schappler, S. Rudaz and J-L. Veuthey, *Trends Anal. Chem.*, 2010, **29**, 15.
49. D. Guillarme, E. Grata, G. Glauser, J-L. Wolfender, J-L. Veuthey and S. Rudaz, *J. Chromatogr. A*, 2009, **1216**, 3232.
50. P. Bennett, *Bioanalysis*, 2011, **3**, 709.
51. M. Mazzarino, S. Turi and F. Botrè, *Anal. Bioanal. Chem.*, 2008, **390**, 1389.
52. O. J. Pozo, W. Van Thuyne, K. Deventer, P. Van Eenoo and F. T. Delbeke, *J. Mass Spectrom.*, 2008, **43**, 394.

53. M. Thevis, *Mass Spectrometry in Sports Drug Testing*, John Wiley & Sons, Hoboken, 2010.
54. M. Thevis and W. Schänzer, *Anal. Bioanal. Chem.*, 2007, **388**, 1351.
55. W. B. Dunn, D. Broadhurst, M. Brown, P. N. Baker, C. W. G. Redman, L. C. Kenny and D. B. Kell, *J. Chromatogr. B*, 2008, **871**, 288.
56. D. Guillarme and J-L. Veuthey, in *Extreme Chromatography: Faster, Hotter, Smaller*, ed. W. C. Byrdwell and M. Holcapek, AOCS, Urbana, 2011, p. 1.
57. O. J. Pozo, P. Van Eenoo, K. Deventer, L. Lootens, S. Grimalt, J. V. Sancho, F. Hernández, P. Meuleman, G. Leroux-Roels and F. T. Delbeke, *Steroids*, 2009, **74**, 837.
58. F. Badoud, E. Grata, L. Perrenoud, M. Saugy, S. Rudaz and J. L. Veuthey, *J. Chromatogr. A*, 2010, **1217**, 4109.
59. F. Badoud, E. Grata, J. Boccard, D. Guillarme, J.-L. Veuthey, S. Rudaz and M. Saugy, *Anal. Bioanal. Chem.*, 2011, **400**, 503.
60. R. Ventura, R. Ramírez, N. Monfort and J. Segura, *J. Pharm. Biomed. Anal.*, 2009, **50**, 886.
61. World Anti-Doping Agency (WADA), The World Anti-Doping Code. Decision Limits for the Confirmatory Quantification of Threshold Substances. Technical Document TD2010DL, Montreal, 2010, www.wada-ama.org (accessed June 2011).
62. F. Marclay, E. Grata, L. Perrenoud and M. Saugy, *Forensic Sci. Int.*, 2011, **213**, 73.
63. H. G. Gika, G. Theodoridis, J. Extance, A. M. Edge and I. D. Wilson, *J. Chromatogr. B*, 2008, **871**, 279.
64. P. A. Guy, I. Tavazzi, S. J. Bruce, Z. Ramadan and S. Kochhar, *J. Chromatogr. B*, 2008, **871**, 253.
65. H. G. Gika, E. Macpherson, G. A. Theodoridis and I. D. Wilson, *J. Chromatogr. B*, 2008, **871**, 299.
66. I. D. Wilson, J. K. Nicholson, J. Castro-Perez, J. H. Granger, K. A. Johnson, B. W. Smith and R. S. Plumb, *J. Proteome Res.*, 2005, **4**, 591.
67. A. Fabregat, O. J. Pozo, J. Marcos, J. Segura and R. Ventura, *Drug Test. Anal.*, 2010, **2**, 630.
68. A. Staub, D. Guillarme, J. Schappler, J-L. Veuthey and S. Rudaz, *J. Pharm. Biomed. Anal.*, 2011, **55**, 810.
69. S. Giraud, N. Robinson, P. Mangin and M. Saugy, *Forensic Sci. Int.*, 2008, **179**, 23.
70. C. Reichel and G. Gmeiner, in *Doping in Sports: Biochemical Principles, Effects and Analysis*, ed. D. Thieme and P. Hemmersbach, Springer-Verlag, Heidelberg, 2010, vol. 195, p. 251.
71. M. Bidlingmaier, J. Suhr, A. Ernst, Z. Wu, A. Keller, C. J. Strasburger and A. Bergmann, *Clin. Chem.*, 2009, **55**, 445.
72. M. Okano, M. Sato, A. Ikekita and S. Kageyama, *Rapid Commun. Mass Spectrom.*, 2011, **24**, 2046.

73. A. Thomas, H. Geyer, M. Kamber, W. Schänzer and M. Thevis, *J. Mass Spectrom.*, 2008, **43**, 908.
74. A. Thomas, M. Kohler, W. Schanzer, P. Delahaut and M. Thevis, *Analyst*, 2011, **136**, 1003.
75. A. Thomas, M. Kohler, W. Schänzer, M. Kamber, P. Delahaut and M. Thevis, *Rapid Commun. Mass Spectrom.*, 2009, **23**, 2669.
76. A. Thomas, W. Schänzer, P. Delahaut and M. Thevis, *Drug Test. Anal.*, 2009, **1**, 219.
77. A. Thomas, M. Kohler, P. Delahaut, W. Schänzer and M. Thevis, Preliminary results for the development of a peptide screening method by means of LC-MS/MS, in *Recent Advances in Doping Analysis*, ed. W. Schänzer, H. Geyer, A. Gotzmann and U. Mareck, Sport & Buch Strauss, Cologne, 2010.
78. M. Kohler, K. Walpurgis, A. Thomas, M. de Maree, J. Mester, W. Schänzer and M. Thevis, *Proteomics Clin. Appl.*, 2010, **4**, 568.
79. M. Kohler, S. Franz, A. Regeniter, A. Ikonen, K. Walpurgis, A. Thomas, W. Schänzer and M. Thevis, *Drug Test. Anal.*, 2009, **1**, 382.
80. A. Staub, D. Zurlino, S. Rudaz, J-L. Veuthey and D. Guillarme, *J. Chromatogr. A*, 2011, **1218**, 8903.
81. E. K. Sugawara, L. M. Ribeiro Neto and I. T. d. N. Verreschi, *Quimica Nova*, 2010, **33**, 447.

CHAPTER 11

UHPLC and UHPLC-MS for the Analysis of Seized Drugs

I. S. LURIE
E-mail: islurie@comcast.net

11.1 Introduction

Seized drugs consist of substances that have potential for abuse and are either banned or controlled by federal, state or local statutes. These substances represent divergent drug classes (*i.e.*, narcotic analgesics, stimulants, depressants, hallucinogens, and anabolic steroids) consisting of acidic, basic and neutral solutes. Seizures by law enforcement personnel consist of powders (*e.g.*, heroin, cocaine and methamphetamine), tablets {*e.g.*, ecstasy [3,4-methylenedioxymethamphetamine (MDMA)] and oxycotin (oxycodone)}, capsules (*e.g.*, barbiturates and benzodiazepines), oils (*i.e.*, anabolic steroids), and plant material (*i.e.*, marijuana and opium). Seized drugs can be complex mixtures such as natural products or powders containing the seized drug, impurities, adulterants, and diluents. The analysis of seized drugs is important for legal and intelligence purposes.[1] Courts frequently require the identity of a banned or a controlled substance. Depending on the jurisdiction, quantitation of the seized drug could also be necessary. For strategic and tactical intelligence, a more in-depth analysis of the drug sample (drug profiling) is usually required. Strategic intelligence involves the determination of a geographical origin and/or manufacturing process. For example, the quantitative analysis of various alkaloids in opium using gas chromatography (GC) is viable for the determination of geographical origin.[2] The determination of the amount of certain basic impurities present in heroin using high performance liquid chromatography (HPLC) and capillary electrophoresis (CE) could be used to

RSC Chromatography Monographs No. 16
UHPLC in Life Sciences
Edited by Davy Guillarme and Jean-Luc Veuthey

Published by the Royal Society of Chemistry, www.rsc.org

ascertain the manufacturing process for this controlled substance.[3–5] Tactical intelligence ascertains whether two or more exhibits came from an identical source. As an example, the HPLC chromatographic profile of acidic and basic impurities present in cocaine could be used to determine the similarity of two cocaine exhibits.[6]

Separation techniques are widely used for the analysis of seized drugs, including both quantitative (amount) and/or qualitative (identity) analysis. GC is widely used for drug analysis due to its ease of use and high peak capacity.[7] However, it has limitations for solutes of forensic interest that are thermally labile, highly polar and non-volatile. Even after derivatization and/or liquid phase extraction, certain solutes may not be amenable to GC analysis. In contrast liquid phase separation techniques, which include CE, HPLC and ultra-high performance liquid chromatography (UHPLC), are amenable to a wider range of solute including non-volatile compounds, and those that are thermally degradable or highly polar. UHPLC, which employs columns packed with sub-2 μm particles or equivalent, is excellent for forensic drug analysis since it affords low H values at relatively high linear velocities. This can result in either high speed or relatively high peak capacity separations. The use of UHPLC for the analysis of seized drugs is described (see Table 11.1 for chromatographic conditions).

11.2 Use of UHPLC for the Analysis of Seized Drugs

11.2.1 Applicability of Reversed Phase Liquid Chromatography (RPLC) with Ultraviolet (UV) or Fluorescence (FL) Detection

Lurie[8] reported on the utility of UHPLC using a C18 column packed with porous hybrid organic–inorganic 1.7 μm particles for the analysis of seized drugs. In this vein, RPLC using 2 mm internal diameter (i.d.) columns of 30–100 mm length with UV detection has been evaluated for the separation of basic and neutral drugs, drug profiling and general screening. UHPLC, in comparison to HPLC and CE, provided significantly better resolution and/or faster analysis.

Phenethylamines such as amphetamine, methamphetamine, 3,4-methylenedioxyamphetamine (MDA), MDMA and 3,4-methylenedioxyethylamphetamine (MDEA) can be difficult to analyze using GC unless a basic extract and/or derivatization is performed.[9] In contrast, these solutes can be easily analyzed directly using liquid phase separations.[9] UHPLC with a 50 mm column and gradient analysis for the above phenethylamines, plus a structurally related internal standard, n-butylamphetamine, provided up to 12 and 3 times faster analysis compared to HPLC and CE respectively.[8] For UHPLC, run times of 3 min were obtained including a re-equilibration time of 1 min. Excellent precision was obtained for both isocratic and gradient analysis. Relative standard deviations (RSDs) of $\leq$0.1% were obtained for both retention times and peak areas. For the above solutes, large differences in

Table 11.1 UHPLC conditions for the analysis of seized drugs.

Solutes	*Column*	*Mobile phase conditions*	*Detector(s)*	*Reference*
Phenethylamines (0.01 mg mL^{-1})	50 mm × 2.1 mm 1.7 μm Acquity UPLC BEH C18 (Waters)	Isocratic and gradient (100 mM phosphate, pH 1.8, acetonitrile), flow 750 μL min^{-1}, temperature 30 °C, injection 20 μL	UV (205 nm, 254 nm)	8
Anabolic steroids (0.03 mg mL^{-1})	30 mm × 2.1 mm 1.7 μm Acquity UPLC BEH C18	Gradient (water, acetonitrile), flow 750 μL min^{-1}, temperature 30 °C, injection 10 μL	UV (215 nm, 240 nm, 280 mm)	8
Anabolic steroids (0.025–1.0 mg mL^{-1})	100 mm × 2.1 mm 1.7 μm Acquity UPLC BEH C18	Gradient (water, methanol), flow 375 μL min^{-1}, temperature 30 °C, injection 10 μL	UV (215 nm, 240 nm, 280 mm)	8
Opium alkaloids (profiling opium)	100 mm × 2.1mm 1.7 μm Acquity UPLC BEH C18	Gradient (100 mM phosphate, pH 1.8, acetonitrile), flow 375 μL min^{-1}, temperature 30 °C, injection 4 μL	UV (205 nm)	8
MDMA (profiling)	100 mm × 2.1 mm 1.7 μm Acquity UPLC BEH C18	Gradient (0.1% TFA, pH 2.2, acetonitrile), flow 375 μL min^{-1}, temperature 30 °C, injection 20 μL	UV (288 nm)	8
Divergent drug classes (0.10–0.80 mg mL^{-1}) (drug screening)	100 mm × 2.1 mm 1.7 μm Acquity UPLC BEH C18	Gradient (100 mM phosphate, pH 1.8, acetonitrile), flow 430 μL min^{-1}, temperature 30 °C, injection 1 μL	UV (210 nm, 240 nm)	8
Benzene substituted phenethylamines (Internet products)	100 mm × 2.1mm 1.7 μm Acquity UPLC BEH C18	Isocratic [ammonium formate (0.1% TFA, acetonitrile and methanol], flow 200 μL min^{-1}, temperature 40 °C, injection 10 μL	FL (excitation 450 nm, emission 550 nm)	12
Divergent drug classes (0.10 mg mL^{-1}) (drug screening)	50 mm × 2.1mm 1.8 μm Zorbax Rapid Resolution HT StableBond C18 (Agilent)	Gradient (100 mM phosphate, pH 1.8, acetonitrile), flow 800 μL min^{-1}, temperature 95 °C, injection 2 μL	UV (210 nm)	11

Table 11.1 (*Continued*)

Solutes	*Column*	*Mobile phase conditions*	*Detector(s)*	*Reference*
Anabolic steroids (0.10 mg mL^{-1})	100 mm × 2.1mm 1.8 μm Zorbax Rapid Resolution HT StableBond C18 (Agilent)	Gradient (25 mM phosphate, pH 2.4, methanol), flow 800 μL min^{-1}, temperature 95 °C, injection 2 μL	UV (215 nm, 240 nm, 280 mm)	11
Barbiturates, benzodiazepines	30 mm × 2.1mm 1.7 μm Acquity UPLC BEH C18	Gradient (0.1% formic acid, pH 2.0, acetonitrile), flow 600 μL min^{-1}, temperature 30 °C or 90 °C, injection 1 μL	UV	13
MDA, MDMA, cocaine, heroin, LSD, GHB, shikimic acid	50 mm × 2.1mm 1.7 μm Acquity UPLC (BEH HILIC or BEH Amide)	Isocratic [ammonium formate[a] or ammonium acetate[a] (1.25–7.50 mM), acetonitrile], flow 200–900 μL min^{-1}, temperature 25–40 °C, injection 1 μL	UV (205 nm) or MS [positive and negative ESI (SIM)]	26
Divergent drug classes	1.9 μm Hypersil Gold PFP (Thermo Fisher Scientific)	Gradient (0.1% formic acid, pH 2.0, acetonitrile and methanol)	MS (positive and negative ESI)	15
Phenethylamines (over the counter preparations)	100 mm × 2.1 mm 1.9 μm Hypersil Gold PFP	Gradient (0.06% acetic acid, pH 2.0, acetonitrile), flow 1000 μL min^{-1}, temperature 45 °C, injection 1 μL	MS [positive ESI (FS, SIM)	16, 17
Benzene substituted phenethylamine	50 mm × 2.1mm 1.7 μm Acquity UPLC BEH C18	Gradient (10 mM ammonium acetate, acetonitrile), flow 200 μL min^{-1}, temperature 40 °C, injection 5 μL	MS [positive ESI (FS)]	18
LSD (candy)	100 mm × 2.1 mm 1.9 μm Hypersil Gold PFP	Gradient (0.06% acetic acid, pH 2.0, acetonitrile, methanol), flow 1000 μL/min^{-1}	MS [positive ESI (FS)]	19

Table 11.1 (*Continued*)

Solutes	*Column*	*Mobile phase conditions*	*Detector(s)*	*Reference*
Psilocybin, psilocin (mushrooms)	100 mm × 2.1 mm 1.9 μm Hypersil Gold PFP	Gradient (0.06% acetic acid, pH 2.0, acetonitrile, methanol), flow 1000 μL min^{-1}, temperature 45 °C, injection 2 μL	MS [positive ESI (FS)]	**20**
Cannabinoids (baked goods)	100 mm × 2.1 mm 1.9 μm Hypersil Gold PFP	Gradient (0.06% acetic acid, pH 2.0, acetonitrile, methanol), flow 1000 μL min^{-1}, temperature 45 °C, injection 2 μL	MS [positive ESI (FS)]	21
Synthetic cannabinoids (herbal products)	100 mm × 2.1mm 1.8 μm Acquity UPLC HSS T3	Gradient (10 mM ammonium formate, pH 3.5 or 0.1% formic acid, acetonitrile), flow 300 μL min^{-1}, temperature 40 °C, injection 1–5 μL	UV (254, 275, 280, 314 and FS) and MS [positive and negative ESI (FS)]	22–25
MDA, MDMA, cocaine, heroin, LSD, GHB, shikimic acid	50 mm × 2.1mm 1.7 μm Acquity UPLC (BEH HILIC or BEH Amide)	Isocratic [ammonium formate[a] or ammonium acetate[a] (1.25–7.50 mM), acetonitrile], flow 200–900 μL min^{-1}, temperature 25–40 °C, injection 1 μL	UV (205 nm) or MS [positive and negative ESI (SIM)]	26
Phenethylamines including novel analogues PMA, 4-MTA and MBDB, and ketamine	50 mm × 2.1mm 1.7 μm Acquity UPLC BEH C18	Isocratic and gradient (pyrrolidine,[b] methanol), flow 400 μL min^{-1}, temperature 40 °C, injection 1–5 μL	UV (254 nm), MS [positive ESI (product ion and MRM)]	28
Fentanyl and fentanyl homologs and analogs (1 μg mL^{-1})	150 mm × 2.1mm 1.7 μm Acquity UPLC BEH C18	Gradient (1% formic acid, pH 2.0, acetonitrile), flow 270 μL min^{-1}, temperature 25 °C, injection 20 μL	MS [positive ESI (MRM)]	29
MDMA, TFMPP and BZP	50 mm × 2.1mm 1.7 μm Acquity UPLC BEH HILIC	Isocratic [ammonium formate (1.25 mM), acetonitrile], flow 500 μL min^{-1}, temperature 30 °C, injection 10 μL	MS [positive ESI (MRM)]	14

Table 11.1 (*Continued*)

Solutes	*Column*	*Mobile phase conditions*	*Detector(s)*	*Reference*
Phenethylamines (ephedra extract)	150 mm × 2.1mm 1.7 μm Acquity UPLC BEH HILIC	Isocratic [ammonium formate (7.50 mM), acetonitrile], flow 500 μL min^{-1}, temperature 25 °C, injection 10 μL	MS [positive ESI (MRM)]	14
GHB, shikimic acid	50 mm × 2.1mm 1.7 μm Acquity UPLC BEH Amide	Isocratic [ammonium acetate (5 mM), acetonitrile], flow 200 μL min^{-1}, temperature 25 °C, injection 10 μL	MS [negative ESI (MRM)]	14
Heroin and heroin impurities (heroin profiling)	100 mm × 2.1mm 1.7 μm Acquity UPLC BEH C18	Gradient (1% formic acid pH 2.0, acetonitrile), flow 270 μL min^{-1}, temperature 30 °C, injection 10 μL	MS [positive ESI (MRM)]	34
Heroin and heroin impurities (heroin profiling)	50 mm × 2.1mm 1.7 μm Acquity UPLC BEH Shield RP 18	Gradient (10 mM ammoniun formate pH 9, methanol), flow 500 μL min^{-1}, temperature 30 °C, injection 5 μL	MS [positive ESI (MRM)]	35
Fentanyl and fentanyl impurities (fentanyl profiling)	150 mm × 2.1mm 1.7 μm Acquity UPLC BEH C18	Gradient (1% formic acid, pH 2.0, acetonitrile), flow 300 μL min^{-1}, temperature 25 °C, injection 20 μL	MS [positive ESI (MRM)]	36
Methamphetamine and methamphetamine impurities (methamphetamine profiling)	150 mm × 2.1mm 1.7 μm Acquity UPLC BEH HILIC	Isocratic [ammonium formate (7.50 mM), acetonitrile], flow 500 μL min^{-1}, temperature 30 °C, injection 10 μL	MS [positive ESI (MRM)]	14

[a]Prepared from stock solutions of pH 3.0 and pH 6.4 respectively. [b]Mixture of glacial acetic acid and pyrrolidine

selectivity were observed by using a low pH (1.8) *versus* high pH (11.5) buffer in the mobile phase. The good performance of the basic phenethylamines in terms of tailing factors and plate counts was obtained at low pH by employing sample concentrations ≤0.01 mg mL^{-1} and phosphate concentrations of 100 mM. Likewise, the good performance of these solutes in terms of tailing factors and plate counts was obtained at high pH by employing sample concentrations ≤0.10 mg mL^{-1} and a 10 mM 3-(cyclohexylamino)-l-propane-sulfonic acid (CAPS) buffer.

Unlike GC, HPLC and CE, all 26 neutral anabolic steroids could be fully resolved using a 100 mm UHPLC column with a water/methanol gradient over 28 min.[8,10] In practice, most seized exhibits would contain only a single anabolic steroid.[11] Therefore, the use of a fast gradient on a 3 cm column was investigated.[8] Such chromatographic conditions would be particularly useful for providing an additional confirmation of the presence of an anabolic steroid and for quantitative analysis. For a 2.5 min gradient, it was possible to baseline resolve 16 of the above anabolic steroids with RSDs of retention time and peak area of ≤0.16% and ≤0.17% respectively (see Figure 11.1). In contrast, using conventional HPLC, 20 out of possible 24 of the above anabolic steroids were separated over a 30 min gradient at ambient temperature using a C18 column

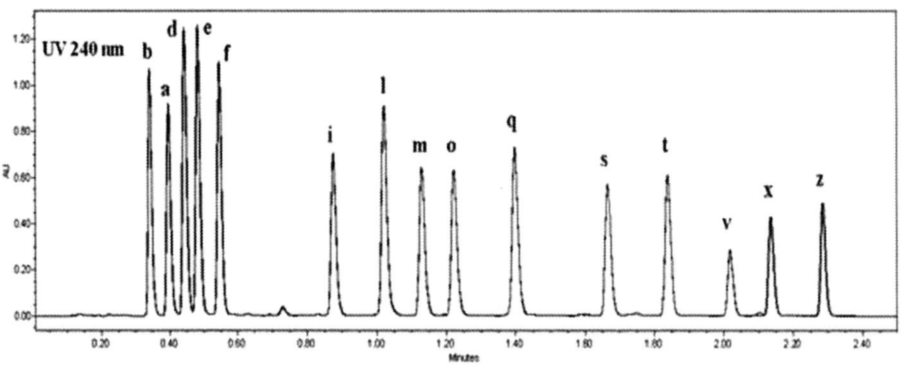

Figure 11.1 Overlay of seven injections of a UHPLC separation of anabolic steroids at solute concentrations of 0.030 mg mL^{-1}. Conditions: injection size, 10 uL (partial fill mode); column, 30 × 2.1 mm, 1.7 μm Acquity UPLC BEH C18. Initial conditions: 35% acetonitrile, 65% water. Final conditions: 100% acetonitrile, 2 min linear gradient, hold for 0.49 min, 0.5 min gradient re-equilibration; flow rate, 750 μL min^{-1}; temperature, 30 °C. Peaks: (a) boldenone, (b) fluoxymesterone, (d) methandrostenolone, (e) testosterone, (f) methyltestosterone, (i) boldenone acetate, (l) testosterone acetate, (m) nandrolone propionate, (o) testosterone propionate, (q) testosterone isobutyrate, (s) testosterone isocaproate, (t) testosterone enanthate, (v) boldenone undecylenate, (x) nandrolone decanoate and (z) testosterone undecanoate. Reprinted from *Journal of Chromatography A*, **1100**, I. S. Lurie, High-performance liquid chromatography of seized drugs at elevated pressure with 1.7 μm hybrid C18 stationary phase columns, 168–175, 2005, with permission from Elsevier.

of 250 mm and a particle size of 5 μm.[10] For complex mixtures such as opium, and MDMA and synthesis impurities, UHPLC using a 100 mm column resolved approximately twice as many peaks as CE and HPLC respectively.[8] Therefore UHPLC would be excellent for drug profiling. A comparison of UHPLC and HPLC for impurity profiling of MDMA is shown in Figure 11.2.

UHPLC, which results in a relatively high peak capacity and is amenable to most drugs of forensic interest, is well suited for general screening In this vein, 24 solutes of different drug classes including narcotic analgesics, stimulants, depressants, hallucinogens, and anabolic steroids were fully separated using a 100 mm column and a 13.5 min gradient.[8] For rapid screening, these same solutes were partially resolved using a 30 mm column and a 2.4 min gradient. In this way, the use of the more specific mass spectrometry (MS) detection with an appropriate buffer would facilitate drug screening.

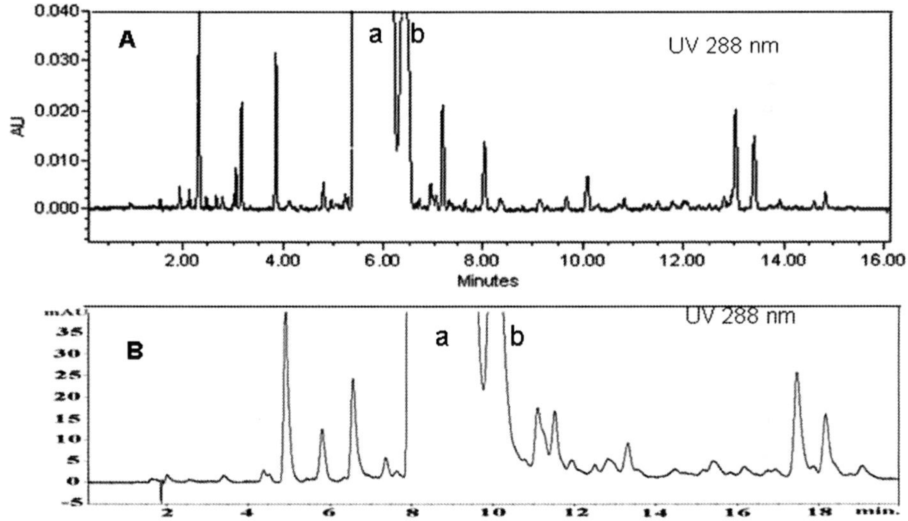

Figure 11.2 Comparison of UHPLC (A), and conventional gradient HPLC (B), for the drug profiling of MDMA (1.0 mg mL^{-1}). UHPLC conditions: injection size, 20 μL (overfill mode); column, 100 mm × 2.1 mm Acquity UPLC BEH C18. Initial conditions: 2% acetonitrile, 98% of 0.1% TFA buffer (pH 2.2). Final conditions: 40% acetonitrile, 60% of 0.1% TFA buffer (pH 2.2), 15 min linear gradient, hold for 1.2 min, 2.0 min gradient re-equilibration; flow rate, 375 μL min^{-1}; temperature, 30 °C. Conventional HPLC conditions: injection size, 100 μL; column, 12.5 cm × 3.2 mm, 5 μm, Partisil 5 ODS3. Initial conditions: 2% acetonitrile, 98% of 0.1% TFA (pH 2.2). Final conditions: 40% acetonitrile, 60% of 0.1% TFA (pH 2.2), 15 min linear gradient, hold for 1.2 min, 12.0 min gradient re-equilibration; flow rate, 375 μL/min; temperature 30 °C. Peaks: (a) MDMA and (b) dimethyl-MDMA. Reprinted from *Journal of Chromatography A*, **1100**, I. S. Lurie, High-performance liquid chromatography of seized drugs at elevated pressure with 1.7 μm hybrid C18 stationary phase columns, 168–175, 2005, with permission from Elsevier.

Min *et al.*[12] reported methodology for the UHPLC/FL determination of 11 "new" hallucinogenic phenethylamines which were recently regulated under Japanese Law. After derivatization with 4-(*N*,*N*-dimethylaminosulfonyl)-7-fluoro-2,1,3-benzoxadizole (DBD-F), the resulting 11 fluorophores were well separated in less than 26 min by RPLC employing an isocratic mobile phase. Intra-day and inter-day precision (%) and accuracy (CV) of ≤10% and ≤±9 respectively were obtained under optimized FL conditions for the various solutes of interest. The methodology was applied to street samples including "BDB" (pale yellow powder), "Jets" (dried brown mushroom) and "Honey Flash 2" (colorless liquid). The levels reported for these hallucinogenic phenethylamines in these exhibits (low-to-mid μg mL^{-1} range) could have been easily performed using UV detection. Analyzing solutes without derivatization is one of the major advantages of liquid phase separation. In addition, four of the reported solutes which contain the methylenedioxy moiety are naturally fluorescent. As reported in this manuscript,[12] the high sensitivity obtained with FL detection would be useful for the determination of these solutes in biological specimens such as blood and urine.

The use of elevated temperature UHPLC−UV was investigated for the different drug classes with 2.1 mm i.d, 1.8 μm, C18 columns of 50–100 mm length (stable at elevated temperature with low pH buffers).[11] As expected for RPLC the retention times of all 20 solutes decreased over the temperature range 15–95 °C (10% for most solutes using a 5 min linear gradient). However, under these same conditions relative retention times (*i.e.*, selectivity) increased, decreased or remained the same. It should be noted that no significant sample degradation at elevated temperature was observed for potentially thermally labile compounds such as diazepam, psilocybin, LSD, heroin, cocaine, phencyclidine (PCP) and clostebol acetate. The use of elevated temperature with concurrent decrease in mobile phase viscosity and increase in solute diffusivity allows for the use of significantly higher flow rates without loss of column efficiency (μ_{opt} for selected parabens increased ~2.4 times from 30 °C to 90 °C).[13] Taking all the above in account, UHPLC at elevated temperature (especially with the higher specificity of MS detection) could be useful for rapid drug screening.

Nguyen *et al.*,[13] using UHPLC-UV, separated a pharmaceutical cocktail of eight solutes, including barbiturates and benzodiazepines, in less than one-third of the time with much better overall resolution by operating at increase temperature (90 °C *versus* 30 °C) and increased flow rate (2000 μL min^{-1} *versus* 600 μL min^{-1}). A C18 column packed with 1.7 μm particles was used with gradient analysis. It should be noted that that positive changes in selectivity with increased temperature played a major role in this separation.

Nineteen out of a possible 27 anabolic steroids (the same compounds as described by Lurie[8] plus one additional anabolic steroid) were separated by Lurie and Li[11] in 6 min using UHPLC at 95 °C with 2.1 × 100 mm, 1.8 μm, C18 column with gradient analysis. The flow rate was limited to 800 μL min^{-1} in order to safely operate under the instrument pressure limitation of 9000 psi. Although anabolic steroids are neutral compounds, an acidic buffer was necessary in the mobile phase to enhance the stability of the C18 stationary phase at elevated temperature.

11.2.2 Applicability of RPLC with MS Detection

The identification of seized drugs, even at trace levels, is facilitated by the use of selective and sensitive MS detection. Technique such as positive electrospray ionization (ESI) is applicable to most drugs of forensic interest including bases, acids, such as cannabinoids, and neutrals, including anabolic steroids. The use of smaller particles has been shown to increase ESI response by 2.2 to 4.7 times.[14] An optimized UHPLC-MS RP separation of 14 illicit drugs and related compounds of divergent drug classes has been reported by Jiang.[15] Baseline separation was obtained in under 12 min using a ternary gradient. For positive ESI detection, linearity correlation coefficients (R^2) ≥ 0.995 and LODs ($S/N = 3$) from 0.29 to 90 ng mL^{-1} were obtained.

Jiang and co-workers[16,17] developed UHPLC-MS (positive ESI) methodology with single ion monitoring (SIM) detection for the determination of the methamphetamine precursor pseudoephedrine in over-the-counter (OTC) cold products. A baseline separation of ephedrine, pseudoephedrine, amphetamine, methamphetamine and MDMA was obtained in under 4 min using RPLC with gradient analysis. No chromatographic data were presented for other drugs possibly present in these OTC preparations, including antihistamines, naproxen, acetaminophen and ibuprofen. For the reported assay, a structurally related internal standard amphetamine was employed. For pseudoephedrine, the linearity R^2 was 0.997, while for three preparations the percentage of recovery for this solute was $\geq$ 87.1% [%RSD ($n = 6$) $\geq$ 5.6%]. The presence of the target compound was confirmed by its MS spectrum. Since psuedoephedrine is not present at trace levels in the OTC preparations, the use of considerably more precise and accurate PDA- UV detection would be more desirable for quantitation.

Kanai *et al.*[18] reported on UHPLC-MS (positive ESI) for the simultaneous analysis of six phenethylamine-type designer drugs. The 2,5-dimethoxy analogues included 2,5-dimethoxyphenethylamine (2C-H), 2,5-dimethoxyamphetamine (2,5-DMA), 4-bromo-2,5-dimethoxyphenethylamine (2-CB), 4-bromo-2,5-methoxyamphetamine (DOB), 4-iodo-2,5-dimethoxyphenethylamine (2C-I) and 4-iodo-2,5-dimethoxyamphetamine (DOI). In the full scan mode, all solutes except for DOB and 2C-I were resolved in less than 11 min using RPLC with gradient analysis. As shown in Figure 11.3, the use of extracted ion monitoring allows for selective detection of these solutes including DOB and 2C-I. For these solutes, the linearity R^2 was $\geq$ 0.982; intra-day precision: 2 μg mL^{-1} %RSD $\leq$ 22.4, 10 μg /mL %RSD $\leq$ 10.6 and 50 μg mL^{-1} %RSD $\leq$ 6.3; and inter-day precision: 2 μg mL^{-1} %RSD $\leq$ 21.8, 10 μg mL^{-1} %RSD $\leq$ 9.1 and 50 μg mL^{-1} %RSD $\leq$ 9.4. These solutes were confirmed by their MS spectra which contained a $[M+H]^+$ ion as their base peak , as well as one other non isotopic ion. In contrast GC-MS (EI) analysis

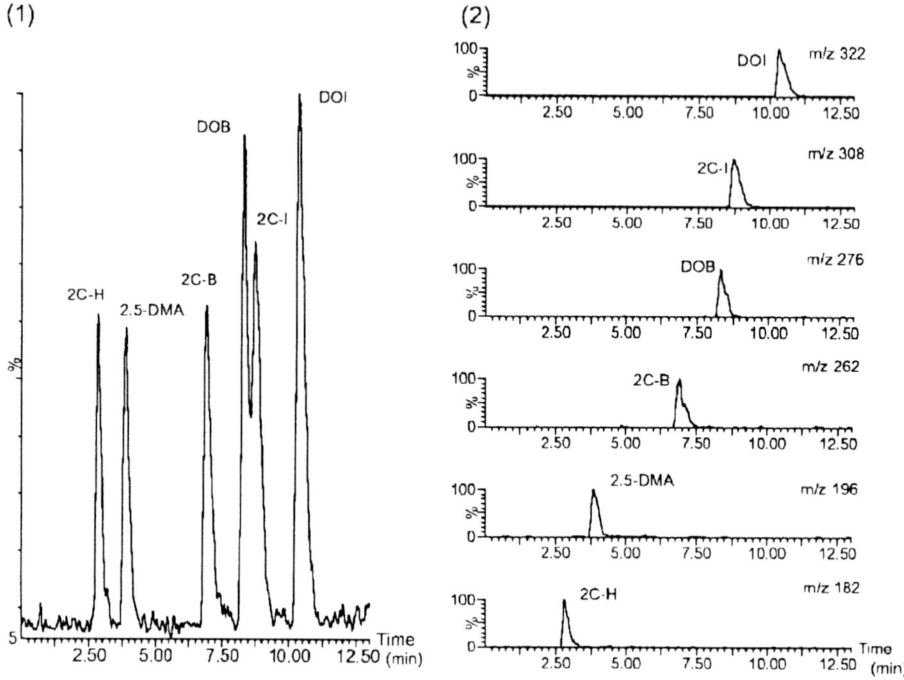

Figure 11.3 UHPLC-MS gradient separation of a mixed standard solution of 2,5- dimethoxyphenethylamine analogs. Total ion chromatogram (1) and mass chromatograms (2) of a standard solution of the six drugs (100 μg mL^{-1} each). $[M+H]^+$ ions were used as monitoring ions for mass spectrometry. UHPLC conditions: column, 50 mm × 2.1 mm Acquity UPLC BEH C18. Initial conditions: 10% acetonitrile, 90% of 10 mM ammonium acetate. Final conditions: 25% acetonitrile, 75% of 10 mM ammonium acetate, 13 min linear gradient; flow rate, 200 μL min^{-1}; temperature, 40 °C. Reprinted from *Forensic Toxicology*, **26**, K. Kanai, K. Takekawa, T. Kumamoto, T. Ishikawa and T. Ohmori, Simultaneous analysis of six phenethylamine-type designer drugs by TLC, LC-MS and GC-MS, 6–12, 2008, with permission from Springer Japan KK.

of the trifluoroacetic acid (TFA) derivatives of these solutes gave rise to multiple peaks, but non-base peak M^+ ions. In fact, 2C-I *via* GC-MS gave rise to a barely discernible M^+ ion, which is not uncommon for drugs of forensic interest. The detection limit for extracted ion monitoring using UHPLC-MS was 2.5 ng per injection for 2C-H, 2,5-DMA, 2C-I, and DOI, and 5 ng per injection for 2C-B and DOB. For the total ion chromatogram the detection limit was 25 ng per injection for all compounds.

The identification of thermally labile lysergic acid diethylamide (LSD) using GC-MS requires derivatization.[191] Stenzel and Jiang[19] employed UHPLC/MS (positive ESI) in the full scan mode for the identification of LSD in candy. For this examination, RPLC with gradient analysis was employed.

Psilocybin mushrooms contain the hallucinogenic compounds psilocin and psilocybin which are difficult to distinguish between by GC due to dephosphorylation of the latter solute at elevated temperature.[20] Stenzel and Jiang[20] employed UHPLC/MS (positive ESI) in the full scan mode for the identification of both solutes *via* retention time and MS spectra. For this analysis, RPLC with gradient analysis was employed.

The GC-MS analysis of marihuana cannabinoids in food matrices can be difficult due to possible interference from solutes found in baked goods.[21] Jiang and Stenzel[21] utilized UHPLC/MS (positive ESI) in the full scan mode for the identification of Δ-9-tetrahydrocannabinol (Δ9-THC) in cookies and brownies. For this investigation, cannabidiol, Δ9-THC and cannabinol were well resolved in less than 8 min using RPLC with gradient analysis.

Uchiyama *et. al.*[22–25] utilized UHPLC-UV-MS along with GC-MS, high-resolution MS (HRMS) and NMR for the identification of eight synthetic cannabinoids (a cyclohexylphenol cannabicyclohexanol, the napthoylindoles JWH-015, JWH-018, JWH-073, JWH-081 and JWH-200, and the phenylacetylindoles JWH-250 and JWH-251) in herbal or chemical products. RPLC with gradient analysis was employed with photodiode array (PDA) detection and negative and/or positive ESI detection in the full scan mode. UHPLC/MS analysis for an extract of a herbal product (including chromatograms, and UV and MS spectral data) is shown in Figure 11.4. Due to the presence of formic acid in the mobile phase (appreciable UV absorbance below 240 nm), quantitative analysis was performed for solutes at UV_{max} higher than 240 nm.[23,24] For this assay, betamethasone valerate was used as an internal standard. Some of these compounds exhibited relatively low extinction coefficients at these high wavelengths. This could explain why for certain solutes poor linearity in terms of UV detection was obtained ($R^2 = 0.998$). For quantitative analysis, the use of the UV transparent phosphate buffer would allow detection of the compounds of interest at lower wavelengths where they exhibit significantly better signal-to-noise than at higher wavelengths. Additional cyclohexylphenol solutes identified in herbal products by GC-MS and UHPLC-UV-MS included the *trans* diastereomer of cannabicyclohexanol and CP-47,497.[24]

11.2.3 Applicability of Hydrophilic Interaction Liquid Chromatography (HILIC) with UV or MS detection

Lurie *et al.*[26] reported on the utility of UHPLC with HILIC for the analysis of seized drugs. The chromatographic properties of eleven solutes of forensic interest were studied using 2 mm i.d., 1.7 μm columns (an unbonded hybrid organic–inorganic particle or the same substrate bonded with an amide moiety) with either a low or moderate pH buffer (ammonium formate *versus* ammonium acetate). For all studies, either UV or MS detection was used. These drugs included weak bases (a primary amine MDA, a secondary amine MDMA, and tertiary amines heroin, cocaine and LSD), a very weakly basic

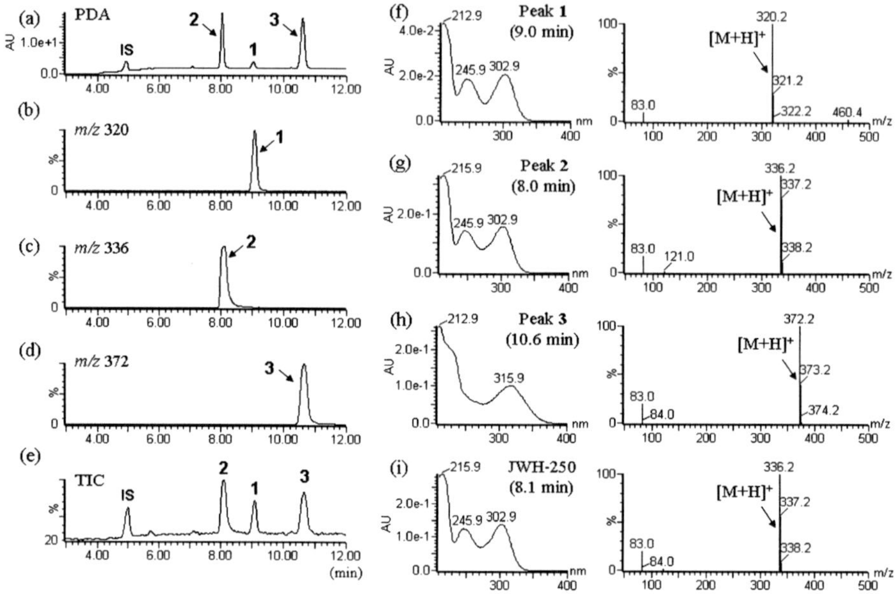

Figure 11.4 UHPLC-UV (a) and UHPLC-MS analysis for the extract of product no. 11. Mass chromatograms obtained at *m/z* 320 (b, 1), 336 (c, 2), and 372 (d, 3) and the TIC ( ); UV spectra and ESI mass spectra for each peak (f–h) and those of the standard of JWH-250 (i). Compounds 1 and 3 refer to JWH-250 and JWH-081 respectively. UHPLC conditions: column, 100 mm × 2.1 mm Acquity UPLC T3. Initial conditions: 50% of 0.1% formic acid in acetonitrile, 50% of 0.1% formic acid (hold for 5 min). Final conditions: 70% of 0.1% formic acid in acetonitrile, 30% of 0.1% formic acid, 2 min linear gradient (hold for 7 min); flow rate, 300 μL min^{-1}; temperature, 40 °C. Reprinted from *Forensic Toxicology*, **29**, N. Uchiyama, M. Kawamura, R. Kikura-Hanajiri and Y. Goda, Identification and quantitation of two cannabimimetic phenylacetylindoles JWH-251 and JWH-250, and four cannabimimetic naphthoylindoles JWH-081, JWH-015, JWH-200, and JWH-073 as designer drugs in illegal products, 25-37, 2011, with permission from Springer Japan KK.

tertiary amine diazepam, weak acids (γ-hydroxybutyric (GHB) and the cocaine biosynthesis precursor shikimic acid), very weakly acidic solutes (phenobarbital and Δ9-THC) and the neutral compound testosterone. Based on this investigation, HILIC can offer highly efficient, selective, and rapid isocratic separations of certain seized drugs and related compounds with excellent peak shapes and low back pressures. In contrast, RPLC often requires gradient analysis, can exhibit poor retention of hydrophilic acids, and there can be extensive overlap between acidic, neutral and basic solutes.[5,8] HILIC, in contrast to RPLC, exhibits a much greater loading capacity, and is therefore an excellent technique for drug profiling.[26,27]

In one of the studies described above,[26] chromatographic variables investigated included the effect of linear velocity on H, and the effect of volume fraction of buffer as well as buffer concentration on k. In addition, factors affecting selectivity and the effect of solute loading on column efficiency, tailing factor T and k were studied. For both types of columns studied minimum H values of ≤ 7 μm were obtained for most solutes. For the basic solutes MDA, MDMA and LSD, relatively flat Van Deemter plots were obtained on both stationary phases, with a significantly higher optimum linear velocity for the unbonded phase. An acidic solute shikimic acid exhibited non-flat Van Deemter plots and lower optimum H values than basic solutes on both stationary phases. For both stationary phases depending on the pK_a of the solute and the buffer employed (low or moderate pH buffer), k increased with the percentage of acetonitrile and either decreased (basic solutes) or increased (acidic solutes) with buffer concentration. Based on the above studies, selective retention of relatively stronger bases and/or acids is possible depending on the stationary phase buffer combination. In addition for retained solutes, factors influencing selectivity include the volume fraction of buffer, buffer concentration, buffer type and stationary phase.

11.2.4 Applicability of MS/MS Detection Including Drug Profiling

Tandem mass spectrometric detection (MS/MS) in the product ion or multiple reaction monitoring mode (MRM), which provides even higher sensitivity and selectivity of detection than the MS mode, is particularly well suited for the identification of seized drugs, even at ultra-trace levels. For this purpose, the use of the product ion spectra or MRM ratios is utilized. The use of MRM monitoring, which allows the measurement of target compounds at trace and ultra-trace levels, is particularly well suited for drug profiling.

Apollonio *et al.*[28] described the application of UHPLC-MS/MS (positive ESI) for the separation and identification of ketamine, common phenethylamine-type drugs (amphetamine, methamphetamine, MDA and MDMA) and novel designer analogues (para-methoxyamphetamine (PMA), 4-methylthioamphetamine (4-MTA) and N-methyl-1-(3,4-methylenedioxyphenyl)-2-butanamine (MBDB)). Using RPLC with an isocratic mobile phase, significant overlap of several of these solutes precluded their identification using UV detection. Positive ESI MS/MS detection, *via* product ion mass spectra, allowed the unique identification of all these solutes under the same chromatographic conditions, with far greater specificity than UV detection. Unresolved solutes pairs (4-MTA, ketamine; PMA, MDA; PMA, amphetamine) all have unique precursor ions.

Lurie and Iio[29] utilized UHPLC with positive ESI tandem MS in the MRM mode for the identification of fentanyl and 16 of its corresponding homologs and analogs. As shown in Table 11.2, these very similar solutes have nominal masses of 336 and 350. In spite of the relatively high peak capacity of UHPLC, several of the compounds of interest were poorly resolved using RPLC with gradient analysis (see Figure 11.5). It was further investigated whether these

Table 11.2 Structure of fentanyl and corresponding homologs and analogs. Reprinted from *Journal of Chromatography A*, **1216**, Ira. S. Lurie, Reiko Iio, Use of multiple- reaction monitoring ratios for identifying incompletely resolved fentanyl ,homologs and analogs via ultra-high-pressure liquid chromatography-tandem mass spectrometry, 1515-1519, 2009, with permission from Elsevier.

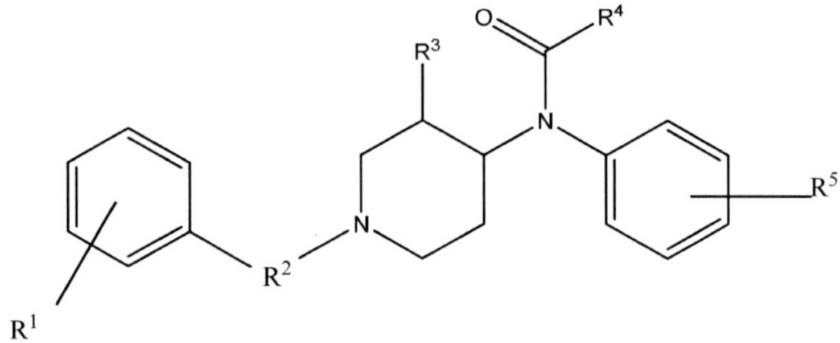

Solute	*Mass*	*R^1*	*R^2*	*R^3*	*R^4*	*R^5*
a	336	----	CH_2CHCH_3	----	CH_3	----
b	336	----	CH_2CH_2	----	CH_3	o-CH_3
c	336	o-CH_3	CH_2CH_2	----	CH_3	----
d (I.S.)	336	----	$CH_2CH_2CH_2$	----	CH_3	----
e (fentanyl)	336	----	CH_2CH_2	----	CH_2CH_3	----
f	336	----	CH_2CH_2	----	CH_3	m-CH_3
g	336	m-CH_3	CH_2CH_2	----	CH_3	----
h	336	p-CH_3	CH_2CH_2	----	CH_3	----
i	336	----	CH_2CH_2	----	CH_3	p-CH_3
j	350	----	CH_2CHCH_3	----	CH_2CH_3	----
k1	350	----	CH_2CH_2	trans-CH_3	CH_2CH_3	----
k2	350	----	CH_2CH_2	cis- CH_3	CH_2CH_3	----
l	350	----	CH_2CH_2	----	CH_2CH_3	o-CH_3
m	350	----	CH_2CH_2	----	$CH_2CH_2CH_3$	----
n	350	----	CH_2CH_2	----	CH_2CH_3	m-CH_3
o	350	m-CH_3	CH_2CH_2	----	CH_2CH_3	----
p	350	----	CH_2CH_2	----	CH_2CH_3	p-CH_3

unresolved compounds could be distinguished based on their dual MRM transitions (see Figures 11.5 and 11.6). Compound **j** was easily distinguished based on its unique precursor ion. Compounds **i**, **n** and **o** could be distinguished based on the different product ions that was produced from compounds with the same precursor ions. The two remaining solutes **g** and **h**, which have identical precursor and product ions, were distinguished based on vastly different dual MRM transitions.

HILIC with its high organic content is highly amenable to mass spectrometric detection. For the same particle size column, HILIC can produce 5.6 to 8.8 times higher MS response than RPLC.[14]

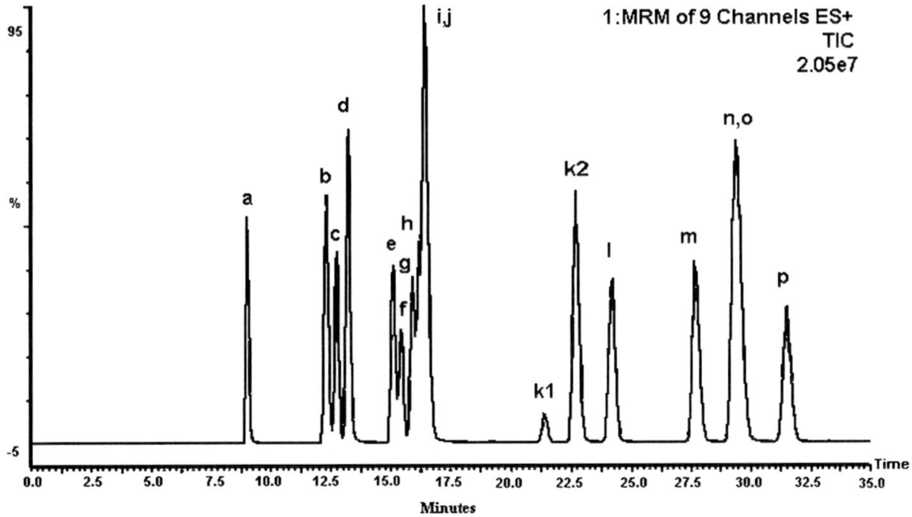

Figure 11.5 UHPLC-MS/MS gradient separation of a standard mixture of fentanyl and corresponding homologs and analogs (1 μg mL^{-1} of each compound). UHPLC conditions: column, 150 mm × 2.1 mm Acquity UPLC BEH C18. Initial conditions: 32% methanol, 68% of 1% formic acid (pH 2.0). Final conditions: 36.7% methanol, 63.3% of 1% formic acid (pH 2.0), 35 min linear gradient, 4.0 min gradient equilibration; flow rate, 270 μL min^{-1}; temperature, 25°C. See compound letters in Table 11.2. Reprinted from *Journal of Chromatography A*, **1216**, I.S. Lurie and R. Iio, Use of multiple-reaction monitoring ratios for identifying incompletely resolved fentanyl homologs and analogs *via* ultra-high-pressure liquid chromatography- tandem-mass spectrometry, 1515–1519, 2009, with permission from Elsevier.

HILIC UHPLC-MS/MS with positive ESI in the MRM mode has been shown to be an excellent alternative to the GC-MS analysis of primary and secondary amines in illicit tablets (MDMA, 3-trifluoromethylphenylpiperazine (TFMPP) and benzylpiperazine (BZP)), and extracts of the plant *Ephedra sinica* (norephedrine, norpseudoephedrine, ephedrine, pseudoephedrine, methylephedrine, and methylpseudoephedrine).[26] The relatively polar primary and secondary amine bases are problematic to analyze by GC-MS, due to poor peak shapes, unless a prior basic extraction is used. Even after such extraction, detection and/or separation can be difficult. Gradient analysis is required for RPLC-MS/MS analysis of these drug mixtures. These solute mixtures were fully resolved in less than 6.5 min with excellent specificity using HILIC UHPLC-MS/MS in the isocratic mode.

In order to monitor the effect of herbicides on cocaine eradication, it is necessary to measure the amount of the cocaine precursor shikimic acid present in the coca plant.[30] Ionized hydrophilic acids such as shikimic acid and the date rape drug GHB can be difficult to retain on a RP HPLC column, even with a stationary phase that is aqueous phase compatible.[31,32] Both these

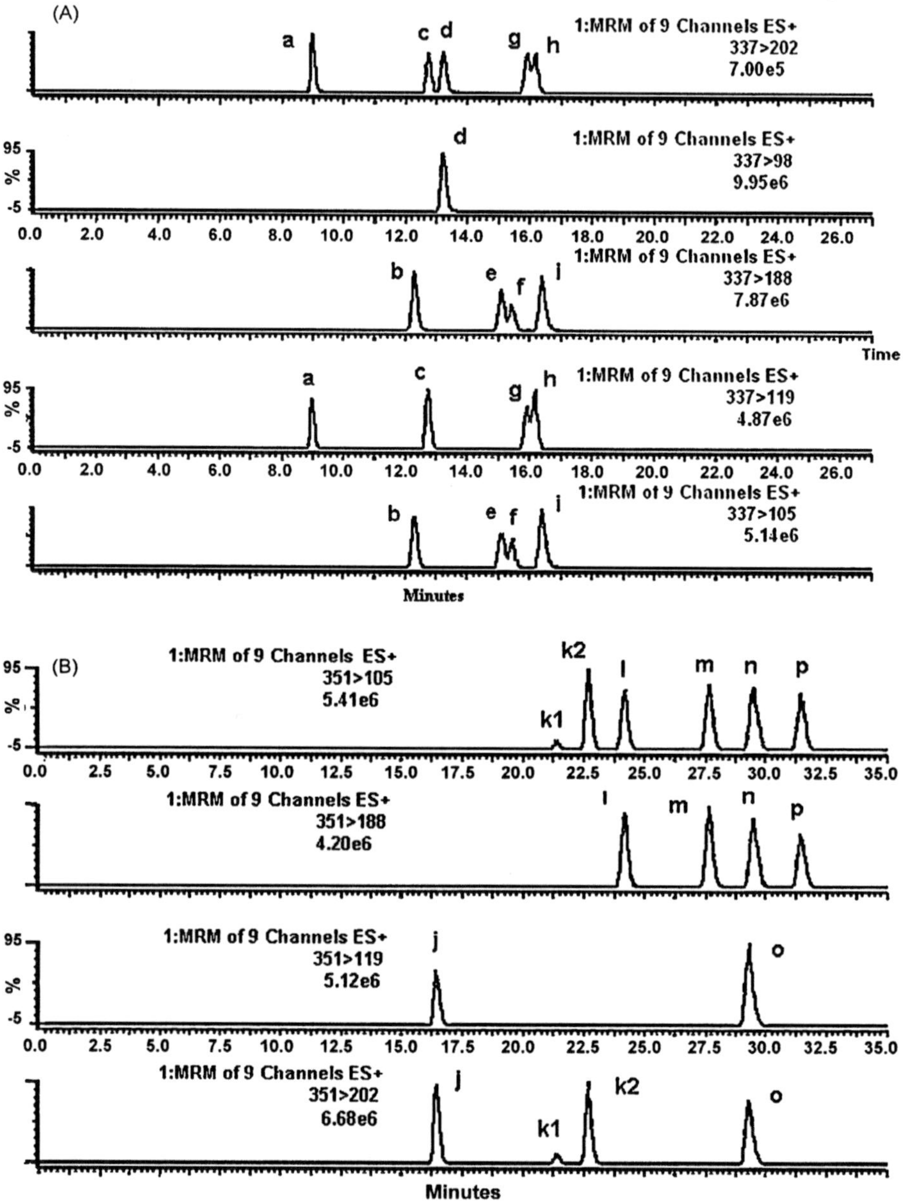

Figure 11.6 UHPLC-MS/MS gradient separation of a standard mixture of compounds with (A) precursor ion *m/z* 337 and (B) precursor ion *m/z* 351 (1 μg mL^{-1} of each compound). Chromatographic conditions are given in Figure 11.5. See compound letters in Table 11.2. Reprinted from *Journal of Chromatography A*, **1216**, I. S. Lurie and R. Iio, Use of multiple-reaction monitoring ratios for identifying incompletely resolved fentanyl homologs and analogs via ultra-high-pressure liquid chromatography–tandem-mass spectrometry, 1515–1519, 2009, with permission from Elsevier.

solutes are easily retained using HILIC, and are detected using negative ESI MS/MS in the MRM mode.[26] UHPLC-MS/MS HILIC, in contrast to GC-MS, does not require derivatization for these polar acids.[33]

The analysis of ultra-trace levels of impurities for enhanced drug profiles of basic drugs such as heroin and fentanyl is complicated by the presence of the large quantity of the seized drugs (and possibly other bases present in an exhibit). UHPLC, in contrast to capillary GC and CE, is more amenable to the injection of a large amount of the major component(s) in order to determine low levels of impurities.

Lurie and Toske[34] demonstrated the applicability of UHPLC-MS/MS for heroin profiling. Using RPLC with positive ESI detection in the MRM mode, heroin basic impurities at levels as low as 10^{-6} % w/w were detected aided by 75 μg of heroin on column. Impurities detected at ultra-trace levels, depending on sample origin, included morphine, codeine, noscapine, papaverine, and the previously unreported compounds reticuline, reticuline monoacetate (two products), reticuline diacetate, narceine, codamine, laudanidine, cryptopine, laudanosine, and norlaudanosine. The UHPLC-MS/MS RP gradient separation of a standard mixture, as well as a Southeast Asian heroin HCl sample, is shown in Figure 11.7. The use of a mobile phase containing ammonium bicarbonate (pH 10.0) *versus* 1% formic acid (pH 2.0) allowed for less band spreading of the heroin peak, major selectivity differences, approximately 10 times higher limits of detection and the analysis of certain neutral impurities as low as 10^{-5} % w/w. Although operating at the higher pH had certain advantages, a major concern was the hydrolysis of heroin to O6-monoacetylmorphine in the basic injection solvent. Using an acidic injection solvent (which inhibits hydrolysis), gave rise to double peaks. The use of the aprotic injection solvent acetonitrile (which inhibits hydrolysis), with a 2 μL injection (instead of a 10–20 μL injection) to prevent peak distortion is an option. However, a higher heroin sample concentration would be required, which could be problematic for cut samples.

Debrus *et al.*[35] also investigated the use of RP UHPLC-MS/MS with positive ESI in the MRM mode for heroin profiling, in particular for tactical intelligence. Employing 200 ng of heroin on a column with gradient analysis allowed the trace level detection of solutes such as noscapine, papaverine, meconine, acetylthebaol and the non-trace level detection of heroin, acetylcodeine and O6-monoacetylmorphine. Sample comparison was accomplished by employing chemometric tools on peak areas from individual runs. In order for the present data to be compatible with an existing GC-MS database, it was necessary to develop a mathematical model to convert LC-MS data into "GC-like" data. This was required because of the different relative peak intensities of the two techniques. A pH 9 ammonium formate buffer was used because it allowed higher selectivity and retention than an acidic pH buffer. In addition, lower MS limits of detection were obtained because of a better desolvation when analytes were eluted with a higher proportion of organic solvents. Lurie and Toske[34] obtained higher limits of detection using a high pH *versus* a low pH buffer, apparently because the

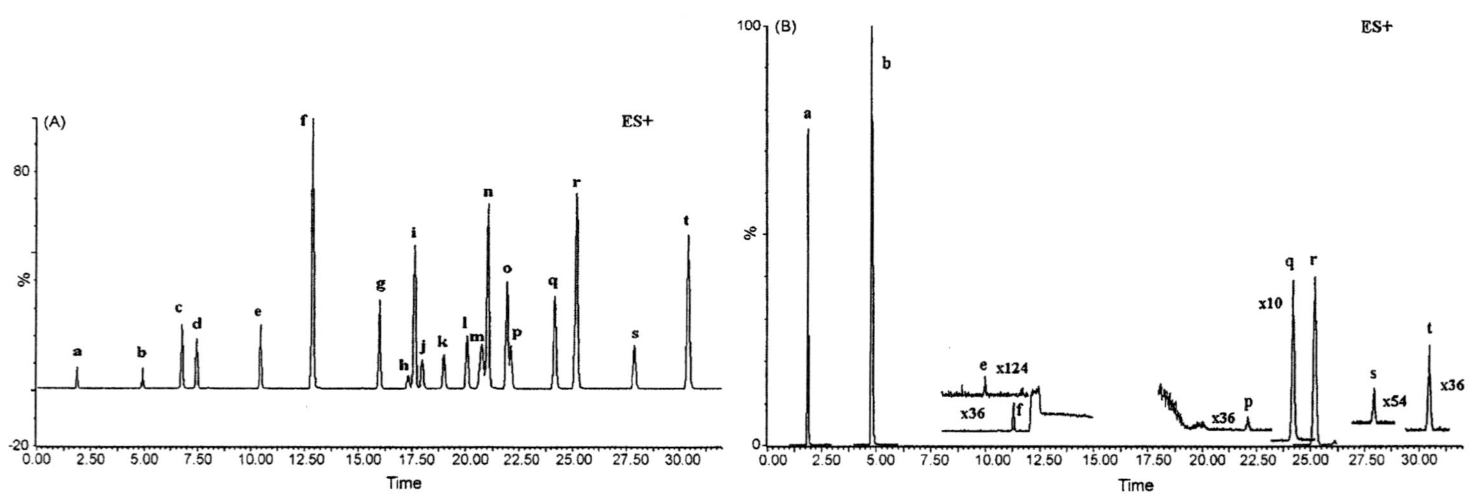

Figure 11.7 UHPLC-MS/MS gradient separation of (A) TIC of MRMs for a standard mixture of heroin, and heroin impurities [1 μg mL^{-1} of each solute; injection size, 20 μL (overfill mode)] and (B) timed programmed MRMS for a Southeast Asian heroin HCl exhibit (heroin concentration 3.75 mg mL^{-1}; injection size, 10 μL). UHPLC conditions: column, 100 mm × 2.1 mm Acquity UPLC BEH C18. Initial conditions: 5% acetonitrile, 95% of 1% formic acid (pH 2.0). Final conditions: 21% acetonitrile, 79% of 1% formic acid (pH 2.0), 32 min linear gradient, 2.0 min gradient equilibration; flow rate, 300 μL min^{-1}; temperature, 30 °C. Peaks: (a) morphine, (b) codeine, (c) O3-monoacetylmorphine, (d) O6-monoacetylmorphine, (e) boldine (heroin exhibit possible bodine isomer), (f) reticuline, (g) codamine, (h) thebaine, (i) laudanidine, (j) acetylcodeine, (k) heroin, (l) reticuline monoacetate (1), (m) cryptopine, (n) laudanosine, (o) norlaudanosine, (p) reticuline monoacetate (2), (q) papaverine, (r) noscapine, (s) narceine, and (t) reticuline diacetate. Reprinted from *Journal of Chromatography A*, **1188**, I. S. Lurie and S. G. Toske, Applicability of ultra-performance liquid chromatography–tandem mass spectrometry for heroin profiling, 322–326, 2008, with permission from Elsevier.

mobile phase containing ammonium bicarbonate generates a considerable background noise due to is relative lack of volatility. A concern of employing basic buffers (pH 9–10) is possible shifts in retention (pH close to pK_a) unless precise pH control is utilized.

Lurie *et al.*[36] presented methodology for the profiling of illicit fentanyl using UHPLC-MS/MS with positive ESI in the MRM mode. RP chromatography was employed with gradient analysis. Target analysis was performed for 40 fentanyl processing impurities, several of which are markers for either a Siegfried or Janssen synthetic route. The coupling of the high separation power and relatively high loading capacity of UHPLC, with highly selective and sensitive detection of MS/MS is amenable to the determination of synthetic route and linking of drug seizures.

Compared to RPLC, HILIC using MS compatible mobile phases offers >100 times increased loading (at or near the highest N value) for retained basic drugs.[26] In addition for these same solutes, HILIC offers significantly higher N values and improved peak shapes *versus* RPLC.[26]

The applicability of HILIC UHPLC-MS/MS with positive ESI in the MRM mode for methamphetamine profiling has been demonstrated.[14] Unlike GC-MS, where liquid–liquid or solid phase extraction is required to remove methamphetamine, HILIC UHPLC-MS/MS allows for 10 μg of methamphetamine on column with minimal interference from the major component.[37,38] Not only does HILIC UHPLC-MS/MS afford excellent limits of detection, but allows for the analysis of solutes removed by liquid–liquid or solid phase extraction.

References

1. B. A. Perillo, R. F. X. Klein and E. Franzosa, *Forensic Sci. Int.*, 1994, **69**, 1.
2. K. Narayanaswami, H. C. Golani and R. D. Dua, *Forensic Sci. Int.*, 1979, **14**, 181.
3. I. S. Lurie and K. McGuinness, *J. Liq. Chromatogr.*, 1987, **10**, 2189.
4. I. S. Lurie, *LC-GC*, 1990, **8**, 454.
5. I. S. Lurie, P. A. Hays, A. D. Garcia and S. Panicker, *J. Chromatogr. A*, 2004, **1034**, 227.
6. I. S. Lurie, J. M. Moore, D. A. Cooper and T. C. Kram, *J. Chromatogr.*, 1987, **405**, 273.
7. H. Huizer in *The Analysis of Drugs of Abuse*, ed. T. Gough, John Wiley and Sons, Chichester, 1991, p. 23.
8. I. S. Lurie, *J. Chromatogr. A*, 2005, **1100**, 168.
9. I. S. Lurie, P. A. Hays, and K. Parker, *Electrophoresis*, 2004, **25**, 1580.
10. I. S. Lurie, P. A. Hays, A. R. Sperling and R. P. Meyers, *J. Forensic Sci.*, 1994, **39**, 74.
11. I. S. Lurie and L. Li, *J. Liq. Chromatogr.*, 2009, **32**, 2615.

12. J. Z. Min, Y. Shimizu, T. Toyo'oka, S. Inagaki, R. Kikura-Hanajiri and Y. Goda, *J. Chromatogr. B*, 2008, **873**, 187.
13. D. T. T. Nguyen, D. Guillarme, S. Heinisch, M. P. Barrioulet, J. L. Rocca, S. Rudaz and J. L. Veuthey, *J. Chromatogr. A*, 2007, **1167**, 76.
14. E. S. Grumbach, D. M. Diehl and U. D. Neue, *J. Sep. Sci.*, 2008, **31**, 1511.
15. G. Jiang, *American Lab.*, 2010, **42**, 40.
16. G. Jiang, R. Chen and C. Loran, *The Application Notebook LC GC*, July 2 2008.
17. G. Jiang, *Forensic Magazine*, February/ March 2009.
18. K. Kanai, K. Takekawa, T. Kumamoto, T. Ishikawa and T. Ohmori, *Forensic Toxicol.*, 2008, **26**, 6.
19. J. R. Stenzel and G. Jiang, *Thermo Fisher Application Notebook*, 2009, **1**.
20. J. R. Stenzel and G. Jiang, *The Application Notebook LC GC*, February 1, 2009.
21. J. R. Stenzel and G. Jiang, *Thermo Scientific Application Note* 433, 2009.
22. N. Uchiyama, R Kikura-Hanajiri, N. Kawahara, Y. Haishima and Y. Goda, *Chem. Pharm. Bull.*, 2009, **57**, 439.
23. N. Uchiyama, R. Kikura-Hanajiri, N. Kawahara and Y. Goda, *Forensic Toxicol.*, 2009, **27**, 61.
24. N. Uchiyama, R. Kikura-Hanajiri, J. Ogata and Y. Goda, *Forensic Sci. Int.*, 2010, **198**, 31.
25. N. Uchiyama, M. Kawamura, R. Kikura-Hanajiri and Y. Goda, *Forensic Toxicol.*, 2011, **29**, 25.
26. I. S. Lurie, L. Li and S. G. Toske, *J. Chromatogr. A*, 2011, **1218**, 9336.
27. D. V. McCalley, *J. Chromatogr. A*, 2007, **1171**, 46.
28. L. G. Apollonio, I. R. Whitall, D. J. Pianca, J. M. Kyd and W. A. Maher, *Rapid Commun. Mass Spectrom.* 2006, **20**, 2259.
29. I. S. Lurie and R. Iio, *J. Chromatogr. A*, 2009, **1216**, 1515.
30. J. Casale and J. Lydon, *J. Forensic Sci.*, 2007, **52**, 573.
31. K. A. Anderson, W. T. Cobb, and B. R. Loper, *Commun. Soil Sci. Plant Anal.*, 2001, **32**, 2831.
32. A. Garcia, I. S. Lurie, presented at 2001 American Academy of Forensic Science Meeting, Seattle, WA, February 2001.
33. R. D. Blackledge and M. D. Miller, *Microgram J.*, 1991, **24**, 172.
34. I. S. Lurie and S. G. Toske, *J. Chromatogr. A*, 2008, **1188**, 322.
35. B. Debrus, J. Broseus, D. Guillarme, P. Lebrun, P. Hubert, J. L. Veuthey, P. Esseiva and S. Rudaz, *Anal. Bioanal. Chem* , 2011, **399**, 2719.
36. I. S. Lurie, A. L. Berrier, J. F. Casale, R. Iio and J. S. Bozenko, Jr, *Forensic Sci. Int.*, in preparation.
37. K. Kuwayama, K. Tsuijkawa, H. Miyaquchi, T. Kananori, Y. Iwata, H. Inoue, S. Saitoh and T. Kishi, *Forensic Sci. Int.*, 2006, **160**, 44.
38. J. S. Lee, E. Y. Han, S. Y. Lee, E. M. Kim, Y. H. Park, M. A. Lim, H. S. Chung and J. H. Park, *Forensic Sci. Int.*, 2006, **161**, 209.

CHAPTER 12

UHPLC-MS for Multi-residue Screening of Pharmaceuticals in Environmental Samples

REBECA LÓPEZ-SERNA[a], SANDRA PEREZ[a], MIRA PETROVIC*[b,c] AND DAMIÀ BARCELÓ[a,b]

[a] Department of Environmental Chemistry, IDAEA-CSIC, c/Jordi Girona 18–26, 08034 Barcelona, Spain; [b] Catalan Institute for Water Research (ICRA), c/Emili Grahit, 101, Cientific and Technologic Park of Girona University, 17003 Girona, Spain; [c] Catalan Institution for Research and Advanced Studies (ICREA), c/Lluís Companys, 23, 08010 Barcelona, Spain
*E-mail: mpetrovic@icra.cat

12.1 Pharmaceuticals as Environmental Contaminants

Pharmaceutically active compounds (PhACs) are a group of chemical substances that have medicinal properties and are produced worldwide on a 100 000 tonne scale. Most modern drugs are small organic compounds with a molecular weight (MW) below 500 Da, which are moderately water soluble, as well as lipophilic, in order to be bioavailable and biologically active. After the oral, parenteral and/or topical administration, PhACs are excreted *via* urine or faeces as a mixture of parent compound and metabolites that are usually more polar and hydrophilic than the original drug. Afterwards, depending on the efficiency of the waste-water treatment and chemical nature of a compound, PhACs can reach surface waters and groundwaters, where they can undergo

RSC Chromatography Monographs No. 16
UHPLC in Life Sciences
Edited by Davy Guillarme and Jean-Luc Veuthey

Published by the Royal Society of Chemistry, www.rsc.org

different chemical, photolytic and biological reactions that modify the structure and physical transport of a compound.

The significance of pharmaceuticals as trace environmental pollutants in waterways, and on land to which treated sewage sludge or wastewater has been applied, is largely unknown.[1] Due to several facts, PhACs deserve special attention because: (i) their continuous introduction *via* effluents from sewage treatment facilities and from septic systems, pharmaceuticals are referred to as "pseudo" persistent contaminants (*i.e.* high transformation/removal rates are compensated by their continuous introduction into environment); (ii) they are developed with the intention of performing a biological effect; (iii) pharmaceuticals often have the same type of physico-chemical behaviour as other harmful xenobiotics (persistence in order to avoid the substance to be inactive before having a curing effect, and lipophilicity in order to be able to pass membranes); and (iv) pharmaceutical substances are used by humans in quantities similar to those of many pesticides.

When released into the environment, pharmaceuticals can have unintended effects on animals and micro-organisms. Although the effects of the pharmaceuticals are investigated through safety and toxicology studies, the potential environmental impacts of their production and use are less understood and have only recently become a topic of research interest. Particularly, there are a lack of data regarding effects on the aquatic ecosystems resulting from long-term low-dose exposure. Such analysis of the chronic toxicity of PhACs on organisms is essential to obtain a realistic environmental risk assessment; however, the direct estimation of effects caused by PhACs on the ecosystems is not a straightforward task.[2]

The first step, as the prerequisite for a proper risk assessment, is the analysis of PhACs (both parent compounds and metabolites or transformation products) in different environmental compartments. Therefore, the availability of a multi-residue analytical method which permits the measurement of pharmaceuticals belonging to different therapeutic classes at low ng L^{-1} levels or even below is the key issue.

12.2 Analysis of Pharmaceuticals in Environmental Samples

In the European Union (EU), approximately 3000 PhACs are approved to be used in human medicine; however, only a small subset of these compounds ($\sim$200) has been investigated in environmental studies. A large number of analytical methodologies have been developed for the determination of pharmaceutical residues in both surface and wastewaters. Generally, liquid chromatography–tandem mass spectrometry (LC-MS/MS) is indicated as the technique of choice to assay pharmaceuticals and their metabolites, and is especially suitable for environmental analysis because of its selectivity.[3] Due to development of hyphenated chromatography–mass spectrometry (MS) techniques, the impressive improvements in detection limits for organic contaminants have pushed the target concentrations from the microgram to the nanogram, or pictogram, per litre range. Consequently, a number of new or previously ignored or unrecognized contaminants, some of them being pharmaceuticals, have been brought under scrutiny.

The need for increased capabilities in environmental analysis is now driving both new strategies and instrument advances. In the recent years, two general trends are observed in the analysis of pharmaceuticals in environmental samples. The first one is the development of multi-residue methods as more efficient alternative to group-specific methods. The multi-residue methods are becoming the preferred and required tools against single group analysis, as they provide wider knowledge about occurrence of contaminants in the aquatic media which are necessary for further study of their removal, partition and ultimate fate in the environment, especially when results for multiple parameters are required. The second general trend is the shortening of the analytical run times as a consequence of a growing demand for high-throughput analysis due to a growing number of samples in laboratories conducting environmental monitoring studies. Run times of several tens of minutes are not tolerable for truly high-throughput analyses and nowadays emphasis is directed towards achievement of maximum chromatographic resolution in a drastically reduced time. Among modern approaches in achieving fast separation without compromising resolution and separation efficiency, the most often used approach, is liquid chromatography (LC) at ultra-high pressures using sub-2 μm particles packed columns [ultra-high pressure liquid chromatography (UHPLC)], while other options for achieving fast chromatography, such as the use of monolith columns, the use of fused core columns, and high-temperature liquid chromatography (HTLC) are less frequently employed.[4]

Majority of UHPLC applications for the analysis of PhACs is focused on the analysis of water matrices, such as waste and river water,[5–10] with fewer methods developed for sea[11,12] and groundwater.[13] Solid environmental matrices such as sediment, soil or sewage sludge have received much less attention and only a few methods have been developed for these samples using UHPLC technology.[14–17]

The majority of methods were developed for the purpose of quantitative determination of target compounds, usually monitoring multi-class pharmaceuticals, while only several methodologies were designed for the screening of non-target pharmaceuticals.[15,18–23] Non-target screening is mainly oriented toward identification of transformation products in bio- or photo-degradation studies. Among the pharmaceuticals, the therapeutic groups most frequently studied by UHPLC are antibiotics, followed by analgesics, anti-inflammatory and cardiovascular drugs. Table 12.1 provides an overview of the most representative methods based on UHPLC-MS for the quantitative determination of target pharmaceuticals in different environmental samples.

Table 12.1 Multi-residue methods for the analysis of pharmaceuticals in environmental samples.

No compounds/ no PhACs	*Matrix*	*Sample preparation*	*UHPLC system*	*Detection*	*Reference*
47/47	SW, WW	SPE (Oasis HLB)	Waters Acquity UPLC	ESI-QqQ	29
27/25	Reverse osmosis retentate	SPE (Oasis HLB)	Waters Acquity UPLC	ESI-QqQ	24
15/13	SW and WW	SPE (Oasis HLB)	Waters Acquity UPLC	ESI-QqQ	48
38/28	SW	SPE (Oasis HLB)	Waters Acquity UPLC	ESI-QqToF	5
10/9	Sea water	Bag-SPE (XAD-2)	Waters Acquity UPLC	ESI-QqToF	11
58/37	DW, SW and WW	Hand-assembled SPE mixed cartridge (Oasis HLB, Strata X-CW, Strata X-AW, Isolute ENV+)	Waters Acquity UPLC	ESI-QqQ	44
55/55	WW	SPE (Oasis HLB)	Waters Acquity UPLC	ESI-QqQ	35
10/10	SW	SPE (Oasis HLB)	Merck L-2160U	TUV	6
20/20	SW and WW	SPE (Oasis HLB)	Waters Acquity UPLC	ESI-QqQ	49
7/7	SW, WW	SPE (MIP)	Waters Acquity UPLC	ESI-QqQ	50
13/13	WW	Bag-SPE (XAD-2)	Waters Acquity UPLC	ESI-QqToF	51
15/15	WW	SPE (Oasis MCX and MAX in series)	Waters Acquity UPLC	ESI-QqToF	52
42/42	SW and WW	SPE (Oasis HLB)	Waters Acquity UPLC	ESI-QqToF	21
54/54	SW and WW	SPE (Oasis MCX)	Waters Acquity UPLC	ESI-QqQ	26
54/36	SW and WW	SPE (Oasis MCX)	Waters Acquity UPLC	ESI-QqQ	9
32/28	DW and WW	SPE (Oasis HLB)	Waters Acquity UPLC	ESI-QqToF	13
28/25	SW	SPE (Oasis MCX)	Waters Acquity UPLC	ESI-QqQ	10
30/8	Sediment	USE [ACN/water (5:3, v/v)]; SPE (Oasis HLB)	Agilent 1200	ESI-QqQ	14
6/6	Sediment	USE (0.1% FA in MeOH); SPE (Oasis HLB)	Agilent 1200	ESI-QqQ	16

Table 12.1 (*Continued*)

No compounds/ no PhACs	*Matrix*	*Sample preparation*	*UHPLC system*	*Detection*	*Reference*
30/8	Sewage sludge	USE [ACN/water (5:3, v/v)]; SPE (Oasis HLB)	Agilent 1200	ESI-QqQ	14
6/6	Sewage sludge	USE (0.1% FA in MeOH); SPE (Oasis HLB)	Agilent 1200	ESI-QqQ	16
11/11	Biota (*Mytilus edulis*)	PLE [1% FA in ACN/water (3:1)]; SPE (Strata-X)	Thermo Accela	ESI-QqQ	45

DW, drinking water; SW, surface water; WW, wastewater.

12.2.1 Sampling and Sample Preparation for UHPLC-MS Analysis

Generally, samples obtained from environmental sources are complex mixtures of organic and inorganic compounds, consisting of thousands of individual components, therefore the analysis of PhACs constitutes a difficult task. Difficulties arise from the complexity of the matrices and because of very low concentrations of the target compounds. A variety of chemical and physical characteristics of PhACs belonging to different therapeutic groups, often present in complex mixtures, requires a comprehensive approach for their analysis and often imposes a compromise in the selection of experimental conditions for their extraction, LC separation and MS detection.

12.2.1.1 Aqueous Matrices

The sampling strategy commonly applied for the monitoring of surface waters (river, lake or sea) is grab sampling over a certain sampling period. When studying the efficiencies of drinking and waste-water treatment plants (DWTP and WWTP), composite sampling is typically applied. Composite sampling of DWTP and WWTP is a technique whereby multiple temporally discrete water samples are collected using automated sampling devices that apply different sampling modes (continuous flow-proportional or constant sampling, discrete time-proportional, flow-proportional or volume-proportional) and thoroughly homogenized, and treated as a single sample. Usually, 24 h composite samples are taken for the influent and the effluent of the water treatment plant, by taking into account the hydraulic retention time (HRT) in order to evaluate properly the efficiency of the treatment.

For the analysis of pharmaceuticals, in the majority of published methods, no preservatives were used, except for references 16 and 24 in which phosphoric acid and NaN_3 was added, respectively. In order to preserve the integrity of the target analytes, the most acceptable practice is to keep the sample under cooling conditions and to perform extraction within 24 h or a maximum 48 h after sampling.

Due to low concentrations in aqueous samples (typically ranging from low ng L^{-1} in environmental water to low μg L^{-1} in raw wastewater), a pre-concentration step, usually using on-line or off-line solid phase extraction (SPE), respectively, is required. Typical problems encountered when performing a multi-residue analysis of PhACs are related to the simultaneous extraction of compounds with polarities varying from less or non-polar to highly polar compounds. Simultaneous extraction of all target analytes in one single SPE step is the approach most widely employed. Another widely employed option consists in the combination of two or more SPE materials operating either in series or in parallel to classify target analytes in two or more groups according to their physico-chemical properties.

For the first approach, the Oasis HLB from Waters and Strata X from Phenomenex, which exhibits both hydrophilic and lipophilic retention characteristics, are the preferred sorbents and were used for simultaneous extraction of neutral, basic and acidic pharmaceutical residues.[25] Both are polymeric sorbents with improved wetting characteristics and mass transfer, and with additional possibilities for the interaction of functional groups of analytes that allows high pre-concentration factors. Due to the high retention capabilities of these sorbents, acidic compounds can be extracted from water samples without previous acidification. This is of great importance when performing a multi-residue analysis, because the risk of acidic hydrolysis of other compounds is not enhanced. Furthermore, no additional clean-up step is needed for the removal of humic and fluvic acid, and also there is a possibility to couple on-line extraction units using large sample volumes. Another often used SPE sorbent is Oasis MCX from Waters that contains both a hydrophobic–lipophilic balanced co-polymer for a reversed-phase interaction and a strong cation exchanger for the selective retention of basic analytes. Less common cartridges employed are C_{18} and Lichrolut ENV+ from Merck, since they generally require pH adjustment prior to extraction depending on the acidic, neutral or basic nature of the analytes.

In multi-residue methods including antibiotics, Na_2EDTA is added to the sample before the SPE in order to prevent complexation of analytes (*i.e.* tetracycline, sulfonamides and polypeptide antibiotics) and therefore improve their recovery.[9,21,24,26] In addition, when analysing tetracycline, it is highly recommended to use polytetrafluoroethylene (PTFE) instead of glass materials, since they tend to adsorb to the glass, resulting in significant losses.[27]

As mentioned before, the simultaneous extraction of compounds from different groups with quite different physico-chemical characteristics requires a compromise in the selection of experimental conditions (*i.e.* sorbent material, pH, elution solvent), which in some cases are not the best conditions for all the analytes studied. For example, in the analysis of 74 pharmaceuticals, López-Serna *et al.*[28] used Oasis HLB and reported recoveries in the range of 50–150% for the 80, 76, 65, 61 and 57% of target compounds in HPLC grade water, groundwater, river water, WWTP effluent and WW influent, respectively. Some polar compounds, such as salbutamol, famotidine, atenolol, cimetidine, sotalol and ranitidine, as well as some strongly hydrophobic compounds such as atorvastatin, tamoxifen, phenylbutazone and mevastatin, showed rather poor recoveries ($<$50%).

Other example is the method of Gracia-Lor *et al.*[29] who developed a multi-class method for the simultaneous quantification and confirmation of 47 pharmaceuticals in environmental and wastewater samples. Using Oasis HLB and after correction with isotopically labelled internal standards, some compounds still yielded quite low recoveries (norfloxacin, 40%; ketoprofen, 55%; naproxen, 55%), whereas for the rest of compounds the recoveries were over 70%.

12.2.1.2 Solid Matrices

The complexity and the presence of interfering substances in solid environmental samples, such as sediment, soil or sewage sludge, demands either a very selective detection or tedious extract clean-up, or even both. Generally, multistep sample pre-treatment aimed at the reduction of the matrix content and the enrichment of the target compounds still remains the most direct means of obtaining maximum sensitivity.

In the case of solid samples, water is usually removed by freeze-drying the sample, *i.e.* by lipophilization or by drying at room temperature in a clean fume hood. Afterwards, there are various procedures described in the literature concerning the extraction of pharmaceutical compounds. In recent years, these methods have usually been based on liquid partitioning with ultrasonication (USE), microwave-assisted extraction (MAE) or the more advanced pressurized liquid extraction (PLE).[30,31] The extraction solvents commonly applied are methanol (MeOH), acetonitrile (ACN), water and acidified water (*e.g.* acetic acid), in different proportions. For the extraction of tetracycline antibiotics by PLE special precaution must be taken, since temperatures higher than room temperature can cause their transformation into epi- or anhydrous forms, and also they tend to form complexes with metal ions. Also, values higher than 100 °C promote the degradation of macrolides.[32]

The most effective clean-up of the extracts of sediment/soil/sludge samples containing pharmaceutical residues have been proved to be SPE using Oasis HLB, Oasis MCX and Strata-X cartridges.

12.2.2 Chromatographic Separation using UHPLC

High-performance liquid chromatography (HPLC) has been the most used separation method for the analysis of non-gas chromatography-amenable organic molecules such as pharmaceuticals. HPLC uses packed columns with typical particle sizes from 3 to 5 μm. It presents many advantages such as robustness, ease of use and good selectivity; however, the main limitation can be seen in the large run time needed when high peak capacity is wanted. In the last decade, to achieve fast separation without compromising resolution, new approaches have emerged as follows: (a) HTLC, to decrease the solvent viscosity and permit faster flow rates with low backpressure; (b) columns packed with sub-3 μm shell particles, allowing the generation of higher plate count than regular HPLC columns, with reasonable backpressure; (c) monolithic columns which permits the increase of the flow rate without generating high backpressure; and (d) the use of columns packed with sub-2 μm porous particles in UHPLC, drastically improving the performance of conventional HPLC but at higher backpressures.[4] The well-known van Deemter equation for HPLC shows that the efficiency of the chromatographic separation is largely governed by the particle size. Small particles provide more uniform flow and less solute diffusion through the column. By decreasing the particle size of the packing material using UHPLC, the analyst can reduce the height of a theoretical plate, making the use of shorter column lengths possible and widening the range of usable flow rates, without sacrificing separation power. Therefore, the analytical method is shortened without losing separation efficiency. UHPLC has been introduced to enhance sample throughput and reduce analysis time and ultimately mobile phase consumption compared to HPLC. For instance, comparing columns packed with sub-2 μm particles for UHPLC with 3 μm particle size HPLC columns, the analysis should be shortened by about three times.[33] However, UHPLC columns packed with sub-2 μm particles lead to higher backpressures. For instance, reducing the particle diameter by a factor of three results in an increase in the backpressure by a factor of nine times for the same plate count.[33] Therefore, specific instrumentation, such as pumps holding 15 000 psi of pressure limit and also specific injection valves with fast injection cycles, are required. Until 2000, the columns used were not sufficiently stable for such purposes. Some researchers[34] started testing UHPLC conditions using fused silica capillaries. Fused silica capillaries with inner diameters of 29 μm and lengths up to 70 cm were packed with non-porous 1.5 μm isohexylsilane-modified (C_6) silica particles using a carbon dioxide-enhanced slurry packing method. They obtained efficiencies as high as 570 000 plates per m at optimum linear velocity. Analysis times were reduced considerably, both by operating at higher than optimum linear velocities and by shortening the column.

In UHPLC methods developed for pharmaceuticals, run times between 2 min[6] and 25 min[9] are achieved depending on the number of compounds analyzed and the used flow rate. Figure 12.1 shows an example of UHPLC separation of 50 pharmaceuticals (among them sic metabolites) and five oestrogens.[35] MeOH and ACN were the most often used organic mobile phases. ACN was probably preferred because, although comparable resolution and reproducibility are acquired by both, ACN usually provides shorter retention times as well as better peak shapes. In addition, the use of ACN results in a 2-fold reduction of backpressure compared to MeOH due to lower viscosity. Modifiers of the organic as well as the aqueous phase were also normally used as they provide ion pairing which helps in the peak resolution and shape. Since most of the detectors used are mass spectrometers (MS) with electrospray ionization (ESI) interfaces, for positive electrospray ionization (PI), formic acid and acetic acid and their corresponding buffers with ammonium salts (ammonium formate and ammonium acetate, respectively) are the most commonly used, since they also provide an acid pH to the mobile phase which will help to the ionization of the analytes in the ESI interface.

12.2.3 MS Detection

The last two decades, LC-MS has been the most used instrument for the qualitative and quantitative analysis of pharmaceuticals in environmental samples. Because of high speed separations in UHPLC, and consequently the analyte elution in very narrow peaks (commonly 2–6 s), they have to be

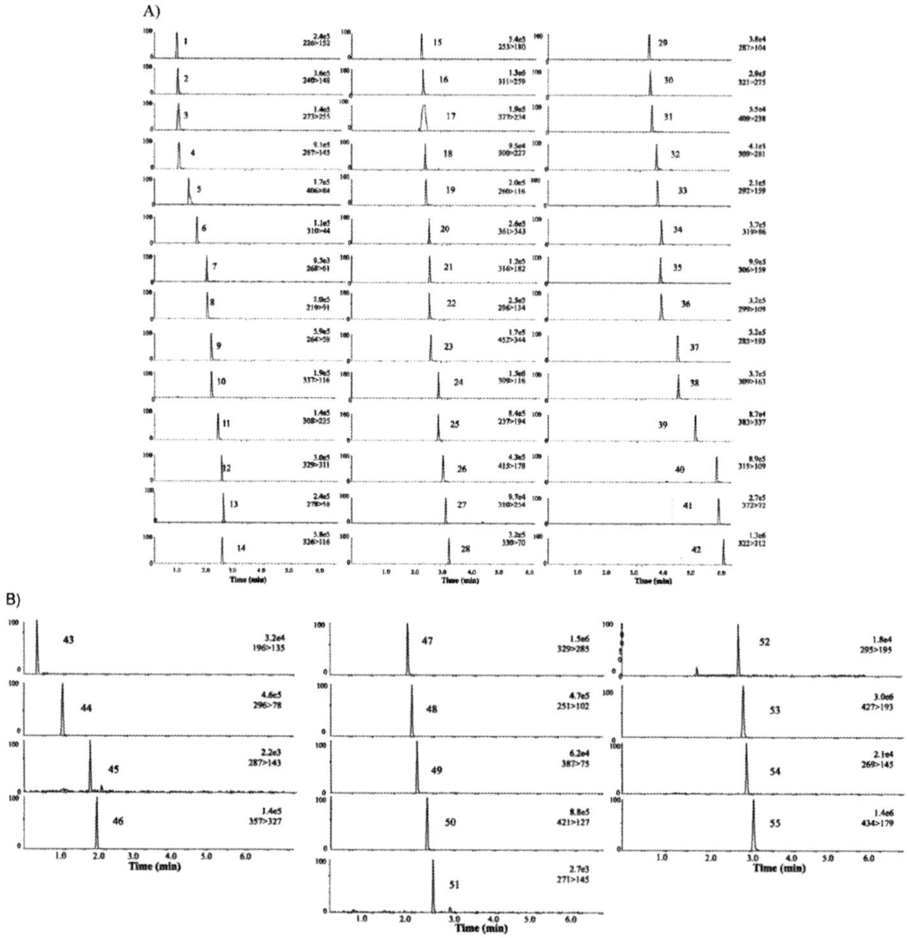

Figure 12.1 Total ion chromatogram of a standard mixture (1.5 μgL^{-1}) in SRM acquisition mode of 50 pharmaceuticals (among them six metabolites) and five oestrogens. (a) Positive ESI; (B) negative ESI. 1, Terbutaline; 2, salbutamol; 3, sotalol; 4, atenolol; 5, lisinopril; 6, fluoxetine; 7, metoprolol; 8, primidone; 9, desmethylvenlafaxine; 10, acebutolol; 11, zolpidem; 12, labetalol; 13, venlafaxine; 14, bisoprolol; 15, carbamazepine epoxide; 16, desloratadine; 17, enalapril; 18, chlordiazepoxide; 19, propranolol; 20, prednisolone; 21, bromazepam; 22, norfluoxetine; 23, doxazosin; 24, betaxolol; 25, carbamazepine; 26, diltiazem; 27, nadolol; 28, paroxetine; 29, oxazepam; 30, lorazepam; 31, amlodipine; 32, alprazolam; 33, desmethylsertraline; 34, chlorpromazine; 35, sertraline; 36, norethindrone; 37, diazepam; 38, warfarin; 39, loratadine; 40, progesterone; 41, warfarin; 42, clopidogrel; 43, Idopa; 44, hydrochlorothiazide; 45, estriol; 46, prednisone; 47, furosemide; 48, phenytoin; 49, cetirizine; 50, losartan; 51, estradiol; 52, ethynyl oestradiol; 53, irbesartan; 54, oestrone; 55, valsartan. Reprinted with permission from Huerta-Fontela *et al.*,[35] © Elsevier 2010.

monitored by detectors that offer acquisition rates high enough to record a reasonable number of data points across the chromatographic peak. MS presents many advantages over light absorption detectors. Apart from retention time matching, MS is capable of measuring the compound mass that allows positive identification of compounds in the environmental samples.[36] For pharmaceutical analysis, the three most frequently instruments used in combination with UHPLC are triple quadrupole (QqQ), (quadrupole-) time-of-flight (ToF) and Orbitrap MS. Whereas older QqQ instrumentation may still be compatible with UHPLC if a limited number of transitions are recorded or if the mass range in full-scan acquisitions is not too wide, multi-analyte methods with monitoring a large number of transitions or methods that involve data-dependent analysis require high acquisition rates and very short inter-channel delays. The most powerful quadrupole-based instruments can nowadays meet these requirements.[37] A mass analyzer that is fully compatible with UHPLC separations is the ToF-mass spectrometer (ToF-MS), which is capable of recording and storing complete mass spectra at rates as high as 100 Hz.[38] In contrast to the scanning mode in QqQ, the duty cycle in ToF-MS is much higher as essentially all ions introduced into the flight tube in each pulsed extraction can be recorded. This feature allows generating mass spectra at a rate much faster than all other mass analyzers. In addition to speed, other characteristics of ToF-MS including high ion transmission, unlimited mass range, simplicity, and ease of peak deconvolution without sacrificing peak shape or separation efficiency, all make ToF-MS an ideal detector for fast separations.[36] Operating the mass analyzer at higher acquisition rates, however, implies reducing the number of transient accumulated to generate mass spectra. Therefore, the high acquisition rate comes at the expense of sensitivity. A further ion analysis technique is the Orbitrap mass analyzer which is becoming more popular as a MS technique.[39] However, the coupling of UHPLC to Orbitrap is not so efficient because the latter suffers from a slower data acquisition rate compared to (Q)ToF instruments, and therefore it is not fully compatible with fast chromatography at resolution of 100 000.[40] For example, LTQ Orbitrap at an acquisition rate of 2.5 Hz provides a mass resolution of 30 000, equivalent to 25 data points across a peak of width at baseline of 10 s.[39] Regarding the acquisition of product ion spectra, the instrument can either be operated in the parallel-scanning linear ion trap at low resolution or in the Orbitrap at a resolution of 7500 with a scan time of 0.1 s. In combination with a full-scan resolution of 30 000 [full width at half maximum (FWHM)], this stills allow the collection of 20 data points across the peak.[38] Regarding the coupling of UHPLC to the stand-alone Orbitrap analyzer Exactive™, it is also slower than ToF instruments and it can deliver 30–40 points across the 3–4 s wide-peak.[39] Setting the resolution of the Orbitrap analyzer to maximum resolution comes at the expense of data acquisition rate. Depending on the objective of the study, this circumstance is less of an issue in qualitative analysis in which spectral information at high resolution is of interest. For quantitative studies, however, well-defined chromatographic peaks are crucial for accurate peak integration. For the successful coupling of the Exactive™ to UPLC, Herrera Rivera *et al.*[41] set the resolution of the mass analyzer to 50 000 (FWHM) in their analysis of 11 pharmaceuticals in different water samples.

12.2.4 Matrix Effect in UHPLC-MS Analysis

In multi-residue methods aimed at analyzing multi-class pharmaceuticals with different polarities, typically, generic SPE protocols with sorbents able to simultaneously extract all analytes (such as Oasis HLB) are used, which also results in the efficient extraction of matrix components. Such extraction and further simplification of the sample clean-up step, in order to reduce the analysis time, results in dirty extracts with high co-extractive substance content which led to a signal enhancement or suppression during MS ionization. Atmospheric pressure ionization (API) interfaces are especially susceptible to matrix interferences, and in the case of very complex matrix such is wastewater, even when using selected reaction monitoring (SRM) detection, both false-negative results (due to "ion suppression") and false-positive results (due to insufficient selectivity, "ion enhancement") can be obtained.[42] The probability and the intensity of this effect, is expected to be more significant for matrixes with a high dissolved carbon content (DOC), such as wastewater and solid samples,[10] or with a high salinity, such as sea water.[12] There are several strategies in quantification to reduce matrix effect.[12,31] They comprise the use of suitable calibration approaches, such as external calibration using matrix-matched samples, standard addition or internal standards, as well as the dilution of sample extracts. Although they correct the matrix effect, they have some limitations or simply disadvantages. Standard addition is a reliable method, but it is time-consuming. On the other hand, an appropriate internal standard (a structurally similar unlabelled compound or isotopically labelled standard) are not always commercially available or they are expensive. In addition, in the case of multi-residue screening, a huge number of standards would be necessary because of the important number of compounds to be monitored, which is unacceptable. In case of lack of matrix-matched materials and isotopically labelled standards, sample extract dilution can be an effective alternative solution.

In addition to above mentioned approaches, several studies[8,21,43] reported that application of UHPLC may contribute to the reduction in ion suppression that result from co-elution of the metabolites and endogenous compounds and its application therefore helps to alleviate problems related to the matrix effect. Using HPLC-MS, matrix effects are generally substantial and often cannot be compensated for with analogue internal standards. Due to the better resolution and more narrow peaks in UHPLC, analytes co-elute less with interferences during ionization, so matrix effects could be lower, or even eliminated.

12.3 Occurrence of Pharmaceuticals in Environmental and Wastewater Samples

Generally, pharmaceuticals are found to be widespread environmental contaminants. The most abundant loads in urban wastewaters, which are main source of pharmaceuticals, are commonly reported for non-steroidal anti-inflammatory drugs (NSAIDs), such as ibuprofen, acetaminophen and naproxen, which is attributed to their wide consumption as over-the-counter (OTC) drugs. Besides these OTC drugs, pharmaceuticals ubiquitous in raw sewage are long-term prescription drugs such as β-blockers (atenolol, metoprolol and propranolol), lipid regulators (bezafibrate, atorvastatine) and psychiatric drugs (carbamazepine). The occurrence of antibiotics in the environment is also considered to be substantial, due to their widespread consumption in human and veterinary medicines.

Maximum concentrations in raw wastewaters can reach levels of several μg L^{-1}, as reported for some compounds such as the antibiotics azithromycin,[44] clarithromycin[44] and metronidazole,[44] the psychiatric drugs carbamazepine,[44] oxazepam[35] and gabapentin,[44] the lipid regulator bezafibrate,[44] and the cardiovascular drugs atenolol,[35,44] hydrochlorothiazide,[35] furosemide[35] and enalapril.[35] Concentrations found in surface waters (rivers, lakes and sea) are normally at ng L^{-1} levels, although in some cases at points close to the WWTPs, levels can reach low μg L^{-1} levels for analgesics such as acetaminophen, ibuprofen and salicylic acid.[29]

Regarding solid samples, such as sediments and sewage sludge, variable concentrations were observed, with maximum levels reaching several μg g^{-1} (d.w.) in sewage sludge for the antibiotics clotrimazole and miconazole.[16]

The only work which used UHPLC for the analysis of pharmaceuticals in biota, in *Mytilus edulis*, a marine mussel,[45] found that salicylic acid was ubiquitous in all samples, found at a maximum concentration of almost 300 ng g^{-1} (d.w.).

As far as screening analysis of non-target compounds, UHPLC coupled to QqToF has successfully been used for the identification of micro-contaminants in surface and wastewaters[21,22] and in fresh water sediments,[15] as well as for the identification and structural elucidation of transformation products in biodegradation experiments,[20,23,46] membrane bioreactors[47] or photo-degradation.[18,19] Figure 12.2 shows an UHPLC-ToF-MS chromatograms of the influent wastewater and effluent from a Zenon membrane bioreactor with annotated peaks of the main transformation products of some antibiotics showing good separation capacity of UHPLC.

12.4 Conclusions

The rapid developments in the field of LC-MS/MS have transformed this technique into a key approach for the analysis of pharmaceuticals as environmental contaminants. Many pharmaceutical compounds that were

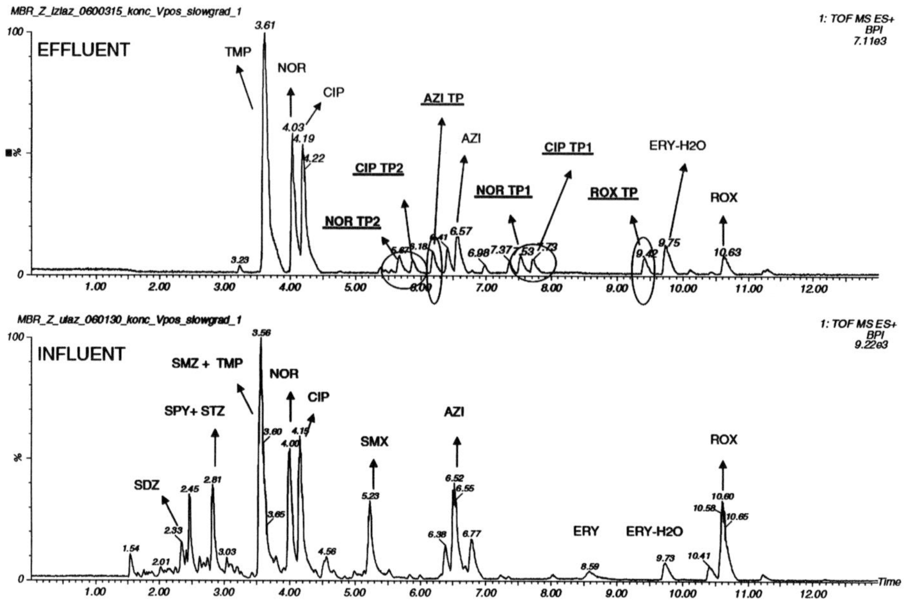

Figure 12.2 UHPLC-TOFMS chromatograms of the influent wastewater and effluent from a Zenon membrane bioreactor with annotated peaks of the main transformation products. SDZ, sulfadiazine; SPY, sulfapyridine; STZ, sulfathiazole; SMX, sulfamethoxazole; TMP, trimethoprim; NOR, norfloxacine; CIP, ciprofloxacine; AZI, azithromycin; ERY, erythromycin; ERY-H2O, dehydrated erythromycin; ROX, roxithromycin; and NOR TP1, NOR TP2, CIP TP1, CIP TP2, AZI TP and ROX TP denote the corresponding transformation products. Reprinted with permission from Terzic *et al.*,[47] © Springer Verlag 2011.

not possible to detect at concentrations lower than the ng L^{-1} level are nowadays part of the routine analysis in environmental laboratories. At the same time, with the progresses in analytical instrumentation, extraction techniques have become more simple, fast and inexpensive, providing the enrichment of analytes of interest from matrices as complicated as wastewater or sewage sludge.

UHPLC coupled to MS has increasingly been used in environmental analysis of pharmaceuticals. It enables elution of sample components in much narrower, more concentrated bands, resulting in better chromatographic resolution and increased peak capacity, and its application is expected to increase.

Acknowledgements

This material is based upon work supported by the Spanish Ministerio de Educación y Ciencia, [projects CGL2007-64551/HID and Consolider-Ingenio 2010 CSD2009-00065]. S.P. acknowledges the contract from the Ramón y

Cajal Program of the Spanish Ministry of Science and Innovation. R.L.S. acknowledges the Spanish Ministry of Education and Science for economical support through the FPI pre-doctoral grant.

References

1. C. G. Daughton, *Environ. Impact Assess. Rev.*, 2004, **24**, 711.
2. A. Ginebreda, I. Muñoz, M. L. de Alda, R. Brix, J. López-Doval and D. Barceló, *Environ. Int.*, 2010, **36**, 153.
3. M. Petrovic, M. Farre, M. L. de Alda, S. Perez, C. Postigo, M. Kock, J. Radjenovic, M. Gros and D. Barcelo, *J. Chromatogr. A*, 2010, **1217**, 4004.
4. D. Guillarme, J. Schappler, S. Rudaz and J. L. Veuthey, *Trends Anal. Chem.*, 2010, **29**, 15.
5. R. López-Roldán, M. L. de Alda, M. Gros, M. Petrovic, J. Martín-Alonso and D. Barceló, *Chemosphere*, 2010, **80**, 1337.
6. I. Baranowska and B. Kowalski, *Water Air Soil Pollut.*, 2010, **211**, 417–425.
7. J. M. Conley, S. J. Symes, S. A. Kindelberger and S. M. Richards, *J. Chromatogr. A*, 2008, **1185**, 206.
8. J. C. Van De Steene and W. E. Lambert, *J. Am. Soc. Mass Spectrom.*, 2008, **19**, 713.
9. B. Kasprzyk-Hordern, R. M. Dinsdale and A. J. Guwy, *Anal. Bioanal. Chem.*, 2008, **391**, 1293.
10. B. Kasprzyk-Hordern, R. M. Dinsdale and A. J. Guwy, *J. Chromatogr. A*, 2007, **1161**, 132.
11. J. Magnér, M. Filipovic and T. Alsberg, *Chemosphere*, 2010, **80**, 1255.
12. J. Wu, X. Qian, Z. Yang and L. Zhang, *J. Chromatogr. A*, 2010, **1217**, 1471.
13. M. Farré, M. Gros, B. Hernández, M. Petrovic, P. Hancock and D. Barceló, *Rapid Commun. Mass Spectrom.*, 2008, **22**, 41.
14. Y. Yu, Q. Huang, J. Cui, K. Zhang, C. Tang and X. Peng, *Anal. Bioanal. Chem.*, 2011, **399**, 891.
15. S. Terzic and M. Ahel, *Environ. Pollut.*, 2011, **159**, 557.
16. Q. Huang, Y. Yu, C. Tang and X. Peng, *J. Chromatogr. A*, 2010, **1217**, 3481.
17. A. Nieto, F. Borrull, R. M. Marcé and E. Pocurull, *J. Chromatogr. A*, 2009, **1216**, 5619.
18. L. Tong, S. Pérez, C. Gonçalves, F. Alpendurada, Y. Wang and D. Barceló, *Anal. Bioanal. Chem.*, 2011, **399**, 421.
19. J. Radjenović, C. Sirtori, M. Petrović, D. Barceló and S. Malato, *Appl. Catal. B Environ.*, 2009, **89**, 255.
20. T. Kosjek, E. Heath, S. Pérez, M. Petrović and D. Barceló, *J. Hydrol.*, 2009, **372**, 109.
21. M. Ibáñez, C. Guerrero, J. V. Sancho and F. Hernández, *J. Chromatogr. A*, 2009, **1216**, 2529.

22. M. Ibáñez, J. V. Sancho, F. Hernández, D. McMillan and R. Rao, *Trends Anal. Chem.*, 2008, **27**, 481.
23. J. Radjenović, S. Pérez, M. Petrović and D. Barceló, *J. Chromatogr. A*, 2008, **1210**, 142.
24. S. B. Abdelmelek, J. Greaves, K. P. Ishida, W. J. Cooper and W. Song, *Environ. Sci. Technol.*, 2011, **45**, 3665.
25. M. Gros, M. Petrović and D. Barceló, *Anal. Bioanal. Chem.*, 2006, **386**, 941.
26. A. L. Batt, M. S. Kostich and J. M. Lazorchak, *Anal. Chem.*, 2008, **80**, 5021.
27. M. Gros, M. Petrović and D. Barceló, *Talanta*, 2006, **70**, 678.
28. M. R. P. López-Serna and D. Barceló, *Chemosphere*, 2011, in press.
29. E. Gracia-Lor, J. V. Sancho and F. Hernández, *J. Chromatogr. A*, 2011, **1218**, 2264.
30. M. S. Díaz-Cruz, M. J. García-Galán, P. Guerra, A. Jelic, C. Postigo, E. Eljarrat, M. Farré, M. J. López de Alda, M. Petrovic and D. Barceló, *Trends Anal. Chem.*, 2009, **28**, 1263.
31. A. Jelić, M. Petrović and D. Barceló, *Talanta*, 2009, **80**, 363.
32. A. M. Jacobsen, B. Halling-Sorensen, F. Ingerslev and S. H. Hansen, *J. Chromatogr. A*, 2004, **1038**, 157.
33. L. Nováková, L. Matysová and P. Solich, *Talanta*, 2006, **68**, 908.
34. N. Wu and A. M. Clausen, *J. Sep. Sci.*, 2007, **30**, 1167.
35. M. Huerta-Fontela, M. T. Galceran and F. Ventura, *J. Chromatogr. A*, 2010, **1217**, 4212.
36. N. Wu, D. C. Collins, J. A. Lippert, Y. Xiang and M. L. Lee, *J. Microcolumn Sep.*, 2000, **12**, 462.
37. D. Guillarme, J. Ruta, S. Rudaz and J. L. Veuthey, *Anal. Bioanal. Chem.*, 2010, **397**, 1069.
38. W. B. Dunn, D. Broadhurst, M. Brown, P. N. Baker, C. W. G. Redman, L. C. Kenny and D. B. Kell, *J. Chromatogr. B,* 2008, **871**, 288.
39. A. Makarov and M. Scigelova, *J. Chromatogr. A*, 2010, **1217**, 3938.
40. A. Tolonen, M. Turpeinen and O. Pelkonen, *Drug Discov. Today*, 2009, **14**, 120.
41. Z. Herrera Rivera, E. Oosterink, L. Rietveld, F. Schoutsen and L. Stolker, *Anal. Chim. Acta*, 2011, **700**, 114.
42. T. A. Ternes, *Trends Anal. Chem.*, 2001, **20**, 419.
43. M. Petrovic, M. Gros and D. Barcelo, *J. Chromatogr. A*, 2006, **1124**, 68.
44. B. Morasch, F. Bonvin, H. Reiser, D. Grandjean, L. F. De Alencastro, C. Perazzolo, N. Chèvre and T. Kohn, *Environ. Toxicol. Chem.*, 2010, **29**, 1658.
45. K. Wille, J. A. L. Kiebooms, M. Claessens, K. Rappé, J. Vanden Bussche, H. Noppe, N. Van Praet, E. De Wulf, P. Van Caeter, C. R. Janssen, H. F. De Brabander and L. Vanhaecke, *Anal. Bioanal. Chem.*, 2011, **400**, 1459.
46. S. Pérez and D. Barceló, *Anal. Chem.*, 2008, **80**, 8135.

47. S. Terzic, I. Senta, M. Matosic and M. Ahel, *Anal. Bioanal. Chem.*, 2011, **401**, 353.
48. Q. Sui, J. Huang, S. Deng, G. Yu and Q. Fan, *Water Res.*, 2010, **44**, 417.
49. E. Gracia-Lor, J. V. Sancho and F. Hernández, *J. Chromatogr. A*, 2010, **1217**, 622.
50. K. Demeestere, M. Petrović, M. Gros, J. Dewulf, H. Van Langenhove and D. Barceló, *Anal. Bioanal. Chem.*, 2010, **396**, 825.
51. J. A. Magnér, T. E. Alsberg and D. Broman, *Anal. Bioanal. Chem.*, 2009, **395**, 1481.
52. M. Lavén, T. Alsberg, Y. Yu, M. Adolfsson-Erici and H. Sun, *J. Chromatogr. A*, 2009, **1216**, 49.

CHAPTER 13

UHPLC in Natural Products Analysis

P. J. EUGSTER AND J.-L. WOLFENDER*

School of Pharmaceutical Sciences, EPGL, University of Geneva, University of Lausanne, Quai Ernest-Ansermet 30, 1211 Geneva 4, Switzerland
*E-mail: jean-luc.wolfender@unige.ch

13.1 Introduction

Natural products (NPs) are known to possess a very high diversity in chemical space,[1] and, as a result, they have a profound impact on chemical biology and drug development.[2] Bioactive NPs can be found in many different biological matrices, such as plants, marine organisms, micro-organisms and animals. In many cases, each organism produces a huge variety of these NPs. Plants, for example, are known to produce NPs that either are essential for their life (primary metabolites) or are not directly involved in their normal growth, development or reproduction, but are necessary for survivability, fecundity or aesthetics (secondary metabolites). The complete composition of a given organism, known as the metabolome, can be extremely large and has been estimated to contain a few thousand constituents; however, the exact size of a plant or fungal metabolome is still unknown.[3]

The high chemical diversity of secondary metabolites can probably be explained by the effects of evolutionary pressure, which provided an impetus for organisms to create biologically active molecules, and/or by the structural similarity of protein targets across many species. This large chemical diversity[4] is also directly linked to a high variability of the intrinsic physicochemical properties of NPs, which causes the separation and universal detection of NPs to be extremely challenging.

RSC Chromatography Monographs No. 16
UHPLC in Life Sciences
Edited by Davy Guillarme and Jean-Luc Veuthey

Published by the Royal Society of Chemistry, www.rsc.org

The analysis of individual NPs in a complex crude extract requires the efficient separation of individual components before detection. In this respect, high-performance liquid chromatography (HPLC) has been recognised since the early 1980s as the most versatile technique for the efficient separation of NPs in crude mixtures without the need for complex sample preparation.[5] HPLC has been greatly developed through the years in terms of its convenience, speed, choice of stationary phases, sensitivity, applicability to a broad variety of sample matrices and its ability to couple to spectroscopic detection methods.[6] The development of HPLC columns with different phase chemistries (especially reversed-phase) enabled the separation of almost any type of NPs.

HPLC is thus widely used and has been adapted to the analysis of a broad range of NPs, generally without the need for complex sample preparation. Because in many cases, the NPs of interest must be isolated from their original biological matrix, other liquid chromatography (LC) preparative techniques that use similar HPLC phase chemistries can be used to isolate milligram amounts of pure NPs. These techniques include low pressure LC (LPLC), medium pressure LC (MPLC), semi-preparative and preparative HPLC.[7]

Crude extracts of natural origin can be separated either by using the raw mixtures or by using samples that are enriched by extraction *via* solid phase extraction (SPE) or liquid–liquid extraction (LLE). These separations are usually performed by reversed-phase chromatography on C18 material with the acetonitrile/water (ACN/H_2O) or methanol/water (MeOH/H_2O) solvent systems in the gradient elution mode. To improve the separation efficiency, various modifiers or buffers can be added to the mobile phase to tune the selectivity of the separation or the sensitivity of detection.

However, the choice of the appropriate HPLC detector is crucial because of the diversity of NPs, and thus there is no universal technique for NP detection. Simple detectors, such as ultraviolet (UV), evaporative light scattering detection (ELSD), fluorescence detection (FD), electrochemical detection (ECD), refractive index detection (RID), flame ionisation detection (FID), chemiluminescence (CL) and charged aerosol detection (CAD), can be used, with UV and ELSD being the most widespread.[8] In addition, the coupling of HPLC with photodiode array (PDA), mass spectrometry (MS), nuclear magnetic resonance (NMR) and infrared (IR) is often of key importance in the dereplication process to collect online preliminary structural information during the HPLC separation.

The latest developments in HPLC technologies, including the recent introduction of phase chemistries that are stable at a wide range of pH values, fully porous sub-2 μm and core-shell sub-3 μm packing particles[9] or monolith columns, have considerably improved the performance of HPLC systems in terms of their resolution, speed and reproducibility. Efficiencies exceeding 100 000 plates and peak capacities over 900 can be attained by coupling columns together.[10] In this Chapter, we will review all aspects related to the introduction of ultra-high-pressure liquid chromatography (UHPLC) for NP analysis and the transition of conventional HPLC profiling methods to this new technology in different fields of plant research, including quality control (QC), metabolite profiling/fingerprinting and dereplication, and metabolomics.

13.1.1 Implementation of UHPLC in NP Analysis

As in other fields of analytical science, the introduction of UHPLC systems that operate at very high pressures and use porous sub-2 μm packing columns in NP research has allowed for a remarkable decrease in analysis time and increases in peak capacity, sensitivity and reproducibility as compared to conventional HPLC. This technology has started being implemented in many laboratories that work in NP research, and it has not only replaced conventional HPLC but also opened new fields of research, such as metabolomics and high-resolution profiling.

Thus, NP chemists have used UHPLC to considerably enhance the throughput of their targeted analyses using very rapid gradients on short columns (*e.g.*, for QC or for crude extract standardisation). In addition, they have also pushed the UHPLC technology to its limit for performing very high-resolution profiling of complex mixtures using slow gradients on long columns (*e.g.*, for detailed metabolite profiling and dereplication). For many applications, important improvements in the overall performance have been reported.

Thus, there is a growing interest in UHPLC in plant science and in other aspects of NP research, such as fungal or bacterial metabolite studies and the standardisation of herbal products. The number of reported UHPLC-related applications for NP analysis has constantly increased since its introduction in 2006, while the reported applications of conventional HPLC methods remain relatively stable.[11] In 2010, 171 articles were published on the applications of UHPLC in NP analysis, while only one paper was reported in 2005.[12] For comparison, in 2010, the number of reported HPLC applications exceeded 3500 [number of papers by year of publication retrieved from Scifinder Scholar (Chemical Abstracts) using the keywords "UHPLC" or "UPLC" or "RRLC" and "plants" or "phytochemistry" or "natural products" in September 2011]. The main factor hindering a faster implementation of UHPLC in NP research laboratories is the need for specific instrumentation, as conventional HPLC systems cannot tolerate the high pressure generated by the sub-2 μm columns.

13.2 Multiple Facets of UHPLC in NP research

As mentioned above, HPLC has been extensively used in many aspects of NP research, and UHPLC is advantageously replacing HPLC for both high-throughput and high-resolution applications. However, to use UHPLC, the acquisition speed of the detectors must be adapted for monitoring thinner LC peaks and the sample extracts must be prepared in a way that meets the requirements of the sub-2 μm columns.

13.2.1 UHPLC Detectors used for NP Analysis

As described for HPLC, different detectors have been used with UHPLC to analyse NPs. Spectroscopic methods are often used in hyphenation with UHPLC, which is important for dereplication in metabolite profiling studies. Compared to HPLC, however, UHPLC imposes some limitations in the choice of the detector, both in terms of the acquisition rate and the loading capacities of the column. For these reasons, UHPLC at present is not compatible with spectroscopic detectors such as NMR (LC-NMR hyphenation) and IR (LC-IR), which is a disadvantage, as these types of detectors are important for the *de novo* structure determination of NPs online or at-line.[13] However, UHPLC can be efficiently hyphenated to MS which is the most useful detector for NP analysis.[8,14] Indeed, although it is expensive, the use of a mass spectrometer as detector for LC systems provides excellent sensitivity and selectivity in the analysis of NPs in complex biological matrices. Furthermore, MS detection provides important online structural information, such as the molecular mass, molecular formula and diagnostic fragments, which are crucial for dereplication and rapid online characterisations of NPs.[13]

Single or triple quadrupole systems have been coupled to UHPLC but have mainly been used for the specific detection of NPs through single ion monitoring (SIM)[15] or multiple reaction monitoring (MRM)[16,17] experiments rather than for full scan acquisitions of MS or tandem MS (MS/MS) spectra. With quadrupole-based analysers, the sampling rate can be problematic, and modern instruments that possess improved acquisition rates should be selected for coupling with UHPLC. Dwell times and inter-channel delays have, however, been reduced down to less than 5 ms in the SIM mode with new analyser generations, for example.[18]

The use of a high acquisition-rate mass analyser, such as the time-of-flight MS (TOF-MS) detector, has considerably boosted the use of UHPLC, providing a powerful UHPLC-TOF-MS platform with high sensitivity and specificity of detection and accurate mass detection is used.[19] Indeed, TOF-MS instruments are well adapted to record and store data over a broad mass range without compromising sensitivity, with high resolving power [generally $>$10 000 full width at half maximum (FWHM)] to be attained in routine analysis at speeds up to 40 spectra s^{-1}.[11] Even higher resolving power, up to 50 000 for the latest generation of TOF instruments have been reported, reducing the risk of false negative results when complex biological matrices are to be analysed.[20]

Furthermore, with hybrid systems, such as hybrid quadrupole TOF mass spectrometer (Q-TOF-MS/MS), acquisition of MS/MS spectra at high frequency provides more online structural information or more specific detection.

Trap systems have given very useful structural information on conventional LC-MS system through MS^n experiments. High-resolution Orbitrap Fourier transform (FT) MS, with a resolving power up to 100 000, is very useful for structural identification and provides high quality spectra when used with infusions of pure NPs.[21] Trap systems have rarely been used in conjunction with UHPLC[22] because the lower peak duration in UHPLC does not match the time needed by the spectrometer to acquire different MS^n spectra with sufficient ion statistics, especially when high-resolution measurements are required. The most recent generation of Orbitrap systems are, however, able to work at up to a 5 to 10Hz acquisition frequency with a reasonable loss in resolution.[23] First attempts of hyphenation of such analysers with UHPLC were successful in proteomics[24] and are also promising for the high-resolution profiling of low-molecular-mass compounds such as NPs.

In conclusion, MS analysers are the most useful detectors coupled to UHPLC for NPs analysis. For specific and sensitive detection, the high acquisition rate of triple quadrupole instruments should be preferred, while non-targeted analysis with full scan spectra acquisition can be efficiently performed on TOF-MS systems. In the years to come, faster Orbitrap systems may compete with the high resolving power of TOF-MS and also provide MS^n structural information online.[25]

13.2.2 Targeted *Versus* Untargeted Analyses of NPs

UV detection has often been used for targeted, quantitative UHPLC analyses of NPs (*ca*. 60% of applications). In addition to simple UV detectors, PDA detectors have also been applied for online UV spectra acquisition.[26] Indeed, UV is the most simple and the most widely used among all LC detectors.[27] It is quite easy to optimise UV-visible (UV-Vis) and UV-diode array detector (UV-DAD) detectors to meet the requirements of UHPLC in terms of the sampling rate. The UV cell volume should be reduced to avoid peak dispersion in UHPLC, while maintaining a sufficient path length of light passing through the UV cell, as the absorbance is directly proportional according to the Beer-Lambert law. Generally, the UV cell in conventional HPLC systems has a volume between 10 and 25 μL and a path length of 10 mm. The UV cell in UHPLC, however, is reduced to 0.25–3 μL with a path length of 3–10 mm, depending on the provider. Although UV suffers from some limitations, particularly for NPs that do not possess UV chromophores, this detection method has the best combination of sensitivity, linearity, versatility and reliability of all of the LC detectors that have been developed. Most NPs adsorb UV light in the range of 200–550 nm, including all substances having one or more double bonds and all substances that have unshared electrons. Thus, even compounds with weak chromophores, such as triterpene glycosides, can be successfully detected by UV at short wavelengths (203 nm).[28] However, in this system, several mobile-phase constituents that exhibit high UV cut-offs should be avoided, as they might inhibit the detection of NPs with weak chromophores.[29] The use of PDA detection provides UV spectra directly online and is particularly useful for the detection of natural products with characteristic chromophores.[30] For example, polyphenols can be efficiently localised by this method because they possess characteristic chromophores.

With this type of compound, PDA-UV spectral libraries can be built and used for dereplication.[26]

ELSD is also compatible with UHPLC. ELSD is a quasi-universal detector for LC, as it can detect any analyte that is less volatile than the mobile phase, regardless of its optical, electrochemical or other properties.[31,32] ELSD is a mass-dependent detector (in contrast to UV, which is a concentration-dependent detector), and the generated response does not depend on the spectral or physicochemical properties of the analyte. Because the detection is based on the measurement of light scattering (using a photomultiplier or a photodiode) produced by the non-volatile residual particles after the evaporation of the mobile phase, the sampling rate is generally not critical (equal to at least 50 Hz in any commercial devices) and thus is sufficient for even ultra-fast experiments. As was recently reported,[33,34] the coupling of UHPLC with ELSD is possible, but the latter remains a non-negligible source of additional dispersion that increases with higher mobile-phase flow rates.

In NPs analysis, HPLC-ELSD has been mainly used for the detection of compounds with weak chromophores, such as terpenes, in both aglycone and glycosidic forms, saponins and some alkaloids.[8] For example, coupling ELSD to UHPLC has been used to quantify triterpenoids in phytopharmaceuticals containing black cohosh (*Actaea racemosa*).[34]

Figure 13.1 shows the complementarity of UHPLC-PDA-ELSD and -MS for profiling a plant extract (crude isopropanol extract of *Arabidopsis thaliana*). The UV detection at 350 nm is rather selective, showing mainly peaks that are related to flavonoids. The trace at 254 nm displays most of the NPs containing an aromatic chromophore or conjugated double bonds. The ELSD provides more peaks than the UV detector, especially for the detection of non-polar compounds (mainly lipids, in this example). Finally, the MS trace [base peak intensity (BPI) positive ionisation (PI) mode] demonstrates that almost all constituents can be ionised, with the exception of the very polar constituents that are detected by ELSD. With LC-MS, however, the response cannot be linked to the quantity of NP detected, as it is compound-dependent.[8] Figure 13.1 demonstrates the use of TOF-MS detection for the selective detection and rapid online characterisation of natural products, which is not possible with UV or ELSD detectors. As shown in the display of the UHPLC-ESI-PI-TOF-MS trace of the ion at *m/z* 741, the TOF-MS system provided a selective detection of this compound in the crude extract of *A. thaliana*. In addition, the corresponding high-resolution spectrum of compound F enabled the precise determination of the molecular formula ($C_{33}H_{41}O_{19}$) of its protonated molecule $[M+H]^+$ (*m/z* 741.2242), allowing for the identification of this compound as a flavonol triglycoside. More details on this dereplication procedure are provided in Section 13.5.3.

As previously mentioned, not all detectors that have been used for HPLC are fully compatible with UHPLC. However, those detectors that can be adapted to this technology, mainly by increasing their acquisition rates, generally demonstrate important improvements in data quality and throughput as

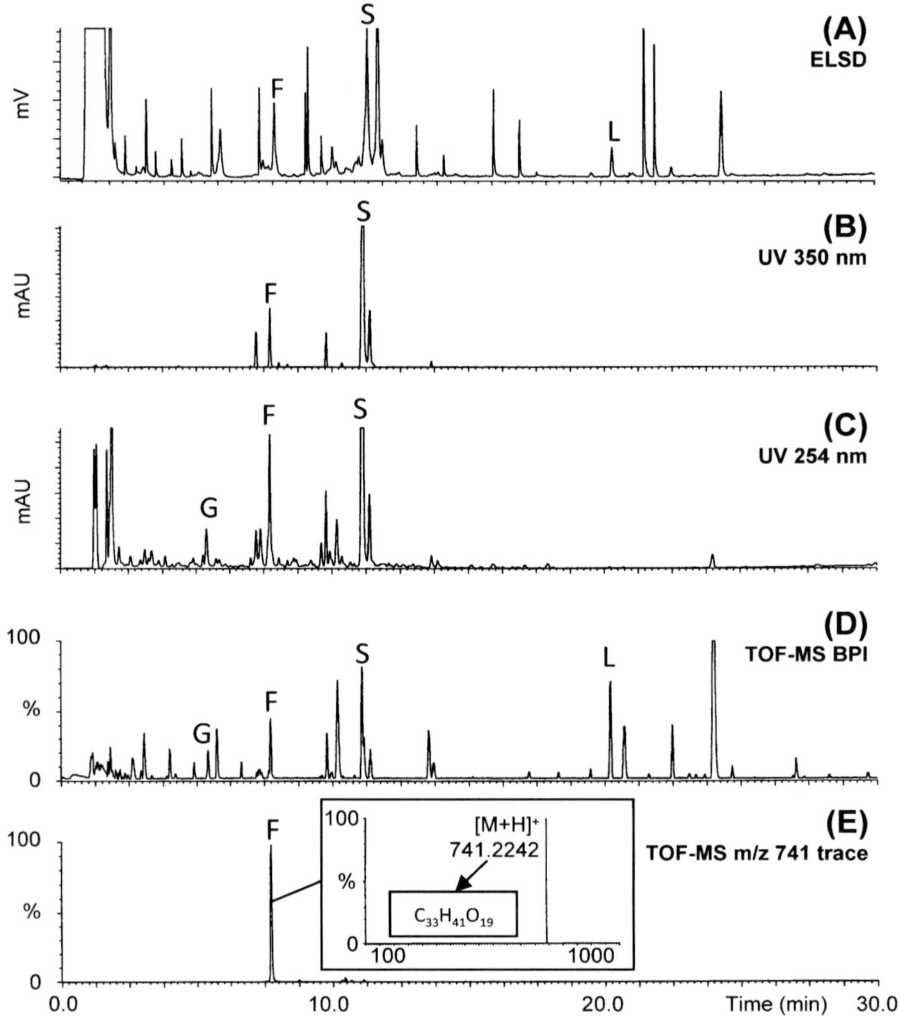

Figure 13.1 Chromatograms of *A. thaliana* (crude leaves isopropanol extract) obtained with different detection techniques: (A) ELSD, (B) PDA trace at 350 nm, (C) PDA trace at 254 nm, (D) TOF-MS BPI in PI mode and (E) extracted ion trace of *m/z* 741.2, TOF-MS in PI mode. Separation was carried out on an Acquity^{TM} BEH C18 column (150 × 2.1 mm; 1.7 μm), with a 5–98% ACN gradient in 45 min; both water and ACN contained 0.1% FA. Compound F is a flavonol glycoside, G is a glucosinolate, S is a synapoyl derivative, and L is a galactolipid. Inset: TOF-MS spectra of F. Adapted from reference 8 with permission from Thieme Verlag

compared to their HPLC counterparts. The use of UV, ELSD or MS enables very fast, targeted analysis to be performed, mainly on a quantitative viewpoint, while PDA and high-resolution MS or MS/MS provide high-quality online spectroscopic information in non-targeted analysis, which is especially useful for metabolite profiling, chemical screening or dereplication applications.

UHPLC has thus conquered domains related to the QC of plants or food extracts, especially for the standardisation and safety assessment of medicinal plants, phytopharmaceuticals or dietary supplements. In this respect, standard HPLC procedures are gradually being replaced by high-throughput, targeted and quantitative UHPLC methods.[35] UHPLC is also being used more commonly for dereplication purposes in drug discovery programmes in conjunction with both PDAs and MS detection. Dereplication is the process of differentiating NP extracts that contain known secondary metabolites from those that contain novel compounds of interest.[36] Here, the high resolving power of UHPLC is required for the deconvolution of closely related metabolites, such as isomers, to obtain high-quality online spectra without interference, which can then be used for database searching or for spectral interpretation. Such a process represents an important step in drug discovery programmes, as the early structural determination of known NPs avoids the time-consuming processes required for their isolation and enables the optimisation of bioactive-guided isolation procedures.[37]

In order to illustrate both the high throughput and the high chromatographic resolution that can be obtained by UHPLC as compared to standard HPLC, the metabolite profiling of a representative crude plant extract (the widely used phytomedicine *Ginkgo biloba*) is displayed in Figure 13.2. By using UHPLC conditions for the profiling of this standardised extract, a 9-fold reduction in analysis time could be obtained by transferring the 60 min gradient from the HPLC column (150 × 4.6 mm; 5 μm) to a short gradient on a 50 mm UHPLC column (50 × 2.1 mm; 1.7 μm). It should be noted that even if chromatographic calculations indicate that the peak capacity should remain constant with such a method transfer, a significant decrease is measured, mainly due to peak broadening in the MS source. The use of the same gradient time on a longer UHPLC column (150 × 2.1 mm; 1.7 μm) provided a notable increase in resolution.

In addition, UHPLC, and especially UHPLC-TOF-MS, is beginning to play an important role in new research fields, such as metabolomics.[3] This holistic approach has recently emerged with other 'omics' technologies in biological research[38] and concerns the large-scale analysis of metabolites in given organisms at different physiological states. Profiling the metabolome has the potential to provide the most "functional" information among the 'omics' technologies that are used in systems biology.[39] Currently, UHPLC-TOF-MS represents a key method for both metabolite fingerprinting and for metabolite profiling from crude extracts. Metabolic fingerprinting consists of high-throughput separations that are used not with the intention of identifying each observed metabolite, but rather to compare patterns or "fingerprints" of metabolites that change in response to disease, nutrition, toxin exposure, or environmental or genetic alterations. In contrast, metabolic profiling focuses on the analysis of a group of metabolites that are either related to a specific

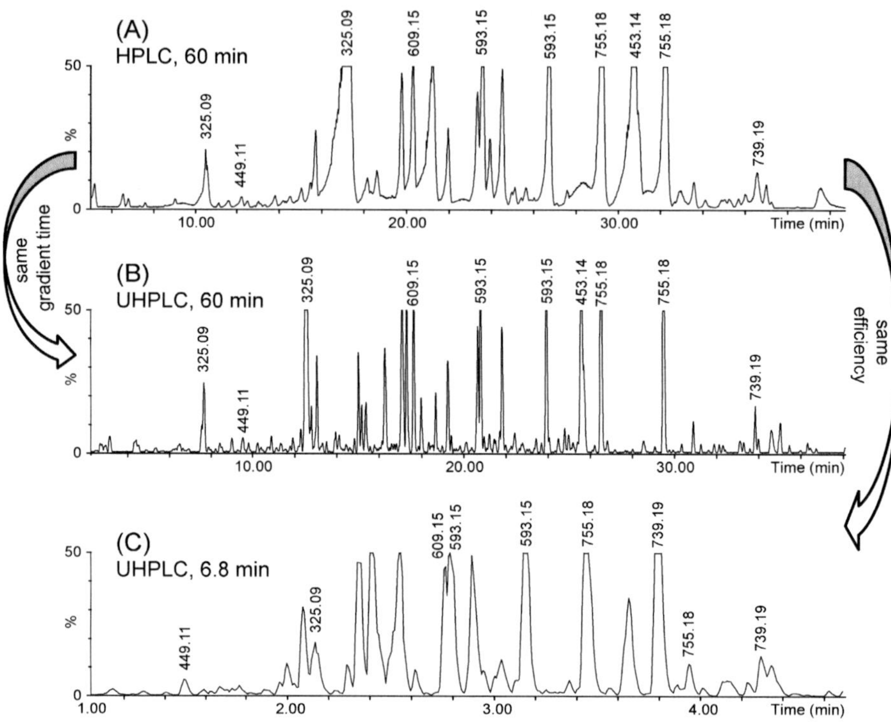

Figure 13.2 Example of an HPLC–UHPLC transfer: chromatograms of a standardised *G. biloba* extract with a 5–40% ACN gradient using (A) a classic HPLC column (150 × 4.6 mm; 5 μm) in 60 min at 1.0 mL min^{-1}, (B) a UHPLC column (150 × 2.1 mm; 1.7 μm) in 60 min at 0.35 mL min^{-1}, and (C) a UHPLC column (50 × 2.1 mm; 1.7 μm) in 6.8 min at 0.60 mL min^{-1}. Detection was carried out by an ESI-TOF-MS analyser in the NI mode in the 100–1000 *m/z* range.

metabolic pathway or to a class of compounds. In most cases, metabolic profiling is a hypothesis-driven approach rather than a hypothesis-generating one.[40]

Metabolic fingerprinting is also becoming frequently used to assess phytoequivalence in untargeted QC methods.[35,41]

13.2.3 Column Phase Chemistries for NP Analysis

As mentioned previously, most separations of crude extracts with complex NP compositions are carried out using a gradient mode on reversed-phase columns. However, as discussed in the Introduction, the chemical space occupied by NPs is very broad, and the analysis of both very polar and very lipophilic compounds is important. The large variety of sub-2 μm phase chemistries that are available can resolve almost all analytical issues: C8 and C18 are used for plant extracts of average polarity; C4 and cyano for the most apolar fractions; hydrophilic interaction liquid chromatography (HILIC) with bare silica or diol, amino bonding for the most polar fractions; and biphenyl, pentaflurorophenyl (PFP) or zirconia for alternative selectivity.[11] Despite this wide variety, relatively few applications have reported the use of reversed-phase columns other than C18 (see Figure 13.3).

Concerning column geometries, in Figure 13.3, it can be also noted that more that 50% of all separations are performed on a 100 mm column, 13% are performed on a 150 mm column, and approximately 30% are performed on a 50 mm or shorter column. The coupling of up to three 150 mm column in series (total column length of 450 mm) has also been reported for very high-resolution profiling of NP.[42]

13.3 Fast Targeted Analysis

Fast targeted analyses are typically used to check the purity of an isolated NP, to quantify a compound in a complex mixture or as QC in the pharmaceutical field. With the introduction of UHPLC, the targeted QC of plant extracts consists of high-throughput methods where only a few constituents that are representative of the plant sample must be evaluated. As described in the first Chapters of the Book and illustrated in Figure 13.2, it is theoretically possible to obtain a 9-fold reduction of the analysis time as compared to standard HPLC methods, while maintaining an equivalent performance with UHPLC. However, it should be noted that method transfers in plant analysis are seldom purely geometric due to the use of different stationary phase chemistries in many applications. Thus, analysis times are often slightly shorter or longer than the predicted 9-fold reduced time.

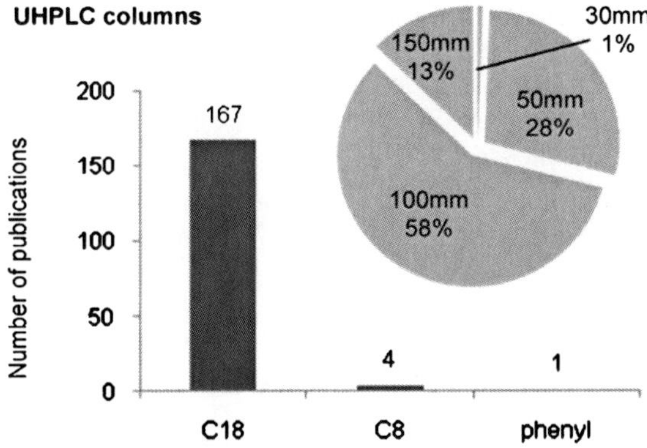

Figure 13.3 Classification of all papers retrieved from a Scifinder Scholar (Chemical Abstracts) search using "UPLC OR UHPLC" and "natural products" or "plants" or "phytochemistry" keywords as related to column phase chemistries (C18, C8 and phenyl). The inset shows the column length distribution. (Compiled in August 2011.)

A targeted analysis using UHPLC can be carried out to check the purity of a given NP during its isolation from a complex extract through a rapid analysis of the collected LC fractions. In this case, the analysis is very similar to that performed for any simple mixture of organic compounds, and the separation is optimised for the target compound. The high throughput of UHPLC is an appreciable advantage for bioactivity-guided fractionation approaches due to the high number of simplified fractions that are generated and the need for these fractions to be individually analysed for pooling and for the final purification of the compound of interest.

In the large majority of literature reports using targeted analysis by UHPLC (mainly UHPLC-UV or UHPLC-MS/MS), the targeted applications of UHPLC have been developed for the specific detection and quantification of a given NP or a set of NPs in a complex mixture. This was applied, for example, for monitoring the biosynthesis of microbial products in fermentation broths or in plant cell cultures, determining phytohormones in model or crop plants, quantifying specific markers in herbal drugs or food, standardising nutraceuticals or phytopharmaceuticals, or detecting NPs with toxic properties.

Using UHPLC for the QC of herbal medicine is of particular interest. Indeed, a suitable standardisation and QC procedure is required to guarantee the botanical identity of the raw material and the quality, safety and efficacy of the final phytopharmaceutical product. Furthermore, NP amounts in a given extract are strongly dependent on the season, time, place of harvest, and extraction method; thus, the NP amounts need to be quantified to guarantee efficacy. The same is also needed for some dietary supplements or functional foods, although to a lesser extent because of less strict regulations. In general, most of the standardisation methods consist of the quantification of one marker (a secondary metabolite characteristic of the plant of interest) and the verification that possible toxic constituents are absent or below a given limit. In an ideal case, the quantified marker is the NP holding the bioactivity, but, in many cases, this compound is unknown, or the activity is the result of several compounds acting synergistically.[41] For the standardisation of such preparations, a complete fingerprinting or the selective quantification of many markers might be required. Because of these characteristics, the QC of plant extracts is difficult, but it nonetheless remains mandatory.[43]

13.3.1 UHPLC-UV

Several plant extracts, mainly phytopharmaceuticals and food, have been analysed by high-throughput UHPLC-UV methods using simple UV or PDA detection. For example, the polyphenols of green tea have been extensively studied for their potential health benefits.[44] In a recent study, a comprehensive profiling of 29 phenolic compounds comprising caffeine in tea preparations, infusions and extracts has been carried out quantitatively. The phenolic compounds were separated in less than 20 min on a C18 column (100 ×

2.1 mm; 1.7 μm) using a gradient elution mode with a 0.1% formic acid and methanol mobile phase.[45] The high-throughput capability of UHPLC was demonstrated in another study on green tea with the baseline separation of seven catechin standards in only 30 s using a short 50 mm column (50 × 2.1 mm; 1.7 μm), although the analysis of enriched extracts required longer columns.[17] In Figure 13.4, we demonstrate that with specific detection at 280 nm, caffeine and the main catechin of green tea, epigallocatechin gallate (EGCG), could be separated and quantified using the same 50 mm column in 2 min (including the reconditioning step) by transferring and re-optimising an existing HPLC method in which the gradient time exceeded 30 min. The linearity of both the HPLC-UV and the UHPLC-UV methods were similar,

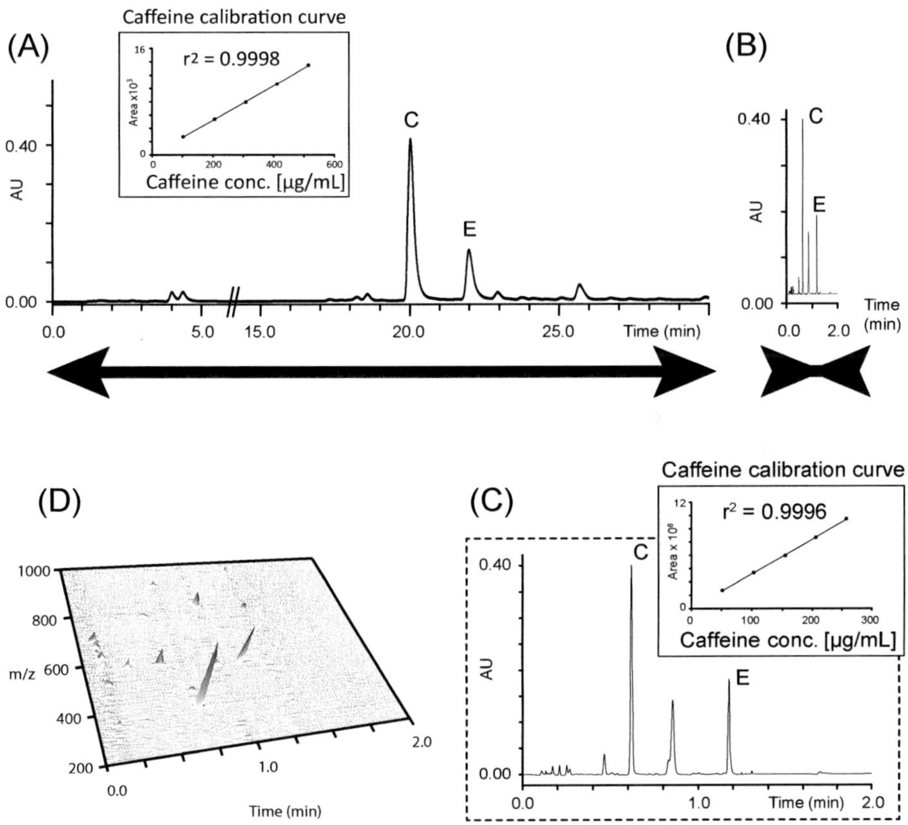

Figure 13.4 Quantitative caffeine analysis of a green tea infusion. (A) 30 min HPLC separation on a C18 column (150 × 4.6 mm; 5 μm) with UV detection at 280 nm. The calibration curve is shown in the inset. (B) and (C) are the same analysis after transfer of the HPLC method on to a short UHPLC column (50 × 2.1 mm; 1.7 μm) with further optimisation. The final analysis time is 2.0 min. (D) A three-dimensional ion map of the same UHPLC separation with TOF-MS detection in PI mode.

and the selectivity of the UHPLC method was demonstrated by further UHPLC-TOF-MS analyses.

For the standardisation of phytopharmaceuticals containing anthraquinones, an UHPLC-PDA method enabled the simultaneous determination of five anthraquinone derivatives in three *Rheum* species.[46] The method was fully validated in terms of precision, accuracy, and linearity according to ICH guidelines, and UHPLC analysis was performed in only 3 min after optimisation using a 50 mm column (50 × 2.1 mm; 1.7 μm). A more recent study also incorporated the ionic liquid-based ultrasonic/microwave-assisted extraction (IL-UMAE) of five anthraquinones (physcion, chrysophanol, emodin, rhein and aloe-emodin) from rhubarb prior to their UHPLC-UV determination.[47] Another example of the quantification of NPs in phytopharmaceuticals is the standardisation of black cohosh by UHPLC-UV and UHPLC-ELSD. In this case, triterpenoids and isoflavonoids were identified and quantified in the rhizomes.[34] The analysis time of the extracts was reduced to 7 min with UHPLC using a 45–65% gradient of ACN/MeOH (7:3). Finally, faster separations have also been developed. For example, a powerful 1 min UHPLC-PDA quantification method of *N*-acyl-D/L-homoserine lactones in *Hordeum vulgare* and in *Pachyrhizus erosus* plants was proposed using a specific sample preparation.[48]

13.3.2 UHPLC-MS

MS or MS/MS analysers providing more specific detection have been used for the quantitation of NPs in various matrices. A good example of fast standardisation of phytopharmaceuticals and dietary supplements is the specific detection of the terpenes (ginkgolides and bilobalide) responsible for the antiplatelet activity of *Gingko biloba* by LC-MS on both a simple quadrupole and a TOF system. These methods involve the use of the $[M+NH_4]^+$ and $[M+H]^+$ ions of Gingko terpene in the PI mode with extractive ion monitoring by HPLC-TOF-MS and selected ion monitoring by UHPLC-MS using a single quadrupole analyser. The limit of detection (LOD) values for ginkgolide J, ginkgolide C, ginkgolide B, and ginkgolide A were in the range of 1–10 ng mL^{-1} with both methods. The LOD for bilobalide was 200 ng mL^{-1} by HPLC-TOF-MS and 35 ng mL^{-1} by UHPLC-MS. The gradient analysis in UHPLC was performed in 7 min.[49] For an even more specific detection, UHPLC-MS/MS was used on a triple quadrupole system operated in the MRM mode for the quantification of eight major alkaloids in extracts of *Coptis chinensis,* a commonly used herbal drug in traditional Chinese medicine (TCM). In this example, the mobile phase comprised an ammonium acetate buffer to optimise the peak shape and the separation of the alkaloids, and the complete analysis was performed in 5 min. This method was used for a rapid authentication and quality evaluation of this TCM obtained from various locations.[50]

Very specific detection methods are also important for the determination of minor plant constituents that have key hormonal effects. In this respect, a validated method has been used for the simultaneous analysis of different phytohormones (auxins, cytokinins,and gibberellins) in vegetables in less than 7 min. UHPLC-MS/MS was performed in both PI and negative ionisation (NI) modes. The sample preparation was reduced to a minimum using a simple and fast extraction procedure in which all extractions and sample preparations are performed in the same tube (QuEChERS-based method). The method was validated, and mean recoveries were evaluated at three concentration levels (50, 100 and 250 mg kg^{-1}), ranged from 75 to 110% at the three levels assayed. Intra-day and inter-day precisions, expressed as relative standard deviations (RSDs), were lower than 20 and 25%, respectively. Limits of quantification (LOQs) were equal to or lower than 10 µg kg^{-1}. The developed procedure was applied to different courgette samples, and naphthylacetic acid, naphthylacetamide, and benzyladenine were found in several of the analysed samples.[16]

QC in routine analysis must be fast and well optimised to ensure high reproducibility. Because plant samples are often complex and several homologous compounds need to be separated in a fast gradient run, chromatographic modelling software has been used to optimise separations. For example, the UHPLC conditions for the QC of *Rhizoma coptidis*, a plant containing different NPs with overlapping LC peaks, were calculated based on the retention time (RT) and peak shape parameters of the target peaks. The calculated chromatograms proved to be well correlated to the experimental ones, and the calculated method was found to be very helpful in obtaining satisfactory separation conditions of target compounds that were rapid and efficient.[51]

Several additional, recent characteristic applications are summarised in Table 13.1. Other applications reported especially prior to 2009 have been previously reviewed.[11]

13.4 Fast Non-targeted Analysis, Fingerprinting and Metabolomics

The high-throughput capacities of UHPLC are not only interesting for targeted quantification methods of NPs but also for non-targeted qualitative analyses. Indeed, coupling UHPLC with a mass spectrometer such as a TOF-MS provides a sensitive MS detection in full-scan mode with a high acquisition rate, resulting in a powerful analytical platform for the non-targeted metabolite fingerprinting of crude plant extracts or other natural matrices.[63,71] When performed in a given series of analytical runs, UHPLC-TOF-MS fingerprinting data are reproducible and accurate. The separation of the different features (*m/z vs.* RT) related to each NP generates ion maps where metabolites are resolved in both the *m/z* scale and RT dimensions. This approach generates metabolite fingerprints that can be used for various purposes such as extract standardisation based on fingerprinting, metabolomics and chemotaxonomic studies.

In the field of metabolomics, UHPLC-MS has been largely used for biomarker discovery in human and animal samples (see related Chapter 14). In applications related to natural products analysis, several studies have demonstrated the usefulness of UHPLC fingerprinting as a rich and valuable source of analytical data for differential metabolomic studies.[3]

13.4.1 UHPLC-MS for Plant Metabolomics

For differential metabolomic studies, in our group, we use both (i) the high-throughput capabilities of UHPLC on short columns to acquire rapid UHPLC-TOF-MS fingerprints of numerous replicates and (ii) slow high-resolution profiling on long columns of pool representative samples for the localisation and determination of biomarkers (see below). The high-throughput analysis of many biological replicates improves the reproducibility of the LC-MS detection, allowing large series of samples to be analysed over a short time period and thereby avoiding drift of the MS detection. The increase in the number of biological replicates gives more significant weight to metabolome variations in relation to a given physiological modification versus the natural biological variation of the samples. The data mining of such fingerprinting data is thus notably improved.[72]

Such a strategy was used to study the stress caused by wounding in the model plant *A. thaliana.* Indeed, wounding is known to mimic the attack of herbivores, and metabolomics has the potential to provide a global picture of all chemical events triggered by this stress for the discovery of new wound biomarkers.[59] In these experiments, UHPLC-MS analyses were performed in a two-step strategy to detect induced metabolites and precisely localise these compounds among the numerous constitutive metabolites from a leaf of *A. thaliana*. In a first step, rapid UHPLC-TOF-MS fingerprints of the isopropanol leaf extracts were acquired on a 50 mm column (50 $\times$ 1.0 mm; 1.7 μm) with a rapid gradient (see Figure 13.5A) and were submitted to multivariate analysis.

For each sample, UHPLC-TOF-MS produces a large amount of three-dimensional information (retention RT $\times$ m/z $\times$ intensity) that can be displayed in the form of two-dimensional (2D) ion maps (see Figure 13.5). Pre-processing of the data was required for data mining.[57] In the first step, noise filtering, peak detection and matching were concomitantly performed, making use of both the UHPLC high peak capacity and resolving chromatographic power and the high mass accuracy of TOF-MS detection. Due to the high reproducibility of the data sets obtained, no alignment of the LC-MS was required. After completing the integration parameters, a report of peaks based on areas was generated for each sample and a comprehensive list of the detected components was created. The final data table consisted of

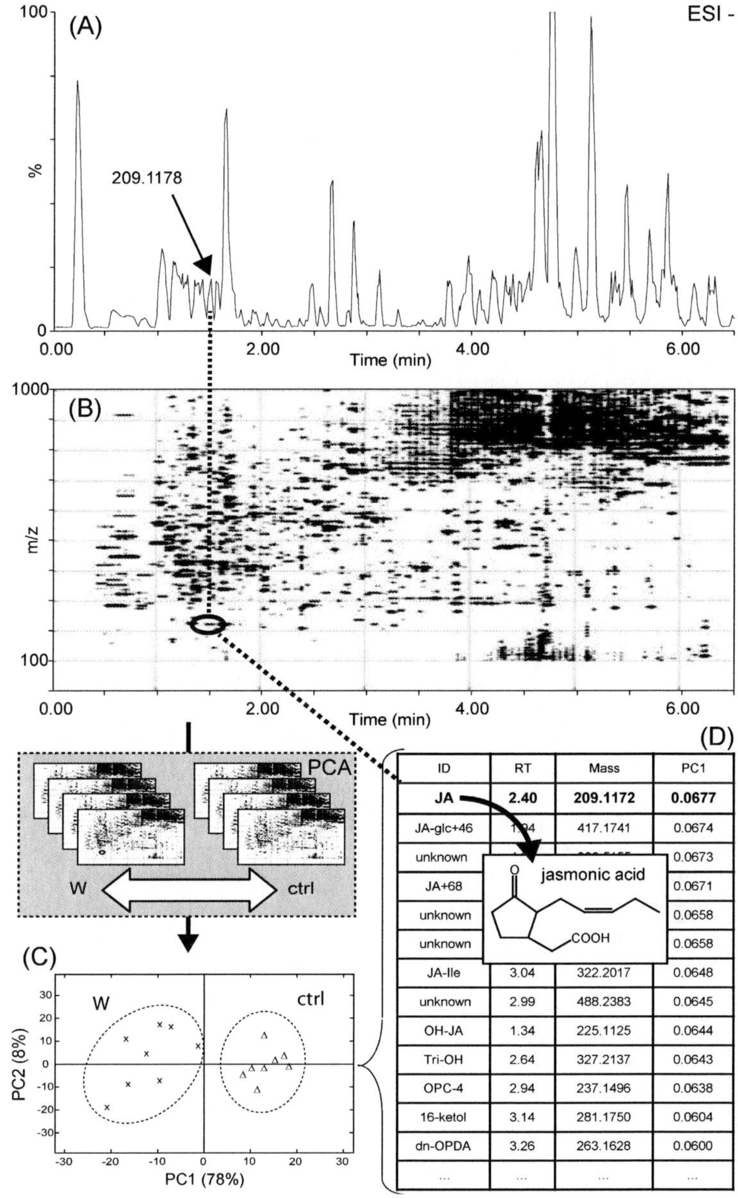

Figure 13.5 UHPLC-TOF-MS based metabolomic analysis of the wound response in *A. thaliana*. (A) Metabolite fingerprint of a wounded *A. thaliana* leaf extract using a fast 6.0 min gradient with TOF-MS detection in NI mode. (B) 2D ion map of the fingerprint (*m/z vs.* RT). (C) PCA score plot of nine wounded (W) and control (ctrl) leaf extracts. (D) Loading of the PCA analysis indicating that jasmonic acid (JA) was the most significant biomarker induced upon wounding of the leaves. All other features (*m/z vs.* RT) ranked according to the PC1 score are related to additional wound biomarkers.

retention times and positive or negative *m*/*z* data pairs as labels; these data were then exported to perform multivariate analyses.

The data were then used to produce interpretable projections of samples in a reduced dimensionality (score plots) (see Figure 13.5C) and to highlight putative biomarkers responsible for the group separation (loading plots) (see Figure 13.5D). Statistical methods, such as principal components analysis (PCA), were initially used and provided an unsupervised data reduction without using class information. Complementary analysis tools and supervised methods were also used for further in-depth investigations of subtle metabolome modifications that occurred at different times after wounding.[73]

After PCA, a clear clustering of plant specimens was demonstrated (control *vs.* wounded plant after 90 min, Figure 13.5C), and the highest discriminating ions given by the complete data analysis were selected, leading to the specific detection of discrete-induced ions (*m*/*z* values) (see Figure 13.5D).

In the generated loading plots, jasmonic acid was found to be the most significant wound biomarker responsible for the PCA clustering between wounded and unwounded plants (see Figure 13.5D). Jasmonic acid is a well-known phytohormone involved in the wound response, and its detection by this non-targeted approach validated the model used. The majority of the other biomarkers highlighted in the loading were then characterised either based on the formula that was detected or in a second, confirmatory step.[57,74,75] For biomarker identification, high-resolution LC profiling was performed on pooled samples by UHPLC-TOF-MS. An example of the type of high resolution profile obtained is illustrated in Figure 13.6. This strategy allowed for a precise localising of the putative biological markers induced by wounding through the specific extraction of accurate *m*/*z* values. The localised markers could then be isolated using semi-preparative LC after method transfer to allow for their subsequent characterisation by capillary NMR.[56]

In addition to our own investigations, the model plant *A. thaliana* has been the topic of many metabolomics studies involving several MS-based metabolomic approaches.[3] Most of these studies are related to the evaluation of responses to different type of biotic or abiotic stresses. For example, non-polar lipids were efficiently analysed by UHPLC-TOF-MS in *A. thaliana* using a C8 column at 60 °C with a relatively long (25 min) aqueous gradient with MeOH and isopropanol as organic modifiers. The aim of this study was to analyse the short-term changes in the *A. thaliana* glycerolipidome in response to temperature and light in a time-resolved manner. This UHPLC-TOF-MS lipidomic approach enabled the monitoring of several glycerolipid species that have been reported in *Arabidopsis* leaves. The exposure of these plants to various light and temperature regimes resulted in two major effects. The first effect was the dependence of the saturation level of phosphatidylcholine and monogalactosyldiacylglycerol pools on light intensity, probably arising from light regulation of *de novo* fatty acid synthesis. The second effect concerns an immediate decrease in unsaturated species of phosphatidylglycerol at high temperatures (32 °C), which could mark the first stages of adaptation to heat stress conditions.[61]

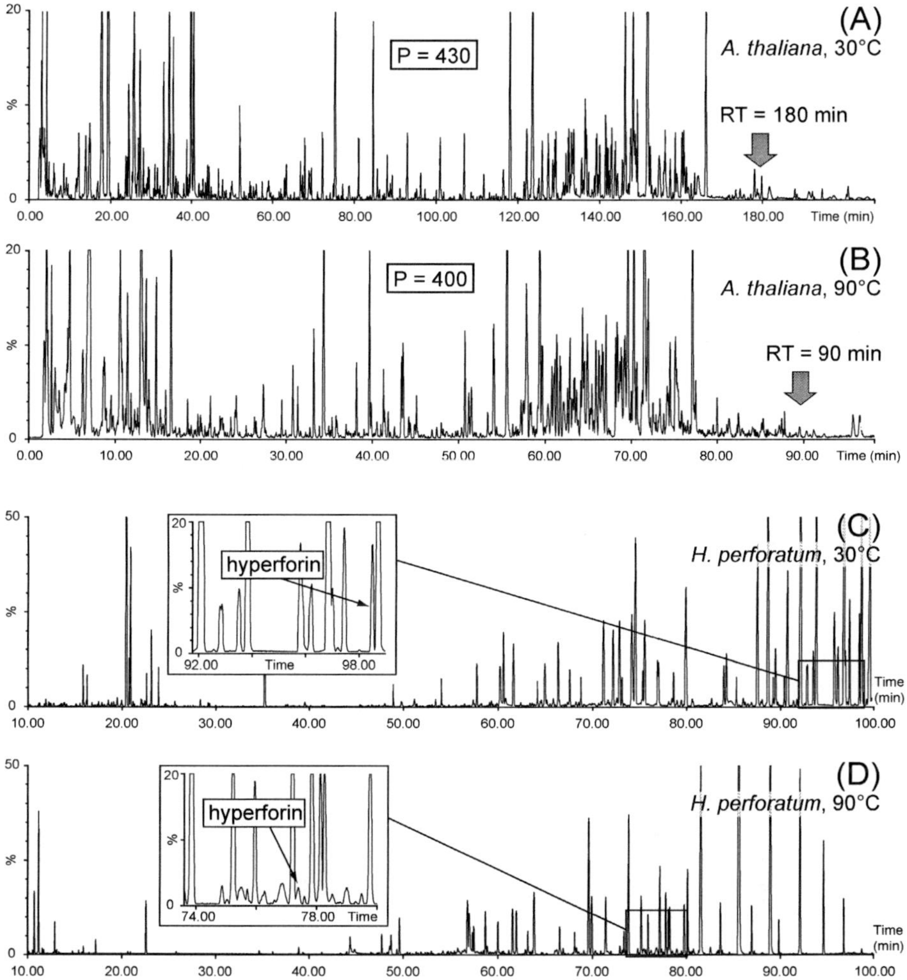

Figure 13.6 (A) *A. thaliana* extract analysed on two Acquity™ BEH C18 columns (150 × 2.1 mm; 1.7 μm) coupled in series at 30 °C with a 240 min gradient. (B) The same separation carried out at 90 °C, with a higher flow rate providing the same pressure with a 120 min gradient. (C) *H. perforatum* extract analysed on a Acquity™ BEH C18 (150 × 2.1 mm; 1.7 μm) column at 30 °C with a 93 min gradient. (D) The same separation carried out at 90 °C with a higher flow rate providing the same pressure. The selectivity is changed, and the peaks are thinner [see inset in (C). Adapted from ref 63 with permission from Elsevier.

Interactions between plant and insects were also studied from a metabolomic perspective. In this case, a comprehensive study of the interaction between feral cabbage (*Brassica oleracea*) and small caterpillars (*Pieris rapae*) was conducted based on a 15 min UHPLC-TOF-MS fingerprinting of the extracts of both participating organisms in this plant–insect herbivore interaction. The metabolomic results provided more insight into the metabolites that were possibly involved in such interactions, and it was finally concluded that the attack history of the plants affects a specific part of the metabolome of the herbivore.[76] Similarly, the plant–insect interface [*Spodoptera* spp. (maize)] was also investigated from a metabolic viewpoint to highlight modifications of bioactive plant secondary metabolites by insect herbivores for understanding animal detoxification processes and plant–insect interactions.[62]

The effect of artificial stresses that are known to induce non-thermal permeabilisation of cell membranes, such as those generated by pulsed electric fields (PEFs), were also assessed based on a UHPLC-TOF-MS metabolomic approach on potato tissue. In this case, the UHPLC-TOF-MS fingerprinting data were complemented by gas chromatography (GC) using a GC-TOF-MS system to obtain a more comprehensive survey of the potato metabolites. Clustering analysis showed that 24 h after the application of PEFs, the potato metabolism shows PEF-specific responses characterised by changes in the hexose pool that may involve starch and ascorbic acid degradation.[58]

UHPLC-QTOF-MS in complement to headspace-solid phase micro extraction-GC-MS (HS-SPME-GC-MS) and 2D gel electrophoresis was also used for an extensive characterisation of the metabolic changes occurring in *Brunfelsia calycina* petals after the flower's opening. In particular, the anthocyanin degradation products were profiled and characterised based on the UHPLC-MS/MS analyses performed. Globally, this multi-level metabolomic study resulted in the identification of nine main anthocyanins in *Brunfelsia* flowers, 146 up-regulated genes, 19 volatiles, seven proteins and 17 metabolites that increased during anthocyanin degradation, suggesting an induction of the shikimate pathway.[60]

In addition to its application in fundamental plant sciences, UHPLC-TOF-MS studies based on metabolomics have been applied to the study of metabolic variations that occur in plants of medicinal value. For example, the effects of the duration of steaming on the metabolome composition of *Panax notoginseng* were monitored by UHPLC-TOF-MS using a 10 min generic water/ACN gradient. A qualitative profiling of multi-parametric metabolic changes of raw *P. notoginseng* during the steaming process was thus obtained. Both the unsupervised and supervised data mining on the fingerprinting results demonstrated strong classification and clear trajectory patterns with regard to the duration of steaming. Using this tool, the minimum duration of steaming providing the maximum production of bioactive ginsenosides could be predicted.[54] Such a methodology can be used for fundamental research and for quality assessment for commercial preparations. For other recent applications, see Table 13.1.

13.4.2 UHPLC-MS/MS-based Targeted Metabolomics

Metabolomics studies can also be performed in a semi-targeted manner by UHPLC-MS/MS when hundreds of previously selected constituents are included in the data sets. This metabolomics methodology has been established to quantify hundreds of targeted plant metabolites by MRM in a high-throughput manner in 14 plant species from Brassicaceae, Gramineae and Fabaceae. As mentioned, the use of MRM after high-throughput UHPLC separation is a well-established method for the targeted analysis of specific NPs. In this study,[77] however, the inclusion of a high number of metabolites provides a rich data set that can be investigated with similar data mining tools as those used for non-targeted metabolomics. Thus, approximately 100 metabolites were quantified in each of the plant extracts investigated, and five transitions were monitored in each 3 min UHPLC gradient run. A hierarchical cluster analysis based on the metabolite accumulation patterns clearly showed differences among the plant families, and family-specific metabolites could be predicted using a batch-learning, self-organising map analysis. Such an automated, widely targeted metabolomics approach represents an interesting alternative method for elucidating metabolite accumulation patterns in plants. It also represents an elegant way to combine the high-throughput potential of UHPLC to the performance of MS/MS for quantitation and appropriate data mining to achieve a comprehensive evaluation of the results obtained.

13.4.3 UHPLC Fingerprinting for QC

All of the examples previously described demonstrate the potential of UHPLC-TOF-MS-based metabolite fingerprinting to obtain a fast overview of an extract metabolic content. Thus, QC procedures are progressively adopting such strategies for identification, categorisation or standardisation purposes. UHPLC is especially useful in these situations because plant extracts are complex and consist, among other things, of numerous metabolites acting synergistically that could not be accurately considered separately. TCM preparations, which often consist of several herbs, require even more extensive rational approaches. Moreover, the identification of plants based on fingerprints is more valuable than identification based on one or few constituents (targeted analysis).[78] NMR metabolomics[79] is well established for this type of global metabolite fingerprinting, but only provides the detection of the main NPs in a given extract. UHPLC-MS is now also starting to be more extensively used for detailed composition comparisons. For example, the QC in commercial preparations of angelica roots (*Angelica acutiloba*) was performed by comparison of high-throughput (10 min gradient) PI and NI UHPLC-TOF-MS fingerprinting using chemometric tools. Partitioning of root samples was effectively achieved by PCA, showing that the cultivation area was one of the most significant parameters for quality determination. This method proved to be an efficient and rapid QC method that can be used on a routine basis.[53] For other recent applications, see Table 13.1.

Table 13.1 Summary of some published NPs UHPLC analyses classified according to the type of studies performed.

Ref.	*Year*	*Analysed compounds or aim of the study*	*Plants or organisms studied*	*Stationary phase chemistry*	*Column size (mm × mm i.d.)*	*Analysis time (min)*[a]	*Detection*[c]
			QC (targeted)				
48	2008	Simultaneous quantitative determination of 10 diterpenes	*Salvia miltiorrhiza*	BEH C18	50 × 2.1	15/10	PDA
42	2008	Quantitative determination of 5 anthraquinone derivatives	*Rheum* spp.	BEH C18	50 × 2.1	3[b]	PDA
30	2009	Quantitative analysis of formononetin and triterpenoid glycosides	*Actaea racemosa*	BEH C18	100 × 2.1	12.5/7	ELSD + Q-MS
45	2009	Quantitative analysis of ginkgolides and bilobalides	*Ginkgo biloba*	BEH Shield RP18	50 × 2.1	12.5/7	Q-MS
16	2010	Quantification of steviol (st) and its glycosides (g); 2 conditions with high temperature of mobile phase	*Stevia rebaudiana*	HSS C18	150 × 2.1 (g), 100 × 2.1 (st)	?/4 (g), 4.4 (st)[b]	QqQ-MS/MS
18	2010	Qualitative analysis of 7 polyphenols in tea samples	*Camellia sinensis*	BEH Shield RP18	50 × 1.0 + 100 × 2.1	?/0.5 and 7.2	UV + QqQ-MS/MS
41	2010	Quantitative analysis of 29 phenolics in tea infusions or extracts	*Camellia sinensis*	BEH C18	100 × 2.1	29/20	PDA
17	2011	Quantitative analysis of phytohormones; use of a QuEChERS-based extraction method	*Cucurbita pepo*, courgette Elena variety	BEH C18	100 × 2.1	6/4	QqQ-MS/MS
43	2011	Quantitative determination of 5 anthraquinones; use of ionic liquid-based ultrasonic/microwave-assisted extraction (IL-UMAE)	*Rheum* spp.	BEH C18	100 × 2.1	?/33	PDA
			QC (untargeted)				
49	2009	Metabolite fingerprinting for untargeted standardised QC; use of chemometric tools	*Angelica acutiloba*	BEH C18	150 × 2.1	?/10.1	TOF-MS
50	2010	Metabolic profiling for the evaluation of raw and steamed *P. notoginseng*	*Panax notoginseng*	BEH C18	100 × 2.1	10/8.3	TOF-MS

Table 13.1 (*Continued*)

Ref.	*Year*	*Analysed compounds or aim of the study*	*Plants or organisms studied*	*Stationary phase chemistry*	*Column size (mm × mm i.d.)*	*Analysis time (min)*[a]	*Detection*[c]
51	2010	Quality evaluation of the *Radix linderae* TCM by metabolite fingerprinting analysis	*Lindera aggregate*	HSS T3	150 × 2.1	?/37	PDA + QqQ-MS/MS
		Metabolomics or signal/biomarker study					
44	2007	Study of bacterial signal molecules in plants	*Hordeum vulgare* + *Pachyrhizus erosus*	BEH C18	100 × 2.1	?/1	PDA
52	2008	Metabolomic study of oxylipins induced by wounding	*Arabidopsis thaliana*	BEH C18	150 × 2.1	?/107	TOF-MS
53	2008	Metabolomic study of oxylipins induced by wounding in a two-step strategy	*Arabidopsis thaliana*	BEH C18	50 × 1.0 + 150 × 2.1	?/10 and 119.7 or 325.8	TOF-MS
54	2009	Study of metabolic changes in potato tissues after pulsed electrical field stress	*Solanum tuberosum* cv. Bintje	BEH C18	100 × 2.1	19/12.5	TOF-MS
55	2009	Signal propagation study in wounded plant leading to jasmonic acid accumulation	*Arabidopsis thaliana*	BEH C18	100 × 2.1	13/6	TOF-MS
56	2010	Study of the metabolic changes after flower opening	*Brunfelsia calycina*	BEH C18	100 × 2.1	26/22	PDA + QTOF-MS/MS
57	2011	Study of the lipid metabolism in plant under light or temperature stress	*Arabidopsis thaliana*	C8	150 × 2.1	25/17	QTOF-MS/MS
58	2011	Monitoring of the dynamic network of benzoxazinoids at the plant-insect interface; 2 methods: (1) metabolomic study, (2) quantification	*Zea mays*	BEH C18 (1,2)	50 × 1.0 (1), 50 × 2.1 (2)	8.0/4.9 (1), 5.0/2.9 (2)	TOF-MS (1), QTOF-MS/MS (2)
		Profiling/dereplication/screening					
59	2009	High resolution profiling optimisation; use of high and low temperature	*Arabidopsis thaliana* + *Ginkgo biloba*	BEH C18	150 × 2.1 + 2x 150 × 2.1	Many, up to 240 min	TOF-MS

Table 13.1 (*Continued*)

Ref.	*Year*	*Analysed compounds or aim of the study*	*Plants or organisms studied*	*Stationary phase chemistry*	*Column size (mm × mm i.d.)*	*Analysis time (min)*[a]	*Detection*[c]
60	2009	Identification of 10 alkaloids	3 *Lycopodiaceae* spp.	BEH C18	100 × 2.1	28/12	TOF-MS
47	2009	Separation optimisation of a TCM; use of computer target optimisation	*Coptis chinensis*	BEH Shield RP18	100 × 2.1	?/9	PDA
61	2010	Study of composition of pine needles in organic acids and antibacterial activity	*Pinus massoniana*	BEH C18	100 × 2.1	?[b]	QqQ-MS/MS
62	2010	Dereplication and identification of 39 bufadienolides	*Bufo bufo gargarizans*	HSS T3	100 × 2.1	?/30	QTOF-MS/MS
63	2011	Fast and comprehensive profiling using a 2D LC (NP × RP) strategy; UHPLC as the 2nd dimension	*Stevia rebaudiana*	Zorbax SB C18	30 × 2.1	0.33/0.27	PDA
64	2011	Dereplication of 28 triterpenoid saponins	*Albizia julibrissin*	RRHD SB-C18	100 × 2.1	?/60	PDA + QTOF-MS/MS
20	2011	Chemical screening of micro-organisms associated with marine invertebrate, in 96-well plates	*Erythropodium caribaeorum*	?	?	?/6	ELSD + IT-MS^n
22	2011	Dereplication of caffeoylquinic acids and flavonoids based on retention time and PDA spectra	*Hemerocallis fulva*	BEH C18	100 × 2.1	?/9	PDA
65	2011	Analysis of anthocynins in red wine	Red wine	BEH C18	2 × 100 × 2.1	?/98	PDA + QTOF-MS/MS
			Chemotaxonomy				
66	2009	Metabolic profiling of *Gentiana* and *Gentianella* spp. for chemotaxonomic study (flavonoids and xanthones)	*Gentianaceae* spp.	BEH C18	150 × 2.1	?/16.1	TOF-MS

[a]Total analysis time with / without wash and equilibration. Unless specified, all analyses are performed in gradient mode. [b]Isocratic mode. [c]Abbreviations: UV, single trace UV; PDA, photo diode array detector; Q-MS, simple quadrupole MS; QqQ-MS/MS, triple quadrupole MS in MS/MS mode; TOF-MS, time-of-flight MS; QTOF-MS/MS, quadrupole-time-of-flight MS in MS/MS mode; IT-MS^n, ion trap MS in MS^n mode; ELSD, evaporative light scattering detector.

13.4.4 Chemotaxonomic Studies

As mentioned above, fingerprinting can be used as a chemotaxonomic tool to discriminate plant species based on their secondary metabolite composition. The *Gentiana* and *Gentianella* genera were distinguished among the Gentianaceae family based on their UHPLC-TOF-MS fingerprints. Separations were carried out on a UHPLC column (150 × 2.1 mm; 1.7 μm) in 15 min with a 5–55% aqueous/ACN gradient. The fingerprints of three *Gentianella* species were strikingly similar. In contrast, fingerprints of the *Gentiana* species were very different from those of *Gentianella* species and from each other. Several compounds were determined as unique to each genus and, therefore, could be used as biomarkers. This result was helpful for an unambiguous classification of plants belonging to these genera.[70] Another study enabled the classification of different Brazilian species of the *Lippia* genus.[80]

13.5 High-resolution Profiling and Metabolite Identification

In most of the fingerprinting studies discussed above, the UHPLC conditions were mainly optimised for a high-throughput comparison of many crude extract replicates for data mining with either very fast separation or with fingerprinting methods, providing peak capacities equivalent to conventional HPLC methods.

UHPLC can also be used to extend the achievable resolution for the separation of NPs in crude extracts with complex compositions. In this case, the goal is to provide the best possible separation of closely related NPs, which often occur as positional isomers or diastereoisomers. Furthermore, the high chromatographic resolution of UHPLC reduces the ionisation suppression problems that often occur with electrospray ionisation in UHPLC-MS; thus, a better detection of minor constituents is also obtained. When coupled with MS or PDA, the high peak purity obtained by this method provides a better deconvolution of the MS or UV spectra recorded for online structural determination or dereplication purposes.[57]

However, the price of this enhancement in the quality of data is a longer analysis time. With similar run times as in conventional HPLC profiling methods, an enhancement of peak capacity of approximately a factor of 3 can be expected with UHPLC systems.[42] This improvement in resolution also depends on the column length and the number of theoretical plates and optimised conditions obtained. Because the profiling of crude extracts metabolites is generally performed in gradient mode and because peak capacity is related both to the plate number and to the column dead time, the improvement in peak capacity is not dependent just on the column length. Thus, an optimum for the column length and the gradient time has to be found. An accepted compromise is that a 150 mm, 1.7μm column should be preferentially selected for gradient lengths up to 60 min at 30 °C, while the columns coupled in series (3 × 150 mm, 1.7 μm) are attractive only for a gradient time higher than 250 min.[42]

The enhancement of peak capacity that can be obtained on crude plant extracts has been well demonstrated by the chromatograms obtained for the extract of *G. biloba* (see Figures 13.2A and 13.2B). In this case, the gradient time was kept similar to the original HPLC method, and the UHPLC columns and conditions were optimised to achieve the maximum peak capacity.

In the case of UHPLC-TOF-MS coupling, this enhancement of chromatographic resolution also provides a much more detailed localisation of the different NPs that constitute the metabolome of a given organism. The ion maps obtained with a high-resolution profiling method transferred from the fast fingerprinting method were used for the metabolomic study of the wound response in *A. thaliana* (see the corresponding high-resolution profile in Figure 13.6A). With such a high peak capacity measurement, several isomers were well separated, and, for example, for this plant, a peak at *m/z* 225 that appeared as a single wound biomarker in the metabolomic study based on the rapid fingerprinting was found to correspond to four different isomers that could be resolved using a high-resolution metabolite profiling method.[56]

13.5.1 Very High-resolution Profiling

In order to push forward the quest for high resolution, it is also possible to increase the column length of the UHPLC column and increase the peak capacity by using gradient times exceeding 60 min.[42] In the case of the metabolite profiling of *A. thaliana*, the use of two 150 mm columns coupled in series provided an increase of 40% in peak capacity as compared to the separation obtained from one 150 mm column. The gradient transfer on this 300 mm column was, however, performed in 240 min, as compared to 60 min on the 150 mm column.[63]

One possible way to decrease this very long gradient time is to perform the separation at high temperature. Indeed, most of the new hybrid silica-based columns are stable at high temperatures, rendering this type of analysis possible. In the case of *A. thaliana*, the gradient time could be reduced from 240 to 120 min while maintaining approximately the same high peak capacity when the separation was performed at 90 °C instead of 30 °C.[63] On the 300 mm column, the maximum flow rate at 30 °C was 200 μL min^{-1}, but this could be increased to 350 μL min^{-1} at higher temperatures. Figures 13.6(A) and 13.6(B) show metabolite profiles obtained at 30 °C and 90 °C, respectively. The baseline separation of more than 300 metabolites could be practically achieved by this means. The potential degradation of NPs during separation was examined, but no apparent degradation was observed for even the longest separations at 90 °C.[63]

As described for short gradients, to a lesser extent, however, a non-negligible loss of resolution may occur due to the extra-column volume and is related to the type of detector used. This parameter has to be taken into account in addition to the gradient time and column length optimisation in order to improve the performance of the analytical platform. The source of some MS detectors, for example, may generate a loss of more than 20% in peak capacity as compared to UV detectors.[63]

The use of high temperatures provides a significant increase in throughput; however, temperature modifications also affect the polarity of the mobile phase and the selectivity of the separation. For the separation of non-polar NPs, the use of HT-UHPLC can represent an advantage, and compounds that would be difficult to elute from C18 columns even with a high percentage of organic solvent may elute much faster in these conditions. For example, this result was seen for hyperforine, a non-polar phloroglucinol derivative found in the standardised extract of *Hypericum perforatum* that is involved in the antidepressant effect of this phytopharmaceutical. As shown in Figures 13.6(C) and 13.6(D), the use of high temperature significantly affects the selectivity and above all the retention for the different *Hypericum* constituents. Hyperforine was found to elute at 78 min at 90 °C, while it did not elute until 98 min at 30 °C.

13.5.2 LC × LC for Improved Resolution

For complex plant extracts, the use of reversed-phase separation alone, even with very high peak capacity, might not be sufficient for the separation of all metabolites in a single profiling analysis. In this case, the use of an orthogonal separation using a column with different phase chemistry might be needed. In this respect, a very recent 2D LC application based on UHPLC has been described for separating all of the components of interest contained in *Stevia rebaudiana*, a plant from Paraguay that is currently used worldwide as a sweetener. For the profiling of this plant, neither RP-HPLC (reversed-phase HPLC) nor NP-HPLC (normal-phase HPLC) alone has been capable of separating all of the components of interest. A combination of 2D LC (LC × LC) for the profiling of this extract was used. The first dimension used a classical polyamine HPLC (250 × 1.0 mm; 5 μm) column in normal-phase mode at ambient temperature with a 100 min gradient at 20 μL min^{-1}. UHPLC was employed for the fast second dimension: the eluate was divided into fractions by a 20 μL loop and then injected online in triplicate in the second dimension. This second dimension consisted of a Zorbax RRHD SB-C18 UHPLC column (30 × 2.1 mm; 1.7 μm) operating in reversed-phase mode at 70 °C in a fast gradient of 20 s (with re-equilibration) at 3.4 mL min^{-1}. This high flow rate allowed a very short wash and re-equilibration times. Thanks to the high throughput of the UHPLC separation, the reduced cycle time allowed 3 to 12 samplings for each peak eluted by the first dimension. Polyphenolic and stevioside compounds were thus efficiently identified by combining the information coming from the position of the compounds in the 2D plot and the UV spectra with that of reference materials.[67]

13.5.3 Metabolite Identification and Dereplication

The high-resolution profiling of UHPLC provides a good separation of NPs in complex mixtures. This baseline separation of analytes is important for quantification if a simple detector, such as UV or ELSD, is used. The deconvolution of LC peaks based on chromatography is also important for recording online UV-PDA and MS spectra of good quality for facilitating the dereplication process. An example of the type of online spectral information that can be obtained for the dereplication of natural products is illustrated in Figure 13.7 for the profiling of *Viola tricolor*, a herbal drug used traditionally for its anti-inflammatory properties. The extract of *V. tricolor* was analysed with an optimised gradient on a BEH C18 column (150 × 2.1 mm; 1.7 μm) in 60 min. As shown, a good separation of most of the metabolites was obtained, and minor compounds were also detected in both PI and NI modes thanks to the sensitivity of the TOF-MS system. The TOF-MS detection provided exact molecular weights (<5 ppm) and retention time information for all of the compounds detected. Different successive filters were applied to extract and ascertain molecular formulas in order to reduce the number of structural possibilities. This filtering is derived from heuristic rules: (1) restrictions for the number of elements; (2) LEWIS and SENIOR chemical rules; (3) isotopic patterns; (4) hydrogen/carbon ratios; (5) element ratio of nitrogen, oxygen, phosphor and sulphur *versus* carbon; and (6) element ratio probabilities.[81] For the most abundant unknown compounds, PDA-UV spectra were recorded and used as a complement to the MS data in the dereplication process. Chemotaxonomic information was then also added for the final selection of putative structures. Based on these structural hypotheses, correlations between retention time, lipophilicity and elution behaviour in a series of related compounds were performed.

As shown in Figure 13.7A, the peak at RT 11.33 min displayed a molecular ion at *m/z* 577.1543 $[M–H]^-$ in NI mode and at *m/z* 579.1741 NI $[M+H]^+$ in PI mode (Figure 13.7C). This information confirmed that the MW was 578 Da. With a 15 ppm tolerance in NI mode, this exact online mass determination gave five possible formulas (Figure 13.7B). The application of the heuristic filter confirmed that the only valid possibility for was $C_{27}H_{30}O_{14}$. A cross-search of this formula with chemotaxonomic information on the *Viola* genus found in a NP database[82] revealed that the only possible hit corresponded to violanthin (see Figure 13.7C). This peak annotation was confirmed by the UV-PDA spectrum that was recorded online, confirming that both maxima matched well with this flavonoid diglycoside, which could be efficiently dereplicated by this means. The same procedure was applied to all LC peaks that were efficiently separated by this high-resolution profiling.[83]

13.6 Conclusions

As shown here, UHPLC presents many advantages for analysing NPs in complex biological matrices, such as crude plant extracts. Indeed, for all of the examples discussed, the efficiency of UHPLC either in terms of its high-throughput (QC, fingerprinting) or in terms of its high-resolution (dereplication, profiling) is very advantageous as compared to classical HPLC. Also,

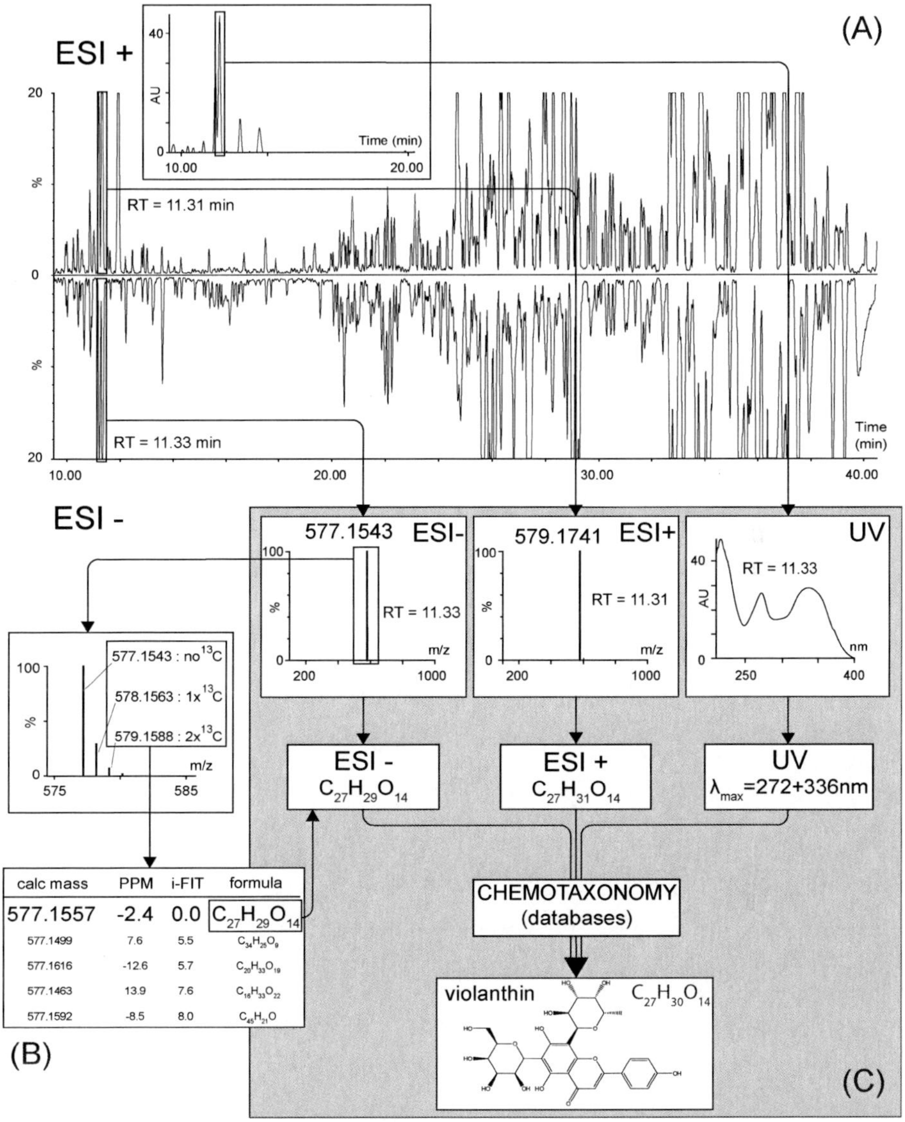

Figure 13.7 Peak annotation procedure for dereplication based on a high resolution *V. tricolor* profiling on a C18 UHPLC column (150 × 2.1 mm; 1.7 μm). (A) PI (upper trace) and NI (lower trace) ESI-TOF-MS BPI chromatograms obtained with a slow gradient (5–95% ACN in 50 min). A UV trace (366 nm) is displayed in the inset. (B) Putative molecular formulas assignment based on the precision (ppm) and isotopic pattern (iFIT) obtained from the NI ESI-TOF-MS spectrum of the LC peak at RT 11.33 min. (C) Annotation of the LC peak at RT 11.33 min based on PI and NI molecular formula assignment and the UV PDA spectrum. Final structural assignment based on a cross search with chemotaxonomic information.

because the diameter of the column used is smaller, there is a significant reduction in solvent and sample consumption. For metabolomics, this technique provides clear advantages in terms of its reproducibility, resolution and throughput, yielding data that could not be attained by conventional HPLC methods in practically achievable analysis times. Such characteristics are essential for the satisfactory comparison of fingerprints with data mining methods.

In NPs research, compounds must often be isolated for either *de novo* structure determination or for bioactivity assessment. In this respect, it is still difficult to find semi-preparative columns with similar phase chemistries as those developed for UHPLC, which might hinder the possibility of performing efficient gradient transfers for all types of applications. However, this problem will likely be solved when the technique spreads more widely to research groups involved in NP research. The number of applications of UHPLC in crude extract analysis is still scarce compared to HPLC, probably due to the required dedicated instrumentation that is needed to work with the high pressures generated by the use of columns with sub-2 μm particles.

The development of highly efficient, sub-2 μm columns has also stimulated the development of other columns that share similar characteristics and are compatible with a conventional HPLC system. An example of this is columns with core-shell particles consisting of a 1.7 μm solid core surrounded by a 0.5 μm porous silica shell, as discussed in Chapter 5. This type of particle shares a similar chromatographic performance to the sub-2 μm particles; however, their backpressure is much lower.[84] These types of columns have already been used with success for profiling crude extracts.[85,86] Core-shell particles can thus be a good low-pressure alternative to columns packed with the sub-2 μm particles for the separation of complex mixtures with only a small sacrifice in peak efficiency.

For detection and dereplication, an MS analyser is the optimal detector to be coupled with UHPLC. Although TOF-MS detectors have an adequate acquisition frequency to cope with the LC peak width obtained by UHPLC, this type of MS detector still requires improvement to achieve faster acquisition rates. Indeed, an ideal system should be able to provide both MS and MS/MS spectra in a single run at high resolution and in both PI and NI mode simultaneously. However, this process is a very demanding one for a MS analyser and further improvements are expected to appear in the coming years.

UHPLC thus represents a very valuable tool for NP chemists. With the increasing requirements for QC, profiling and fingerprinting, dereplication and metabolomics, it is very likely that UHPLC will gradually replace most of the HPLC applications developed for NP research in years to come.

Acknowledgments

The authors are indebted to the Swiss National Foundation for supporting their plant metabolomic studies (grant no. 205320-124667/1 to J.-L. W.).

References

1. C. M. Dobson, *Nature*, 2004, **432**, 824.
2. J. Hong, *Curr. Opin. Chem. Biol.*, 2011, **15**, 350.
3. J. L. Wolfender, G. Glauser, J. Boccard and S. Rudaz, *Nat. Prod. Commun.*, 2009, **4**, 1417.
4. J. Larsson, J. Gottfries, S. Muresan and A. Backlund, *J. Nat. Prod.*, 2007, **70**, 789.
5. D. G. I. Kingston, *J. Nat. Prod.*, 1979, **42**, 237.
6. T. K. Natishan, *J. Liq. Chromatogr. Relat. Technol.*, 2004, **27**, 1237.
7. K. Hostettmann, A. Marston and M. Hostettmann, *Preparative Chromatography Techniques: Applications in Natural Product Isolation*, 2nd edn, Springer-Verlag, Berlin Heidelberg, 1997.
8. J. L. Wolfender, *Planta Med.*, 2009, **75**, 719.
9. D. T. T. Nguyen, D. Guillarme, S. Rudaz and J. L. Veuthey, *J. Sep. Sci.*, 2006, **29**, 1836.
10. F. David, G. Vanhoenacker, B. Tienpont, I. Francois and P. Sandra, *LC-GC Eur.*, 2007, **20**, 154.
11. P. J. Eugster, D. Guillarme, S. Rudaz, J. L. Veuthey, P. A. Carrupt and J. L. Wolfender, *J. AOAC Int.*, 2011, **94**, 51.
12. M. I. Churchwell, N. C. Twaddle, L. R. Meeker and D. R. Doerge, *J. Chromatogr. B*, 2005, **825**, 134.
13. J-L. Wolfender, G. Marti and E. F. Queiroz, *Curr. Org. Chem.*, 2010, **14**, 1808.
14. K. W. Cheng, F. Cheng and M. Wang, in *Bioactive Natural Products: Detection, Isolation, and Structural Determination,* 2nd edn, ed. S. M. Colegate and R. J. Molyneux, CRC Press, London, 2008, pp. 245.
15. C. Gardana, M. Scaglianti and P. Simonetti, *J. Chromatogr. A*, 2010, **1217**, 1463.
16. M. I. Alarcon Flores, R. Romero-Gonzalez, A. Garrido Frenich and J. L. Martinez Vidal, *J. Sep. Sci.*, 2011, **34**, 1517.
17. D. Guillarme, C. Casetta, C. Bicchi and J. L. Veuthey, *J. Chromatogr. A*, 2010, **1217**, 6882.
18. J. Schappler, R. Nicoli, D. Nguyen, S. Rudaz, J. L. Veuthey and D. Guillarme, *Talanta*, 2009, **78**, 377.
19. D. Guillarme, J. Schappler, S. Rudaz and J-L. Veuthey, *Trends Anal. Chem.*, 2010, **29**, 15.
20. A. Pelander, P. Decker, C. Baessmann and I. Ojanpera, *J. Am. Soc. Mass Spectrom.*, 2011, **22**, 379.
21. J. J. J. van der Hooft, J. Vervoort, R. J. Bino, J. Beekwilder and R. C. H. de Vos, *Anal. Chem.*, 2011, **83**, 409.
22. F. Berrue, S. T. Withers, B. Haltli, J. Withers and R. G. Kerr, *Mar. Drugs*, 2011, **9**, 369.
23. J. V. Olsen, J. C. Schwartz, J. Griep-Raming, M. L. Nielsen, E. Damoc, E. Denisov, O. Lange, P. Remes, D. Taylor, M. Splendore, E. R. Wouters,

M. Senko, A. Makarov, M. Mann and S. Horning, *Mol. Cell. Proteomics*, 2009, **8**, 2759.
24. T. Koecher, R. Swart and K. Mechtler, *Anal. Chem.*, 2011, **83**, 2699.
25. Y. O. Tsybin, L. Fornelli, A. N. Kozhinov, A. Vorobyev and S. M. Miladinovic, *Chimia*, 2011, **65**, 641.
26. Y-L. Lin, C-K. Lu, Y-J. Huang and H-J. Chen, *J. Agric. Food Chem.*, 2011, **59**, 8789.
27. R. P. W. Scott, in *Chrom-Ed Book Series*, Library4Science, http://www.library4science.com/, 2003.
28. X. F. Gao, M. Dan, A. H. Zhao, G. X. Xie and W. Jia, *Biomed. Chromatogr.*, 2008, **22**, 244.
29. S. A. Ozkan, *Chromatographia*, 2007, **66**, S3.
30. T. O. Larsen and M. A. E. Hansen, in *Bioactive Natural Products: Detection, Isolation, and Structural Determination,* 2nd edn, ed. S. M. Colegate and R. J. Molyneux, CRC Press, London, 2008, pp. 221.
31. N. C. Megoulas and M. A. Koupparis, *Crit. Rev. Anal. Chem.*, 2005, **35**, 301.
32. N. Vervoort, D. Daemen and G. Torok, *J. Chromatogr. A*, 2008, **1189**, 92.
33. R. Russo, D. Guillarme, S. Rudaz, C. Bicchi and J-L. Veuthey, *J. Sep. Sci.*, 2008, **31**, 2377.
34. B. Avula, Y. H. Wang, T. J. Smillie and I. A. Khan, *Planta Med.*, 2009, **75**, 381.
35. X. M. Liang, Y. Jin, Y. P. Wang, G. W. Jin, Q. Fu and Y. S. Xiao, *J. Chromatogr. A*, 2009, **1216**, 2033.
36. G. Lang, N. A. Mayhudin, M. I. Mitova, L. Sun, S. van der Sar, J. W. Blunt, A. L. J. Cole, G. Ellis, H. Laatsch and M. H. G. Munro, *J. Nat. Prod.*, 2008, **71**, 1595.
37. K. Hostettmann, J. L. Wolfender and C. Terreaux, *Pharm. Biol.*, 2001, **39**, 18.
38. L. W. Sumner, P. Mendes and R. A. Dixon, *Phytochemistry*, 2003, **62**, 817.
39. O. Fiehn, J. Kopka, P. Dormann, T. Altmann, R. N. Trethewey and L. Willmitzer, *Nat. Biotechnol.*, 2000, **18**, 1157.
40. W. B. Dunn, *Phys. Biol.*, 2008, **5**, 011001.
41. Y. Z. Liang, P. S. Xie and K. Chan, *J. Chromatogr. B*, 2004, **812**, 53.
42. D. Guillarme, E. Grata, G. Glauser, J. L. Wolfender, J. L. Veuthey and S. Rudaz, *J. Chromatogr. A*, 2009, **1216**, 3232.
43. J. B. Calixto, *Brazilian J. Med. Biol. Res.*, 2000, **33**, 179.
44. S. Sang, J. D. Lambert, C-T. Ho and C. S. Yang, *Pharmacol. Res.*, 2011, **64**, 87.
45. L. Novakova, Z. Spacil, M. Seifrtova, L. Opletal and P. Solich, *Talanta*, 2010, **80**, 1970.
46. J. B. Wang, H. F. Li, C. Jin, Y. Qu and X. H. Xiao, *J. Pharm. Biomed. Anal.*, 2008, **47**, 765.
47. C. X. Lu, H. X. Wang, W. P. Lv, C. Y. Ma, P. Xu, J. Zhu, J. Xie, B. Liu and Q. L. Zhou, *Chromatographia*, 2011, **74**, 139.

48. C. Gotz, A. Fekete, I. Gebefuegi, S. T. Forczek, K. Fuksova, X. Li, M. Englmann, M. Gryndler, A. Hartmann, M. Matucha, P. Schmitt-Kopplin and P. Schroder, *Anal. Bioanal. Chem.*, 2007, **389**, 1447.
49. B. Avula, Y. H. Wang, T. J. Smillie and I. A. Khan, *J. AOAC Int.*, 2009, **92**, 645.
50. J. H. Chen, F. M. Wang, J. Liu, F. S. C. Lee, X. R. Wang and H. H. Yang, *Anal. Chim. Acta*, 2008, **613**, 184.
51. G. W. Jin, X. Y. Xue, F. F. Zhang, Y. Jin and X. M. Liang, *Talanta*, 2009, **78**, 278.
52. P. Li, G. Xu, S. P. Li, Y. T. Wang, T. P. Fan, Q. S. Zhao and Q. W. Zhang, *J. Agric. Food Chem.*, 2008, **56**, 1164.
53. S. Tianniam, T. Bamba and E. Fukusaki, *J. Sep. Sci.*, 2009, **32**, 2233.
54. D-F. Toh, L-S. New, H-L. Koh and E. C-Y. Chan, *J. Pharm. Biomed. Anal.*, 2010, **52**, 43.
55. Y-j. Wu, Y-l. Zheng, L-j. Luan, X-s. Liu, Z. Han, Y-p. Ren, L-s. Gan and C-x. Zhou, *J. Sep. Sci.*, 2010, **33**, 2734.
56. G. Glauser, D. Guillarme, E. Grata, J. Boccard, A. Thiocone, P-A. Carrupt, J-L. Veuthey, S. Rudaz and J-L. Wolfender, *J. Chromatogr. A*, 2008, **1180**, 90.
57. E. Grata, J. Boccard, D. Guillarme, G. Glauser, P. A. Carrupt, E. E. Farmer, J. L. Wolfender and S. Rudaz, *J. Chromatogr. B*, 2008, **871**, 261.
58. F. G. Galindo, P. Dejmek, K. Lundgren, A. G. Rasmusson, A. Vicente and T. Moritz, *Planta*, 2009, **230**, 469.
59. G. Glauser, L. Dubugnon, S. A. R. Mousavi, S. Rudaz, J-L. Wolfender and E. E. Farmer, *J. Biol. Chem.*, 2009, **284**, 34506.
60. A. Bar-Akiva, R. Ovadia, I. Rogachev, C. Bar-Or, E. Bar, Z. Freiman, A. Nissim-Levi, N. Gollop, E. Lewinsohn, A. Aharoni, D. Weiss, H. Koltai and M. Oren-Shamir, *J. Exp. Bot.*, 2010, **61**, 1393.
61. A. Burgos, J. Szymanski, B. Seiwert, T. Degenkolbe, M. A. Hannah, P. Giavalisco and L. Willmitzer, *Plant J.*, 2011, **66**, 656.
62. G. Glauser, G. Marti, N. Villard, G. A. Doyen, J.-L. Wolfender, T. C. J. Turlings and M. Erb, *Plant J.*, 2011, **68**, 901.
63. E. Grata, D. Guillarme, G. Glauser, J. Boccard, P. A. Carrupt, J. L. Veuthey, S. Rudaz and J. L. Wolfender, *J. Chromatogr. A*, 2009, **1216**, 5660.
64. R. Ho, N. Marsousi, P. Eugster, J. P. Bianchini and P. Raharivelomanana, *Nat. Prod. Commun.*, 2009, **4**, 1349.
65. S. Feng, W. Zeng, F. Luo, J. Zhao, Z. Yang and Q. Sun, *Food Sci. Biotechnol.*, 2010, **19**, 35.
66. Y. Liu, Y. Xiao, X. Xue, X. Zhang and X. Liang, *Rapid Commun. Mass Spectrom.*, 2010, **24**, 667.
67. F. Cacciola, P. Delmonte, K. Jaworska, P. Dugo, L. Mondello and J. I. Rader, *J. Chromatogr. A*, 2011, **1218**, 2012.
68. L. Han, G. Pan, Y. Wang, X. Song, X. Gao, B. Ma and L. Kang, *J. Pharm. Biomed. Anal.*, 2011, **55**, 994.

69. A. de Villiers, D. Cabooter, F. Lynen, G. Desmet and P. Sandra, *J. Chromatogr. A*, 2011, **1218**, 4660.
70. A. Urbain, A. Marston, E. Marsden-Edwards and K. Hostettmann, *Phytochem. Anal.*, 2009, **20**, 134.
71. I. D. Wilson, J. K. Nicholson, J. Castro-Perez, J. H. Granger, K. A. Johnson, B. W. Smith and R. S. Plumb, *J. Proteome Res.*, 2005, **4**, 591.
72. J. Boccard, J. L. Veuthey and S. Rudaz, *J. Sep. Sci.*, 2010, **33**, 290.
73. J. Boccard, A. Kalousis, M. Hilario, P. Lanteri, M. Hanafi, G. Mazerolles, J-L. Wolfender, P-A. Carrupt and S. Rudaz, *Chemom. Intell. Lab. Syst.*, 2010, **104**, 20.
74. G. Glauser, E. Grata, L. Dubugnon, S. Rudaz, E. E. Farmer and J. L. Wolfender, *J. Biol. Chem.*, 2008, **283**, 16400.
75. G. Glauser, J. Boccard, S. Rudaz and J-L. Wolfender, *Phytochem. Anal.*, 2010, **21**, 95.
76. J. J. Jansen, J. W. Allwood, E. Marsden-Edwards, W. H. van der Putten, R. Goodacre and N. M. van Dam, *Metabolomics*, 2009, **5**, 150.
77. Y. Sawada, K. Akiyama, A. Sakata, A. Kuwahara, H. Otsuki, T. Sakurai, K. Saito and M. Y. Hirai, *Plant Cell Physiol.*, 2009, **50**, 37.
78. P. S. Xie, S. B. Chen, Y. Z. Liang, X. H. Wang, R. T. Tian and R. Upton, *J. Chromatogr. A*, 2006, **1112**, 171.
79. F. van der Kooy, F. Maltese, Y. Hae Choi, H. Kyong Kim and R. Verpoorte, *Planta Med.*, 2009, **75**, 763.
80. P. J. Eugster, C. Funari, F. Mattioli, G. Durigan, S. Martel, P. Carrupt, D. Silva and J-L. Wolfender, *Planta Med*, 2011, **77**, PA49.
81. T. Kind and O. Fiehn, *BMC Bioinformatics*, 2007, **8**, 20.
82. C. Hall, Dictionary of Natural Products on CD-ROM, Chapman & Hall, 2008.
83. J. L. Wolfender, P. J. Eugster, D. Guillarme, H. Kratou, G. Glauser, S. Martel, S. Rudaz and P. A. Carrupt, *Planta Med.*, 2010, **76**, 1178.
84. J. M. Cunliffe and T. D. Maloney, *J. Sep. Sci.*, 2007, **30**, 3104.
85. P. L. Yang, G. R. Litwinski, M. Pursch, T. McCabe and K. Kuppannan, *J. Sep. Sci.*, 2009, **32**, 1816.
86. N. Manchon, M. D'Arrigo, A. Garcia-Lafuente, E. Guillamon, A. Villares, A. Ramos, J. A. Martinez and M. A. Rostagno, *Talanta*, 2010, **82**, 1986.

CHAPTER 14

Application of UHPLC-MS to Metabolomic/metabonomic Studies in Man

HELEN G. GIKA[a], GEORGIOS A. THEODORIDIS[b] AND IAN D. WILSON*[c]

[a] Department of Chemical Engineering Aristotle University Thessaloniki, 541 24 Thessaloniki, Greece; [b] Department of Chemistry Aristotle University Thessaloniki, 541 24 Thessaloniki, Greece; [c] Drug Metabolism and Pharmacokinetics IM, Mereside, Alderley Park, Macclesfield, Cheshire SK10 4TG, UK

*E-mail: ian.wilson@astrazeneca.com

14.1 Introduction

Whilst the terms metabonomics and metabolomics have been defined in various ways (*e.g.* refs. 1–3), in current practice they tend to be used interchangeably and the term used by different investigators tends to be a matter of personal choice (for a commentary on the evolution of these terms the interested reader is directed to ref. 4). Most often these studies aim to provide a comprehensive global (or holistic) profile of the metabolites present in a set of samples. From examination of the resulting data, usually *via* a multivariate statistical approach, differences are sought between the samples forming the experimental group *vs.* an appropriate control group. From these differences, the expectation is that a deeper mechanistic understanding of the system under study may be obtained (see below). Initially the practice of global

RSC Chromatography Monographs No. 16
UHPLC in Life Sciences
Edited by Davy Guillarme and Jean-Luc Veuthey

Published by the Royal Society of Chemistry, www.rsc.org

metabolic profiling of the type employed in metabolomic or metabonomic investigations was dominated by the use of analytical methodology based on either NMR spectroscopy or gas chromatography-mass spectrometry (GC-MS). However, with the development of reliable liquid chromatography-MS (LC-MS) systems it was inevitable that they would also be employed for this type of analysis and, indeed, in the last decade, the field has benefitted enormously from the application of LC-MS-based techniques (reviewed in refs. 5–7). However, the introduction of LC separations using stationary phases based on sub-2 μm particles, the so-called ultra-high performance LC (UHPLC, or UPLC) technique, to global metabolic profiling[8] propelled the usefulness of LC-MS in this type of application to an even higher level, such that it is now becoming the standard approach. As indicated above, this methodology seeks to provide a better metabolic understanding of biological phenomena based on the discovery of new knowledge *via* an "omic", hypothesis-free (but necessarily hypothesis generating), approach. Thus, it is expected that variations in the metabolic composition of biofluids and tissues, in animals and humans, will provide insights into, for example, normal physiological processes (growth, development, aging, diurnal variation, exercise *etc.*), normal and abnormal nutrition, and biomedical areas such as the onset and progression of disease and responses to therapeutic intervention (both drug efficacy and toxicity).[9–15] The production of these comprehensive, non-targeted metabolite profiles, requiring as it does the analysis of complex mixtures of the very diverse metabolites present in biofluids or tissue extracts, depends upon analytical platforms, such as NMR spectroscopy and MS. NMR is generally applied directly to the sample to be profiled. Whilst MS-based methods can also be applied directly to some biofluids, without the concomitant use of a separation technique (such as, for example, GC, capillary electrophoresis (CE) or LC), there are disadvantages to such approaches. Thus, whilst direct introduction does provide the potential for very rapid analysis of samples there is the potential for matrix interferences and an inability to distinguish between structural isomers. Hyphenating the mass spectrometer to a high resolution chromatographic separation provides the opportunity to minimize ion suppression/enhancement effects and may allow the resolution of isomeric/isobaric substances. The use of a separation is, of course, not without associated costs (both financial and experimental). For example, difficulties that hyphenation can add include additional complexity (of both the instrumental set-up and data analysis), increased analysis time, the requirement to optimise both separation and MS detection and a need for expertise in both separations and MS *etc.*, and these must be balanced against the experimental requirements of the analysis. Currently, however, the vast majority of the MS-based metabonomic/metabolomic studies undertaken on mammalian systems (animal and human) are performed using hyphenated techniques.

Very many global metabolic profiling studies have now been performed, investigating both animals and human subjects, and a significant proportion of these have used LC-MS-based analysis, either alone or in combination with other analytical platforms. Increasingly, we see the application of LC-MS, and particularly UHPLC-MS, to global metabolic profiling studies in man (both small investigative studies into, for example, the understanding of disease, or larger epidemiological investigations). This makes it an appropriate time to examine the conduct of such UHPLC-MS-based analysis, particularly for plasma/serum and urine, which represent the most widely accessed samples. Whilst it is possible to perform sample analysis using very short run times (*ca.* 1–2 min),[16] this is at the cost of a reduced metabolome coverage and an increased risk of ion suppression. Thus, the methods described here generally use separations taking less than $<$ 20 min, as these provide medium sample throughput combined with good chromatographic resolution. Clearly, in some instances longer analysis times may provide advantages in terms of increased metabolome coverage, but in that case sample throughput suffers. Whilst aimed at describing the utility of UHPLC-MS for human samples, exactly the same methodology can usually be applied, with little or no modification, and with an equal expectation of success, for the analysis and profiling of similar animal-derived samples.

14.2 Pre-analysis Considerations: Protocol Design, Sample Collection, Storage and Preparation

Whilst UHPLC-MS provides an excellent analytical platform for the generation of high-quality metabolic data from human-derived samples, it cannot compensate for inadequate samples resulting from poor study design, sample storage or preparation. It is therefore perhaps worthwhile devoting space to these topics prior to discussing the use of the methodology.

14.2.1 Protocol Design

Whilst not specific for LC-MS-based studies, it goes without saying that any investigation to be performed on humans requires that both ethical approval and the consent of the participants must be obtained. An important variable in any metabolic profiling study is diurnal variation and, in the case of urine, if "spot" samples rather than 24 h collections are being made then thought should be given to attempting to collect samples at the same time of day (*e.g.* see ref. 17). Diet is another important variable and where this cannot be controlled subjects should be asked to provide details of food consumption as this can greatly affect urinary profiles (*e.g.* see ref. 18). The metabolome of blood-derived samples such as plasma or serum can also be expected to be affected by both diurnal and dietary factors, especially proximity to meal times. In addition, it is good practice to ascertain and record the use of medications *etc.* For human studies, the selection of appropriate age-, gender- and disease-matched controls is also something that needs careful consideration as, depending upon the study, these may also be important confounding factors.

14.2.2 Sample Collection and Storage

In any global metabolite profiling, the correct collection of the sample set is critical to success. Whilst a failure to correctly analyze the sample set due to, for example, instrument failure or human error can be addressed by re-analysis, any failure in the initial sample collection from patients/volunteers (at the start of the study) cannot be compensated for. At best, this may compromise the study, and at worst may completely invalidate it.

14.2.2.1 Collection and Storage of Blood-derived Samples

The most common samples used for the analysis of the blood metabolome are plasma and serum, although, recently, the use of so called "blood spots" has also begun to receive attention. Serum results from the process of allowing the blood sample to clot naturally and removal of the clot then provides the serum. The main advantage of serum is that it is relatively easy to prepare without the need for sophisticated equipment. It is thus ideal for multicentre collection facilities such as hospital wards, GP surgeries *etc.* Plasma results from adding an anticoagulant to the blood on collection (usually contained within the sampling tube) and then centrifuging the sample (3000 ***g*** at *ca.* 4 °C) to separate the plasma from the cells and platelets *etc.* Various anticoagulants are available (EDTA, citrate and lithium heparin). Given that citrate is an important endogenous metabolite, and EDTA can cause problems with metabolic profiling *via* the introduction of interferences, lithium heparin is preferred. A similar conclusion was reached in an investigation designed to develop a method to analyze human plasma *via* UHPLC-MS for application in studies on the effect of diet.[19] The disadvantage of plasma compared to serum is clearly the need for refrigerated centrifuges in order to prepare plasma. This need may limit the sites where plasma can be obtained.

To date no detailed studies have been performed that have rigorously assessed the stability of the plethora of metabolites that make up the serum or plasma metabolomes. However, the current best practice is that, as soon as possible after preparation, both serum and plasma samples should be rapidly frozen and stored at the lowest available temperature until analysis.[20] Typically, this would be −80 °C. It would also seem to be good practice to divide samples into appropriate sub-aliquots (*e.g.* 0.5 ml) so that multiple freeze/thaw cycles are not applied to the whole sample in the case that re-analysis is required. In this way, if multiple aliquots of the sample are available, each aliquot need only be analyzed once. This approach also allows samples from the same subject to be stored in different freezers, which may be important in the case of a failure, and allows the samples to be shared between laboratories.

To an extent, the choice of either serum or plasma as the matrix is driven by the practicalities of the investigation reflecting both the resource/facilities available and the investigators' personal preferences. It should be recognized, however, that the different methods used to remove blood cells are likely to result in the measurement of slightly different metabolomes and best practice would seem to be not to mix the analysis of plasma and serum samples in the same study. Further, in the absence of data showing method and sample comparability, caution is necessary when trying to compare results obtained between different studies on the same topic where one data set is derived from the analysis of plasma and the other from serum (see, for example, the results of recent comparisons of serum and plasma by GC-MS[21] and HPLC-MS[22]).

As final point, it must be emphasised that a critical factor that must be addressed before any study is undertaken is the need to carefully screen, using the analytical methodology that will be used for routine analysis, the containers that will be used both for collection and sample storage (if different) in order to demonstrate their suitability. This is for the simple reason that these containers can be a rich source of interfering contaminants (such as polyethylene glycol (PEG) or plasticizers *etc*.,) that can interfere with accurate metabolome analysis. Once selected sufficient containers should be purchased from the same batches to collect and store all of the samples that will be obtained.

As mentioned above, as well as liquid samples such as plasma and serum, a possible method of sample collection is the use of blood spots. These samples, most often taken *via* simple finger or heel pricks are widely used clinically, especially in paediatric work looking for, for example, inborn errors of metabolism. The blood is spotted on to an adsorbent paper (either treated with chemicals to promote sample stability or untreated, as required). The blood is allowed to dry at ambient temperature and the card is then stored until analysis, at which point a segment from the centre of the spot is punched out and the sample extracted into a solvent such as methanol. The approach has many potential advantages including ease of collection and transport etc. In a preliminary study where we compared this methodology to analysis using conventional solvent protein precipitated human plasma, both blood and plasma spots gave comparable profiles and numbers of ions *etc*., using both positive and negative electrospray ionization (ESI).[23] Whilst promising these initial studies were undertaken on samples spotted and then analyzed within a relatively short timescale and further work is required to ensure appropriate sample stability *etc.* before the method can be used routinely as an alternative to conventional methods.

14.3 Sample Preparation for Serum/Plasma

Plasma and serum are complex samples that contain both low molecular mass metabolites and macromolecules such as proteins. The metabolites comprise both non-polar compounds, such as lipids, steroids *etc*., and polar metabolites (*e.g.* aminoacids, sugars *etc*.). For LC-MS-based methods some type of pre-chromatographic sample preparation to remove proteins is essential to prevent an otherwise catastrophic loss of system performance resulting from their presence in the sample. Removal of proteins generally employs the addition of an organic solvent to the sample and numerous methods have been described for global metabolite profiling, with a range of compositions and temperatures used to achieve precipitation. For example, in an investigation of sample preparation for human blood with UHPLC-TOF-MS as the analytical end step the utility of the organic solvents acetonitrile, acetone, methanol and ethanol for protein precipitations were evaluated for their effects on factors such as, for example, the number of components detected, data quality/reproducibility and column lifetime.[24] In addition, based on the initial experiments a further set of conditions were examined on selected precipitations to optimise the vortex time and temperature. These authors selected two solvent compositions [methanol/ethanol (1:1, v/v) and methanol/acetonitrile/acetone (1:1:1, v/v/v)], added to plasma in a ratio of 4:1, as being optimal for the preparation of human plasma samples.[24] We have found that methanol, added in a ratio of 3:1 (v/v), followed by centrifugation at 14 000 ***g*** represents a simple and efficient process that provides a sample that is suitable for preparing human serum for UHPLC-MS[20,25] (although depending on the chromatographic conditions and the injection volume, it may be necessary to alter the sample composition by, for example, dilution with water or freeze drying and reconstitution in an alternative solvent may be required to ensure suitable analyte peak shape).

To minimize the effects of serum/plasma enzymes on the composition of the metabolome, the temperature at which sample preparation is performed needs to be considered.[20] Clearly, performing sample preparation at low temperatures will reduce the activity of these enzymes and samples should therefore be thawed on ice and be kept cold (0–4 °C), on ice, until the precipitating solvent has been added (which should be as soon as possible after thawing).

14.3.1 Solid Phase Extraction (SPE)

Another method of sample preparation that might have utility in global metabolic profiling of the type required for metabonomic/metabolomic investigations is SPE. A limited study of the applicability of SPE for profiling human plasma compared the technique against the standard methodology described above or solvent-based protein precipitation.[26] Plasma samples were therefore either subjected to SPE on a C18-bonded phase or treated with 3 volumes of either methanol or acetonitrile for protein precipitation. Once loaded on to the C18 SPE cartridge the metabolites were recovered by elution with methanol followed by analysis by gradient reversed-phase (RP) UPLC-MS. The eluent from the column was subjected to analysis by positive ESI using a Q-TOF mass spectrometer and the overall conclusion was that the repeatability of SPE was significantly better than that of solvent precipitation.[26] It is also arguable that SPE could be used to concentrate samples and enrich the extracts with low abundance metabolites. However, against this is the slightly increased complexity of the method (although automation is readily available), the increased cost of the consumables compared to solvent precipitation and the potential for loss of metabolites *via* non-retention on the stationary phase (or incomplete recovery). The effect of non-retention, or poor recovery, can of course be minimized by the selection of an SPE material that matches the retention properties of the analytical column to be used for analysis as the same effects on the analytes will no doubt apply to both extraction and chromatography.

As well as SPE, an in-line method of sample preparation that has shown some potential for global metabolite profiling of plasma/serum, which has so far been demonstrated for HPLC-MS but not UHPLC-MS, is the use of the so-called turbulent flow chromatography methodology.[27] Turbulent flow conditions are achieved using high flow rates combined with the use of relatively large particles (25–50 μm) for the chromatographic stationary phases, contained within small internal diameter (i.d.) columns (0.5–1 mm). Under these turbulent flow conditions plasma/serum samples can be injected directly into the system without any need for pretreatment. This is possible because, unlike in conventional LC, where laminar flow occurs and where the direct injection of plasma would result in proteins being retained by the stationary phase, leading to a rapid loss in column performance, employing turbulent flow conditions prevents this, and proteins pass through the column unretained,[28,29] whilst the low molecular mass metabolites are retained. The chromatographic performance of such turbulent flow columns is, however, somewhat limited in terms of chromatographic resolution and to overcome this they are generally coupled to a more conventional analytical column using column switching. In these systems the eluent containing the unwanted proteins is directed to waste and, once this part of the separation is complete, the metabolites are then eluted on to the analytical column for profiling. This metabolic profiling study demonstrated that turbulent flow chromatography could indeed be used as a form of on-line SPE for human plasma analysis in this type of application, with the significant benefit that off-line sample handling was significantly reduced. However, the metabolite profiles obtained in this way showed substantial differences compared to methanol-precipitated HPLC-MS due, at least in part to the *ca.* 10-fold reduction in the amounts of phospholipids detected using the turbulent flow methods compared to protein-precipitated samples.

14.4 Collection and Storage of Urine Samples

The composition of urine is not carefully controlled as its function is to help maintain homeostasis in the organism and dispose of waste products. The highly variable composition of urine, however, is valuable as it can often highlight metabolic deregulation and can provide a rich source of biomarkers, thereby providing insights into changes in response to disease, toxicity or physiological challenges. Points to be considered include the potential for great variability in ionic strength, pH and osmolarity, with concentrations of metabolites encompassing wide dynamic ranges. Whilst there is considerable diversity in the types of metabolites present in urine, these tend to be polar water soluble compounds and, unlike, for example, plasma/serum, large amounts of non-polar lipids are generally absent. Urine is also a major route for the elimination of many exogenous compounds, including environmental pollutants, food additives, drugs/drug metabolites and other xenobiotics (dealt with under the topic of the "xenometabolome"[30]). Human urine represents a readily obtained biofluid for metabolic profiling studies that can usually be obtained non-invasively, and that, in general, is unlikely to be volume-limited, unlike other types of samples such as plasma/serum. Urine can easily be sampled in a serial fashion, allowing temporal metabolic changes to be studied and both timed/spot and 24 h samples can be obtained. For 24 h sample collection the specimens should, as a minimum, be stored in the fridge at 0–4 °C between collections. The best practice for spot collections is to collect a "mid-stream" sample directly into a suitable container. Whilst few studies have been undertaken in order to ascertain the optimum for sample storage conditions, and despite finding little variation in the metabolic profiles of samples stored at either −20 or −80 °C (analyzed by LC-MS[31]), independent of the number of freeze/thaw cycles, it would seem sensible to use the lowest available temperature for storage. A further sensible precaution is to store the sample sub-divided into convenient aliquots (thereby minimizing subsequent freeze/thaw cycles) in appropriate containers for long-term storage (and transport if the sample is to be analyzed at a site distant from where it has been collected). To avoid the decomposition of unstable analytes, or bacterial gradation, which would result in unwanted changes in metabolic profiles, samples should be frozen as rapidly as possible following collection. The best practice would then be to store the samples at, at least, −20 °C, and preferably at the lowest available temperature. As for plasma/serum collection, the containers used both for collection and sample storage (if different) should be carefully screened to ensure their suitability for sample collection and storages.

14.4.1 Sample Preparation for Urine

Sample preparation for urine from healthy human subjects containing only low concentrations of protein can often be limited to centrifugation to remove particulates that would otherwise block the LC column, and dilution with water (1:1, v/v). Samples can then be aliquoted in to either maximum recovery vials or 96−well plates (with cap mats) and placed in an LC autosampler maintained at 0–4 °C throughout the analysis. If analysis is not to be performed immediately after aliquoting, the vials/96-well plates can be stored frozen or, if analysis is to be undertaken shortly, on ice or refrigerated at 0–4 °C. In LC-MS-based studies we have found samples of human urine to be relatively stable for up to 48–72 h at 0–4 °C in the LC autosampler.[32]

14.5 UHPLC-MS-based Metabolite Profiling

Carefully undertaken, with appropriate attention to factors that affect repeatability, UHPLC-MS can provide a sensitive, accurate and specific analytical methodology for the detection and determination of the myriad metabolites present in biofluids such as urine or plasma/serum. UHPLC provides an efficient and relatively rapid separation combined with the good dynamic range of MS-based detection for a range of structural types. The ability of UHPLC-MS to detect amines, amino acids, organic acids, lipids, sugars, hormones, vitamins, small peptides *etc*. provides a platform that is well suited to untargeted analysis of biofluids for biomarker discovery. However, excellent performances from both parts of the analytical system are essential to obtain these benefits. Whilst good results have been obtained using conventional HPLC-MS methods, the introduction of UHPLC marked a step change in the ability of the analyst to perform global metabolic profiling.[8] In the case of these UPHPLC separations, gradient RP chromatography has been the most widely employed separation mode for metabolic profiling. This reflects the fact that RPLC is easily compatible with aqueous samples and well suited to the chromatography of many of the mid-polar to non-polar, water soluble, metabolites found in urine and serum/plasma without any requirement for derivatization. This ability to separate a wide range of metabolites *via* a carefully optimized separation renders the determination of complex metabolic profiles possible. The RP columns used for UHPLC are typically C18-bonded silicas, with sub-2 μm particle sizes, contained in 2.1 mm i.d. columns of between 5 and 15 cm in length. Increased metabolome coverage for certain polar compounds can be obtained using high strength silica (HSS) columns.[32] However, irrespective of the choice of stationary phase, a decision has to be made concerning the length of the chromatographic analysis as this will directly affect metabolome coverage. Therefore, UHPLC gradients for the analysis of plasma/serum or urine can range from a few minutes (for example, see ref. 16 for a *ca.* 1–2 min separation for urine analysis) up to 30 min or more. There is no doubt that, in general, short runs will detect fewer ions than longer ones and thus the latter will provide a more comprehensive metabolic profile. However, if a large number of samples have to be analyzed, a compromise clearly has to be made with respect to required throughput *versus* maximizing metabolome coverage. Pragmatically, investigators have tended to recommend separations in the region of 15–20 min for both urine and plasma samples (including column re-equilibration times) (see, for example, refs. 20, 33). Of course, one major difference between urine and plasma/serum lies in the presence of large amounts of lipid in the latter. Recently, a "lipidomic" methodology, designed for the analysis of chloroform/methanol extracts of plasma, has been described.[34] Whilst based on a RP gradient separation, this method begins the gradient at a relatively high organic content of 50% methanol (plus 50% 20 mM aqueous ammonium formate). The gradient rises to 70% methanol over 14 min, followed by a slower gradient to 90% methanol over 50 min, then an isocratic phase of 15 min and finally 5 min at 100%

methanol. Both positive and negative ESI data were described. The method was investigated qualitatively and quantitatively using human blood plasma as the matrix and was shown to be able to determine some 65% of the detected features with an relative standard deviation (RSD) of below 10%. The detected features were shown to be derived mainly from (lyso-)phospholipids, sphingolipids, mono-, di- and tri-acylglycerols, and cholesterol esters. The separations achieved by this type of analysis are shown in Figure 14.1 which provides a total ion chromatogram (TIC; top panel) and base peak chromatogram (BPC; bottom panel) acquired in positive and negative ionization mode respectively. Negative ionization mode exhibited significant ion suppression in the retention time region where neutral cholesterol esters and triacylglycerols elute.

A typical gradient RP-UHPLC separation for urine performed on a 2.1 × 150 mm Acquity BEH C18 in our hands might employ a mobile phase consisting of water/0.1% formic acid as solvent A and acetonitrile/0.1% formic acid as solvent B. The gradient profile would begin with a short isocratic period of 95 or 100% solvent A for 0.5 min followed by a linear increase to 20% B at 5 min, and then a further linear increase to 45% B at 7.5 min followed by a final linear increase to 100% B at 9.8 min. These conditions are then held for a further 1.2 min to wash the column before returning to the starting conditions with re-equilibration then allowed for a further 3 min prior to the next injection. Typically, flow rates would be 300–400 μl min^{-1} with the column temperature maintained above ambient (to minimize changes in retention *etc.*, due to fluctuations in the laboratory temperature) at between 40–60 °C. For the analysis of plasma/serum samples, typical gradient profiles employ much higher percentages of organic solvents to elute lipids. We have found that

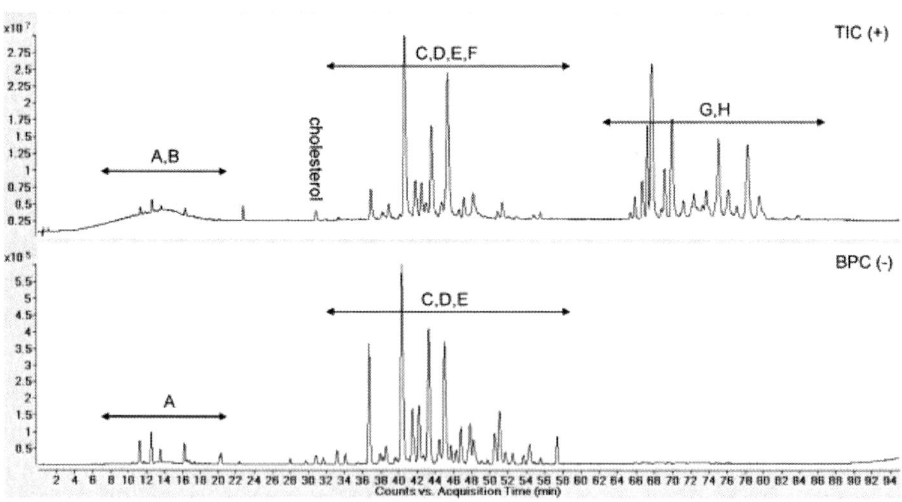

Figure 14.1 TIC (top panel) and BPC (bottom panel) acquired in positive and negative ESI mode respectively of UHPLC-MS of blood plasma lipid extract. Reproduced from reference 34 with permission from Elsevier.

solvent systems based on the use of methanol as the organic modifier are more useful for the analysis of blood-derived samples.[20] Whatever the chromatographic elution profile, the samples are generally injected on to the column under conditions of low eluotropic strength where the mobile phase is predominantly aqueous to ensure good retention of polar metabolites. To further enhance the profiling of polar substances, formic acid is often added to the eluents to suppress the ionization of weak organic acids, and thereby

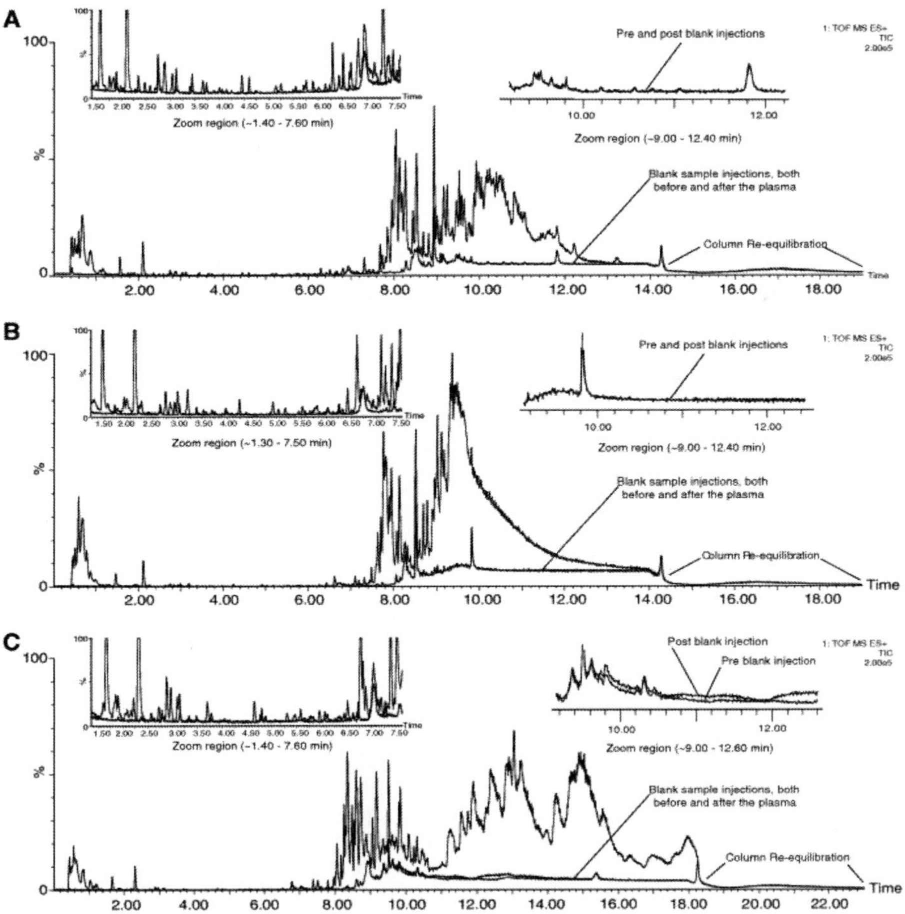

Figure 14.2 UHPLC-TOF-MS analysis of plasma extracts and blank samples injected on to a C8 (A), a phenyl (B), and a C18 (C) UHPLC column. Pooled plasma was extracted with 80% methanol solution prior to injection on a UHPLC column (2.1 × 100 mm). Zoomed regions show minute differences in the early part of the chromatogram (zoomed region left 1.4–7.6 min), but significant differences in the lipophilic part of the chromatogram (zoomed region right 9.0–12.0 min). The C8 column provided the best performance and was selected. Reproduced from reference 35 with permission from Elsevier.

improve retention. In our hands, C18 stationary phases have provided a good "workhorse" for the metabolic profiling of samples such as urine and plasma, whereas other workers who have evaluated C8, C18 and phenyl-based materials have concluded, in experiments on plasma samples obtained from both male and female subjects, that the C8 column provided the best results.[35] It was seen that all three columns performed in a similar manner during the first steps of the gradient program, however, in the plasma phospholipid region (8–11 min) the C8 and C18 columns (Figures 14.2A and 14.2C) display better separation compared to the phenyl column (Figure 14.2B). The C18 and phenyl column retained lipophilic compounds, resulting in increased background in the late (lipophilic) part of the chromatogram (10–12min), as observed in blank injections (zoom region showing the pre- and post-blank injection samples). The C8 column was better "washed" and made ready for the subsequent analyses and for this reason was selected for this study.

The excellent repeatability of both retention and peak height that can be obtained using UPLC-MS for the analysis of biological samples is illustrated in Figure 14.3. This example shows the results for a single ion present in a pooled urine sample that was analyzed at regular intervals during the investigation of a large number of human urine samples using gradient UPLC-Q-TOF-MS. As Figure 14.3 shows, there was little change in either retention or peak height over the entire run.

Whilst it is normal practice to perform these RP-UHPLC separations at temperatures close to ambient it is also possible to use elevated temperatures either with isothermal conditions[36] with a solvent gradient or isocratic solvent conditions with a thermal gradient.[37] The major benefit of the use of higher temperatures is the reduced solvent viscosity which results in reduced back pressures. An example of the high temperature separation of human urine, with a solvent gradient for elution,[36] is shown in Figure 14.4, exhibiting high chromatographic resolution in two gradient program lengths: 10 min (Figure 14.4A) and 60 min (Figure 14.4B).

Whilst UHPLC with C18-bonded phases represents an excellent and dependable workhorse for the global metabolic profiling of a wide range of sample types, it has limitations with respect to polar and polar ionic substances that elute largely unretained from such systems. Such polar substances (*e.g.* sugar phosphates, amino acids *etc.*) represent important classes of metabolites that clearly need to be determined as part of any comprehensive metabolome evaluation. The current approach to increasing the amount of the metabolome accessed by LC-MS-based methods is to use so-called hydrophilic interaction chromatography (HILIC). HILIC, performed on stationary phases prepared from silica or polar derivatized silica, whilst not a complete solution to the problem analyzing of polar compounds, does provide complementary data that can be used in addition to that obtained *via* conventional RP-UHPLC.[38–42] HILIC-UHPLC-MS has had limited application to the analysis of urine[40–42] and been demonstrated to have a different selectivity to that of RP separations on the same samples,[41] thereby providing complementary metabolite information and

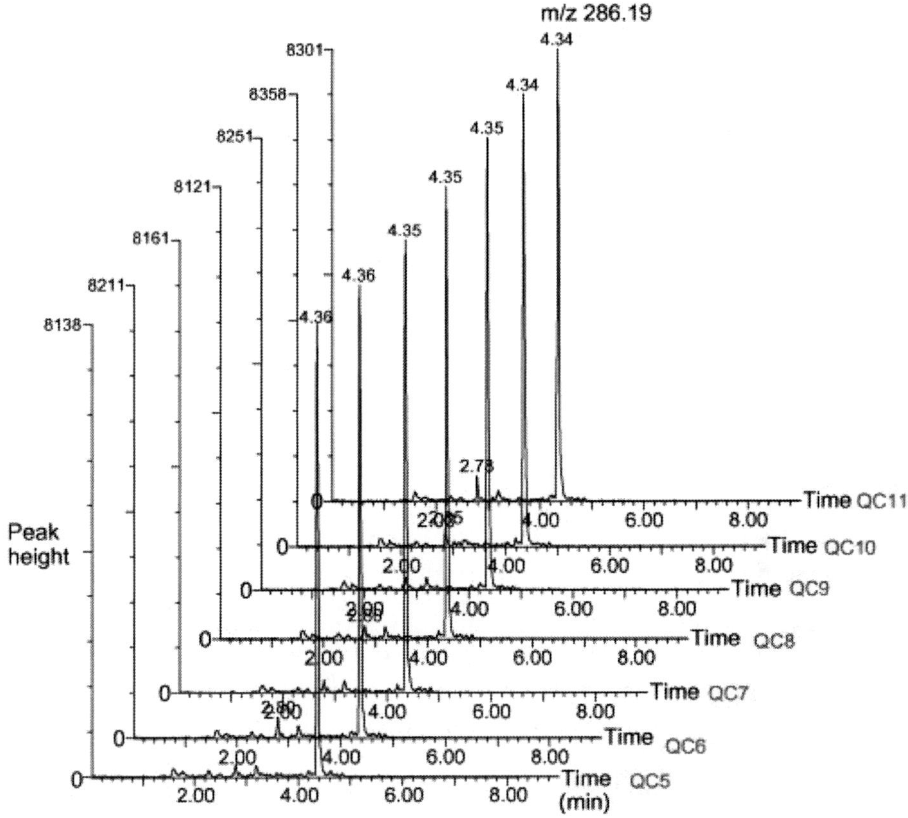

Figure 14.3 Extracted ion chromatograms of a selected ion (286.19 amu, retention time 4.3 min) in seven sequential QCs samples analysed in between 67 test samples. Very high repeatability in peak height and retention time is observed. Reproduced from reference 37 with permission from Elsevier.

thus enhanced metabolome coverage.[41] The use of a single two-dimensional HPLC separation combining RP and HILIC modes of chromatography has also been described.[43]

The greater metabolome coverage provided by the combination of RP and HILIC leads inevitably to the question of which represents the optimal profiling strategy that should be adopted when faced with a set of samples. Should analysis be performed by RP-UHPLC alone, RP-UHPLC plus HILIC or RP-UHPLC followed by HILIC if there seem to be differences between the samples concentrated towards the polar metabolites? Pragmatically, we would currently tend towards the last approach as being the most conservative of resources, as we believe that conventional RP-based separations represent a convenient and robust means of obtaining a metabolite profile for many applications, and a good starting point for the initial phases of a metabonomic investigation.

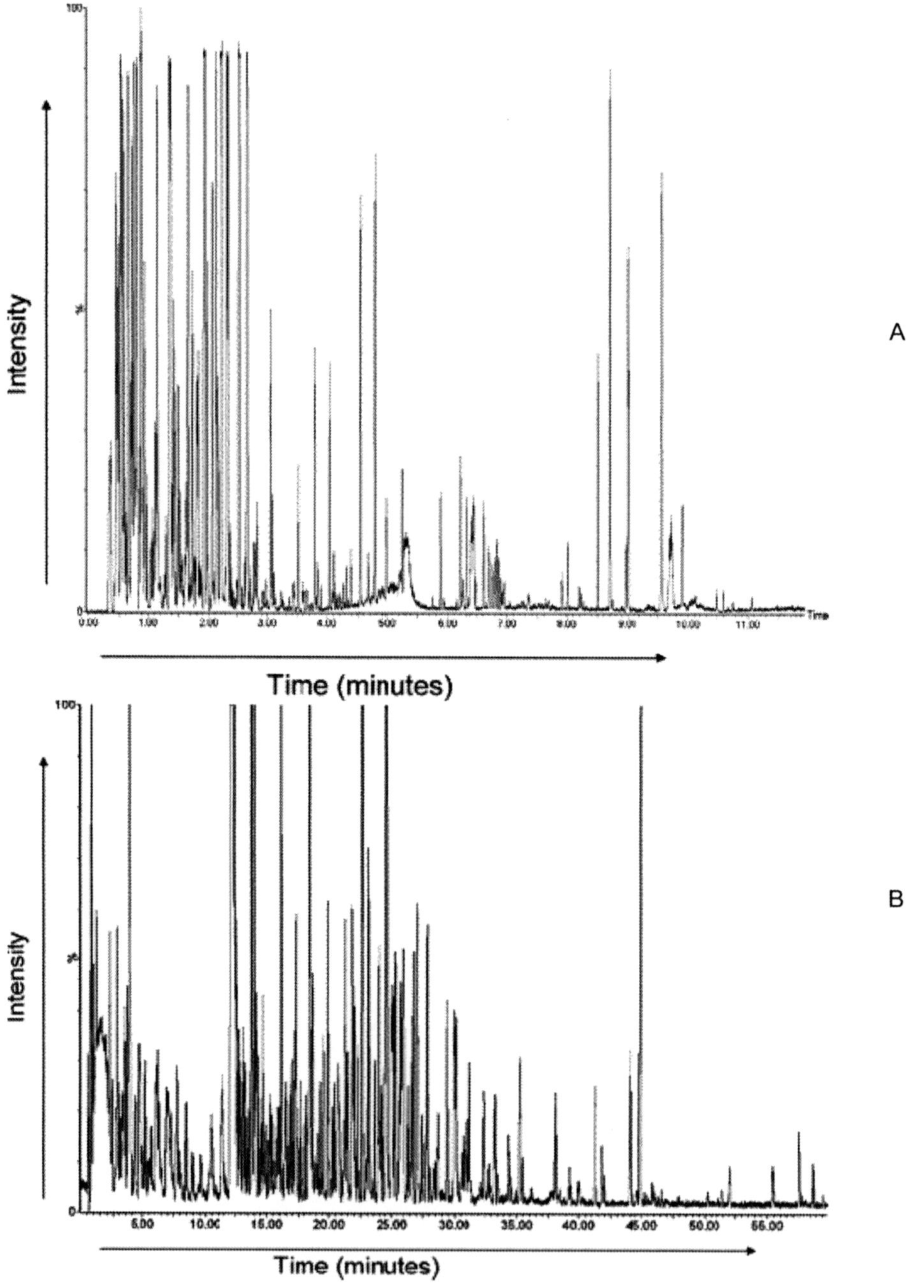

Figure 14.4 (A) RP-UHPLC gradient program separation of rat urine on over 10 min at 90 °C. (B) RP-UHPLC gradient program separation of rat urine over 60 min at 90 °C. In both cases separations were performed on a 2.1 × 150 mm, 1.7 μm, Acquity BEH C18 column. Reproduced from reference 36 (Plumb *et al.*, 2006) with permission from American Chemical Society.

14.5.1 The Use of Quality Control Samples

A major practical problem with any LC-MS-based metabolomic study is the need to ensure that the data generated are reliable. In conventional bioanalytical practice this is normally achieved by the use of appropriate internal standards and thorough validation. However, in the world of global metabolic profiling the composition of the sample is unknown before the analysis commences, and in any case contains, potentially at least, thousands of analytes. The use of internal standards is therefore, in our opinion, somewhat problematic. The approach that we, and many others, have adopted is to make up a "quality control" (QC) sample based on the matrix to be analyzed (urine, plasma serum, tissue extract *etc.*). These QC samples are then run at the beginning and end of the analytical run and at regular intervals (*e.g.* every 5 or 10 samples) throughout the analysis. For small scale studies the QC sample can be prepared by mixing a small aliquot of each of the samples to be analyzed to give a pooled sample,[20,33,44–46] which has the advantage of effectively providing a sample containing a mean concentration of all of the analytes that will be present in the samples to be analyzed. In longer term epidemiological/population studies such as, for example, the HUSERMET project,[20,25] the logistics of sample collection and analysis make such a simple approach to QC generation impractical and instead a bulk QC sample is prepared from a suitable source of the matrix (in this case a commercial serum sample) for use when the study samples are assayed. Once the sample batch has been subjected to UHPLC-MS statistical analysis, *via* principle component analysis (PCA), it can be used to show that the QC samples which, if the methodology was perfect would appear as a single point on the scores plot, cluster closely and that there are no obvious outliers or time-related trends[47] (see Figure 14.5A). Should clustering of the QCs when the run is analyzed using PCA be poor (see Figure 14.5B), it suggests that there might be a problem somewhere with the analysis (*e.g.* retention time drift or changes in MS response) and further analysis of the data is recommended to try to determine the cause(s).

Based on this examination of the data, decisions can then be made on (qualified) acceptance of the run or re-analysis of the batch. If the QCs cluster tightly and the analytical data appear acceptable, then further analysis to look for biomarkers can be attempted. The great value of the QC data is that, once potential biomarkers have been found, the performance of those individual ions in the QCs can be examined to see if they are sufficiently repeatable for further evaluation (pragmatically, we take a CV of less than 30%, and preferably below 20% over the whole run as a minimum for taking a ion forward). In large scale studies, in which numerous analytical runs are required, the use of a common QC also allows the day-to-day performance of the method to be monitored, and the QC data can also be used to normalize the between run data and compensate for "drift" in the method over time (*e.g.* see ref. 20). The use of QCs is described in a number of articles on HPLC and UHPLC.[20,33,44–46] As well as QCs, test mixtures, made up of a limited number

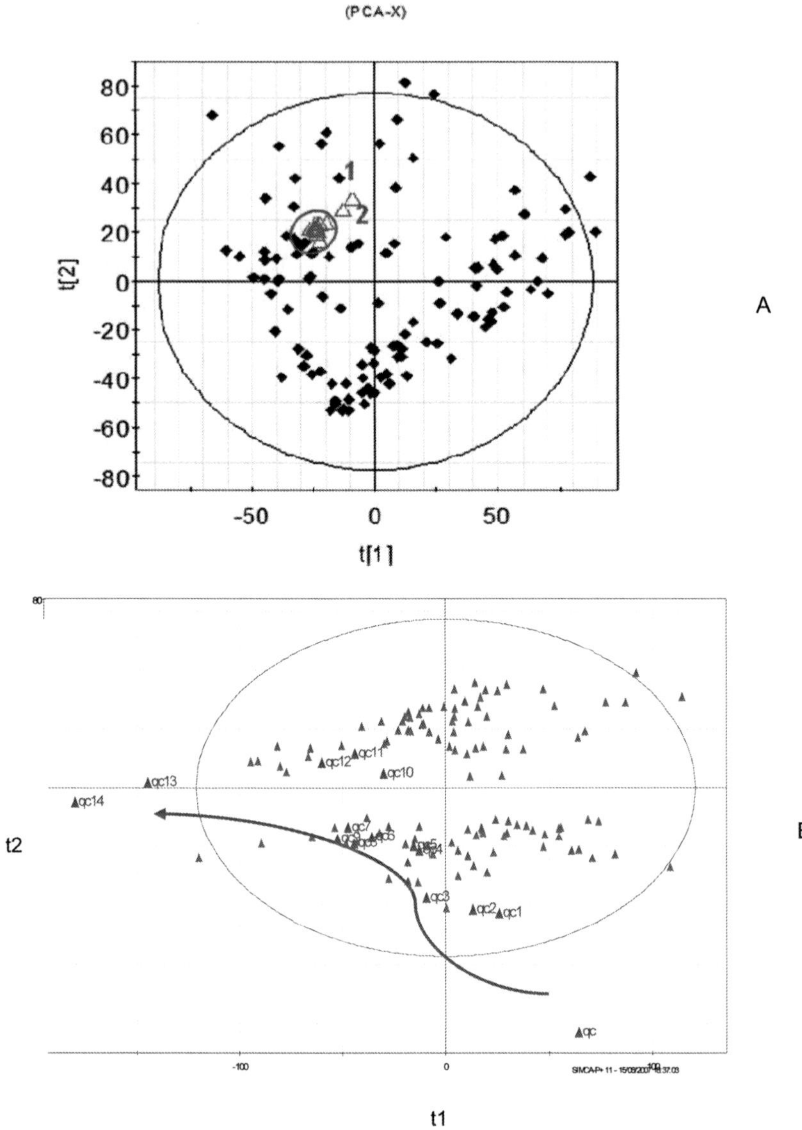

Figure 14.5 (A) PCA scores plot generated from UHPLC-TOF-MS data from the analysis of human urine. Individual test samples are indicated as black diamonds. Open light grey triangles (representing QC samples) are tightly clustered, within the small, light grey circle, indicating a worthy data set. The two triangles labelled 1 and 2 represent conditioning injections. (B) PCA scores plot generated from UHPLC-TOF-MS data from the analysis of urine from medium sized cohort (*ca.* 120 subjects). Individual test samples indicated as grey triangles. Darker grey triangles represent QC samples; these are spread throughout the scores plot, indicating a significant time trend shown with a grey arrow in PC1. Further examination of data as extracted ion chromatograms indicated retention time changes along the run. The data was not used further.

of pure metabolites can be used to provide a rapid means of assessing chromatographic and detector performance for factors such as retention time stability, peak shape, detector response and mass accuracy. Experience suggests that these metabolites should perhaps best be employed by "overspiking" them into the QCs for the urine or plasma/serum samples, rather than use aqueous solutions, to avoid adversely affecting the conditioning of the column (see Section 14.5.2) achieved by the injection of the matrix. The test mixture should be injected at the beginning, middle and end of the analytical run and brief examination of the data (retention time, peak shape *etc.*) should be enough to decide if the analysis is likely to be acceptable, even before the PCA to examine the QCs.

14.5.2 Column Conditioning

Clearly, there is a need to ensure that factors such as stable retention times, signal intensities and mass accuracies for the compounds present in the complex mixtures found in all types of biological matrix represents an important analytical aim for metabolite profiling studies. One of the types of experimental best practice that has emerged from the attempts to make LC (HPLC and UHPLC)-based global metabolic profiling into a reliable and repeatable method has been the need to "condition" the system to ensure stability for the duration of the analysis. In particular, it has been noted in a number of studies that the first few injections of the sample show small changes in both the chromatographic retention time and signal intensity[20,25,33,44–46] that make the data from these samples unrepresentative of the rest of the set. Once a number of injections of the matrix have been made the system stabilizes, presumably as active sites on the stationary phase are masked *etc.*, and retention times then show little variability through the remainder of the run (*e.g.* see Figure 14.3). In practice, we have found it useful to "condition" the system by chromatographing at least five urine samples and a minimum of 10–15 plasma/serum samples (the pooled "quality control or QC" samples described above in Section 14.5.1. provide a convenient sample for column conditioning) before starting the analysis of the test batch. Post analysis, the data from these conditioning samples can be analyzed together with the analytical batch and the achievement of system stability should be evident from the scores plots. Once the analyst is satisfied that the system has attained equilibrium as a result of the injection of the conditioning samples, the data relating to them are then discounted from further statistical analysis. In addition to the QCs, it may also be worthwhile to analyze some, or all, of the samples more than once in the same run as technical replicates in order to provide a further demonstration of the quality of the analysis. However, it is our view that, with a robust method, the QC samples should provide sufficient reassurance of satisfactory method performance, therefore the need for technical replicates is reduced.

14.5.3 MS

As currently practiced, almost all types of mass spectrometer, including single and triple quadrupoles, time-of-flight (TOF), "Orbitrap" and Fourier transform (FT) ion cyclotron instruments, have been used in LC-MS-based metabolomic/metabonomic profiling experiments on both animal- and human-derived samples. Probably the most widely used mass spectrometers to be combined with UHPLC separations have been the TOF, particularly the quadrupole TOF (or Q-TOF), instruments. In general the bulk of the reported applications have employed ESI, in both positive and negative modes, as this type of ionization is well suited to the polar/ionic analytes generally encountered in biological fluids.

ESI is useful because this type of ionization generally provides molecular ion information and minimal fragmentation (although this is obviously analyte dependent). Solvent and salt adducts are often also formed, and there is also the potential for some analytes, *e.g.* peptides, to form multiply charged ions. Care therefore has to be taken when analyzing LC-MS data to ensure that these factors (and isotope peaks) are taken into account during data processing.

Different classes of metabolites show different ionization characteristics and, for example, compounds composed only of C, H and O will probably be best detected using negative ESI, whilst those which contain N in addition to C, H and O will generally be more prone to ionize in positive ESI mode. A problem for global metabolic profiling is thus the need to obtain data in both positive and negative ESI modes to ensure the maximum coverage of the metabolome. For some mass spectrometers this means that the UHPLC separation must be performed twice, once in positive and then again in negative ESI, although using other types of instrument can allow this to be performed in a single analysis with an obvious saving in instrument time and avoidance of retention time variation between runs. However, the need for polarity switching can result in the loss of some sensitivity. In addition to ESI, there is also the potential to use the other commonly available ionization technique of atmospheric pressure chemical ionization (APCI), which is perhaps best suited to non-polar analytes. In targeted analysis it is often seen that ESI and APCI ionize different classes of compound with different efficiencies (albeit that there is significant overlap as a large number of analytes can be analyzed using either mode). Hence, there is an argument for analyzing samples with both ESI and APCI, albeit at the expense of increased analysis time per sample. As well as ionization modes, there are clearly a whole range of mass analyzers available offering mass resolution ranging from low (1000) to very high (>100 000). The low mass resolution MS instruments, such as the widely available single and triple quadrupoles (and quadrupole ion traps), provide mass spectrometric data with unit mass resolution. Triple quadrupole MS instruments are useful for metabolite identification based on their ability to perform tandem MS (MS/MS) experiments that provide diagnostic fragmentation data, whilst the quadrupole ion traps can be used for both MS, MS/MS and MS^n experiments for metabolite identification. The TOF and Q-TOF instruments have wide mass ranges, provide rapid fast scanning capabilities and mass accuracies of *ca.* 5 ppm. The fast scanning speeds provided by TOF and Q-TOF mass spectrometers make them well suited for use in combination with UHPLC separations where, because of their great chromatographic efficiency, peak widths can be narrow. The accurate mass capability provided is also of great utility in providing atomic composition data that can be of considerable value in metabolite identification, whilst the newly introduced MS^E method allows accurate mass data to be obtained on both precursor and product ions simultaneously. The very highest mass resolutions and accuracies (sub-ppm) are provided by the use of FT mass spectrometers. The widely used Orbitrap instrument employs an electrostatic ion trap and fast FT (FFT) to yield mass accuracies in the order of 1–2 ppm and resolution of up to 100 000 or higher.

Whilst the low mass resolution mass analyzers can be used with UHPLC for global metabolic profiling studies, the higher resolution instruments provide distinct advantages in being able to distinguish between co-eluting metabolites of the same nominal mass. On the other hand, the triple quadrupole instruments, operated in multiple reaction monitoring (MRM) mode, are ideal for more targeted quantitative approaches once biomarkers have been identified (*e.g.* see refs. 48–50).

14.5.4 Sample Run Order

Once the UHPLC and mass spectrometric conditions have been optimized for the biofluid/sample set that is to be investigated, then serious analysis can begin. One factor that then becomes essential is to ensure that biases due to small drifts in performance (of the type that the QCs are designed to monitor) do not adversely affect the outcome of the experiment. An important means of ensuring that such minor changes do not adversely affect the subsequent data treatment is to ensure that the samples are carefully randomized prior to the start of the analysis. This ensures that the analysis, and therefore the resulting data, for all of the experimental and control groups are equally affected and that the subsequent multivariate statistical analysis is unbiased.

Taking all of these considerations into account, a typical UHPLC-MS analysis for biomarker discovery *via* global metabolic profiling would be structured as follows. The column, mass spectrometer inlet and source should be meticulously cleaned prior to the run and before the samples are committed a system suitability test should be performed using a test mixture to check chromatographic and mass spectrometer performance. If these pre-run tests are satisfactory then, depending upon the sample type, they should be followed by 5–10 conditioning injections (using the QC samples), the first QC sample and then batches of 5–10 randomized test samples, each followed by a QC sample. At the end of the run a repeat of the system suitability test mixture is recommended. This should be followed by column washing with a strongly eluotropic solvent such as methanol or acetonitrile, or a mixture of these, to remove strongly retained contaminants. Consideration should also be given to the need for the mass spectrometer inlet and source to be decontaminated before the next run is undertaken (this will be dependent on the type of sample, and the total number of samples analyzed as, for example, plasma analysis will result in a more rapid build-up of contaminants than urine).

14.6 Applications of UHPLC-MS to Human Metabolic Profiling Studies

An important area that can be anticipated for the application of UHPLC-MS based studies of the metabolome is the field of human disease and, to date, this has indeed been the area where the largest number of published investigations in man have been performed. Hopefully, such initiatives will provide specific biomarkers for the early detection, diagnosis of disease, together with patient stratification and disease prognosis, as well as monitoring of therapeutic outcomes. These studies employing UHPLC-MS, which are still at an early stage, are considered below, together with applications in the area of nutrition.

14.6.1 UHPLC-MS Applications in Cancer

As an appropriate biofluid for prospecting for biomarkers of kidney disease urine would seem to be uniquely appropriate and, indeed, this sample type was employed in an early UHPLC-MS-based study to investigate the effects of renal cell carcinoma (RCC) on the urinary metabolome.[51] As well as RP gradient UHPLC-MS, both HILIC and GC-MS (with derivatization) were used to profile the samples which were obtained from a relatively small number (6) of diseased subjects of various ages and genders, ethnicities and disease states, and an equivalent number of controls. The use of all three techniques, with their complementary selectivities for different types of metabolite, ensures a wider coverage of the metabolome than using a single technique alone and provides a means of covering lipophilic and hydrophilic metabolites present in urine. An important observation from this study was that the pre-treatment of the urine samples with urease, commonly employed to remove urea from the samples prior to analysis, in addition to removing urea also resulted in major changes to the observed metabolic profiles. As such, use of this sort of treatment should clearly be avoided. For data analysis MZmine and XCMS, two freely available peak alignment methods, were compared for peak detection and retention time alignment, with the number of features somewhat dependent on the program used. Multivariate statistical analysis [partial least squares (PLS)] showed that all three analytical methods were able to distinguish between the healthy controls and diseased subject. Thus, from several thousand features detected in the samples, a number of significant components were highlighted that discriminated between the two groups, and this number was then reduced to less than 30, as yet unidentified, components *via* a feature selection process to be concentrated on in future work using a larger patient cohort. The authors expressed the hope that such investigations would lead to "clinically applicable assays for earlier diagnosis of RCC, as well as other malignancies, and thereby improved patient prognosis".[51] More recently, work on metabolic profiling, using a range of analytical techniques including UHPLC-MS, for investigation into human renal cancer has described changes in the urinary profiles of acylcarnitines in subjects with the disease.[52]

In a similar vein, a more recent study reported the investigation of the urinary metabolic profiles of groups totalling 24 pre- and post-operative colorectal cancer (CRC) patients and 80 healthy volunteers.[53] The metabolic profiling was performed using a conventional RP gradient UHPLC-TOF-MS (in positive ESI) method combined with statistical analysis of the resulting data using partial least squares to latent structure–discriminant analysis (PLS-DA). The authors reported the detection of "at least several thousands of peaks" in the samples and the results of the statistical analysis of the data obtained in the study showed that, in the pre-operative samples from CRC patients, two unidentified metabolites (with masses of 283 and 294 amu) were elevated in comparison to the group of healthy volunteers. After surgery, the amounts of these two components were seen to decrease. The authors suggested that their preliminary results indicate that the UHPLC-MS-based analysis, when combined with pattern recognition, could lead to methods that have clinical utility for the diagnosis of CRC and, indeed, this may be the case. In addition, the identification of these molecules would enable their biochemical significance to be determined, which would greatly increase confidence in the value of these potential markers. Another study using UHPLC-TOF-MS, in combination with GC-TOF-MS, for profiling the serum metabolomes of CRC patients compared to normal healthy subjects for the detection of discriminating metabolites has been reported.[54] In this study samples of serum from 32 patients with colon and 32 with rectal tumours were obtained, together with those of 65 normal subjects. For UHPLC-MS analysis serum was mixed with an equal volume of water containing an internal standard, and then two volumes of a 5:4 mixture of methanol/acetonitrile. After vortexing and centrifugation, the supernatant was filtered through a syringe filter (0.22 μm) and analyzed using a RP gradient based on water and acetonitrile, both modified with 0.1% formic acid, with data collection in both positive and negative ESI. Multivariate statistical analysis using orthogonal partial least squares-discriminant analysis (OPLS-DA) resulted in the detection of 16 potential marker metabolites, of various classes with some, such as glycerol phosphate, pyruvate, lactate and "carnitines", higher in concentration in CRC patients compared to controls, whilst others, such as, for example, arginine, phenylalanine, tryptophan palmitate and dopamine were present in smaller amounts.

Although urine and plasma are perhaps the most easily accessible and easily analyzed samples that can be obtained from humans; other sample types, such as saliva, biopsy tissues and feaces, are also potentially rich sources of biomarkers. An example of the analysis of saliva samples from patients suffering from oral squamous cell carcinoma (OSCC) or leukoplakia (OLK) has been reported[55]. The study provided an investigation of the metabolic profiles of saliva obtained from 37 OSCC, 32 OLK and 34 healthy subjects. Samples were analyzed with RP gradient UHPLC-MS. Prior to analysis the samples were first mixed with water and acetonitrile (at ratio of 1:1:1, v/v/v), followed by centrifugation and then filtration through a syringe filter (0.22 μm). A water/acetonitrile gradient was employed, with both solvents containing 0.1% formic acid. Analysis of the resulting UHPLC-MS data by PCA and OPLS-DA showed metabolites such as γ-aminobutyric acid (GABA), phenylalanine, valine, *n*-eicosanoic acid and lactic acid, as discriminating between patients and controls, and between the OSSC and OLK groups.

An example of the investigation of faecal metabolomes is provided by a study comparing those of normal healthy individual to subjects with liver cirrhosis or hepatocellular carcinoma (HCC).[56] Faecal extracts prepared by mixing weighed samples with methanol (at a ratio of 3 ml g^{-1}), centrifuging and filtering through a 0.2 μm pore size membrane, were then analyzed directly using RP UHPLC-Q-TOF MS (positive ESI) with subsequent data analysis to enable discrimination between healthy and diseased states performed *via* PLS-DA-derived models. The same statistical approach was also used to compare the metabolic profiles of the faecal extracts of cirrhotic and HCC subjects so as to identify potential biomarkers for cirrhosis and HCC that were differentially expressed in the respective faecal metabolomes. Detailed statistical examination of the UHPLC-MS data suggested that liver cirrhosis and HCC patients could be distinguished from healthy humans on the basis of six metabolites. The potential biomarkers, which showed a significant increase in patients with cirrhosis and HCC in comparison with healthy subjects, included lysophosphatidylcholines (lysoPCs), whilst bile acids and bile pigments were correspondingly decreased. Figure 14.6 shows UHPLC-MS profiles of faecal extracts together with an "S-plot" generated from these data in order to better visualize the contribution of the different features in group clustering (the biomarkers are highlighted with red boxes). Table 14.1 provides the identities of the six key metabolites, the VIP value (SIMCA *P* parameter), *P* value (T test), retention time, measured *m*/*z* and major fragments, and the concentration trend in liver cirrhosis and HCC states. The authors concluded on the basis of their study that UHPLC-MS had potential for use in the screening of faecal samples as an aid to the early diagnosis of the two disease states.

In addition to biofluids such as plasma and urine (or faecal samples), there is also a clear opportunity to employ UHPLC-MS-based methods to investigate cancer biology *via* profiles of both tissue biopsies and using *in vitro* incubations of suitable cell types.

A recent contribution to the evolving use of UHPLC-MS for cell-based studies investigated the development of sample extraction methods to enable the intracellular metabolite profile of preparation methods for MCF-7 cells, an immortalized human breast cancer-derived cell line.[57] This methodology

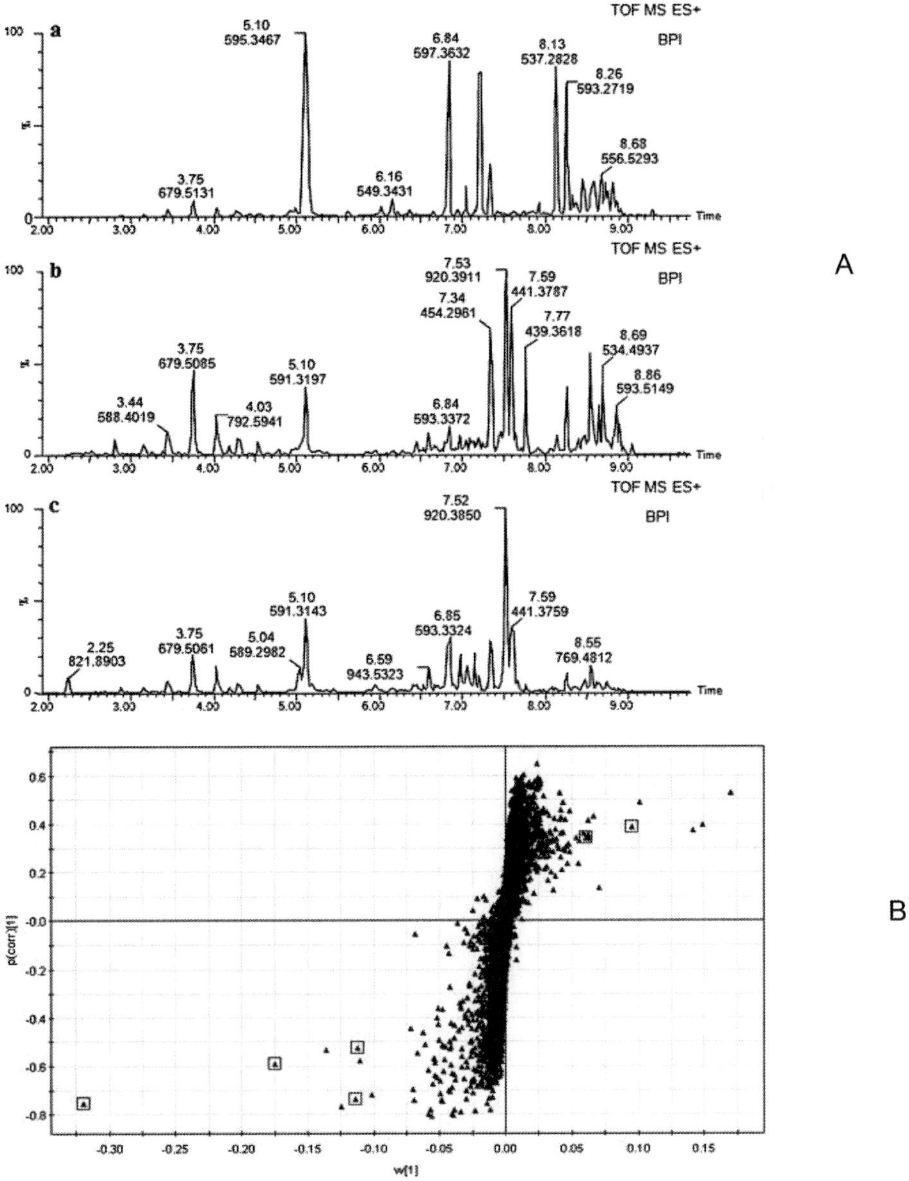

Figure 14.6 (A) UHPLC-MS profiles of faecal extracts of (a) a healthy human, (b) a patient with liver cirrhosis and (c) a patient with HCC. (B) An S plot that visualizes the contribution of the different features in group clustering; biomarkers are highlighted with black boxes. Table 14.1 provides the identities and statistical measures of six key metabolites.

Table 14.1 The most important metabolites discriminating the faecal profiles of liver cirrhosis and HCC patients compared to healthy humans.

Metabolite	*VIP value*	P *value*	*Retention time (min)*	*Measured* m/z *	*Major fragments*	*Liver cirrhosis*	*HCC*
Chenodeoxycholic acid dimer	37.34	< 0.01	7.201	785.5898	357.28	↓	↓
Urobilin	28.38	< 0.01	5.011	595.3467	180.10, 303.16	↓	↓
Urobilinogen	17.13	< 0.01	6.834	597.3632	180.10, 303.16	↓	↓
7-ketolithocholic acid	13.37	< 0.01	7.044	391.2837	355.26, 373.27	↓	↓
LPC C18:0	11.60	< 0.05	7.620	524.3707	104.11, 184.07	↑	↑
LPC C16:0	10.42	< 0.05	7.354	496.3410	104.11, 184.07	↑	↑

*$[M+H]^+$ or $[2M+H]^+$

involved detaching the cells from the Petri dishes in which they were grown by scraping followed by metabolite extraction. Two methods of cell extraction were compared, the first based on lysis with water, and the second using methanol and chloroform. The extracts resulting from these procedures were then re-dissolved in acetonitrile/water (1:1, v/v) for analysis. The extracts were analyzed *via* RP gradient UHPLC-TOF-MS in both positive and negative ESI modes, and the authors concluded that the multiple solvent extraction method, perhaps not unexpectedly, gave superior detection of features in the non-polar region of the chromatogram compared to water, resulting in *ca.* 2000 and 800 features detected in positive and negative ESI respectively (compared to *ca.* 1000 and 600 for water). The authors then went on to apply the methodology to an array of human cell lines commonly used in cancer research (including breast, prostate colorectal and pancreatic derived cell lines), in order to demonstrate the generic applicability of the method.

14.6.2 Infection

In a small scale pilot study, based on the analysis of plasma and urine from 11 children plus community controls from the Gambia in West Africa with World Health Organization (WHO)-defined severe pneumonia,[58] an attempt was made to use UHPLC-MS to find disease markers. The authors defined the objective of the investigation as the application of metabolomic analysis to childhood pneumonia "to explore its potential to improve pneumonia diagnosis in a high burden setting". For this, they used RP gradient UHPLC-TOF-MS, in both positive and negative ESI, with the data analyzed using a range of multivariate statistical techniques as well as with the machine-learning algorithm "Random Forests". From this, they were able to determine statistically significant differences ($P < 0.05$) between the plasma of control and infected children in metabolites such as uric acid, hypoxanthine and glutamic acid, which were increased in concentration in samples from sick children compared to the controls. In contrast L-tryptophan and adenosine-5′-diphosphate (ADP) were found to be present in lower amounts. In urine samples, reduced amounts of uric acid and L-histidine were seen for pneumonia cases compared to healthy controls. Metabolomic analysis clearly distinguished severe pneumonia patients from community controls. Whilst noting the limitations in the study with respect to the sample size, the clear ability to detect differences between the patients and the controls was seen as promising with larger studies required to validate if these identified markers were disease-specific responses *etc.*

14.6.3 Interstitial Cystitis (IC)

UHPLC-MS-based global metabolic profiling of urine has also been used to search for biomarkers of IC.[59] IC is a chronic, non-bacterial, clinical syndrome affecting the lower urinary tract associated with urinary frequency and urgency and/or pelvic pain. The diagnosis of the condition currently requires cystoscopy, which is both painful and expensive and therefore methods that could non-invasively diagnose IC would assist in the initiation of therapy and improve the patient's quality of life. In order to find such biomarkers, urine studies were performed on samples from 15 patients with IC, a further 10 with bacterial cystitis and 15 healthy volunteers. Initially, urine samples from 10 subjects from each group were profiled by RP gradient UHPLC-TOF-MS in positive ESI to identify suitable candidates. Some 7700 features were seen in the data with two ions, with *m/z* values of 265.118 and 279.161 amu and retention times of *ca.* 2.0 and 9.6 min respectively, found to be characteristic of IC patients *via* unsupervised PCA. Of these two ions, the former (nominal mass 265 amu) appeared to be the most promising candidate, being elevated in IC patients compared to both normals and those suffering from bacterial cystitis. Database searches, followed by attempts at confirming identity with standards based on "hits", proved unsuccessful and the investigators then resorted to preparative scale isolation of the metabolite and structural identification by MS, MS/MS and ^{1}H and ^{13}C NMR spectroscopy. These studies indicated that the metabolite was phenylacetylglutamine (PAGN); having confirmed this identification, the authors then went on to perform quantitative analysis *via* UHPLC-UV to determine urinary concentrations of this metabolite in a further five IC subjects and five healthy volunteers. These measurements showed that PAGN, when measured relative to creatinine (CR), was, indeed, present in significantly greater amounts in the urine of IC patients (mean 0.47 mg mg^{-1} Cr) compared to both bacterial cystitis patients (mean 0.25 mg mg^{-1} Cr) and healthy controls (mean 0.11 mg mg^{-1} Cr). However, some complexity is provided by the observation that the urinary PAGN/Cr ratios in patients with both mild and moderate IC (grades 1 and II) respectively were higher than for patients with the severe form of the disease (grade III). Moreover, urinary PAGN/Cr ratios with mild and moderate IC patients (mean 0.30 mg mg^{-1} Cr) were higher than healthy controls (mean 0.059 mg mg^{-1} Cr), in the validation set. The authors concluded that, although the reason for patients excreting higher concentrations of PAGN in their urine was not clear, the urinary PAGN/Cr ratios could nevertheless be used as a novel urinary marker of IC, and might also have a role to play in the early diagnosis of IC.

14.6.4 Renal Failure

In another application of gradient RP UHPLC-TOF-MS-based metabonomic investigations, serum samples obtained from 32 patients suffering from chronic renal failure (CRF) and 30 age-matched healthy control subjects were compared in an attempt to find biomarkers of the disease.[60] Samples were prepared for analysis by mixing serum with acetonitrile at a volume ratio of 1:2, followed by centrifugation, freeze drying and reconstitution in water/acetonitrile (1:1, v/v) for analysis, which employed negative ESI for detection. Some 9600 features were initially detected, reduced to *ca.* 2000 after the removal of background and non-biologically derived ions. Using PLS-DA, a number of metabolites were identified which correlated with CRF, including creatinine, tryptophan, phenylalanine, kynurenine and three lysoPCs (LPC C16, LPC 18:0 and LPC18:1) and were identified as either increased (*e.g.* creatinine, kynurenine, phenylalanine, LPC 18:0) or decreased (*e.g.* tryptophan, LPC C16, LPC 18:1).

14.6.5 Alzheimer's Disease (AD)

AD is becoming an increasing problem as Western populations age, and there is clearly a major requirement for biomarkers that provide a non-invasive and accurate diagnosis of the condition. Equally, markers that give an indication of the likely rate of disease progression and response to therapy would be very welcome additions to the clinician's armoury of diagnostic tools. In a recent UHPLC-MS study, plasma was analyzed for potential biomarkers of AD in patients compared to a control group of age- gender- and BMI-matched healthy subjects.[61] Prior to analysis, the plasma was treated with acetonitrile (in a ratio of 1:2) to precipitate proteins, followed by centrifugation, evaporation of the supernatant to dryness under nitrogen and then prepared for UHPLC by re-dissolving the dry residue in water/acetonitrile (15:85). For sample analysis, RP elution, with a water/acetonitrile (with both components acidified with 0.1% formic acid) gradient and detection in positive ESI was performed. PCA of the resulting data revealed metabolic differences between the two groups with reductions in nine potential biomarkers, including lysoPCs, phytosphingosine, dihydrosphingosine, hexadecasphinanine and tryptophan detected in subjects with AD compared to the control subjects.

14.6.6 Pregnancy

A major complication of pregnancy is pre-eclampsia (PE), a multi-system disorder that is thought to be mediated by factors released into the circulation as a result of damaged placental villous trophoblast. In a study employing UHPLC-MS, the metabolites consumed from and released into serum-conditioned culture medium were analyzed using UHPLC-MS following the *in vitro* culture of factors released from villous trophoblast tissue obtained from normal pregnancies ($n = 6$) and those with PE ($n = 6$).[62] A total of four experimental groups were set up using explanted placental villous fragments from these subjects. Tissue from each group was cultured for 96 h in either 1% O_2 (hypoxia) or 6% O_2 (placental normoxia). The analysis of the resulting data showed that the relative concentrations of 154 features of the metabolic profile (or "footprint") changed in culture medium obtained from normal pregnancies cultured in normoxic and hypoxic conditions, whilst 21 and 80 features were found to be different in the culture medium of PE *vs.* normal pregnancies cultured in hypoxic and normoxic conditions, respectively. This investigation revealed evidence of changes in the villous tissue in PE, some of which could be replicated by culturing villous tissue in hypoxic conditions. When all four groups were compared some 47 features showed similar relative concentrations in the media of PE-derived tissue medium cultured in normoxic conditions compared to conditioned medium from normal villous tissue cultured in hypoxic conditions. Areas of metabolism highlighted by this work for further investigation included glutamate and glutamine, tryptophan metabolism and leukotriene or prostaglandin metabolism. The authors suggested on the basis of their *in vitro* data that hypoxia might have a role in the placental pathogenesis of PE.

In addition to the *in vitro* study described above, in an investigation using UHPLC-LTQ-Orbitrap-MS, blood serum from 20 Caucasian women with PE were compared with a control group matched for age, parity and BMI.[63] RP gradient chromatography was performed using both positive and negative ESI on serum samples precipitated with methanol in a ratio of 1:3 (v/v) serum/methanol. The resulting data were normalized using XCMS and, whilst this study was primarily undertaken to assess the suitability of the analytical approach for serum analysis rather than biomarker detection, the resulting data nevertheless indicated a number of possible biomarkers of PE, whilst demonstrating that both the instrumentation and XCMS software were suitable for the production of reproducible and valid data, including the detection of disease biomarkers in serum.

In an *in vitro* investigation similar to that for PE, a study by the same group examined whether there were differences in the metabolic footprint of placental villous explants cultured at different oxygen (O_2) tensions between women who had delivered a baby defined as being born small for gestational age (SGA) compared to those resulting from normal controls.[64] This condition is associated with a significantly increased risk of perinatal morbidity and mortality and may be the result of a poorly perfused and abnormally developed placenta. As with PE, it appears that some of the features seen in SGA (*e.g.* abnormal cell turnover, impaired nutrient transport *etc.*,) can be reproduced by placing placental explants under hypoxic conditions. Differences were therefore sought in the metabolic footprints, determined by UHPLC-MS (using a hybrid LTQ-Orbitrap mass spectrometer) of placental villous explants from women delivering an SGA baby ($n = 9$) and cultured at different oxygen (O_2) tensions and those obtained from normal controls ($n = 8$). Tissue explants from both groups were cultured for 96 h in 1% (hypoxic), 6% (normoxic) and 20% (hyperoxic) O_2. From the UHPLC-MS analysis of the medium, 574 metabolite features were found to have significant differences between SGA- and normal-derived explants at one or more of the oxygen tensions studied. Data analysis on a univariate level showed that, in the case of 49% of the metabolites of interest, the medium from SGA explants that had been cultured under hypoxic conditions showed the same metabolic signature as control explants cultured under normoxic conditions. No such behaviour was observed under hyperoxic culture conditions. The authors hypothesized that perhaps that SGA tissue was acclimatized to hypoxic conditions *in vivo*. In terms of metabolic pathways, glycerophospholipids and tryptophan were highlighted.

14.6.7 Gastrointestinal Tract

An investigation of irritable bowel syndrome (IBS), using both UHPLC-MS-based lipidomics and a GC × GC-TOF metabolomic approach, was applied to the global metabolic profiling of mucosal biopsies obtained from the ascending colon of patients and control subjects.[65] Samples were obtained from 15 IBS patients fulfilling the Rome II criteria and nine healthy volunteers. The lipid analysis of these mucosal biopsies showed that concentrations were higher for IBS sufferers compared to the controls, with the greatest up-regulation observed for the pro-inflammatory lysoPCs. In addition, there was significant up-regulation of metabolites such as lipotoxic ceramides, glycosphingolipids and di- and tri-acylglycerols in the IBS group compared to the controls. The metabolite 2(3H)-furanone was almost 14-fold up-regulated in IBS patients compared to healthy subjects. These results enabled the investigators to conclude that the mucosa of IBS patients was "characterised by a distinct pro-inflammatory and lipotoxic metabolic profile".

UHPLC-TOF MS-based metabonomics has been used for the investigation of the metabolic effects of the development of intestinal fistula. Such fistula can result as a complication of surgery or can develop spontaneously as a result of antenatal anomalies, trauma, inflammation and severe acute pancreatitis.[66] In this investigation, serum from 40 patients with intestinal fistulas was compared with that of 17 healthy individuals in an attempt to discover potential biomarkers of the condition. Prior to UHPLC, the serum was mixed with acetonitrile in a ratio of 1:4 (v/v) serum/acetonitrile and, after removal of the protein pellet by centrifugation, the supernatant was freeze dried and then re-dissolved in water/acetonitrile (1:4, v/v). RP gradient UHPLC-MS was performed with both positive and negative ESI used for metabolite detection. Data analysis indicated that *ca.* 4900 features could be detected in positive ESI and a further 3600 in negative ESI. Following UHPLC-MS multivariate statistical analysis using PLS-DA and coefficient of correlation analysis were performed for marker selection and identification. This resulted in nine metabolites being highlighted as potential biomarkers. These included the bile acids glycochenodeoxycholic acid, glycodeoxycholic acid, taurochenodexycholic acid and taurodeoxycholic acid, together with and two types of lysophosphatidyl choline (C16:0 and C18:2); all of which were found in greater amounts in patients with intestinal fistulas compared to controls. Conversely, the metabolites phenylalanine, tryptophan and carnitine were present in decreased amounts in patients compared to controls. The authors hypothesis was that these changes were perhaps indicative of a subclinical hepatic injury combined with abnormal metabolism of the essential amino acids phenylalanine and tryptophan as well as carnitine in patients with an intestinal fistula.

As well as studies on subjects with intestinal fistulas, UHPLC-MS has also been undertaken on the plasma of a patient recovering from an intestinal transplant,[67] in this case a small bowel transplant, who was compared to 10 healthy individuals. The samples were collected over a period of some 138 days post-surgery, during which time the patient was treated with immunosuppressants (tacrolimus) and prednisolone during periods of acute rejection. Once again, RP gradient UHPLC-TOF-MS, with both positive and negative ESI, was used for the analysis of protein-precipitated plasma (plasma/acetonitrile, 1:4, v/v). Potential biomarkers of rejection were identified as lysophosphatidyl choline, phenylalanine and tryptophan.

14.6.8 Liver Disease

Liver disease, whether caused by disease, alcoholism or toxicity, represents a major cause of human ill health. As the major metabolic organ in the body, disruptions in function of the liver might reasonably be expected to be accompanied by major changes in metabolic pathways and the metabolome. As a result, many studies have been undertaken in animal models to search for potential biomarkers for a range of different liver-related applications and these investigations are now also being undertaken to studies in man.

The condition known as non-alcoholic fatty liver disease (NAFLD) represents the commonest type of chronic liver disease amongst Western populations. Currently, the diagnosis of NAFLD requires either invasive methods such as liver biopsies or expensive techniques such as imaging. A UHPLC-MS-based investigation examined the utility of metabolic analysis of serum as a source of biomarkers for the condition, investigating samples obtained from non-diabetic, morbidly obese, biopsy-proven NAFLD patients.[68] In addition, samples from a rodent model, based on the glycine *N*-methyltransferase knockout (GNMT-KO) NAFLD mouse model, were also investigated. As well as global metabolite profiling, the experiments were also complemented with conventional histopathological and clinical analysis. Serum samples for global metabolic profiling were treated with 4 volumes of methanol to precipitate proteins after which the samples were left at −20 °C overnight before centrifugation and analysis. The UHPLC-MS method employed a RP gradient composed of water and acetonitrile, both containing 0.05% aqueous formic acid, with detection using a Q-TOF mass spectrometer operated in both positive and negative ESI.

Statistical analysis of the data, performed using PCA and OPLS-DA, enabled a number of biomarkers to be identified as significantly altered in the GNMT-KO mouse; these were common to human subjects with NAFLD. Alterations were seen in the profiles of creatinine, organic acids, free fatty acids, phophatidylcholine, lysoPC, bile acids and sphingomyelin compared to normal profiles. The authors noted that some of these compounds could be readily associated with biochemical perturbations found with liver dysfunction and concluded that such "differential metabolic phenotyping" may have a future role in supplementing the information available to clinicians treating NAFLD and aiding in the adaption of clinical practice to "more individualized treatment protocols".

Recently, a UHPLC-MS metabolomics study, employing variable selection and PLS-DA analysis, has been evaluated to determine its utility for discriminating between non-steatotic and steatotic human liver metabolite profiles.[69] A range of different chemometric approaches were studied with the result that the metabolites such as glutathione, lysophospholipids and bile acids were found to be the most important altered metabolites in the metabolic profiles obtained.

Clearly, the study of liver disease and toxicity in humans is complicated by the invasive nature of sampling the organ itself and hepatocytes are often used as a surrogate, especially for the investigation of the actual, or potential, toxicity of drugs and other xenobiotics. UHPLC-TOF-MS was used in an investigation of the effects of three model drugs on human hepatocytes.[70] Both of the intracellular and extracellular metabolomes were analyzed following exposure of the cell to well-characterised activators of the xenoreceptors CAR (NR1I3) and PXR (NR1I2). The compounds used were rifampicin, phenobarbital and 6-(4-chlorophenyl)imidazo[2,1-b][1,3]thiazole-5-carbaldehyde O-3,4-dichlorobenzyl) oxime (CITCO). The aim of the study was to discover endogenous metabolites that were potentially related to a PXR or CAR induction mechanism and also to identify metabolites of the drugs themselves. Cell extracts were obtained *via* re-suspending the cell pellet, obtained after centrifugation of the hepatocytes in phosphate buffer, in a 1:1 (v/v) mixture of water and acetonitrile followed by cell disruption by orbital agitation. Centrifugation of the cell lysate provided a suitable sample for analysis. RP gradient UHPLC-TOF-MS, using both positive and negative ESI, was undertaken on the samples followed by multivariate statistical analysis. The authors noted that a significant source of variability in the experiments was the source of the cell themselves rather than the drug treatments. It was also noted that when multivariate statistical approaches such as PCA and OPLS were used, the effect of the presence of drug metabolites, which were only present only in treated hepatocytes, complicated the interpretation of the data with respect to changes in endogenous metabolites. This required the application of a new statistical approach, called the shared and unique structure (SUS) plot, that then allowed a comparison of different treatments, and enabling the separation of variables such as drug metabolites from the endogenous biomarkers. This allowed the identification of endogenous components (either up- or down- regulated) related to the effects of rifampicin, phenobarbitol and CITCO on the biochemical pathways of human hepatocytes.

As well as liver disease, some drugs can cause profound changes to liver biochemistry, and indeed in the case of fibrates, where there is a reduction in cholesterol production, that is the intention. The activation of the peroxisome proliferator-activated receptor (PPAR) by such drugs results in increased fatty acid catabolism, making these compounds useful for treating conditions such as hyperlipidaemia. Fenofibrate was administered orally at 200 mg day^{-1} for 2 weeks to three healthy male and seven healthy female volunteers with the aim of identifying biomarkers of PPARα activation and hence increased fatty acid β-oxidation.[71] Urine from these volunteers was profiled by RP UHPLC with detection *via* Q-TOF-MS. Data were analyzed using the machine learning algorithm "Random Forests" with the results suggesting that by 2 weeks of dosing that there were notable reductions in pantothenic acid and acetylcarnitine (greater than 5- and 20-fold respectively). In addition to global metabolic profiling, more conventional targeted analysis LC-MS/MS was also performed on pantothenic acid and a range of urinary acyl carnitines, in addition to acetylcarnitine, which were also seen to decline in concentration, including propylcarnitine (>10-fold), isobutyrylcarnitine (>2.5-fold), (*S*)-(+)-2-methylbutyrylcarnitine (5-fold) and isovalerylcarnitine (>5-fold) by day 14 of the study. In addition to this work in human subjects, the presence of these biomarkers as indicators of PPARα activation was confirmed in an animal model *via* examination of samples from *ppar*α-null mice and wild-type mice fed 0.1% fenofibrate in their diet. In this animal study, there was a 40-fold depletion in urinary pantothenic acid and a 88-fold reduction in acetylcarnitine in the wild-type animals and no change in the *ppar*α-null mice. The authors therefore concluded that these compounds may well provide useful biomarkers for PPARα-induced fatty acid β-oxidation in humans.

14.6.9 Obesity

Whilst not necessarily considered a disease, it is clear that obesity is very often a precursor to Type II diabetes, cardiovascular disease and host of other conditions. Once again, given the involvement of metabolism in the condition, it might be anticipated that obesity might have effects on the metabolome. In a UHPLC-Q-TOF-MS-based study[72] comparing the metabolic profiles of 30 overweight/obese men and 30 age-matched, normal weight men, the overweight/obese men were seen to have higher rates of homeostasis model assessment-insulin resistance (HOMA-IR) as well as higher concentrations of triacylglycerols, total cholesterol and low-density lipoprotein (LDL)-cholesterol, and lower concentrations of high-density lipoprotein (HDL)-cholesterol and adiponectin, than the normal control subjects. Overweight/obese men showed higher proportions of stearic acid and lower proportions of oleic acid in serum phospholipids. Additionally, overweight/obese individuals showed higher fat intake and a lower ratio of polyunsaturated fatty acids to saturated fatty acids. Three lysoPCs were identified as potential plasma markers, whilst eight known metabolites for overweight/obese men were confirmed. In particular, overweight/obese subjects showed higher levels of lysoPC C14:0 and lysoPC C18:0 and lower levels of lysoPC C18:1 than lean subjects. Results confirmed abnormal metabolism of two branched-chain amino acids, two aromatic amino acids, and fatty acid synthesis and oxidation in overweight/obese men. Additionally, the amount of dietary saturated fat may influence the proportion of saturated fatty acids in serum phospholipids and the degree of saturation of the constituent acyl group of plasma lysoPC.

14.6.10 Diabetes

Together with obesity, and indeed associated with it, Type II diabetes is becoming an increasing problem for Western societies. This has resulted in a number of investigations of the metabolic consequences of diabetes, and therapy, in human subjects using UHPLC-MS-based analysis. For example, UHPLC-TOF-MS, in both positive and negative ESI, was used for the analysis of serum profiles obtained from 33 Type II diabetics, eight subjects suffering from diabetic nephropathy and 25 healthy controls in order to find potential biomarkers.[73] RP chromatography was performed on protein-precipitated serum (mixed 1:3, v/v, with methanol to precipitate the proteins) using a linear water/acetonitrile gradient (with both solvents containing 0.1% formic acid) from 0 to 95% acetonitrile over 15 min (see Figure 14.7A for a UHPLC-MS profile). Multivariate statistical analysis of the resulting positive ESI data using PCA showed clear clustering of the samples into control, Type II diabetic and diabetic nephropathy subjects (Figure 14.7B). Metabolites contributing significantly to this clustering were identified as leucine, dihydrosphingosine and phytoshpingosine, which the authors interpreted as being indicative of perturbations in amino acid and phospholipid metabolism in diabetic diseases (Figure 14.7C–E).

It is also obviously of interest to monitor the response of patients to therapy. A major first line drug for the treatment of Type II diabetes is metformin, and a combined UHPLC-MS and ^{1}H NMR study of the biochemical differences observed between the serum of patients suffering from this condition treated with the drug and untreated subjects has been reported.[74] The metabonomic analysis was undertaken on samples obtained from 20 untreated Type II diabetics and 15 age-matched patients administered metformin hydrochloride for 3 months. PCA and PLS-DA were then used to interrogate the resulting data. Both methods of analysis detected differences in the samples obtained from drug-treated and untreated patients. Thus, analysis using ^{1}H NMR spectroscopy identified increases in trimethylamine-*N*-oxide and 3-hydroxybutyrate coupled with a reduction in glucose, lactate, acetoacetate and unsaturated lipids (as well as reduced *N*-acetyl glycoprotein and lipoprotein) in the serum obtained from metformin-treated subjects relative to those left untreated. RP UHPLC-MS, in positive ESI mode, was undertaken on serum precipitated with 2 volumes of acetonitrile using a water/acetonitrile gradient. Both solvents were acidified with 0.1% formic acid. The multivariate statistical analysis of the UHPLC-MS data indicated that in the drug-treated group, the amounts of the amino acid tryptophan were increased and that this was combined with reductions in the quantities of phenylalanine and the C16:0, C18:0 and C18:2 lysoPCs compared to the untreated patients.

A more recent study investigated the effects of the traditional Chinese medicine berberine on subjects with Type II diabetes and dyslipidaemia.[75] The research was undertaken on serum samples obtained from 60 patients, with profiling undertaken before and after treatment with either berberine or placebo. Prior to analysis, the plasma was mixed with acetonitrile (1:4, v/v) to precipitate proteins and following their removal by centrifugation, the supernatant was reduced to dryness under vacuum. After redissolving the residue in acetonitrile/water (4:1), samples were analyzed *via* UHPLC-Q-TOF-MS using a RP gradient based on water (containing 0.1% formic acid)/acetonitrile. Both positive and negative ESI were employed. Changes in the profile of serum metabolites were then sought using OSC-PLS-DA (orthogonal signal correction filtered partial least-squares discriminant analysis). This data analysis demonstrated that berberine-treated patients could be distinguished from the placebo group based on changes in the serum metabolite profile. Notable effects of berberine administration on the concentrations of circulating free fatty acids were seen, and these were subsequently confirmed by quantification *via* UHPLC-MS with a single quadrupole mass spectrometer. This showed a highly significant reduction in the concentrations of some 13 fatty acids following drug treatment, suggesting that the mode of action of berberine in the treatment of Type II diabetes might be *via* reducing the concentrations of free fatty acids in the blood.

As well as investigations on Type II diabetes itself using UHPLC-MS, there has also been a study on subjects displaying pre-diabetic traits.[76] Thus, the risk of developing Type II diabetes can be predicted, to some extent, based on an individual's response to an oral glucose challenge or tolerance test. In this study, plasma and spot urines were obtained from a total of 51 non-diabetic

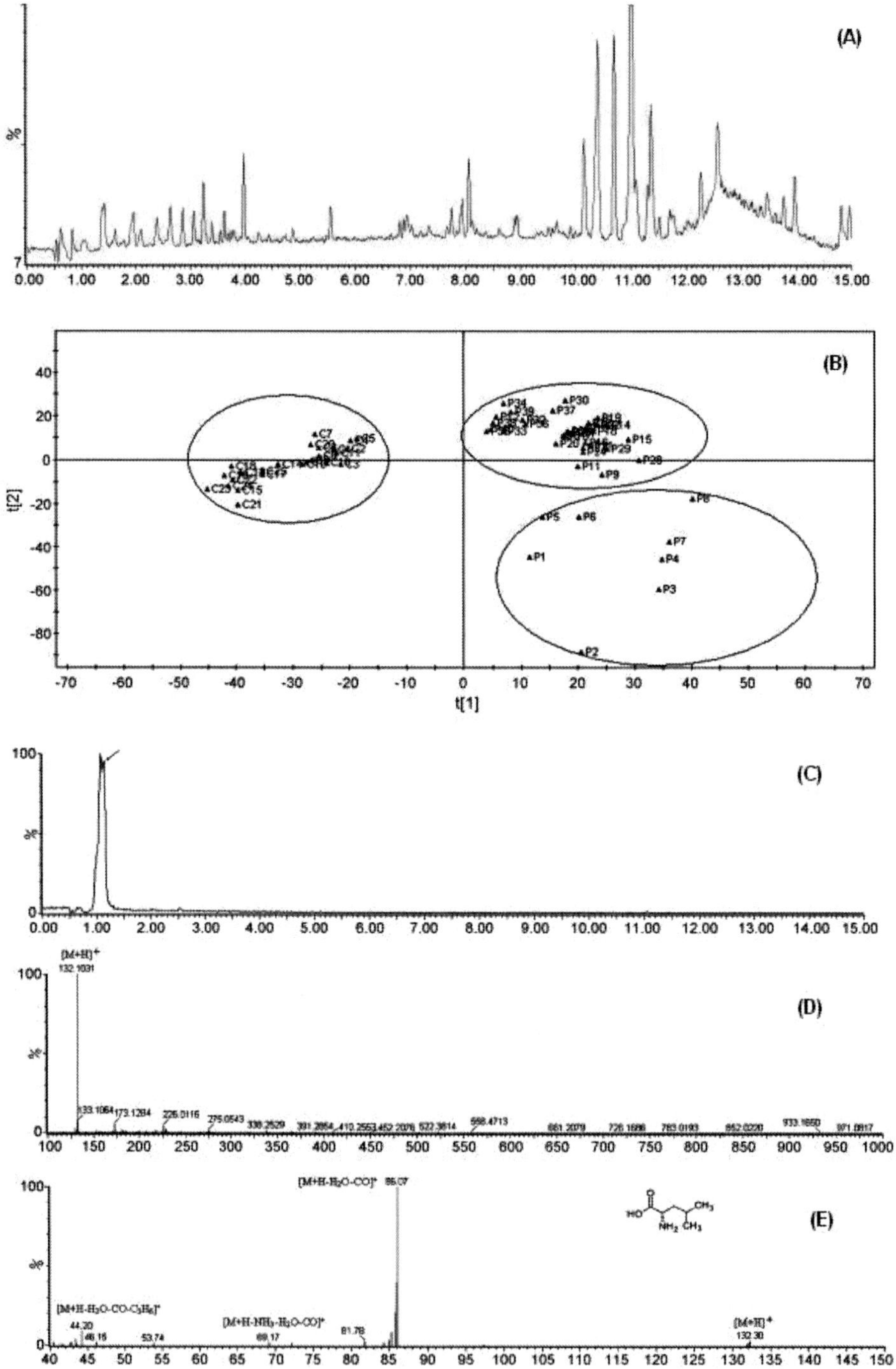
(A)
%
(B)
t[2]
t[1]
(C)
%
(D)
[M+H]+
132.1031
%
(E)
[M+H-H2O-CO]+ 86.07
[M+H-H2O-CO-C3H6]+
[M+H-NH3-H2O-CO]+
[M+H]+
132.30
%

subjects, 39 of whom gave a normal response to the glucose tolerance test and 12 of whom proved to be glucose intolerant, following an overnight fast. These samples were then analyzed by RP UHPLC-Q-TOF-MS. Solvents for gradient elution were water containing 0.1% formic acid and acetonitrile, with different gradient profiles used for urine and plasma. Both sample types were deproteinized by mixing with 2 volumes of acetonitrile, followed by centrifugation and reduction to dryness under vacuum. For analysis, plasma was taken up in acetonitrile in a ratio of 4:1 (v/v), whilst urine was re-dissolved in the same solvents but at a ratio of 1:4 (v/v). Multivariate statistical analysis OSC-PLS-DA was undertaken with the aim of separating those subjects who demonstrated an impaired glucose tolerance from those with a normal response so as to identify those pathways that were altered in the pre-diabetic state. On the basis of their data, the authors were able to show that glucose tolerant subjects could be separated from those subjects with impaired glucose tolerance for both sample types. In urine differences between glucose intolerant subjects and normal individuals included increased concentrations of tryptophan, xanthine and C8:2-OH and C10:2-OH carnitines and decreased quantities of uric acid, methyluric acid, 3- and 7-methylxanthine, hippuric acid, 3-hydroxyhippuric acid and phenylacetylglutamine. With respect to plasma, glucose intolerance was associated with raised concentrations of a range of free fatty acids and the bile acid glycochendeoxycholic acid and lower concentrations of lysoPCs compared to normal subjects.

14.6.11 Nutrition Research

As well as applications in the investigation of human disease there are obvious opportunities to study other aspects of human life such as the effects of food stuffs in nutritional research.[77–79] Many of the current publications in the area of nutrition research concern the development of UHPLC-MS methods for matrices such as urine and plasma, and the conclusions mirror much of the work that has been performed for biomedical applications of UHPLC-MS. So, for example, methods for plasma analysis in nutritional studies have also been developed and validated based on UHPLC-MS of samples collected using heparin as an anticoagulant and methanol for protein precipitation, as this gave the most reproducible results with the best metabolome coverage.[19]

Figure 14.7 (A) Typical UHPLC-MS TIC chromatogram obtained from the serum sample of a diabetic nephropathy patient in positive ESI mode. (B) PCA scores plots derived from the UHPLC-TOF-MS data. Samples from controls (C1–C25) cluster on the left part of the plot separated from samples from patients of diabetic nephropathy (P1–P8) and Type II diabetes (P9–P43). (C) Extracted ion chromatogram of potential biomarker (*m/z* 132) in positive mode, marked with an arrow; (D) corresponding mass spectrum; and (E) MS/MS spectrum (collision energy 15 eV). Adapted from reference 73 with permission from Elsevier.

Similarly, an optimized method for urine profiling in nutritional studies has been described based on RP UHPLC-MS,[77] whilst another urine-based study investigated the effects of sample preparation and chromatographic method differences on the classification and recovery of metabolic biomarkers from UHPLC-MS measurements on the urine of subjects consuming three high-fat meals (rich in saturated, monounsaturated and polyunsaturated fatty acids, respectively).[78] The meals were administered to eight volunteers in a randomized order, and with a washout period in between samples and each subject provided a urine sample before each meal and at three timed intervals subsequently. The urines were then either simply diluted 1:4 (v/v) with water, or subjected to liquid–liquid extraction (LLE). LLE was performed on 2 ml aliquots of the sample, acidified to pH 1, and following the addition of 1.5 g of sodium chloride, using 5 ml of ethylacetate. The organic layer was then taken, alkalified by the addition of 50 μl of 5 M sodium hydroxide in methanol and then reduced to dryness under reduced pressure using nitrogen gas. Following drying the samples were re-dissolved in water/acetonitrile (1:1, v/v), both of which contained 0.1% formic acid. Following sample preparation, the samples were analyzed using RP gradient UHPLC-TOF-MS using two different gradient conditions, with solvents systems based on water and acetonitrile acidified with 0.1% formic acid. The data were analyzed using PLS-DA. Perhaps unsurprisingly, although all three dietary classes could be distinguished whichever method of sample preparation and chromatographic gradient had been employed, the metabolites responsible for this discrimination differed depending upon the method of sample preparation and the separation used.

Another example of the application of UHPLC-MS to nutrition research is the use of the technique for global metabolic profiling of urine samples following the ingestion of pu-erh tea.[79] Urine samples were analyzed by RP gradient UHPLC-QTOF-MS, with acidified water and acetonitrile (0.1% formic acid) as mobile phases and data analysis was performed *via* OPLS-DA. The investigators noted an increase in the amounts of urinary 5-hydroxytryptophan, inositol and 4-methoxyphenylacetic acid excreted during, and after, pu-erh tea ingestion, together will a concomitant reduction in the amount of 3-chlorotyrosine and creatinine excreted.

Given the importance of nutritional research, it is to be expected that, as these methods mature, there will be a large increase in the number of published applications of global metabolic profiling in this area.

14.7 Current Challenges for UHPLC-MS in Global Metabolic Profiling Studies

14.7.1 Biomarker Characterization using UHPLC-MS

As will be clear from the examples provided in the preceding applications, with appropriate study design and samples, the detection of metabolic differences between groups of, for example, diseased subjects and the controls is often possible. The difficulty, however, that often occurs once this process of discovering potential biomarkers has taken place is that of identification of unknown biomarkers. The identification of these ions can often represent a significant challenge in MS-based metabolic profiling. The whole range of MS/MS techniques can be brought to bear on these problems and can be used to provide structural information based on fragmentation, and accurate mass measurements. These can be used to derive part structures and to generate empirical formulae of both parent and, if performed on, for example, a Q-TOF MS, on the fragment ions as well. Clearly, the use of instruments such as the "Orbitrap" or an Fourier transform-ion cyclotron resonance-mass spectrometry (FT-ICR-MS) will provide even better levels of mass accuracy, and therefore greater confidence in the resulting molecular formulas. Generating such data can, of course, can be quite time consuming, and the identification of unknowns is not a trivial undertaking. Although the data obtained from such experiments can significantly reduce the metabolic "search space" for unknowns, they do not always provide an unequivocal identification of the analyte. Where a tentative identification has been made, perhaps *via* interrogation of a database, there still remains no substitute for confirmation of identity *via* comparison of retention time and MS/MS fragmentation patterns with an authentic standard. In the event that these strategies fail, there may be no alternative to the preparative isolation of the unknown for in-depth structural analysis using additional techniques such as, for example, ^{1}H and ^{13}C NMR spectroscopy.

14.7.2 Databases for Metabolite Identification

Whilst LC-MS studies may well generate putative structures that can be followed up experimentally, the metabolome is so large that this can often be a matter of luck. Where such pragmatic approaches fail then the interrogation of one of the increasing number of web-based databases represent an excellent search strategy. In such instance, searches of the following online databases are to be recommended: HMDB[80,81] and METLIN,[82] as well as the KEGG database.

14.7.3 Biomarker Validation

If the aim of all of these global metabolic profiling studies is the detection and identification of biomarkers and the generation of new mechanistic hypotheses based on them, the final proof of their utility must be based on the development and deployment of targeted and specific bioanalytical methods. These "global" profiles are generated using separations and detection strategies that are optimized to detect largest possible range of structures, and are therefore not optimized in any way for individual compounds. Having identified a metabolite, or a group of metabolites, that may indeed represent biomarkers, it is incumbent on the investigator to develop a fully validated method to enable these potential biomarkers to be converted into actual biomarkers suitable for clinical use. In this respect, the guidance issued by the US FDA provides a useful framework for bioanalytical method validation.[83,84]

14.8 Conclusions

UHPLC hyphenated to MS is already making significant contributions to the exploration of the human metabolome in health and disease, as it already has in animal models. Hopefully, this work will result in the production of new insights into biological processes and provide, for example, clinical tools and biomarkers for the understanding of human disease and the monitoring of therapy. Standard methods are now available for the global metabolic profiling of specimens such as urine and serum/plasma[20,33] and further, robust and standardized experimental UHPLC-MS protocols, for example, for tissues and cells *etc.*, will be developed in the near future. The undeniable improvements in performance provided by this combination, compared to conventional HPLC-MS, for the analysis of complex biological samples means that it will inevitably come to dominate the field for this type of work. The major challenge that remains concerns the need to continue the development of tools and strategies that will provide rapid and unambiguous identification of the ions that are identified by the various multivariate statistical methods employed to analyze the data. Such identifications require not only appropriate expertise on the part of the investigator but also both robust databases and, often, the availability of authentic standards.

References

1. J. K. Nicholson, J. C. Lindon and E. Holmes, *Xenobiotica*, 1999, **29**, 1181.
2. J. K Nicholson, J. Connelly, J. C. Lindon and E. Holmes, *Nat. Rev. Drug Discov.*, 2002, **1**, 153.
3. O. Fiehn, *Plant Mol. Biol.*, 2002, **48**, 155.
4. J. K Nicholson and J. C. Lindon, *Nature*, 2008, **455**, 1054.
5. I. D. Wilson, R. Plumb, J. Granger, H. Major, R. Williams and E. M. Lenz, *J. Chromatogr. B.*, 2005, **817**, 67.
6. E. M. Lenz and I. D. Wilson, *J. Proteome Res.*, 2007, **6**, 443.
7. X. Lu, X. Zhao, C. Bai, C. Zhao, G. Lu and G. Xu, *J. Chromatogr. B.*, 2008, **866,** 64.
8. I. D. Wilson, J. K. Nicholson, J. Castro-Perez, J. H. Granger, K. A. Johnson, B. W. Smith and R.S. Plumb, *J. Proteome Res.*, 2005, **2**, 591.
9. D. S. Wishart, *Drugs RD*, 2008, **9**, 307.
10. C. J. Clarke and J. N. Haselden, *Toxicol. Pathol.*, 2008, **36**, 140.
11. C. Lindon, E. Holmes and J. K. Nicholson, *FEBS J.*, 2007, **274**, 1140.

12. M. Coen, E. Holmes, J. C. Lindon and J. K. Nicholson, *Chem. Res. Toxicol.*, 2008, **21**, 9.
13. N. G. A. Gowda, S. Zhang, H. Gu, V. Asiago, N. Shanaiah and D. Raftery, *Expert Rev. Mol. Diagn.*, 2008, **8**, 617.
14. E. M. Lenz, J. Bright, R. Knight, I. D. Wilson and H. Major, *J. Pharm. Biomed. Anal.*, 2004, **35**, 599.
15. E. Holmes, R. L. Loo, J. Stamler, M. Bictash, I. K. S. Yap, Q. Chan, T. Ebbels, M. De Iorio, I. J. Brown, K. A. Veselkov, M. L. Daviglus, H. Kesteloot, H. Ueshima, L. Zhao, J. K. Nicholson and P. Elliott, *Nature*, 2008, **453**, 396.
16. R.S. Plumb, J. H. Granger, C. L. Stumpf, K. A. Johnson, B. W. Smith, S. Gaulitz, I. D. Wilson and J. Castro-Perez, *Analyst*, 2005, **130**, 844.
17. E. M. Lenz, J. Bright, I.D. Wilson, S.R. Morgan and A. F. P. Nash, *J. Pharm. Biomed. Anal.*, 2003, **33**, 1103.
18. E. M. Lenz J. Bright, I. D. Wilson, A. Hughes, J. Morrisson, H. Lindberg and A. Lockton, *J. Pharm. Biomed. Anal.*, 2004, **36**, 841.
19. H. Pereira, J-F. Martin, C. Joly, J-L. Sébédio and E. Pujos-Guillot, *Metabolomics*, 2010, **6**, 207.
20. W. B. Dunn, D. Broadhurst, P. Begley, E. Zelena, S. Francis-McIntyre, N. Anderson, M. Brown, J. D. Knowles, A. Halsall, J. N. Haselden, A. W. Nicholls, I. D. Wilson, D. B. Kell and R. Goodacre, The Human Serum Metabolome (HUSERMET) Consortium, *Nat. Prot.*, 2011, **6**, 1060.
21. K. Dettmer, M. F. Almstetter, I. J. Appel, N. Nurnberger, G. Schlamberger, W. Gronwald, H. H. D. Meyer and P. J. Oefner, *Electrophoresis*, 2010, **31**, 2365.
22. J.R. Denery, A. A. K. Nunes and T. J. Dickerson, *Anal. Chem.*, 2011, **83**, 1040.
23. F. Michopoulos, G. Theodoridis, G. J. Smith and I. D. Wilson, *J. Proteome Res.*, 2010, **9**, 3328.
24. S. J. Bruce I. Tavazzi, V. Parisod, S. Rezzi, S. Kochhar and P. A. Guy, *Anal. Chem.*, 2009, **81**, 3285.
25. E. Zelena, W. B. Dunn, D. Broadhurst, S. Francis-Mckintyre, K. M. Carroll, P. Begley, S. O'Hagan and J. D. Knowles, A. Halsall, I. D. Wilson and D. B. Kell, HUSERMET Consortium, *Anal. Chem.*, 2009, **81**, 1357.
26. F. Michopoulos, L. Lai, H. Gika, G. Theodoridis and I. D. Wilson, *J. Proteome Res.*, 2009, **8**, 2114.
27. F. Michopoulos, A. M. Edge, G. Theodoridis and I. D. Wilson, *J. Sep. Sci.*, 2010, **33**, 1472.
28. A. Edge, in *Handbook of Bioanalytical Separations*, ed. I. D. Wilson, Elsevier, The Netherlands, 2003, vol. 4, p. 91.
29. P. Turnpenny, D. Fraier, C. Chassaing and J. Duckworth, *J. Chromatogr. B*, 2007, **856**, 131.
30. D. J. Crockford, A. D. Maher, K. R. Ahmadi, A. Barrett, R. S. Plumb, I. D. Wilson and J. K. Nicholson, *Anal. Chem.*, 2008, **80,** 6835.
31. G. Gika, G. Theodoridis and I. D. Wilson, *J. Chromatogr. A*, 2008, **1189,** 314.

32. L. S. New and E. C. Chan. *J. Chromatogr. Sci.*, 2008 **46**, 209.
33. E.J. Wantl, I. D. Wilson, H. Gika, G. Theodoridis, R. S. Plumb, J. Shockcor, E. Holmes and J. K. Nicholson, *Nat. Prot.*, 2010, **5**, 1005.
34. K Sandra, A.D.S. Pereira, G. Vanhoenacker, F. David and P. Sandra, *J. Chromatogr. A*, 2010, **1217**, 4087.
35. S.J. Bruce, P. Jonsson, H. Antti, O. Cloarec, J. Trygg, S. L. Marklund and T. Moritz, *Anal. Biochem.*, 2008, **372**, 237.
36. R.S. Plumb, P. Rainville, B.W. Smith, K. A. Johnson, J. Castr-Perez, I. D. Wilson and J. K. Nicholson, *Anal. Chem.*, 2006, **78**, 7278.
37. H. G. Gika, G. Theodoridis, J. Extance, A. M. Edge and I. D. Wilson, *J. Chromatogr. B*, 2008, **871**, 279.
38. V. V. Tolstikov and O. Fiehn, *Anal. Biochem.*, 2002, **301**, 298.
39. H. Idborg, L. Zamani, I. Schuppe-Koistinen and S. Jacobsson, *J. Chromatogr. B*, 2005, **828**, 9.
40. S. Cubbon, T. Bradbury, J. Wilson and J. Thomas-Oates, *Anal. Chem.*, 2007, **79**, 8911.
41. H. G. Gika, G. A. Theodoridis and I. D. Wilson, *J. Sep. Sci.*, 2008, **31**, 1598.
42. K. Spagou, I. D. Wilson, P. Masson, G. Theodoridis, N. Raikos, M. Coen, E. Holmes, J. C. Lindon, R. S. Plumb, J. K. Nicholson and E. J. Want, *Anal. Chem.*, 2011, **83**, 382.
43. Q. Yang, X. Shi, Y. Wang, W. Wang, H. He, X. Lu and G. Xu, *J. Sep. Sci.*, 2010, **33**, 1495.
44. T. Sangster, H. Major, R. Plumb, A. J. Wilson and I. D. Wilson, *Analyst*, 2006, **131**, 1075.
45. H. G. Gika, G. A. Theodoridis, J. E. Wingate and I. D. Wilson, *J. Proteome Res.*, 2007, **6**, 3291.
46. H. G. Gika, E. Macpherson. G. Theodoridis and I. D. Wilson, *J. Chromatogr. B*, 2008, **871**, 299.
47. G. Theodoridis, H. G. Gika and I. D. Wilson, *Trends Anal. Chem.*, 2008, **27,** 251.
48. A. Wijeyesekera, P. A. Clarke, M. Bictash, I. J. Brown, M. Fidock, T. Ryckmans, I. K. S. Yap, Q. Chan, J. Stamler, P. Elliott, E. Holmes and J. K. Nicholson. *Anal. Methods,* 2012, **4**, 65.
49. S. Geenen, F. Michopoulos, J. G. Kenna, K. L. Kolaja, H. V. Westerhoff and I. Wilson, *J. Pharm. Biomed. Anal.*, 2011, 54, 1128.
50. S. Geenen, C. Guallar-Hoyas, F. Michopoulos, J. G. Kenna, K. L. Kolaja, H. V. Westerhoff, P. Thomas and I. D. Wilson, *J. Pharm. Biomed. Anal.*, 2011, **56,** 655.
51. T. Kind, V. Tolstikov, O. Fiehn and R. H. Weiss, *Anal. Biochem.*, 2007, **363**, 185.
52. S. Ganti, S. L. Taylor, K. Kim, C. L. Hoppel, L. Guo, J. Yang, C. Evans and R. H. Weiss, *Int. J. Cancer*, 2011, doi: 10.1002/ijc.26274.
53. Y-L. Ma, H-L. Qin, W-J. Liu, J-Y. Peng, L. Huang, X-P. Zhao and Y-Y. Cheng, *Dig. Dis. Sci.*, 2009, **54**, 2655.

54. Y. Qiu, G. Cai, M. Su, T. Chen, X. Zheng, Y. Xu, Y. Ni, A. Zhao, L. X. Xu, S. Cai and W. Jia, *J. Proteome Res.*, 2009, **8**, 4844.
55. J. Wei, G. Xie, Z. Zhou, P. Shi, X. Xeng, T. Chen, M. Su, A. Zhao and W. Jia, *Int. J. Cancer*, 2011, **129**, 2207.
56. H. Cao, H., Huang, W. Xu, D. Chen, J. Yu, J. Li and L. Li, *Anal. Chim. Acta*, 2011, **691**, 68.
57. K. D. Sheikh, S. Khanna, S. W. Byers, A. Fornace Jr and A. K. Cheema, *J. Biomol. Tech.*, 2011, **22**, 1.
58. E. C. Laiakis, G.A. Morris, A. J. Fornace and S. R. C. Howie, *PLoS ONE*, 2010, **5**, e12655.
59. Y. Fukui, M. Kato, Y. Inoue, A. Matsubara and K. Itoh, *J. Chromatogr.*, *B*, 2009, **877**, 3806.
60. L. Jia, J. Chen, P. Yin, X. Lu and G. Xu, *Metabolomics*, 2008, **4**,183.
61. N-J. Li, W-T. Liu, W. Li, S-Q. Li, X-H. Chen, K-S. Bi and P. He, *Clin. Biochem.*, 2010, **43,** 992.
62. W. B. Dunn, M. Brown, S. A. Worton, I. P. Crocker, D. Broadhurst, R. Horgan, L. C. Kenny, P. N. Baker, D. B. Kell and A. E. P. Heazell, *Placenta*, 2009, **30**, 974.
63. W. B. Dunn, D. Broadhurst, M. Brown, P. N. Baker, C. W. G. Redman, L. C. Kenny and D. B. Kell, *J. Chromatogr. B.*, 2008, **871**, 288.
64. R. P. Horgan, D. Broadhurst, W. B. Dunn, M. Brown, A. E. P. Heazell, D. B. Kell, P. N. Baker and L. C. Kenny, *Placenta*, 2010, **31**, 893.
65. K. Kajander, E. Myllyluoma, S. Kyrönpalo, M. Rasmussen, P. Sipponen, I. Mattila, T. Seppänen-Laakso, H. Vapaatalo, M. Oreşiç and R. Korpela, *World J. Gastroenterol.*, 2009, **15**, 6068.
66. P. Yin, X. Zhao, Q. Li, J. Wang, J. Li and G. Xu, *J. Proteome Res.*, 2006, **5**, 2135.
67. Q. Li, Q. Zhang, G. Xu, P. Yin, N. Li and J. Li, *Digestion*, 2008, **77**, 122.
68. J. Barr, M. Vázquez-Chantada, C. Alonso, M. Pérez-Cormenzana, R. Mayo, A. Galán, J. Caballería, A. Martín-Duce, A. Tran, C. Wagner, Z. Luka, S. C. Lu, A. Castro, Y. Le Marchand-Brustel, M. L. Martínez-Chantar, N. Veyrie, K. Clément, J. Tordjman, P. Gual and J. M. Mato, *J. Proteome Res.*, 2010, **9**, 4501.
69. G. Quintás, N. Portillo, J. C. García-Cañaveras, J. V. Castell, A. Ferrer and A. Lahoz, *Metabolomics*, 2012, **8**, 86.
70. V. Croixmarie, T. Umbdenstock, O, Cloarec, A. Moreau, J-M. Pascussi, C. Boursier-Neyret and B. Walther, *Anal. Chem.*, 2009, **81**, 6061.
71. A. D. Patterson, O. Slanač, K. W. Krausz, F. Li, C. C. Höfer, F. Perlík, F. J. Gonzalez and J. R. Idle, *J. Proteome Res.*, 2009, **8**, 4293.
72. J. Y. Kim, J. Y. Park, O. Y. Kim, B. M. Ham, H-J. Kim, D. Y. Kwon, Y. Jang and J. H. Lee, *J. Proteome Res.*, 2010, **9**, 4368.
73. J. Zhang, L. Yan, W. Chen, L. Lin, X. Song, X. Yan, W. Hang and B. Huang, *Anal. Chim. Acta*, 2009, **650**, 16.
74. T. Huo, S. Cai, X. Lu, Y. Sha, M. Yu and F. Li, *J. Pharm. Biomed. Anal.*, 2009, **49**, 976.

75. Y. Gu, Y. Zhang, X. Shi, X. Li, J. Hong, J. Chen, W. Gu, X. Lu, G. Xu, G. Ning, *Talanta*, 2010, **81**, 766.
76. X. Zhao, J. Fritsche, J. Wang, J. Chen, K. Rittig, P. Schmitt-Kopplin, A. Fritsche, H-U. Haring, E, D. Schleicher, G. Xu and R. Lehmann, *Metabolomics*, 2010, **6**, 362.
77. P. A. Guy, I., Tavazzi, S. J. Bruce, Z. Ramadan and S. Kochhar, *J. Chromatogr. B*, 2008, **871**, 253.
78. C. Legido-Quigleya, C. Stella, F. Perez-Jimenez, J. Lopez-Miranda, J. Ordovas, J. Powell, F. Van-der-Ouderaa, L. Ware, J. C. Lindon, J. K. Nicholson and E. Holmes, *Biomed. Chromatogr.*, 2010, **24,** 737.
79. Xie, M. Ye, Y. Wang, Y. Ni, M. Su, H. Huang, M. Qiu, A. Zhao, X. Zheng, T. Chen and W. Jia, *J. Agric. Food Chem.*, 2009, **57**, 3046.
80. D.S. Wishart, D. Tzur, C. Knox, R. Eisner, A. C. Guo, N. Young, D. Cheng, K. Jewell, D. Arndt, S. Sawhney, C. Fung, L. Nikolai, M. Lewis, M-A. Coutouly, I. Forsythe, P. Tang, S. Shrivastava, K. Jeroncic, P. Stothard, G. Amegbey, D. Block, D. D. Hau, J. Wagner, J. Miniaci, M. Clements, M. Gebremedhin, N. Guo, Y. Zhang, G. E. Duggan, G. D. MacInnis, A. M. Weljie, R. Dowlatabadi, F. Bamforth, D. Clive, R. Greiner, L. Li, T. Marrie, B. D. Sykes, H. J. Vogel and L. Querengesser, *Nucleic Acids Res.*, 2007, **35**, D521.
81. D.S. Wishart, C. Knox, A. C. Guo, R. Eisner, N. Young, B. Gautam, D. D. Hau, N. Psychogios, E. Dong, S. Bouatra, R. Mandal, I. Sinelnikov, J. Xia, L. Jia, J. A. Cruz, E. Lim, C. A. Sobsey, S. Shrivastava, P. Huang, P. Liu, L. Fang, J. Peng, R. Fradette, D. Cheng, D. Tzur, M. Clements, A. Lewis, A. De Souza, A. Zuniga, M. Dawe, Y. Xiong, D. Clive, R. Greiner, A. Nazyrova, R. Shaykhutdinov, L. Li, H. J. Vogel and I. Forsythe, *Nucleic Acids Res.*, 2009, **37**, D603.
82. C. A. Smith, G. O'Maille, E. J. Want, C. Qin, S. A. Trauger, T. R. Brandon, D. E. Custodio, R. Abagyan and G. Siuzdak, *Ther. Drug. Monit.*, 2005, **27**, 747.
83. FDA Guidance for Industry, Bioanalytical Method Validation, Food and Drug Administration, Centre for Drug Evaluation and Research (CDER), May 2001. A Guidance.
84. C.T. Viswanathan, S. Bansai, B. Booth, A. J. Stefano, M. J. Rose, J. Sailstad, V. P. Shah, J. P. Skelly, P. G. Swann and R. Weiner, *Pharm. Res.*, 2007, **24,** 1962.

Subject Index

References to tables and charts are in **bold** type.